# Writing in the Sciences

# THE ALLYN AND BACON SERIES IN TECHNICAL COMMUNICATION

## Series Editor: Sam Dragga, Texas Tech University

**Thomas T. Barker**
*Writing Software Documentation: A Task-Oriented Approach,* Second Edition

**Carol M. Barnum**
*Usability Testing and Research*

**Deborah S. Bosley**
*Global Contexts: Case Studies in International Technical Communication*

**Melody Bowdon and Blake Scott**
*Service-Learning in Technical and Professional Communication*

**R. Stanley Dicks**
*Management Principles and Practices for Technical Communicators*

**Paul Dombrowski**
*Ethics in Technical Communication*

**David Farkas and Jean Farkas**
*Principles of Web Design*

**Laura J. Gurak**
*Oral Presentations for Technical Communication*

**Sandra W. Harner and Tom G. Zimmerman**
*Technical Marketing Communications*

**TyAnna K. Herrington**
*A Legal Primer for the Digital Age*

**Richard Johnson-Sheehan**
*Writing Proposals: Rhetoric for Managing Change*

**Dan Jones**
*Technical Writing Style*

**Charles Kostelnick and David D. Roberts**
*Designing Visual Language: Strategies for Professional Communicators*

**Victoria M. Mikelonis, Signe T. Betsinger, and Constance Kampf**
*Grant Seeking in an Electronic Age*

**Ann M. Penrose and Steven B. Katz**
*Writing in the Sciences: Exploring Conventions of Scientific Discourse,* Second Edition

**Carolyn Rude**
*Technical Editing,* Third Edition

**Gerald J. Savage and Dale L. Sullivan**
*Writing a Professional Life: Stories of Technical Communicators On and Off the Job*

# Writing in the Sciences

## Exploring Conventions of Scientific Discourse

Second Edition

**Ann M. Penrose**

*North Carolina State University*

**Steven B. Katz**

*North Carolina State University*

PEARSON

Longman

New York  Boston  San Francisco
London  Toronto  Sydney  Tokyo  Singapore  Madrid
Mexico City  Munich  Paris  Cape Town  Hong Kong  Montreal

Senior Vice President and Publisher: Joseph Opiela
Vice President and Publisher: Eben W. Ludlow
Executive Marketing Manager: Tamara Wederbrand
Production Manager: Charles Annis
Project Coordination, Text Design, and Electronic Page Makeup: Nesbitt Graphics, Inc.
Cover Designer and Manager: John Callahan
Manufacturing Buyer: Al Dorsey

**Library of Congress Cataloging-in-Publication Data**

Penrose, Ann M.
    Writing in the sciences : exploring conventions of scientific discourse/Ann M. Penrose, Steven B. Katz.— 2nd ed.
        p. cm.
    ISBN 0-321-11204-0
    1. Technical writing. I. Katz, Steven B., 1953– II. Title.
T11.P393 2003
808' .0665—dc22                                    2003021559

Please visit our website at http://www.ablongman.com/penrose

ISBN 0-321-11204-0

        9 10—HT—09 08

# CONTENTS

## PART 1 Scientific Conventions   1

### 1   Science as a Social Enterprise   3

### 2   Forums for Communication in Science   24

### 3   Reading and Writing Research Reports   40

**4**     Reviewing Prior Research                                        83

**5**     Preparing Conference Presentations                        102

**6**     Writing Research Proposals                                   128

**7** ## Documenting Procedures and Guidelines 153

**8** ## Communicating with Public Audiences 176

**9** ## Considering Ethics in Scientific Communication 203

## 12    Research on Supernova Remnants: From Proposal to Publication    356
REYNOLDS AND COLLEAGUES

## 13    Research on the Oracle at Delphi: From Ancient Myth to Modern Interdisciplinary Science    400
HALE AND DE BOER AND COLLEAGUES

## List of Stylistic Features

Chapter sections indicated here. See index for page references.

| Feature | Section |
|---|---|
| Active voice | 3.4; 7.2; 7.7; 9.6 |
| Adverbial phrases as qualifiers | 3.6 |
| Analogy | 8.7 |
| Analysis | 8.6 |
| Apposition | 8.7 |
| Citation form | 4.7 |
| Definition | 8.5 |
| Direct vs. indirect citation | 4.7 |
| Examples | 8.4 |
| Figure captions | 3.5 |
| Graphics | 3.5; 5.6; 5.7; 8.8 |
| Headings | 3.2; 4.6; 6.5; 8.9 |
| Hedges | 3.6 |
| Imperative mood | 7.2; 7.3; 7.5 |
| IMRAD form | 3.1; 3.2 |
| Instructional steps | 7.2; 7.3 |
| Lists | 7.2 |
| Metaphor | 8.7 |
| Modal auxiliary verbs | 3.6 |
| Narration | 8.3 |
| Parallelism | 7.2; 7.3 |
| Parenthetical citation | 4.7 |
| Passive voice | 3.4; 7.7; 9.5 |
| Personal pronouns | 3.4; 7.2; 7.3; 7.6; 9.6 |
| Qualifiers | 3.6 |
| Quotation marks | 8.7 (note 5) |
| Quotations | 4.2; 8.9 |
| Rebuttals | 3.6 |
| Reference form | 4.8 |
| Reporting formulae | 3.5; 4.7 |
| Rhetorical appeals | 6.4; 8.2 |
| Simile | 8.7 |
| Synonym | 8.7 |
| Table titles | 3.5 |
| Topic sentences | 3.5 |
| Verb tense | 3.1; 3.7; 4.5 |
| Works cited list | 4.8 |

# FOREWORD
## by the Series Editor

The Allyn and Bacon Series in Technical Communication is designed to meet the continuing education needs of professional technical communicators, both those who desire to upgrade or update their own communication abilities as well as those who train or supervise writers, editors, and artists within their organization. This series also serves the growing number of students enrolled in undergraduate and graduate programs in technical communication. Such programs offer a wide variety of courses beyond the introductory technical writing course—advanced courses for which fully satisfactory and appropriately focused textbooks have often been impossible to locate.

The chief characteristic of the books in this series is their consistent effort to integrate theory and practice. The books offer both research-based and experience-based instruction, describing not only what to do and how to do it but explaining why. The instructors who teach advanced courses and the students who enroll in these courses are looking for more than rigid rules and ad hoc guidelines. They want books that demonstrate theoretical sophistication and a solid foundation in the research of the field as well as pragmatic advice and perceptive applications. Instructors and students will also find these books filled with activities and assignments adaptable to the classroom and to the self-guided learning processes of professional technical communicators.

To operate effectively in the field of technical communication, today's technical communicators require extensive training in the creation, analysis, and design of information for both domestic and international audiences, for both paper and electronic environments. The books in the Allyn and Bacon Series address those subjects that are most frequently taught at the undergraduate and graduate levels as a direct response to both the educational needs of students and the practical demands of business and industry. Additional books will be developed for the series in order to satisfy or anticipate changes in writing technologies, academic curricula, and the profession of technical communication.

**Sam Dragga**
*Texas Tech University*

# PREFACE

In the Preface to the first edition of *Writing in the Sciences*, we reported that Science Citation Index (SCI®) listed over 3000 peer-reviewed science and technical publications, and that the National Institutes of Health and the National Science Foundation received more than 40,000 proposals from scientists each year (Seiken 1992). Today, only five years since the first edition, the Science Citation Index (SCI®) lists 5800 peer-reviewed print and electronic journals that are available through its parent company *ISI Web of Science*® and the online version, *SciSearch*® (ISI 2003). In 1999, NIH alone received 26,000 proposals and funded about 32 percent of them (NIH 2000). NSF receives about 30,000 proposals per year, of which it also funds about one-third (NSF 2002). During the two-year period from 2000 to 2001, the National Sea Grant College Program reviewed 2249 proposals and funded 520 projects, or 22 percent (National Sea Grant 2003). Numbers like these begin to illustrate both the sheer volume of scientific research and the intensity of the competition. With numbers like these, the need for a clear understanding of communication in science has never been greater.

*Writing in the Sciences* is intended as a resource for scientists and technical communicators and is designed as a primary text for upper division and graduate courses in scientific writing. It is also well suited as a supplemental text for courses in the sciences that seek to provide explicit support for writing. Material in the text has been field-tested in our own classes over the past thirteen years and at other institutions around the country and abroad.

The book presents a rhetorical, multidisciplinary approach to teaching the major genres of writing for science and research: research reports, grant proposals, conference presentations, and a variety of forms of public communication. We introduce general rhetorical concepts and analytical tools to help students recognize the conventions used by scientists in their own fields and use these conventions effectively. Our approach is descriptive rather than prescriptive—the goal is to teach students to recognize distinctive features of common genres by examining how, why, and for whom such texts are created by scientists.

These features are illustrated in a series of research cases in Part 2, which include documents addressing a range of audiences for scientific research and illustrating consequent differences in substance, focus, arrangement, style, and other rhetorical dimensions. Examples are drawn from a variety of research disciplines, enabling students to recognize practices in their own fields, as well as to compare these practices to those of other disciplinary communities. Such comparisons provide the opportunity for instructor and student to identify and investigate common conventions in science, examine variations across fields, recognize relationships between the nature of inquiry in a discipline and the logic and form of the discipline's written and spoken texts, and explore the social and ethical dimensions of those communication practices.

Readers familiar with the first edition of *Writing in the Sciences* will notice some significant changes and additions in this second edition.

1. As in the first edition, the sample research cases in Part 2 ground the analysis of rhetorical principles and conventions in Part 1. We have revised and expanded these cases to illustrate several distinctive research dynamics: the progression from initial theory to clinical application in Chapter 10, from local environmental event to state and federal policy in Chapter 11, from initial research proposal to presentation and publication in Chapter 12, and from myth to scientific validation in Chapter 13.

2. We have thoroughly updated the chapters in Part 1 to reflect continuing developments in rhetorical scholarship on scientific communication as well as changing trends in editorial policy, grant review procedures, and other disciplinary communication practices.

3. We have incorporated a new focus on communication technologies throughout the book, providing an examination of the impact of electronic media on communication and publication processes in science.

4. We have developed a new chapter exploring scientific communication in industry and government contexts. Focusing on procedures and guidelines, Chapter 7 examines how procedural information is shaped for a range of purposes, including quality assurance within an organization, regulatory oversight, commercial application, and governmental protection of public health and safety.

5. A directory of stylistic features has been added to provide easy reference to the stylistic devices and conventions treated in the text.

As this list indicates, one of our primary goals in the second edition is to broaden the scope of our discussion to include applied contexts for scientific research, where many of our science students will pursue their careers. Chapter 10 on the "Ulcer Bug" now includes text from the pharmaceutical industry, illustrating the application of clinical research. Chapter 11 on *Pfiesteria* now includes environmental policies and procedures resulting from the research. These new documents, along with the new Chapter 7 on applied contexts, broaden the range of audiences covered in our book to include contexts beyond those of the lab or field site. In providing windows into the drug development and environmental regulation processes, the text allows instructors to explore the rhetorical challenges of communicating with stakeholders outside the scientific community.

Five new documents also have been added to the sample case on supernova research in Chapter 12, and the case has been reshaped to follow the research process chronologically from proposal to publication. Students can now trace the development of a line of research through its various text forms, from NASA's initial call for proposals, to the research proposal itself, to a series of progress and final reports, to conference proceedings and publication. Both the NASA case and the new case in Chapter 13 help students explore research paradigms beyond the more familiar empirical model.

While the first edition examined research in a range of individual fields, Chapter 13 examines interdisciplinary research that looks across fields. This case foregrounds the collaborative nature of scientific inquiry by documenting how scientists from geology, archeology, and clinical toxicology together are unraveling the mystery behind the legendary oracle at Delphi. Sample published texts integrate historical and literary evidence with the analysis of new geologic and chemical data, thus illustrating interdisciplinary argumentation and offering opportunities to explore the complementarity of research in science and the humanities.

These updates both preserve and enhance one of the most distinctive features of the first edition: its emphasis on exploring actual rhetorical practices in actual research communities.

## Instructor's Manual

The Instructor's Manual presents a quick overview of the purpose and rationale for each chapter in Part 1, along with sample course outlines, and extensive notes on exercises and activities. Most exercises in the text are exploratory, many focusing on texts from students' own research fields; the Manual includes sample responses and/or advice about particular text features that could be highlighted, problems students may encounter, and patterns they may notice across a range of texts. This Manual also includes some commentary on figures when these are not self-explanatory or explained in the text. There are special notes on some of the writing activities suggested at the end of each chapter as well.

## Companion Website
## *http://www.ablongman.com/penrose*

The Companion Website for *Writing in the Sciences*, developed by Katz, contains resources and features that both instructors and students will find useful. Self-correcting questions allow students to test their understanding of each chapter in Part 1. Online exercises help students further focus on rhetorical conventions and apply the organization and audience adaptation skills discussed in the book. Additional writing activities and assignments complement those in the text while also taking advantage of the interactive environment of the World Wide Web. The Companion Website includes many URLs to important sites, providing instructors and students with the opportunity to actively explore research journals, professional organizations, and funding agencies. Additional URLs provide supplemental resources to refine and improve students' writing skills.

## Acknowledgments

In a recent newspaper column celebrating the collaborative nature of science, North Carolina State University emeritus professor of physics Jasper Memory remarked that the ability to work and play well with others may be as important a trait for budding scientists as good grades in math (Memory 2003). Our research into the nature of communication in science supports this observation, and our experience in writing this book confirms it. We could not have completed this project without the help of many others.

First and foremost, we acknowledge an enormous debt to our colleague, research assistant, and friend, Daun Daemon, without whom this edition would not yet be finished. Beyond helping us update samples throughout the book, she tracked down sources and the sources' sources; served as liaison with researchers; and handled copyrights, permissions, and index, for a time with one hand literally tied (in a splint). It was Daun's curiosity and journalistic savvy that led us to the Delphic oracle case that plays an important role in this edition. At times we were motivated solely by the desire to do justice to the quality and depth of her research.

We are greatly indebted as well to the scientists who graciously provided the sample texts and research stories on which much of this text is built: botanist JoAnn Burkholder and physicist Steve Reynolds from North Carolina State University, archeologist John Hale of the University of Louisville, and biophysicist Carl Blackman of the EPA's National Health and Environmental Effects Research Laboratory. The willingness of these researchers to share their work, time, and expertise with us illustrates what we consider the most valuable dimension of science: the ethic of collaboration and sharing.

We also would like to thank Ann Roy, a graduate of North Carolina State University's M. S. Program in Technical Communication, whose research during an independent study with Katz in the Summer of 2002 proved highly valuable in updating this book; Heather Bell and Kathleen Walch for their earlier work on clinical protocols; Gary Blank of North Carolina State's College of Natural Resources for his willingness to share his expertise in forestry writing; and Edward Tick, for his late night translations of the Greek permit—*Efkaristow poli!*

The first edition of this book was adopted by science writing instructors in a variety of courses and fields across the United States, many of whom have offered comments and suggestions that have shaped the new edition. In addition to the authors, ten other instructors at North Carolina State University have used the text in English 333, Communication for Science and Research, an advanced writing course required of science majors at NC State. In particular, Linda Rudd provided valuable feedback on individual exercises, assignments, text passages, and organizational features of the book. We also thank Trish Watson, our research assistant for the first edition, for her thoughtful commentary, and Carolyn Miller, our long-time mentor and colleague, for her constant attention to things rhetorical and scientific.

This book also benefited from the attentive stewardship it received in its new home. We would like to express our appreciation to Sam Dragga, Series Editor, for his invitation to submit a proposal for this second edition; to Eben Ludlow, Vice President of Longman, for his professional guidance; and to Susan McIntyre, Senior Project Manager at Nesbitt Graphics, for the expertise and care she always brought to her work with our manuscript.

Once again, we thank our spouses, David and Alison, who have sustained us in this effort, as they do every day. We had a choice to take on this project; they did not. Our appreciation of their support and encouragement is beyond all words.

And finally, we reiterate our thanks to our science writing students, past, present, and future, who are both observers and participants in scientific discourse.

Ann M. Penrose and Steven B. Katz
Raleigh, North Carolina

# Scientific Conventions

# Science as a Social Enterprise

## 1.1 The Shaping of Knowledge in Science

We begin this introductory chapter with the case of a young scientist who made a revolutionary discovery. In September, 1983, Barry Marshall, an unknown internist at Australia's Royal Perth Hospital, presented his and pathologist J. R. Warren's findings at the Second International Workshop on *Campylobacter* Infections in Brussels. They had already published separate, technical "letters" together in the *Lancet* (Warren and Marshall 1983). In these letters they reported the presence of bacteria (later classified as *Helicobacter pylori [H. pylori]*) in the stomach lining of patients with gastritis. Following the presentation of their joint paper at this international conference, Marshall stood before doctors, researchers, and specialists in microbiology, gastroenterology, and infectious diseases. In response to a question from one of the experts, Marshall declared that he believed this bacterium was the cause of all stomach ulcer disease and that chronic ulcer recurrence could be eradicated in most, if not all, patients with a treatment of common antibiotics and bismuth such as Pepto Bismol (Chazin 1993; Monmaney 1993).

Marshall and Warren's theory that ulcers were the result of bacterial infection in the stomach lining was greeted with both intense interest and skepticism by experts in the field. After all, the claim that *anything* could live in the intensely acidic environment of the stomach was as unbelievable as it was revolutionary. If the claim were true, the implications for ulcer sufferers—and for the major drug companies manufacturing acid blocker treatments—would be substantial. Although experts in the field were polite and respectful in the professional journals, they later told the popular press that when Marshall first presented the theory, they thought he was "brash" (SerVaas 1994, p 62), "a madman" (Chazin 1993, p 122), "a medical heretic" (Monmaney 1993, p 65), "a crazy guy saying crazy things" (Monmaney 1993, p 66). The experts were intrigued, however, and the problem was important enough to demand attention, as evidenced by letters and editorials in scientific journals devoted to gastroenterology and internal medicine (see Chapter 10 for examples of these, and some of the studies mentioned below).

By 1993, more than 1500 studies around the world had lent support to Marshall's theory (Monmaney 1993). In a major shift in policy in February, 1994, a panel of the National Institutes of Health released a consensus statement in which it accepted that there is a relationship between *H. pylori* and ulcer disease and recommended that antibiotics be used in the treatment of stomach ulcers where the bacterium is present (NIH 1994). In April, 1996, the FDA approved the first drug specifically designed for treating ulcers based on Marshall's findings. Today, the testing for *H. pylori* and its treatment have been extended not only to some forms of stomach cancer, but also to the investigation of common dyspepsia (e.g., Chiba et al. 2002). Methods for more accurately and easily detecting the presence of *H. pylori*, such as a breath test for *urease*, continue to be developed (see Alimenterics 1999). And, a sure sign of the acceptance of the theory, at the time of this writing, several books, including a handbook, have already been published on the nature, identification, and manipulation of new species of *H. pylori* and their physical and genetic structure (Harris and Misiewicz 1996; Clayton and Mobley 1997; Heatley 1999; Mobley et al. 2001). Marshall himself has edited a collection of firsthand accounts, spanning almost a century, of discoveries of *H. pylori* worldwide (Marshall 2002). And one of Marshall's most outspoken critics, David Graham, now edits *Helicobacter*, a journal devoted exclusively to the subject.

Dr. Marshall, who later was tenured at the University of Virginia School of Medicine and is now Director of the Helicobacter Foundation, started a revolution in the field of gastroenterology (Carey 1992). Yet, it took more than a decade for his theory to be accepted. And most of the development and support came from other physicians and scientists. Although Marshall continued to publish technical letters, abstracts, reports, and some retrospective studies, he had great difficulty getting his clinical research published and felt his message about ulcer treatment was not being heeded (Chazin 1993). The decade following his presentation in Brussels was a period of trial and tribulation for him.

Why was there such negative reaction to Marshall's presentation in Brussels? Despite all the positive editorials by scientists expressing interest in the theory and the subsequent research and confirmation of the claim, why did Marshall have difficulty getting his research published? Why wasn't his theory immediately embraced, as he, the press, and the public thought it should have been? A close examination of the answers to these questions can provide insight into how knowledge is created and shaped in science and into the role communication plays in that process.

Certainly, part of the answer has to do with the somewhat shaky nature of Marshall's methods—with his science itself. In 1982, Warren, the Royal Perth pathologist, had observed the presence of the bacterium in the stomach and showed Marshall the pathology slides. Together they studied a hundred patients suffering from peptic ulcers and found that *H. pylori* was present in 87 percent of the cases. This was the study they presented in Brussels and later published as a research article in the *Lancet* in 1984 (Marshall and Warren 1984). But by his own accounts and those of experts such as David Graham, chief of gastroenterology at Houston's Veterans Affairs Medical Center, Marshall was "not the greatest researcher of all time" (Chazin 1993, p 123). One scientific editorial pointed to the

1984 Marshall and Warren study in the *Lancet* as "well planned" (Lancet 1984, p 1337), but in 1988 Marshall's large-scale study was rejected by the *New England Journal of Medicine* as "inconclusive" (Chazin 1993, p 123). Other editorials in scientific journals criticized the validity of his methods and conclusions (see Lam 1989), and Marshall admitted to the press that he was more interested in curing patients than in developing adequate methods and conducting large clinical experiments needed to support his claim; he had hoped that clinical success with patients would be enough to convince his colleagues (Chazin 1993). Thus, Marshall was faced with the problem that to some degree all scientists, not just the lucky ones who wind up doing important research, face in their career: How do you get other scientists to listen to you?

To gain the acceptance he thought his theory deserved, Marshall did something that most scientists would—and should—never do. Unable to convince his colleagues that *H. pylori* caused stomach ulcers, in 1984 Marshall created a potent mixture containing the bacteria and drank it, inducing a case of acute gastritis in himself (Marshall et al. 1985). Marshall's methods were not only unorthodox but also radical and potentially dangerous. Although Marshall's experimenting on himself led to further research, the incident made his critics even more skeptical of his professionalism and less accepting of his theory (Morris et al. 1991; Carey 1992).

Another reason Marshall's theory wasn't readily accepted had to do with the prevailing assumptions about, and treatment of, gastrointestinal diseases. Warren and Marshall were not the first to observe bacteria in conjunction with gastric inflammation. Medical researchers as far back as the late 1800s had reported and even published images of the bacteria living in the stomach lining (Blaser 1987; Marshall 2002). But these findings had been dismissed either as contaminants introduced during biopsy or as unrelated agents existing near ulcers; because of the presence of hydrochloric acid, gastroenterologists assumed that the stomach was "a sterile organ" (Warren and Marshall 1983, p 1273). As Monmaney (1993) reported, the causes of ulcers had been attributed to weak stomach linings and/or to an increase in stomach acids caused by emotional trauma, tension, nervousness, or modern life itself: "Marshall's theory challenged widely held and seemingly unassailable notions about the cause of ulcers. No physical ailment has ever been more closely tied to psychological turbulence" (p 64).

The contrast between the initial response to Marshall's hypothesis and the subsequent success of his theory suggests several important points about how knowledge is shaped in science and the importance of communication in that process. We will highlight these briefly here by way of introduction and then will examine them in more depth in this chapter and throughout this book as you begin to explore them in your own field.

- Scientific experimentation and knowledge are governed by tacit beliefs and assumptions about what is factual, valid, and acceptable; these beliefs and assumptions are "social" in nature.
- Communication is central to the growth of scientific knowledge in each discipline, and thus to the advancement of science itself.
- Persuasion is an integral part of scientific communication; it includes the use of sound arguments and an appropriate style of presentation, as well as acceptable scientific theories, methods, and data.

- As social enterprises, scientific fields are also to some degree governed by explicit conventions and rules about how and what to communicate, conventions and rules that professional scientists expect each other to follow; failure to follow these can result in a failure to communicate and thus can hamper the advancement of scientific knowledge.
- Collaboration and cooperation both within and across disciplines and professions are essential to the development of scientific theories, research, and knowledge. As you will learn in your exploration of communication in your field, collaboration and cooperation are central to research and to the actual writing of research papers and proposals. Scientific knowledge is built and shared through collaboration and cooperation.

## 1.2 The Social Nature of Science

As you will see when you explore your own field, science is a social enterprise. In one sense, this means that science is a part of the larger society in which it is situated. Science is shaped by the values of the dominant culture in which scientists participate and live, sharing many of its assumptions, goals, biases, and problems (NAS 1995; Lyne 1998). Conversely, science also exerts a powerful influence on society. Think about decisions you've made recently concerning such practical things as medical treatment, diet, energy and fuel consumption, and weather. Indeed, at a deep level, our very way of thinking about the world is rooted in current scientific practices and beliefs.

Our primary focus here, however, is on the social nature of the activity within scientific communities and related settings, rather than on the general social context in which these communities are embedded; the challenges and implications of communicating science in the public realm will be considered in Chapter 8. Scientists in a discipline constitute a *community* in which knowledge is built, is validated, and has meaning. Robin Warren needed Barry Marshall's clinical knowledge to understand how the bacterium he was observing was related to gastric symptoms (Chazin 1993). Marshall needed Warren in order to understand the biology of the bacterium he was observing in his patients. Both Marshall and Warren needed other scientists and the NIH to further test, validate, accept, and extend their work in theory and practice. In providing healthcare to their patients, physicians applying this research needed the pharmaceutical industry to develop tests and treatments for *H. pylori*, and the drug companies required FDA oversight and approval. We saw earlier that this can be a slow process. Like any society, scientific communities operate by a system of assumptions and beliefs that govern the perception and understanding of phenomena, the methods used, the research that is conducted, and the kinds of conclusions that can be drawn and treatments developed.

Before reading further, try the classic "nine-dot problem" (Adams 1976) presented in Exercise 1.1. For those of you who are having trouble solving this problem, don't worry; most people do. The frustration you are feeling is the creative tension, which scientists, like everybody else, experience when they're trying to solve a problem and can't find the solution. The tension makes you want to keep working on the problem and is one of the driving forces of science. Often the diffi-

To get an idea of how assumptions and beliefs operate in science, try the following experiment. In the next five minutes, connect all the dots below with only four straight lines and without lifting your pencil from the page. If you have done this problem before, do what Einstein did: a *Gedanken* (thought) experiment: connect all the dots with only one line; there are several ways of doing this.

```
    •   •   •


    •   •   •


    •   •   •
```

culty is that the solution cannot be found within your current cognitive framework; your assumptions about the problem, the experimental and cultural context in which you are working, and your expectations about the solution all influence your perception. Thus, according to Kuhn (1996), finding the solution involves something like a "gestalt" switch, where you see the problem in a new context, a context that allows you to find the solution.

Notice that solving the nine-dot problem involves going outside the conceptual space you perceived and believed to be a square. The perception that dots form lines is, of course, one of the basic principles of geometry, which defines a *line* as the track made by a moving point, and so in some sense this perception is based on a social belief or convention. But the belief that the solution must fall within the square created from the dots prevents the perception of the easy solution. To solve the nine-dot problem, you must "violate" or ignore conventional beliefs and assumptions. This is what leading gastroenterologists had to do to accept Marshall's theory that stomach ulcers are caused by bacteria rather than by stress.

The important point here is that assumptions and beliefs influence scientists' perception of phenomena. In *The Structure of Scientific Revolutions*, Thomas Kuhn (1996) calls these sets of assumptions in science *paradigms*. Paradigms are learned tacitly through observation and imitation of scientists and practitioners in the field (Polanyi 1958); and they are learned explicitly through education, textbooks, and specific practices (Kuhn 1996). Thus some of the assumptions in paradigms are only implicit or subconscious, and some are explicit in the form of rules and accompanying examples. As students of science, you are currently engaged in learning the paradigms of your fields. As we will explore in this book, paradigms include conventions, assumptions, and rules about communication in science as well.

Paradigms constrain thinking, as illustrated in the nine-dot problem. Yet, at the same time, paradigms provide the support and context for discovery. Without the background of the square created by the pattern of the dots, there would be no clear way to connect the dots, nothing to break out of, no problem to solve. In the case of *H. pylori,* the discoveries that had been made—and the assumptions, examples, and knowledge about bacteria and about ulcers that already existed in the field—provided both the scaffolding and the backdrop against which to discover the connection between the bacterium and gastritis. Although Warren, Marshall's coworker, stated that the discovery of the ulcer bacteria "was something that came out of the blue," he also admitted he "happened to be there at the right time, because of the improvements in gastroenterology in the seventies" (Monmaney 1993, p 68). We observe here that the discovery by an individual scientist that takes place outside the dominant paradigm nevertheless depends on that paradigm for the perception of it. Without paradigms, we would notice nothing at all. As a set of assumptions, rules, and examples, the prevailing paradigm in a field helps define the problems in that field, specify methods that are allowed to solve the problem, and predict what results can be expected.

This is not to deny the role of the individual scientist's perception and judgment. Some assumptions in science are personal, subjective, and even aesthetic (the scientific values of simplicity and elegance, for example, are aesthetic). The National Academy of Sciences calls attention to this dimension of science:

> Researchers continually have to make difficult decisions about how to do their work and how to present that work to others. Scientists have a large body of knowledge that they can use in making these decisions. Yet much of this knowledge is not the product of scientific investigation, but instead involves value-laden judgments, personal desires, and even a researcher's personality and style. (1989, p 1)

In general practice, this subjective dimension is balanced by social paradigms. Paradigms as social mechanisms act as a check on personal judgment and individual error, and so they make science as we know it possible (Merton 1973a; NAS 1995; White 2001). In fact, what is personal, private, and subjective must be validated socially by the community of scientists to count as "objective" scientific knowledge. Thus, while paradigms that operate in a particular discipline tend to influence perception and constrain scientific research and thought, they also serve to correct and enhance it.

Sometimes, as in the case of Drs. Marshall and Warren, the individual scientist or group of scientists opposing a dominant paradigm in their field turn out to be right. When other scientists begin to shift their beliefs and assumptions and work on the problems created by the new theory, the field undergoes a paradigm shift—what Kuhn (1996) calls a "scientific revolution." Scientific revolutions, such as the one that resulted in the shift from Ptolemaic to Copernican astronomy, or from the belief that the earth is flat to the belief that the earth is round, are very well-known paradigm shifts. Other scientific debates in contemporary society that may involve conflicting paradigms include the discussion of whether evolution proceeds gradually or by leaps (gradualism versus punctuated equilibrium); whether the universe is expanding and contracting or is a steady-state universe; the cause of dinosaur extinction (volcanic activity and/or global climate change versus meteor storm). Can you think of any debates that may involve conflicting paradigms in your field?

In these special instances of revolutionary or extraordinary science, as in normal science, scientific knowledge must be validated by the community of scientists (Toulmin et al. 1984; Kuhn 1996; Good 2000). As we will discuss next, communication is essential to that process and thus is central to science itself.

# 1.3 The Centrality of Communication in Science

*The centrality of communication in science?* To the casual observer, the phrase may seem somewhat nonsensical if not patently false. After all, what matters in science is the science itself—hypotheses, research methods, results. Certainly, these are fundamental to good science. Yet, without the communication of those hypotheses, methods, and results to other scientists, no science would be possible. Scientists in and across fields would not be able to share or build knowledge on the results of other scientists. Science would become a private, redundant, and ultimately futile endeavor.

Some external factors exert enormous pressure on scientists *not* to share. Scientists who work for industry or for government agencies, for instance, often find there are restrictions on what they can talk about. But scientists generally agree that secrecy is bad for science (NAS 1989, 1995; see Huizenga 1992). One of the fundamental principles of science is free and open communication. The National Academy of Sciences puts it bluntly: "If scientists were prevented from communicating with each other, scientific progress would grind to a halt" (1989, p 10).

Barry Marshall's sharing of his theory prompted a rethinking of the field of gastritis, ulcer treatment, and the curing of patients. His presentation in Brussels did "arouse much interest" (Lancet 1984, p 1336), and the desire and necessity to check out his theory propelled scientists in the fields of microbiology, gastroenterology, and internal medicine into a flurry of research. Hundreds of articles investigating the existence, nature, classification, detection, and treatment of *H. pylori* followed, eventually lending support to Marshall's contention.

For the National Academy of Sciences, communication is the engine that drives the "social mechanism" of science (NAS 1989, p 10). And as communication scholar William White (2001) puts it, "[C]onceptual innovation is both a fundamental scientific activity and essentially a communication phenomenon" (p 290). We have already noted that the communication of hypotheses, research methods, and results—in journals, at conferences, over email—is essential for the growth of science. *Sharing ideas is essential to the evolution of every scientific field.* In this sense, it is communication that binds any discipline into a community, which makes science social. Language is the basis of any society. Without language, there could be no communication, no cooperation, no concerted research effort that we note in the investigation of *H. pylori* and other scientific endeavors.

Indeed, communication in science is so important that *the credit for a scientific discovery is awarded not to the scientist who discovers a phenomenon, but to the scientist who publishes its discovery first.* (Actually, even more telling, it is the date the paper is received by the journal that determines historical priority.) This is another way in which communication, and writing itself, is central to the conduct of science. Credit can be awarded only to the scientist(s) who write up their findings—and who write first! The practice of attributing originality to the scientist whose paper first reaches the offices of the journal, along with the promise of quick turnaround time for publication and support by the institution and/or journal in disputes concerning the ownership of ideas and discoveries, was begun by the Royal Society of London in the 18th century. By protecting the rights of the author, the Royal Society hoped to ensure open communication and the sharing of ideas in science by alleviating the (real) fear among scientists that their ideas or results would be stolen by others. The National Academy of Sciences cites the example of Isaac Newton, who wrote in Latin anagrams so his findings could be on record but not publicly available (NAS 1989). Marshall and Warren were not the first to observe the presence of *H. pylori* in the stomach, but they were the first to recognize and write about its role in gastritis, and their joint letters to the *Lancet* thus mark the historical point of discovery of that phenomenon (Warren and Marshall 1983).

Since historical priority is awarded to the scientist(s) whose manuscript reaches a publication first, scientists who do original work and want that work recognized and used by the field must write and publish as quickly as possible. Marshall presumably felt this need, for he, like many other scientists, submitted initial reports of his important discovery in the form of brief technical letters or preliminary notes rather than as fully elaborated research reports (the Watson and Crick letter in *Nature* about the double-helix structure of DNA is a famous example of this genre). Speed of publication is necessary not only because of the rapid advance of the field but also because of scientists' need to ensure their claim of originality (Merton 1973b; see Miller and Halloran 1993).

Thus, while scientific journals protect the science that is submitted and published, the necessary speed of publication in most sciences also creates competition: "researchers who refrain from publishing risk losing credit to someone else who publishes first" (NAS 1989, p 9). It is a matter of historical fact that eminent scientists such as Darwin, Watson and Crick, and others were pressured *to write up* and publish their results to beat the competition. Darwin didn't want to write at all, and only did so when he heard that Alfred Russell Wallace was about to publish a theory of evolution (Campbell 1975); Watson and Crick were hotly competing with another lab to be the first to announce the structure of DNA (Watson 1968; Halloran 1984). Learning to write quickly and well is important in science.

There are other, less obvious but absolutely crucial ways in which communication has come to play a key role in science. The processes of writing and submitting papers, of giving presentations, and of writing grant proposals in a real sense *define the nature and activity of the field and the state of knowledge within it.* The acceptance or rejection of conference abstracts, presentations, papers, grant proposals, and the like by conference organizers, journals, funding agencies, and peers becomes a vehicle not only for the dissemination but also for the control of scientific research. As the National Academy of Sciences states, "At each stage, researchers must submit their work to be examined by others with the hope that it will be accepted. This process of public, systematic skepticism is critical in science" (1989, p 10).

*Peer review is the primary mechanism through which such gatekeeping is accomplished.* In addition to the dissemination of scientific research, it is the function of research journals to ensure "quality control" by deciding what is acceptable to publish in the field. These decisions are typically accomplished via a "peer review" system in which journal editors send the manuscripts they receive from researchers to other experts working in the same field for evaluation. After soliciting evaluations on a given manuscript from several such experts, an editor uses their assessment to make a decision about whether the paper merits publication in the journal. (Authors also receive the comments of the reviewers, which influence subsequent revisions.) Through the process of peer review, journal editors and reviewers determine what gets published and thus influence what scientists read and, to some extent, what scientists work on (Bazerman 1983; NAS 1989; Rowland 1997; Relman 1999). (Marshall's three-year clinical study did not pass peer review because it was inconclusive [Chazin 1993], but his smaller studies and technical notes were allowed into print.) Similarly, research funds are typically allocated using a peer review process (Seiken 1992). Funding agencies such as the National Science Foundation, the Department of Energy, and the National Institutes of Health evaluate proposals or requests for funding by assigning them to appropriate groups of experts for peer review; then, based on those recommendations, the funding agency decides which studies to fund, thus again determining what kind of research can proceed in the field (Myers 1985; Seiken 1992). (We discuss mechanisms of peer review in more detail in Chapters 3, 5, and 6, and consider related ethical issues in Chapter 9.)

Most scientists believe that peer review, as the social system of checks and balances that works against personal bias and ensures quality control, is at the heart of good science. Scientists therefore take this system of checks and balances very seriously and believe that violating or circumventing it makes for risky science. One of the ways the system can be circumvented is by releasing scientific studies or results to the public prior to peer review and publication in professional journals; this prior release has been dubbed "pre-publication" (Ingelfinger 1977). Except for presentations at scientific meetings and conferences (Angell and Kassirer 1991), journal editors become concerned when scientific results are released prematurely or independently of scientific publication. The National Academy of Sciences summarizes this concern:

> Bypassing the standard routes of validation can short-circuit the self-correcting mechanisms of science. Scientists who release their results directly to the public—for example, through a press conference called to announce a discovery—risk adverse reactions later if their results are shown to be mistaken or are misinterpreted by the media or the public. (1989, p 10)

It is because of this gatekeeping function of the peer review process, the need and desire to control the flow of information and ensure "quality control," that fax, email, and Internet discussion lists, which have no quality control mechanisms in place, are a concern for professional scientific organizations (NAS 1995). We will return to this concern in a discussion of electronic publication in the last section of this chapter.

These traditional gatekeeping mechanisms, and the entire consensus process by which theories and results are verified and accepted as knowledge in the scientific community, depend on the fair, accurate assessment of research. *Papers must be written in a way that makes the science accessible, testable, and acceptable* to journal editors and other colleagues in the field. As proposed many years ago by philosopher of science Karl Popper (1959) and still widely believed, a scientific hypothesis or theory must be susceptible to falsification to be valid; that is, a hypothesis or theory must be wholly testable (as opposed to tested) before it can be accepted by the scientific community as a valid hypothesis or theory. One way in which findings are tested is through replication, in which scientists in a later study repeat or build on the methods and results of an earlier one. The practice of replication as standard procedure was recommended at the beginning of modern science by Francis Bacon to address the untrustworthiness of the senses and mind in interpreting what we see, playing tricks with our perception and understanding of reality (Bacon 1605).

In addition, *the development of knowledge in science depends on the willingness and ability of scientists to share information* after *publication.* Research reports must provide enough information for readers to evaluate the plausibility and rigor of the researchers' theory, methods, and results; but scientists can't and don't include every detail of their experiments in their reports (Berkenkotter and Huckin 1995). Researchers must be willing to make further details available to others working in the field. This is especially important when you consider that scientists cannot be present at each other's observations and experiments. Rather, scientists must rely on how observations and experimental results are presented in writing. Here again we see the centrality of writing and communication in science. As Popper (1959) pointed out, the statements contained in a research report come to embody and represent the science itself. Hypotheses, theories, experiments, and results are primarily presented, obtained, and critiqued through publication (Bazerman 1988; Winsor 1993; White 2001).

**EXERCISE 1.2**

1. Carefully read the summary contained in Figure 1.1. Describe the sequence of events. Now think about the discussion in this chapter. Based on the information given here, why do you think Pons and Fleischmann did what they did? Explain the reaction of the scientific community. What principles of science were involved? What principles of scientific communication did Pons and Fleischmann violate? Were they justified in doing so? Can you think of a scenario in which scientists would be justified in doing so?

2. Now compare the story of Pons and Fleischmann with that of Marshall and Warren. What are the similarities between these two cases? What are the differences? Do you think the differences between these two cases have anything to do with the eventual acceptance of Marshall and Warren's theory, or with the rejection of Pons and Fleischmann's? Explain. Why do some scientists still argue in favor of cold fusion and seek to duplicate the results that Pons and Fleischmann said they achieved? What should Pons and Fleischmann (or Marshall) have done differently? Why?

# 1.4 The Role of Persuasion in Scientific Communication

We have already discussed the importance of sharing information in science. But facts do not speak for themselves. Rather, facts are interpreted and presented as evidence in scientific arguments contained, for example, in research reports, conference presentations, or grant proposals. In the two cases we have examined, that of Marshall and Warren and that of Pons and Fleischmann, the initial failure to gain acceptance for a theory can be directly attributed to a failure to convince

On March 23, 1989, at a Salt Lake City press conference called by the University of Utah, two electrochemists, Dr. B. Stanley Pons (University of Utah) and Dr. Martin Fleischmann (University of Southhampton, England) announced to the world that they had achieved cold fusion. They claimed that their electrolysis experiment produced four times the amount of energy required to run the experiment—not by the tremendous heating and smashing and splitting of atoms (fission), but by bringing together positively charged atomic (deuterium) nuclei at normal room temperatures (fusion). The benefits of their method of achieving cold fusion would be that deuterium is available in seawater and produces much less dangerous radioactivity, and thus that the process would not require nuclear reactor facilities. Pons and Fleischmann, both chemists, thought they had made a major breakthrough in nuclear physics, where research into the possibility of cold fusion had been going on for years without much hope of success; they thought their breakthrough would benefit the entire world (Crease and Samios 1989; Maddox 1989). A flurry of experiments followed, with major research labs around the world (MIT, Cal Tech, Harwell in Britain, for example) diverting attention and money to cold fusion projects.

On March 24, the day after Pons and Fleischmann's press conference, *Nature* received a paper by a team of physicists, led by Stephen E. Jones, working on cold fusion at Brigham Young University. This paper made much more modest claims about cold fusion. Pons and Fleischmann apparently had a prearranged agreement with Jones, made at a March 6 meeting between the scientists and the presidents of their respective universities, to submit their papers simultaneously to *Nature* on March 24 (Huizenga 1992). However, on March 11, unknown to Jones, Pons and Fleischmann submitted a paper to the *Journal of Electroanalytical Chemistry* (Fleischmann and Pons 1989); the revised version was received on March 22, the day before the press conference on March 23, and appeared in the *Journal of Electroanalytical Chemistry* on April 10, 1989. The paper had been faxed around the world so many times that only the words "Confidential—Do Not Copy" were legible

(Huizenga 1992, p 24). This paper was later followed by the publication of extensive corrections, called "errata," including the omission of the third author, Marvin Hawkins. Contrary to what was widely believed and reported in the press, Pons and Fleischmann never submitted a paper to *Nature* (Huizenga 1992). On April 26, 1989, the University of Utah asked for $5 million from the Utah state legislature, and Pons and Fleischmann appeared before the U.S. Congress to ask for an additional $25 million immediately (and $125 million later) for a Cold Fusion Institute to continue their research. The paper by Jones et al. was published in *Nature* on April 27. At the American Physical Society meeting in Baltimore on May 1 and 2, other groups reported negative results from cold fusion experiments, and at the American Electrochemical Society meeting in Los Angeles on May 8, Fleischmann reported flaws in some of Pons and his original results. On May 18, the first full-fledged critique of Fleischmann and Pons's paper appeared in *Nature*. Petrasso et al. (1989a) criticized the research on the grounds that it lacked adequate controls and that the equipment may have been miscalibrated, and attributed the reports of energy production and other by-products to those errors. Even though it was "only" a "preliminary" or "technical note," many scientists thought that Fleischmann and Pons's paper should have been revised again before being published. As Huizenga comments:

> When the paper was finally available for examination by an anxious scientific community, most readers were shocked by the blatant errors, curious lack of important experimental detail and other obvious deficiencies and inconsistencies. David Bailey, a physicist at the University of Toronto, said the paper was "unbelievably sloppy." He was quoted as saying, "If you got a paper like that from an undergraduate, you would give it an F." (1992, p 24)

Scientists also complained that there wasn't enough information in the paper for others to replicate the experiment. Pons and Fleischmann refused to answer criticisms directly or to provide crucial details of their experiment (Petrasso et al.

*(continued on page 14)*

**FIGURE 1.1** Chronology of communication events in cold fusion. Among other sources, we are indebted to Huizenga (1992) for the basis of the chronology here.

1989b; Huizenga 1992). As reported by the *New York Times:*

> Drs. Pons and Fleischmann offered little help to the unfortunates struggling to repeat their work; they declined to provide details of their techniques and refused to send samples of their equipment to laboratories for analysis. . . . When someone claimed that it was not possible to produce cold fusion, the two Utah [sic] scientists would add more instructions. As Robert Park, head of the Washington office of the American Physical Society, remarked, "Anytime someone did the experiment with no results they would say, 'You didn't do the experiment right,' and offer up another tidbit." (Crease and Samios 1989, p 3D)

When asked for more information, Pons and Fleischmann claimed "that they preferred to press on with more urgent work rather than stop to handle the reviewers' criticisms" (Crease and Samios 1989, p 3D).

As researchers failed in their attempts to test or reproduce Fleischmann and Pons's results, many of the big laboratories terminated their expensive cold fusion experiments. By early July, a special advisory panel to the Department of Energy, co-chaired by John Huizenga, had recommended against awarding special funds for cold fusion research. Undeterred, Pons and Fleischmann and other supporters (see Moore 2000) continue to stand by the discovery and to work on cold fusion. According to other scientists, however, subsequent research, including a paper published by Pons and Fleischmann in 1993 (Pons and Fleischmann 1993), has added little to what is already known (Amato 1993; Dagani 1993). The field is now split between "believers" and "nonbelievers," both of whom continue to try to convince each other of their positions (Dagani 1993; Greenland 1994).

**FIGURE 1.1**    *(continued)*

colleagues of the validity of the work. The problem is not only a matter of methods and data; it is also a matter of the accessibility, quality, and presentation of evidence—the persuasiveness and style of the argument made.

Neither Marshall and Warren nor Pons and Fleischmann did a particularly good job of persuading their colleagues. In the end, Marshall's critics were persuaded by the results of other researchers, made possible in part because Marshall and Warren provided enough information to make their theory testable; Pons and Fleischmann apparently did not. While Marshall and Warren's theory has revolutionized the field of gastroenterology, Pons and Fleischmann's theory is still hotly disputed, as is their professional credibility.[1]

Persuasion is central to scientific communication. Persuasion tends to be a dirty word in our culture, and a tricky subject in science, which traditionally prides itself on objectivity. But in addition to acceptance by editors and reviewers associated with journals and funding agencies, the work of scientists must ultimately be accepted by the scientific community at large. As mentioned previously, science consists of those findings that have survived the scrutiny of the community and the test of time. Individual findings take on the status of scientific

---

[1]In 1996, a judge ruled against Pons and Fleischmann in a libel suit they had brought against an Italian journalist who had reviewed their work in a book on scientific fraud. In his decision, the judge ruled that the journalist's review was justified, citing "important opposition from the scientific community, not just against the theory of the research and the way the experiments were conducted, but also the way the data were divulged and the conclusions reached about the future direction of research" (Abbott 1996). Yet, cold fusion research does still continue and has some outspoken advocates (see Moore 2000; also see the conference abstracts in Chapter 5, page 107).

knowledge as they are accepted by more and more members of the field. Thus, the process of building scientific knowledge is best described not through individual facts, but through the achievement of consensus about what counts as fact (Kuhn 1996; Good 2000).

And this consensus is created through scientific argument (Prelli 1989; Lyne 1998). In the cases of Marshall and Warren and of Pons and Fleischmann, we get a glimpse of the importance of persuasion, argumentation, and debate in the construction of scientific knowledge. In later chapters you will explore this dimension of science in your own field. To briefly illustrate the role of argumentation in science here, let's return to the case of Barry Marshall. In the exchange of technical letters in the *Lancet* that followed Warren and Marshall's initial letters, scientists focused not only on methods and data, but also on proving or disproving the validity of Marshall's argument. Debate raged not so much about data, but about reasoning from the data. Commenting on a 1989 paper by the Marshall team, Walter Peterson (1989, p 509) pointed to "a number of problems *with this paper* that compel me to urge that *its recommendations* not be accepted" (emphasis ours). Once published, the letters themselves became the object of critique.

## EXERCISE 1.3

In Chapter 10 we have reprinted five letters commenting on developments following Warren and Marshall's *H. pylori* announcement in 1983: Veldhuyzen van Zanten et al. (1988); Lam (1989); Marshall, Warren, and Goodwin (1989); Loffeld, Stobberingh, and Arends (1989); and Bell (1991). In each letter, what is being debated? Is it the facts of the case or the writers' argument? Note all the places where the *reasoning* of the scientists—and thus the persuasiveness of their argument—is being questioned. What is being questioned? Their logic? Their evidence? Their terminology? Their beliefs? Their judgment? Look for instances of each of these categories and create additional categories if necessary.

Persuasion is created not only by the logic of arguments but also by presentation and style (Myers 1985; Montgomery 1996; Fahnestock 1999). Earlier, we said Marshall had trouble getting people to listen to him because of the way he answered questions at the conference in Brussels. Certainly, a part of the problem was Marshall's position as a young internist speaking before seasoned experts in gastroenterology, an outsider working against the dominant assumptions in the field. Marshall himself admitted to the press that the odds were stacked against him. But critiques by his colleagues in the popular press indicated that they were skeptical not only because of his youth and casual appearance but also because of how he presented himself. Beyond one's past "reputation," the persona one projects through language—what Aristotle called *ethos*, the persuasive character of the speaker or writer created in and through language—can be understood to operate in scientific communication as well (see Halloran 1984; Miller and Halloran 1993; Constantinides 2001). Although the contents of his conference paper may have been appropriately qualified and cautious, Marshall struck listeners as brash and reckless because of his presentation style and the way he answered questions:

> Unschooled at such presentations and filled with boyish eagerness, he refused to respond to questions in the measured, cautious manner of most researchers. Asked whether he thought the bacteria were responsible for some ulcer disease, Marshall replied, "No, I think they're responsible for *all* ulcer disease." Such blanket statements, backed only by small studies and anecdotal case histories, alarmed many researchers. (Chazin 1993, p 121–22)

Pons and Fleischmann too, were criticized for overstating their claims. Physicist Stephen Jones, on the other hand, made more modest claims in his report in *Nature*, and so he was more believable. As the *New York Times* reported, Jones's "colleagues took him seriously not because he was one of their own, nor even because he showed up at all the important meetings to defend his work. Rather, it was because his work betrayed an awareness of potential pitfalls" (Crease and Samios 1989, p 3D). As we will see in subsequent chapters, that awareness is reflected not only in what is said, but also in how it is said. Sociological research has shown that the kinds of arguments and styles employed in formal scientific communication often differ from those in informal settings. Much gets said in the lab that would not be said in more formal forums such as the research report or grant proposal. In formal communication, scientists employ a style that subordinates their personal preferences and professional allegiances (Merton 1973a; see Couture 1993). Regardless of the validity of his claims, Marshall's enthusiasm seemed inappropriate in this formal context. Even Walter Peterson, one of Marshall's most staunch opponents, says, "We scientists should have looked beyond Barry's evangelical patina and not dismissed him out of hand" (Chazin 1993, p 124). But the question is: Can scientists look beyond style of argument, appearance, and delivery when this is how science is presented?

The reaction of scientists to Marshall's and to Pons and Fleischmann's presentations of their research illustrates the central role of argument and style in scientific communication. To be persuasive, scientists must make the claims of their research believable in the context of the previous research and the existing paradigm of the field; and they must present these arguments in professional forums and styles that are acceptable in the scientific community.

# 1.5 Scientific Communication and Convention

As illustrated by the preceding discussion, the forums and styles a scientist chooses can make a difference in how well the results of his or her research are heard and understood. The conferences you attend and in which you participate, the publications to which you submit your work, the funding agencies to which you apply, even the institutions for which you work—all can make a difference in how well your research is received and whether it is used by other scientists.

As you will discover in working through subsequent chapters, different types or genres of writing follow different conventions and rules, both implicit and explicit. Understanding what these conventions and rules are and how to use them to demonstrate the nature and significance of your research to other scientists in your field is one of the things that distinguishes a professional scientist from a student scientist. Learning to be a professional scientist means learning both the science and the conventions of communication in your discipline.

While we can't go deeply into these conventions and rules in this introductory chapter, we can talk about them in more general terms. For example, although scientific research is systematic, the process of doing science is really a much more creative and sometimes even haphazard process than is commonly thought (the National Academy of Sciences published *On Being a Scientist*, a booklet intended to attract more students to science, in part to demonstrate this creative dimension of science). Yet, when science is written up and published in a typical research report, it is not *presented* to the scientific community as a personal narrative or story of "what happened" (including missteps and mistakes) in the actual order that it happened. Although the report may include a chronological description of methods, this description is embedded in a broader argument in which a claim or hypothesis is supported or refuted. Thus, several scholars have pointed out that the traditional report does not accurately represent the processes of scientific research (Medawar 1964; Bazerman 1988; Gross 1990). In writing a report, the actual process of scientific imagination and discovery is reconceptualized along the empirical and mathematical lines of the argument needed to justify the science to the scientific community (see Carnap 1950; Holton 1973; cf. Feyerabend 1978; Fuller 2000). The four-part structure of the conventional research report (introduction, methods, results, discussion) requires the writer to begin not with the first step in the experiment but with an argument for the significance of the hypothesis. Personal narrative does play a role in science; it is often employed to communicate science to general audiences (see Katz 1992a; Jorgensen-Earp and Jorgensen 2002), particularly in general-interest magazines and television programs, and often on email, where discussions are more informal. But in a formal research report in most fields, the narrative form probably would be considered inappropriate in convincing other scientists of the validity of research and might actually undermine that attempt.

Throughout this book you will be learning the communication conventions accepted by your field. We also will talk about some of the ethical considerations of this socialization process in Chapter 9. It is important for students contemplating becoming professional scientists to know the conventions of their field, to understand the underlying assumptions and attitudes that give rise to those conventions, and to understand how to work within them. As you will explore in more detail in subsequent chapters, these conventions include not only the structural features of major genres, such as research reports and grant proposals, but also the styles in which these are presented. For now, we would point out that all these genres and styles are social conventions that create the channels through which, and that influence the ways in which, scientists interact and collaborate with one another.

## 1.6 The Role of Collaboration in Scientific Communication

If science is a community and communication in science is social, it should be no surprise that the process of writing in science is also social in nature. Science itself tends to be more collaborative than work in some other fields "because of the specialization and sophistication of modern research methods," and "the increased emphasis on interdisciplinary research being prompted by funding agencies" (Macrina 2000, p 157, 158). Given this trend, the role of communication becomes even more important. (In our citing of scientific research in this book, note the number of times "et al.," designating multiple authorship, has been used; even the shorter technical letters to the editor are often written collectively or represent a group of scientists.) In modern science, collaboration is necessary for the development of theory and the conduct of research. The discovery of the role of *H. pylori* in the development of stomach ulcers clearly has a collaborative history. Marshall drew on earlier studies to support his claim that there is a connection between bacteria and ulcers by showing that bacteria in the stomach lining already had been noted but had been overlooked as a cause of ulcers; Marshall and Warren wrote joint letters and a research report together to argue this claim; finally, other scientists validated, confirmed, and extended Marshall and Warren's research and began testing and developing treatment. The creation of scientific knowledge is truly a collaborative process.

In the Marshall and Warren case, we see how scientific progress involves collaboration within the discipline of gastroenterology. In addition to in-field work, collaboration increasingly occurs across disciplines. A good example of this kind of interdisciplinary collaboration can be found in the case we present in Chapter 13 on the Delphic oracle. In this research, an archeologist, a geologist, a chemist, and a clinical toxicologist eventually teamed up to work on a theory that the legendary trances and visions and occasionally violent frenzies of the prophesying priestesses at the Temple of Apollo at Delphi were caused by vapors that rose from a subterranean fault beneath the temple floor. First postulated by writers in antiquity, this theory was virtually dismissed in the early 20th century by archeologists and geologists who first excavated the site and found no large faults or volcanic activity in the region (Spiller et al. 2002).

Also an example of how science is "rooted in serendipity, hard work, and productive dreaming" (Broad 2002, p D1), two faults—one of them exposed by a widening of the road to accommodate the need for tourist buses to turn around!— were first discovered at the Delphi site by Jelle Zeilinga de Boer, a geologist hired by the Greek government to determine the geological conditions for constructing nuclear reactors in the area (Broad 2002, p D4). When De Boer discussed his puzzling finding over a bottle of wine in Portugal with archeologist John Hale, the two hypothesized that, contrary to accepted belief in their respective fields, these fissures may have been the source of the gas that inspired the Delphi priestesses' prophetic pronouncements and visions. De Boer and Hale applied for and received permission from the Greek government to take geological samples from the site. These samples were analyzed by a geochemist, Jeffrey Chanton, who confirmed the presence of ethane, methane, and, most important, ethylene gases trapped in the rock. These three experts from relatively different fields were joined by a fourth, toxicologist Henry Spiller. Spiller compared the pharmacological effects of the gases when used in 20th century anesthesiology to those reported in ancient texts, and confirmed that the effects were similar, if not identical, and thus affirmed the likelihood that ethylene was responsible for the behavior of the priestesses at Delphi (Spiller et al. 2002, p 193).

In the case of the Delphic oracle, we once again see how scientists, this time from different fields, needed each other's expertise to develop and test a hypothesis. One dimension of this case that is particularly interesting is that this collaboration occurred not only across the boundaries of disciplines, cultures, and languages, but across the geography of time as well. This diverse team of scientists relied not only on each other, but also on the evidence provided by the testimony of another group of diverse authors, separated by time and space: ancient Greek and Latin philosophers, historians, poets, orators, geographers, travel writers, and biographers (Spiller et al. 2002, p 190), particularly Plutarch (see De Boer et al. 2001, on page 406). The keen observations and reports of these ancient writers provided not only the questions that guided the hypothesis of the team of contemporary researchers, but also some compelling evidence (e.g., see De Boer et al. 2001, p 710).

Scientific discovery and investigation may and often do depend on the successful interaction of researchers from diverse scientific fields. This is especially true in "big science," which involves teams of experts from varied fields, including administrators, technicians, politicians, and other nonscientists. For example, imagine the range of experts and professionals involved in the construction, deployment, repair, and maintenance of the Hubble Space Telescope, and in the continuous processing and publishing of images from it. Or the number of scientists and engineers from a variety of countries involved in the construction of the International Space Station. Or the Human Genome Project! In physics, biomedicine, and other fields, the effect of collaboration is vividly illustrated by the growing (and problematic) number of authors appearing on articles (McDonald 1995; CBE Task Force on Authorship 2000).

John Ziman (1968) has called the schools of thought that emerge when researchers regularly work and/or publish with each other "invisible colleges." But not all collaborations have to be large, and sometimes the professional interaction can be quite varied. Thomas Edison relied on an array of lab assistants and workers in all sectors of society (including the press) to develop and promote his inventions, especially electric power and light (see Bazerman 1999). Uglow (2002) describes the close and productive friendship of James Watt, Erasmus Darwin (Charles' grandfather), Josiah Wedgwood, and Joseph Priestly, who together created an informal "society" to share ideas and provide mutual support. As a result of this alliance, for example, Wedgwood sculpted the ceramic equipment Priestly needed to eliminate contamination in his work on gases.

In all of these examples we see the benefits of multiple skills and perspectives that scientists from different fields can bring to an exploration or problem. In the case of the Delphic oracle, geologist De Boer had read Plutarch but was not as familiar with the literature in the field of archeology (Broad 2002, p D4), so he needed John Hale to frame his geological observations, just as Warren needed Marshall's clinical experience to frame his lab results. De Boer and Hale, in turn, needed chemist Jeffrey Chanton to analyze the site samples and toxicologist Henry Spiller to compare and interpret the possible physical effects of the gases on humans. We also see in this case that interdisciplinary collaboration can stretch not only across disciplinary space but also across the vaster reaches of time. In the work of Drs. De Boer, Hale, Chanton, and Spiller, contemporary scientists used ancient observations to develop research questions and hypotheses. But they also attempted to answer a question posed by ancient thinkers and writers using modern methods and experimental techniques (see the last sentence of De Boer et al. 2001, on page 409), just as scientists do indirectly in one way or another every day.

Although perhaps less obviously than in the case of the Delphic oracle, the use of historical research occurs in other fields as well. For example, note how Marshall's letter in the *Lancet* (page 233) situates his study of bacteria in the stomach lining in the context of prior research going back more than half a century, and how he uses that history to build a case for the significance of his own findings. (This is one of the purposes of Marshall's 2002 book as well.) In Figure 1.2 below, we see the relevance and use of ancient Chinese history in what may at first appear to be a most unlikely place: astrophysics.

In recognizing the role history can play in scientific collaboration, we also more clearly see the "situatedness" of collaborative research. A research program develops in a specific time and place. History, politics, economics, and culture all influence the shape of science, just as science influences them (e.g., see Bazerman 1983, 1999; Lewontin 1993; Lyne 1998; Fuller 2000). To begin to test the feasibility of their hypothesis, De Boer and Hale had to request that the Greek government allow their team to take samples from this ancient site (see pages 402–403). For the Greek government and the Greek state, this is not exactly a context-free request: throughout its history, and often without permission, explorers and collectors from other countries have carried away pieces of ancient Greece. This is yet another kind of collaboration, between scientists and governments—in this case, a foreign government. Just as specific research projects are planned and carried out in the context of a paradigm and fields, science as a social enterprise is always situated. History is ongoing and contemporaneous.

Collaboration, like science itself, is also "situated" in technological developments. Technology constitutes a context as well as a means for scientific collaboration. This is not only a matter of the development of new scientific machines, apparatus, and techniques, but of technological developments in communication as well. Innovations in the computer and communication industries obviously facilitate and increase collaboration in science. Much collaborative work in science, especially but not exclusively, international collaboration, is taking place over email, as well as on electronic discussion boards and newslists. Scientific societies are forming and invisible colleges emerging on the World Wide Web. BioMedNet, for

---

Clark and Stephenson (1977) describe the historical records reporting the supernova of 386 AD, during the Chin dynasty. Little information is available, except the asterism (locating the object to within about 15°) and the fact that the supernova disappeared after about 3 months (not for seasonal reasons). They judge the uncertainty in duration to be about 1 month; they also conclude, from lack of Chinese observations of Mira, that to be noticed an object would require an apparent magnitude $\lesssim 1.5$ mag, consistent with a supernova at a distance $d \sim 5$ kpc.

The positional and distance information led Clark and Stephenson to associate SN 386 AD with the Galactic plane supernova remnant G11.2–0.3. . . . [T]he surface brightness and structure have been used to argue for a *smaller* age than 1607 yr (Downes 1984). It seems unlikely that a second supernova remnant, within a few hundred years of SN 366 AD, should occur in the same region of the sky and go unnoticed by the Chinese. We shall henceforth adopt the hypothesis that G11.2-0.3 is the remnant of SN 386 AD, and examine the consequences.

**FIGURE 1.2**   From the Introduction to "X-ray evidence for the association of G11.2–0.3 with the supernova of 386 AD," Reynolds et al. (1994, p L1).

example, bills itself as "The world wide club for the biological community," and as of the end of 2002 boasted a membership of 1,133,819 (http://www.bmn.com).

Open-source databases that collect and provide electronic access worldwide to multiple journals and articles also are proliferating on the Internet. Part of PubMed Central initiated by NIH in 2000, the BioMed Central homepage states: "This commitment [to open access] is based on the view that open access to research is central to rapid and efficient progress in science and that subscription-based access to research is hindering rather than helping scientific communication" (http://www.biomedcentral.com/start.asp).[2] The Public Library of Science (PLoS) also was created and is managed by a nonprofit organization of scientists who want to make the "world's science and medicine literature a public resource" (http://www.publiclibraryofscience.org/). And of course, libraries and other information services around the world are always developing databases, as well as new information systems, to store, retrieve, and otherwise handle the massive amounts of scientific research contained in them. Many journals, such as the *New England Journal of Medicine* and the *Journal of the American Medical Association*, are taking advantage of the flexibility of the electronic medium by publishing in both print and electronic formats, often adding features online that are not available in print. In addition, electronic journals that publish scientific research exclusively online continue to emerge. "arXiv.org" was started by Paul Ginsparg at the Los Alamos National Laboratory in 1991, not only to archive and disseminate print journals, but also to create journals and publish articles on high-energy physics in electronic-only form (Ginsparg 1994); "arXiv.org" has since expanded to create e-journals in other areas of physics, as well as in mathematics and computer science (lanl.arXiv.org). In 2002, Ginsparg, who is now at Cornell University, was named a fellow of the John D. and Catherine T. MacArthur Foundation for his pioneering work in the dissemination of scientific knowledge (MacArthur Fellows 2002).

Proponents of e-journals as *the primary vehicle* for scientific research point out that e-journals publish scientific advances faster and more easily, cost less to produce, and afford greater access to scientists around the world than do print publications (Harnad 1997). In fact, both dual media and e-journals have been heralded as a boon to so-called third world nations that traditionally have not had easy (or in some cases any) access to cutting-edge research (Samarajiva 1990; WHO 2000). Some scientists even believe that by publishing and reaching more scientists, e-journals in particular will help "break down social barriers between central and peripheral scientists and thereby . . . increase the size of 'invisible colleges'" (Matzat 1998). At the same time, there is some concern that not only email and discussion lists, but also e-journals will have fewer quality controls built in because of the emphasis on speed and availability of publication, and so might undermine the system of checks and balances of the traditional peer review system (Rowland 1997; Relman 1999). We will explore these issues in more depth later. Suffice it to say here that issues of scientific collaboration and the development of new technologies are closely intertwined.

---

[2]To read about the NIH proposal for BioMed Central and some of the surrounding controversy, see Ingelfinger 1977; Angell and Kassirer 1995; Butler and Wadman 1999; Day 1999; Delamothe and Smith 1999; Hogan 1999; Lancet 1990; Pear 1999; Relman 1999; Varmus 1999.

This brief overview of collaboration illustrates several important features of the conduct of scientific research: that professional scientists help each other with both the thinking and the writing of science; that readers as well as authors can come from a variety of fields, countries, and even times, a fact made all the more likely by newer electronic forms of communication; and that readers, whether peer reviewers or other professionals, also influence scientific communication (see Gragson and Selzer 1993). Communication is fundamental in science. Indeed, reviewing findings from Latour and Woolgar (1979), Bazerman characterizes the "entire laboratory activity [as] a process of inscription, gradually turning the materials under study into the words and symbols that appear in an article or other scientific communication" (Bazerman 1983, p 165).

## Activities and Assignments

1. To expand your ability to shift paradigms, reclassify, reorganize, and solve problems, represent the information in Figure 1.3 in at least three different ways. This exercise does not necessarily entail any great artistic skill or knowledge in graphics, though you will want to think visually here.
2. Working in a group of students from your field, brainstorm about a theory that has been accepted by your discipline. Prepare to discuss what you know about its history: previous theories to explain the phenomenon, the personalities of the scientists involved, the debates surrounding the newer theory, and the evidence that made it acceptable or unacceptable in your field. What does your discussion tell you about how knowledge is constructed in your field?
3. To demonstrate the value of collaboration as a means of generating ideas and discovering insights, do the NASA simulation exercise in Figure 1.4.
4. Warren and Marshall's joint letters to the *Lancet* announcing their *H. pylori* discovery are reprinted at the beginning of Chapter 10. Write a one- to two-page analysis in which you compare and contrast Warren's letter with Marshall's. How are the letters related? Which arguments are similar and which

**Directions:** Make the information below more accessible and easier to read by classifying and reorganizing it. There are several different ways of formatting the information.

When time is limited, travel by Rocket, unless cost is also limited, in which case go by spaceship. When only cost is limited, an astrobus should be used for journeys of less than 10 orbs, and a satellite for longer journeys. Cosmocars are recommended when there are no constraints on time or cost, unless the distance to be traveled exceeds 10 orbs. For journeys longer than 10 orbs, when time and cost are not important, journeys should be made by superstar.

**FIGURE 1.3**  Breaking paradigms: Representing information in different ways. From Wright and Reid, "Written Information: Some Alternatives to Prose for Expressing the Outcomes of Contingencies," *Journal of Applied Psychology* 57 (1973): 160–66.

Your spaceship has just crash-landed on the lighted side of the moon. You were scheduled to rendezvous with a mother ship 200 miles away on the lighted surface of the moon, but the rough landing has ruined your ship and destroyed all the equipment on board, except for the 15 items below.

Your crew's survival depends on reaching the mother ship, so you must choose the most critical items to take on the 200-mile trip. Working alone, your first task is to rank the 15 items in terms of their importance to your crew in reaching the rendezvous point. In the column labelled "Your Rank" place a number 1 by the most important item, number 2 by the second most important, and so on through number 15, the least important.

When you have finished, your instructor will give you additional directions for working in groups, and then for calculating errors based on the NASA rank.

| | Your Rank | Group Rank | NASA Rank | Your Error | Group Error |
|---|---|---|---|---|---|
| Box of matches | | | | | |
| Food concentrate | | | | | |
| 50 feet of nylon rope | | | | | |
| Parachute silk | | | | | |
| Solar-powered portable heating unit | | | | | |
| Two .45 caliber pistols | | | | | |
| One case of dehydrated Pet milk | | | | | |
| Two 100-pound tanks of oxygen | | | | | |
| Stellar map (of the moon's constellations) | | | | | |
| Self-inflating life raft | | | | | |
| Magnetic compass | | | | | |
| 5 gallons of water | | | | | |
| Signal flares | | | | | |
| First-aid kit containing injection needles | | | | | |
| Solar-powered FM receiver-transmitter | | | | | |

**FIGURE 1.4**    NASA simulation exercise. Adapted from Hall (1971).

are different? Do you detect any difference in emphases and purpose between them? Why do you think Warren and Marshall decided to submit these letters together?

# CHAPTER 2

# Forums for Communication in Science

## 2.1 The Socialization Process: Entering a New Community

As we join a research community, we gradually acquire knowledge of the ways in which that community develops and communicates knowledge. This happens naturally, and usually without conscious effort, as we read papers written by other researchers, attend conferences, and talk informally with new colleagues in the classroom, the field, the development lab, or other research workplace. We are gradually *socialized* into the communication patterns the community has adopted.

You have probably already started thinking as a biologist or soil scientist or wildlife researcher. If you've taken a course or two in your area of science or have pursued a personal interest in the field outside your formal studies, the socialization process has begun. Not only have you begun developing *content knowledge*— the principles, concepts, and terminology that members of the field take for granted; you've also begun to acquire *procedural knowledge*—knowledge of how to do things in this research area: how to solve problems, how to test hypotheses, how to use the basic methods of your field, and how to communicate your concerns, questions, and findings to others in the community. The more familiar you become with the ways of thinking, speaking, and writing in your field, the easier it is for you to quickly understand written texts, to grasp important concepts and recognize issues that matter to the field, and to contribute to written and spoken conversations—that is, to participate in the development of new knowledge.

Your content and procedural knowledge develop together, though not necessarily at the same rate, and are mutually reinforcing. You cannot write like a physicist if you don't know anything about physics. And you cannot acquire knowledge of physics if you're not able to read the texts written by physicists, apply the experimental procedures and theorems that physicists have developed, and communicate with others about your questions and your developing understanding. Thus, the process of socialization involves becoming more sensitive to the *context* of activities in your discipline (Arminen 2000): learning the subject matter of your particular branch of science and also learning how to reason and communicate as a member of your research community.

Theoretically, the more you know about the community you want to join and about the socialization process itself, the easier your initiation into the community will be. Thus, you can simply let the socialization process take its course, or you can actively seek out the distinctive patterns of communication and interaction that characterize your field, rather than waiting to stumble upon them. The discussion and activities in this book are intended to help you recognize the communication patterns that have developed in the scientific community you are joining.

## 2.2 Research Journals and Their Readers

An important first step for any new researcher is to become familiar with the primary journals in the field, for these professional publications represent the principal medium through which individual scientists share their theories and results with others in the scientific community. As Chapter 1 illustrates, the progress of science rests upon the exchange of knowledge among scientists. Indeed, Winsor (1993) describes scientific knowledge as "constructed . . . not just in labs or at field sites, but in arguments that scientists conduct through the medium of scientific papers" (p 128). In journal articles, researchers do much more than describe methods and report results; they present and defend a particular interpretation of those results in light of the field's developing knowledge. The rigorous peer-review system used by most scientific journals ensures that this interpretation is carefully examined by experts before it appears in print. It is through the publication of such arguments that scientific knowledge advances, as results and theories become available to be verified, challenged, refuted, and refined. Journal articles thus represent the heart of the scientific enterprise, and much professional effort is devoted to writing them.

The easiest way to identify the essential journals in your field is to notice those around you. Start by paying attention to what other researchers—your professors, teaching assistants, other students—are reading. When you are assigned research articles to read in your classes, take time to notice where and when the papers were published, what institutions the researchers represent, and what granting agencies provided funding for the research. Identifying the professional associations or institutions that sponsor the journals will help you understand who reads them. Most journals are quite explicit about their audiences and goals, posting editorial goals and guidelines on their websites and publishing them in their print issues as well (in every issue or in one issue annually).[1] For example, the American Institute of Physics publishes or cosponsors dozens of journals, each intended for a specific segment of the physics community. See Figure 2.1 for a sampling of these journals and their intended audiences. Notice as you read the editorial profiles of these journals that the target audiences vary along a number of dimensions, not just in their areas of specialization.

---

[1]For a large sampling of editorial guidelines, see http://www.mco.edu/lib/instr/libinsta.html. At this site, the library of the Medical College of Ohio has assembled links to the editorial websites of more than 3500 journals in the health and life sciences.

### American Journal of Physics

Published by AIP for the American Association of Physics Teachers. Devoted to meeting the needs and interests of college and university physics teachers and students by focusing on the instructional and cultural aspects of physics. Contains feature articles that describe novel approaches to laboratory and classroom instruction and other areas of physics pedagogy.

### The Astrophysical Journal

Published by the University of Chicago Press for the American Astronomical Society. Published three times monthly (in two parts), this publication contains articles on all aspects of astrophysics, astronomy, and related sciences that are primarily applicable to astronomical objects.

### Journal of Applied Physics

Largest general publication medium for research results in applied physics. Contains material applying physics to industry and other sciences. Active fields represented include semiconductor properties and devices, lasers and their applications, magnetic and dielectric materials, plasmas, high-polymer physics, and many more.

### Physical Review A: Atomic, Molecular, and Optical Physics

### Physical Review B: Condensed Matter

### Physical Review C: Nuclear Physics

### Physical Review D: Particles and Fields

### Physical Review E: Statistical Physics, Plasmas, Fluids, and Related Disciplinary Topics

Published by AIP for the American Physical Society, these journals publish original research in the specified areas.

### Physical Review Letters

Published by the American Physical Society. Contains short, original communications in active and rapidly developing areas of physics. The reports are written so that their significance can be appreciated by physicists working outside the area while remaining stimulating to physicists in the same field.

### Physics Today

A semipopular publication that contains articles and news of interest to the reader with only a general interest in physical science, as well as to the professional physicist. Contains feature articles that cover a wide range of topics and interests.

**FIGURE 2.1**   Statements of editorial policy from selected journals published by the American Institute of Physics and affiliated societies (AIP 1990, p 31–33).

We saw in Chapter 1 that the study of science is dynamic: as fields develop and diversify, and as the interests of different fields begin to converge, new areas of research become active and important. Consequently, these new communities, often comprising researchers from different fields, establish specialized journals devoted to their emerging interests. The field of marine ecology, for example, includes researchers in such diverse fields as microbiology, botany, oceanography, conservation, and resource management. The journal *Marine Ecology Progress Series* was created to meet the needs of these researchers, who, despite their different disciplinary backgrounds, have a common set of research interests. The purpose and scope of this journal are clearly announced by the editors at the journal's website and inside the front cover of each issue. This statement is reproduced in Figure 2.2.

# MARINE ECOLOGY PROGRESS SERIES
### Companion Journal to Aquatic Microbial Ecology

MEPS and AME have received international recognition as the leaders in their respective fields of science

**HISTORY:** Marine Ecology Progress Series (MEPS) was founded by Professor Otto Kinne. Its original concept was based on 'Marine Ecology' – the first comprehensive, integrated treatise on life in oceans and coastal waters – conceived, contributed to, organized and edited by Otto Kinne, and published by John Wiley & Sons.

**MEPS is the Citation King 2000.** Of all the journals listed under 'Marine & Freshwater Biology' (including review and society journals) in Journal Citation Reports (JCR) MEPS features the highest number of Total Cites: 14 206.

**AIM:** MEPS serves as a worldwide forum for all aspects of marine ecology, fundamental and applied. The journal covers: microbiology, botany, zoology, ecosystem research, biological oceanography, ecological aspects of fisheries and aquaculture, pollution, environmental protection, conservation, resource management. Ecological research has become of paramount importance for the future of humanity. The information presented here should, therefore, encourage critical application of ecological knowledge for the benefit of mankind and, in fact, of life on earth. MEPS strives for

○ complete coverage of the field of marine ecology
○ the highest possible quality of scientific contributions

○ quick publication
○ a high technical standard of presentation.

**SCOPE:** MEPS is international and interdisciplinary. It presents rigorously refereed and carefully selected Research articles, Reviews and Notes, as well as Comments/Reply Comments*, Theme sections and Discussion forums. Main topics are:

**Environmental factors:** Tolerances and responses of marine organisms (microorganisms, autotrophic plants, and animals) to variations in abiotic and biotic components of their environment; radioecology.

**Physiological mechanisms:** Synthesis and transversion of organic material (mechanisms of auto- and heterotrophy); thermo-, ion-, osmo- and volume-regulation; stress resistance; non-genetic and genetic adaptation; population genetics and ecological genom research; orientation in space and time; migrations; behavior; chemical ecology; molecular basis of ecological processes.

**Cultivation:** Maintenance and rearing of, as well as experimentation with, marine organisms under environmental and nutritional conditions which are, to a considerable degree, controlled; analysis of the physiological and ecological potential of individuals, populations and species; determination of nutritional requirements; ecological aspects of aquaculture; water-quality management; culture technology.

**Dynamics:** Production, transformation, and decomposition of organic matter; flow patterns of energy and matter as these pass through organisms, populations and ecosystems; biodiversity; variability of ecosystem components; trophic interrelations; population dynamics; plankton ecology; benthos ecology; estuarine and coastal ecology; wadden-sea ecology; coral-reef ecology; deep-sea ecology; open-ocean ecology; polar ecology; theoretical ecology; ecological methodology and technology; ecological modelling and computer simulation.

**Ocean management:** Human-caused impacts: their role as modifiers and deformers of living systems, their biological consequences and their management and control; inventory of living resources in coastal areas, estuaries and open oceans; ecological aspects of fisheries; pollution of marine areas and organisms; protection of life in the seas; management of populations, species and ecosystems; management of coastal zones and sea areas; biotechnology.

**Molecular marine ecology**

**Eco-ethics:** Marine ecological research immediately relevant to human thought and conduct oriented to what is right or wrong, beneficial or destructive for the total system 'Homo sapiens plus nature'. For details see the editorial in MEPS 153:1–3 and the Eco-Ethics International Union (*www.eeiu.org*).

*Comments are usually not peer-reviewed. They may include *personal* opinions. For details on Comments/Reply Comments consult MEPS 228:1

**FIGURE 2.2**  Statement of editorial policy for *Marine Ecology Progress Series* (MEPS 2002).

A journal like *Marine Ecology Progress Series* unites researchers from a variety of fields who have a common interest. Most researchers belong to several such research communities or subcommunities, each of which is linked by one or more specialized journals. As scientists' research interests and needs develop, they seek out those journals that are most pertinent to their research agendas, often reading in more than one field.

To stay active in a research field or interdisciplinary area, then, scientists need to be aware of the purposes and intended audiences of the journals published in that area. This knowledge helps them decide which journals they will want to find time to read (of the thousands in publication); and, equally important, it helps them decide where they should submit their own work for publication. Choosing an appropriate journal is a critical step in the publication of a scientific paper. As we will discuss in Chapter 3, "appropriateness for the journal" is the primary criterion used by reviewers in evaluating papers for potential publication. It is important to note that the ultimate influence of a paper in the scientific community may be significantly determined by the journal in which it appears. In science as in other domains, the amount of attention an announcement receives is to some extent determined by the context in which it appears, that is, by where, when, how, and to whom the announcement is made (Miller 1992). A researcher's decision about where to submit his or her paper ultimately determines who will have access to the work.

As a case in point, Chapter 11 includes reports of a research project headed by Dr. JoAnn Burkholder, an aquatic botanist at North Carolina State University. In 1988, Burkholder's team discovered a new genus of algae, a "predatory" dinoflagellate that kills fish by releasing a lethal neurotoxin into the water. In a series of studies over the next few years, Burkholder and her colleagues determined that the alga was responsible for massive fish kills in estuaries along the southeastern coast of the United States and was very likely associated with fish kills in other regions as well. The striking nature of the finding (this was the first known instance of predatory behavior among such organisms), along with its potential international ramifications, warranted submission of the results to the British journal *Nature*, which has a broad, interdisciplinary, and international readership. Published in *Nature* in 1992, this first major paper from the project announced the discovery of the new organism, described its behavior, and briefly listed related fish kills that had been observed. In contrast, subsequent papers focused more narrowly on the results of individual studies of the organism in the lab and in particular estuarine locations. For these later reports, the researchers targeted more specialized audiences within the marine ecology community, publishing in such forums as the *Journal of Plankton Research* and the *Marine Ecology Progress Series* described earlier.

The publications from the Burkholder project also illustrate another important feature of scientific research journals, the variety of *genres,* or types of articles, that they publish. The team's original announcement in *Nature* (included in Chapter 11) was presented as a research letter, an important and increasingly common genre in scientific writing, which is designed to enable researchers to make quick announcements of important findings to the relevant scientific communities. According to *Nature*'s Guide to Authors (2003a), *letters* are "short reports of original research focused on an outstanding finding whose importance means that it will be of interest to scientists in other fields."

In contrast to full-length reports, which often run to 6000 words or more, letters and research notes may be limited to 1000 or 1500 words, enabling researchers to write them more quickly and making it possible for reviewers to read and respond to them quickly, with the goal of getting important information into print as soon as possible. Despite the obvious appeal of this genre for writers, the more traditional full-length report—with its extensive discussion of goals, methods, results, and implications—remains the most common type of journal article published in the sciences. This genre is discussed at length in Chapter 3.

In addition to the primary scholarship presented in letters and full-length reports, many journals, including *Nature*, publish secondary scholarship, or review articles, in which the authors survey recent research on a particular topic of interest to the journal's readership. Often, journal editors will ask a prominent expert to write a review article on a current topic, but some journals accept unsolicited reviews as well. These reviews of previously published results generally provide a quick history of recent research, comparing and contrasting results from a range of studies in an effort to help readers understand where research consensus is forming and what issues are still open for exploration. This genre will be discussed in more detail in Chapter 4.

## 2.3 Research Conferences and Professional Associations

Like research journals, conferences vary in size, scope, and audience. Most are sponsored by professional organizations whose membership and areas of interest vary widely. For example, botany student Alexander Krings reports that researchers in plant community ecology may belong to a small, highly specialized group such as the International Association for Vegetation Science (about 1200 members) and/or to the larger Ecological Society of America (7000 members), whose members come from a wide array of ecology subfields. (Krings's profile of this community is presented in Figure 2.3.)

Conferences represent another formal mechanism that, like the letter genre discussed earlier, has developed to speed up the exchange of scientific knowledge. Most groups meet annually, allowing researchers to present recent results directly to others in their field even before studies appear in print. The conference thus provides opportunities for immediate feedback from other researchers, often on work in progress.

Local, regional, and national conferences play different roles in different fields, and learning about these roles is an important part of the socialization of a new researcher. An easy way to find out about important conferences in your research field is to take notice of conference announcements posted in your department and to pay attention to the details when your professors, lab directors, or teaching assistants go out of town to present papers. Ask where they went, whom they spoke to, what kind of work they presented. Borrow their conference programs or find the program at the sponsoring association's website. Read these materials to get a sense of the kinds of topics the meeting covers and the kinds of sessions scheduled. In addition to panels of formal oral presentations, you'll discover that most organizations also use other meeting formats, such as poster sessions, invited symposia, and workshops. Chapter 5 will discuss the preparation of conference presentations and posters.

# The Plant Community Ecology Research Community

Plant community ecology, or phytosociology, is a study of interactions, of structure, and of composition. The field is broad, with roots in the past. Its history in the Western world, some say, began with early plant geographers such as Alexander von Humboldt (1769–1859). Humboldt traveled extensively in the New World, prolifically describing the natural world around him. Indeed he is credited with coining the word "association" that is still in use among ecologists today. After a period of continued development of the science in the 19th century, it finally grew apart from plant geography in the early part of the 20th century. Today, the field is largely dominated by professionals and academicians studying the causes of community structure as well as the relationships and dynamics of communities. The methods used in the field have, since Day 1, been largely observational. As the root drive of the science is to understand and describe natural plant communities, most efforts have concentrated on strict observation of, and data gathering from, existing communities. There is, however, a growing trend toward controlled experiments in certain subfields, such as those dealing with physiological efficiency research, where observational data no longer prove reliable or accurate.

As with any scientific field of study, the channels of communication within plant community ecology are broad and diverse. They may involve the casual conversation with a colleague in the hallway or the professional presentation of research results as part of an international conference. As the channels of communication are so diverse, they may be grouped easily according to the different levels on which they function—local/intimate, local/formal, and regional/international.

The local/intimate category seems by far the most time-consuming. Forms of communication that fall into this category include casual hallway conversations with colleagues, meetings with graduate students about their respective research proposals/projects, participating on various committees, etc. This is the most intensive level for the individual researcher, as this level builds the ground-

work for the science. Basic questions such as sampling design, matters of statistics, and objectives of studies are addressed on this level. It is this level that sets the stage for the data gathering, the experiments, and the general additions to the body of scientific knowledge. Without the intimate, one-on-one interactions with colleagues, students, and other professionals, there would be no foundation to the science.

The local/formal category closely follows the local/intimate category in terms of time consumption. Forms of communication on this level include the preparation and delivery of lectures, attendance at thesis defenses, and the attendance or delivery of seminars. Frequently, despite the "formality" accorded to the preceding events, there is ample opportunity for one-on-one questioning and discussion. In cases such as seminars, new ideas can have their first chance to be examined while at the same time being distributed to various professionals in a research community. Lectures in themselves play a critical role in furthering the science in that they prepare students for the questions that they are expected to ask in their field. Lectures also serve as gateway communication channels between students eager to enter a field and already established professionals.

The third category of communication channels is the regional/international level. Regional, as used here, more aptly refers to interinstitutional/interagency communication. This level encompasses the justification for various projects, future research needs, peer review of present research, and necessary funding sources. There are various subchannels of communication at this level. Practicing professionals communicate through journals as well as through professional organizations and their conferences.

Of all the various journals in ecology, few deal exclusively with plant community ecology. Of those that do, the most notable examples are the *Journal of Vegetation Science, Vegetatio,* the *British Journal of Ecology,* and *Castanea.* A good example of the breadth of the field and the magnitude of its

**FIGURE 2.3**   Profile of the field of plant community ecology by Alexander Krings. Student sample.

undertakings can be seen in the diversity of the publishing centers. The first three journals mentioned are European creations, whereas the last is a regional, Appalachian-based endeavor. Consequently, the information published in the differing journals may be quite distinct and tailored to a specific audience. The large European journals usually report on worldwide research focusing on an academic audience, whereas *Castanea* concentrates on more regional issues and research, targeting both an academic and a professional audience. In general, the Council of Biology Editors [now Council of Science Editors] style manual is used regarding general matters of style for journal articles, but most researchers simply follow the guidelines issued at least once a year in the journals in which they want to publish.

The publication and review of journal articles are probably the most common communication channels that involve the members of the field as a whole. Other inclusive channels at the regional level include conferences, where practitioners have the opportunity of meeting the "faces behind the publications" and a chance, as Dr. Tom Wentworth (North Carolina State University—Botany) puts it, "to catch up on what is new in the field without having to read the entire spectrum of publications available." Most often it is the professional organizations and agencies that organize conferences. In the field of plant community ecology, individuals may belong to broad-based organizations such as the Ecological Society of America (estimated 7000 members) or more field-specific organizations such as the International Association for Vegetation Science (estimated 1200 members). Depending on individual interest, or even regional interest, societies such as the Torrey Botanical Club, the Association of Southeastern Biologists, or the American Association for the Advancement of Science fit almost everyone's needs. Of crucial importance as well is the fact that such a varying array of journals and organizations can cover many aspects and details that no one journal could cover by itself. This fact is crucial because it furthers the already existing hierarchy of levels of communication and adds to the stability of the field as a whole.

Also within the regional level of communication channels in plant community ecology lies the ever important source of funding. A good deal of time is spent by professionals, especially researchers, procuring adequate funding to carry out their scientific investigations. Grant proposals may be written and rewritten, often being tailored to specific funding sources and their interests, before final success. Often individual researchers' interests may be modified by the funding agencies' interests. It seems better to at least carry out part of your original inquiry with some modifications than not to be able to carry out any investigation at all. Funding sources in plant community ecology include, but are not limited to, the National Science Foundation, the U.S. Forest Service, the U.S. Fisheries and Wildlife Service, the U.S. Department of Agriculture, the National Biological Survey, and the National Park Service.

As the forms of communication on the regional and international level largely overlap, they have been included together for the purpose of discussion, the main difference being the larger scope in terms of institutional backgrounds, geographical distances, and interests over which the communication takes place. Although peer review and the publishing of journal articles, as discussed before, are probably the primary channels of communication internationally, conferences also play a key role. In fact, international conferences are probably the primary means of direct, fact-to-face communication on this level. Although an international perspective is often intriguing, the drawback is a certain lack of depth to the discussion. Primarily, international forms of communication address the field as a whole, its context in society and science, its needs, and its position in shaping international trends. Region-specific details are discussed not so much for their individual validity but for their relation to the whole. Nonetheless, international conferences provide an important service in that they broaden the horizons of individual researchers, allowing them not only to utilize shared concepts, but to continue working with the idea of being part of a greater community of researchers that transcends all national boundaries.

**FIGURE 2.3**    *(continued)*

## 2.4 Research Proposals and Their Audiences

Research proposals represent another critical channel of communication in science and research. As illustrated in the profile in Figure 2.3, the quest for research funding is a central activity in science. Just as journal editors and conference organizers exert control over what information is made available to the community, so funding agencies influence what kinds of research are undertaken in the first place. Berkenkotter and Huckin (1995) illustrate the interrelationships of these gatekeeping processes in a diagram we've reproduced in Figure 2.4. As the diagram indicates, peer review plays a critical role in funding decisions, just as it does in decisions about the publication of journal articles. We saw in Chapter 1 that in denying funding to Pons and Fleischmann for their work on cold fusion, the U.S. Department of Energy dealt a major blow to their research as well as to

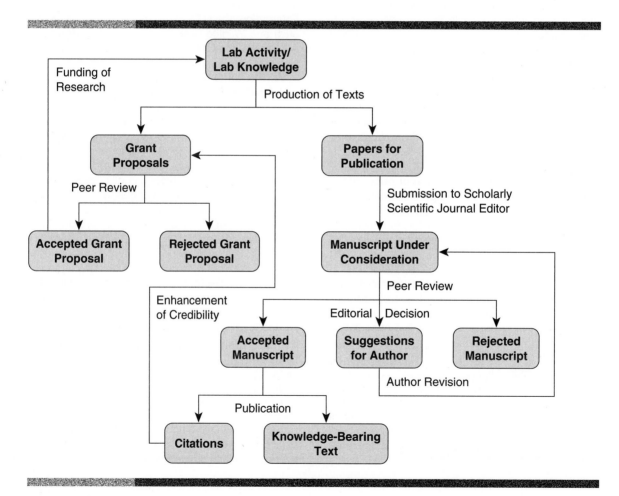

**FIGURE 2.4**   Life cycle of lab knowledge in scientific publication system from Berkenkotter and Huckin (1995, p 62).

the credibility of their research team. Because these all-important decisions are based on reviewers' assessments of the quality and persuasiveness of the proposal document, many scientists consider proposal writing the most important writing they do.

As a new member of a research community, you'll want to find out which public and private sources provide funding for work in your field. Notice where your professors or employers apply for funding, and pay attention to the funding agencies acknowledged in journal articles you read. Like the professional associations that sponsor journals, funding agencies have specific interests; most will accept proposals only on specified topics and may limit their funding to projects with certain types of applications (e.g., industrial, educational, or environmental).

Consider, for example, two proposals written by the Burkholder team for their research on toxic algae, both of which were awarded funding. The first proposal, for basic research on the algae's place in the estuarine food web, was submitted to the National Science Foundation. A second proposal (included in Chapter 11, page 303) described a more applied project, the development of gene probes to help detect the organism in water samples. This project was submitted to the National Sea Grant College Program in response to a specific call for proposals in the area of marine biotechnology.

Large federal agencies like the National Science Foundation encompass a range of specific interests, making it important to target an appropriate division *within* the agency as well. Burkholder and co-investigator Alan Lewitus sent their proposal on the estuarine food web to NSF's Biological Oceanography division. Another large federal agency, NASA (the National Aeronautics and Space Administration), supports a range of research projects in astrophysics and other fields through its many subdivisions. Stephen Reynolds, an astrophysicist whose reports of research on supernova remnants are included in Chapter 12, has submitted proposals to several different NASA programs, including one to the ROSAT (ROentgen SATellite) Guest Observer Program in 1993, proposing a set of observations to be carried out by the ROSAT satellite; and others to NASA's Astrophysics Theory Program in 1994, 1997, and 2000, in which Reynolds and two colleagues proposed work on a new theoretical model of supernova dynamics to be used in interpreting data gathered from the ROSAT and other sources (the first proposal in this series is included in Chapter 12).

In sum, before sitting down to write a proposal, researchers select a target agency carefully and become thoroughly familiar with its funding history and preferences. Funding agencies publish guidelines for proposal writers as well as formal *requests for proposals (RFPs)*. Much can be learned about a funding agency's interests and purposes by analyzing these materials, which are typically available online. We will take a closer look at how to do this in Chapter 6.

## 2.5 Communicating Beyond the Research Community

In this book we focus on how scientists communicate with others in their research communities, overviewing the primary forums through which researchers participate in the development of new knowledge in their fields. But of course, scientific knowledge is generated in other contexts as well and is of interest to audiences well beyond the research community. These broader interactions are conducted through a wide array of specialized communication channels. Scientists who conduct their research in applied settings—for example, as employees of pharmaceutical companies, environmental consulting firms, or regulatory agencies such as the EPA or the FDA—will need to become familiar with other types of internal and external communication as well as the basic genres described above. In addition to scientific colleagues, audiences for this research might include policy specialists, marketing personnel, field technicians, clinicians, resource managers, legal staff, transportation personnel, construction monitors, production line inspectors, local government planning committees, and consumers, among many others. Each of these groups has specific needs for scientific research and applies different perspectives in interpreting it. Chapter 7 will examine a sampling of these specialized forms of communication and explore the rhetorical challenges of communicating with stakeholders in industry and government.

The general public, perhaps the ultimate stakeholder for all scientific work, presents still more communication challenges. In Chapter 8, we will examine the broader issues involved in adapting scientific information for general audiences—readers and listeners in the public realm who bring widely varying interests, experiences, and knowledge to the task of understanding science.

# 2.6 Electronic Communication in Science

As we conclude this overview of communication forums, it is important to recognize that most of the day-to-day work of science takes place through much less formal channels, such as conversations in the lab or field, informal memos within a research group, and departmental presentations and colloquia. In addition, it is important to recognize that increasingly, the work of science also takes place in communication via phone, fax, video conferences, websites, and especially email, online discussion and news groups, databases, and online journals.

The use of websites, email, and online discussion groups is, of course, exploding in all sectors of society. Constructed by scientific organizations, libraries, and scientists themselves, databases that collect published scientific periodicals and/or articles and make them accessible to the worldwide scientific community are becoming a common feature of scientific communication (Day 1999). Many of these databases, such as BioMedNet, are open source sites; the material in them is available, usually in PDF form, without charge, to anyone who wants to read it.

The number of print journals that also make their articles available online (e.g., the *New England Journal of Medicine*) is rising dramatically too. Often these dual-media journals don't merely "republish" what appeared in print, but take advantage of the versatility of the electronic medium to publish additional features, such as news items pertinent to the research community, additional editorials, or titles of articles accepted for publication in future issues. More and more e-journals are emerging as well, that is, journals that publish in electronic format only, such as the *HMSBeagle: The BioMedNet Magazine,* or *AgBioForum,* a journal devoted to the social and economic issues of biotechnology. At the time of this writing, some e-journals are free; others are subscription-based or for members only, in which case the audience for the electronic publication is narrower and more clearly defined. The accessibility of the journal is thus another factor for scientists to take into account when deciding where to submit their work for publication. This consideration adds a literal dimension to our earlier point that the choice of a journal ultimately determines who will have access to a scientist's work.

As we touched on in Chapter 1, more and more researchers believe electronic communication is making scientific information more quickly and readily available to an ever-widening community of scientists, helping them keep up with the latest advances in their fields and facilitating the interaction of researchers throughout the world while reducing the astronomical cost of paper publication (Samarajiva 1990; Odlyzko 1995; Matzat 1998; WHO 2000). Obviously, greater accessibility will be more likely if the databases and journals remain free of charge and available to everyone, but the need for organizations and journals to earn income may make free and open access difficult to maintain in many cases.

At the same time, some scientists question whether electronic journals are a panacea. (Okerson and O'Donnell [1995] reproduce an online debate about these issues—ironically, in book form.) Electronic channels of communication such as email, online news groups, and discussion boards still raise concerns about the appropriateness of "prepublication." And fears that e-journals may undermine the peer-review process (Rowland 1997; Relman 1999) have led experts in a variety of fields to question the credibility and professionalism of publication in e-journals; these questions have been noted as the major obstacle to the quicker adoption of e-journals (Harnad 1997). However, more and more scientists are coming to believe that if rigorous quality control and the permanence of articles in e-journals are ensured, the benefits of electronic publication will outweigh the disadvantages (Harnad 1997).

Communication media are developing rapidly in a number of directions, and reliance on these technologies varies across and within scientific disciplines. The importance of these technologies is underscored by the fact that the American Association for the Advancement of Science (2002) created a set of guidelines to help scientific associations keep records and to publish and disseminate information to the public in various paper and electronic forms (see Macrina 2000, p 231). As a student, it will be important for you to become familiar with electronic forms of communication common in your field and to understand the attitudes of professional scientists toward these forms.

Electronic forms of communication no doubt will continue to emerge and reshape the landscape of scientific communication. As access to technologies increases within a given field, new communication channels will become navigable. Their purposes and audiences will gradually become defined, as will the other conventions governing their use. Students entering your research field a decade from now will no doubt find an even more complex set of communication channels to negotiate.

# Activities and Assignments

The following activities are designed to help you begin to investigate the communication channels in your research community. Your instructor may ask you to work alone or with others in your field.

1. Write a one-page introduction to the "business" of your field for outsiders who are unfamiliar with that branch of science. What do botanists (or soil scientists or organic chemists) do? Where do they work? What do they study? What kinds of questions do they ask? What kinds of methods do they use? Why is their research important? Who uses the results of this research? Who is affected by the results?

2. Interview a member of your research community to learn the kinds of writing, reading, speaking, and listening he or she does in professional life. A good way to elicit this kind of information is to ask your interviewee to describe a typical day or couple of days in the lab or office: With whom did he or she talk? What meetings did he or she attend? In what professional reading or writing activities did she or he engage, in either paper or electronic formats (e.g., recording data, writing notes, drafting part of a report or proposal, revising an article, reading or skimming journal articles or abstracts, reviewing manuscripts or proposals, participating in internet discussions)? Your interview subject also should be able to help you identify the major research journals, professional conferences, funding agencies, online forums, and databases in your field. If you are working with a group, decide together on an appropriate interview subject, develop a set of questions, and arrange for one or more members of the group to conduct the interview. Feel free to interview more than one person, particularly if your group includes students from different subfields. Your instructor may ask you to summarize your findings for the class in an oral presentation, or to present them in writing, as described in Activity 3.

3. Use the research you've conducted in Activities 1 and 2 as the basis for a written profile of communication patterns in your research community, as illustrated in Figure 2.3. Your goal is to describe the distinctive practices that have developed in your community. Include the names of the most important journals, professional associations, funding agencies, databases, and discussion lists as examples, but also focus more broadly on how, why, and with whom scientists in your field communicate. Include both written and spoken channels of communication, and note any journals or funding agencies that have converted partly or wholly to electronic form. Tell your readers what kinds of knowledge or information are exchanged via the various channels, to whom that information is directed or from whom it is received, and what form it must take in each case.

   Your profile should be written for readers who are not familiar with your particular field of study. Begin the paper with a brief introduction to the field (as described in Activity 1) to give readers some idea of the type of research that is carried out in this field. Then go on to describe the channels of communication that have developed to facilitate this research.

4. As an alternative or addition to the written profile in Activity 3, create a graphic representation of communication processes in your research field. Choose one of the following:

   a. Berkenkotter and Huckin (1995) outlined the "life cycle" of research in general terms in the diagram reproduced in Figure 2.4. Create a life cycle diagram for a specific project in your field, in which you identify the journals and funding agencies that served as "gatekeepers" for this line of research.

   b. Draw a more comprehensive map of your field. Present the information you gathered in Activity 3 in visual form. Your map should identify the major journals, associations, and funding agencies that are active in your field and illustrate specific interrelationships among them. A sample diagram of the field of chemistry is presented in Figure 2.5.

5. In groups, exchange your written or graphic profiles with other students in the class. (If profiles were presented orally, use your notes from the oral presentations.) Ideally, each member of the group should represent a different research field or subfield. After all members of the group have read the set of profiles, develop a list of similarities and differences across fields. Prepare a brief oral presentation for the class in which you compare and contrast the conventional ways of communicating in this set of research fields. For example, consider whether scientists in these fields ask similar or different types of questions, whether their objects of study are similar or different in significant ways, and whether and how their methods differ. Determine whether differences in research goals or methods have led to different patterns of communication in these communities. For example, you might examine the importance of conferences versus publications in different fields, paper versus electronic publication, the amount of interaction researchers in each area have with outside audiences, the amount of collaborative work in each field and the mechanisms for supporting that work, and so forth.

6. Go to the library, and find the major research journals you've identified. Working from the editors' statement of purpose or guidelines for authors (as well as information gathered in your interview), annotate your list with a brief summary of the purpose and audience of each journal, as in Figure 2.1. Include any significant electronic journals that may have emerged.

7. Select one research journal in your field, either print or electronic, for closer study. Examine several issues of the journal, and read the instructions for authors and other editorial information posted at the journal's website. Write a three- to four-page rhetorical analysis in which you discuss the following basic elements of communication:

   a. *Author.* Who publishes this journal? Whose research is reported there?

   b. *Audience.* Who are the intended readers? What types or levels of expertise do they have? Are they primarily researchers, policy makers, resource managers, educators, or other experts?

   c. *Subject.* What topics or kinds of topics does the journal cover? How general or specialized are these topics?

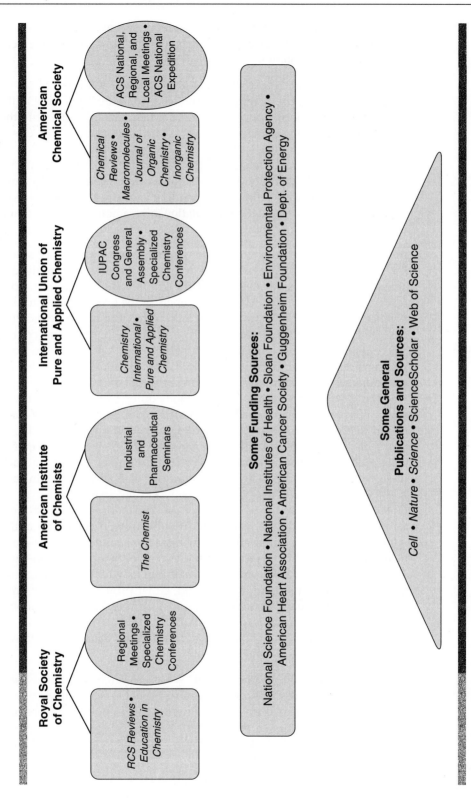

**FIGURE 2.5**   Map of the field of chemistry by Amy Haw-meei Yap. Student sample.

d. *Text.* Describe the distinctive features of the text that helped you make inferences about author and audience. Do the articles have a formal or informal tone? Does the journal include more than one type of article (e.g., research reviews, letters, technical notes)? If so, how do these texts differ in style, tone, format, and purpose? (*Hint:* Don't overlook obvious features of the journal that might provide important information about its primary purpose and audience: the journal's title, institutions represented on the editorial board, announcements and advertisements, types of books reviewed, and so forth. If it is an online journal, who has access?)

8. Write a letter to a real or a hypothetical research collaborator in which you explain why you'd like to submit the results of your collaborative project to a particular research journal. Assume the two of you had narrowed your choices to two journals in the field. You've done some more investigating of the two journals, and you're ready to argue that one rather than the other will be the better place to report your findings. Your argument should be based on a clear comparison of the purpose, scope, and target audiences of the two journals.[2]

---

[2]We are indebted to Christina Haas for this assignment idea.

# Reading and Writing Research Reports

## 3.1 Argumentation in Science

Despite the proliferation of other forms of communication in the sciences, the published research report remains the most common medium used by individual scientists to communicate their findings to the research community at large—and thus to contribute to the developing knowledge of the field. Given the critical role of the research report in the development and exchange of scientific knowledge, it is important to understand its distinctive features and the common strategies scientists use in writing and reading these documents. Research reports in scientific journals are variously referred to as *journal articles, papers,* or *reports.* In this text, we use these terms interchangeably to designate the class of texts reporting original results or theoretical developments in professional research journals. Our purpose in this chapter is to describe some of the general conventions research reports follow and to help you analyze the specific conventions governing reports in your field.

The communication of research results is far from straightforward. It is traditional to think of scientific reports as purely factual or explanatory, but as we saw in Chapter 1, the report also serves an important interpretive and persuasive function. Scientists publish descriptions of their research not simply to tell others what they've done but also to persuade readers that the work is valid and useful. In terms of form, then, the research report is more than a narrative; it is a careful argument. The authors of a research report find themselves in the position of building a case for their research, not simply recounting actions and observations.

From this perspective, the research report represents an extended argument in which researchers seek to convince readers that their research questions are important, their methods were sensibly chosen and carefully carried out, their interpretations of their findings are sound, and their work represents a valid contribution to the developing knowledge of the field. These basic goals are clearly reflected in a

**Sections of a Research Report: Typical Headings and Functions**

| Heading for section | Function of section; comments |
|---|---|
| Introduction | Describes the state of knowledge that gave rise to the question examined by, or the hypothesis posed for, the research. States the question (not necessarily as an explicit question) or hypothesis. |
| Methods and Materials | Describes the research design, the methods and materials used in the research (subjects, their selection, equipment, laboratory or field procedures), and how the findings were analyzed. Various disciplines have highly specific needs for such descriptions, and journals should specify what they expect to find in a methods section. |
| Results | Findings in the described research. Tables and figures supporting the text. |
| Discussion | Brief summary of the decisive findings and tentative conclusions. Examination of other evidence supporting or contradicting the tentative conclusions. Final answer. Consideration of generalizability of the answer. Implications for further research. |
| References | Sources of documents relevant to elements of the argument and describing methods and materials used. |

**FIGURE 3.1**    Sections of a research report as described by the Council of Biology Editors (CBE 1994, p 590).

common research report format, which consists of four standard parts or sections: introduction, methods, results, discussion. Each section of the report contains an argument, and each section plays a part in supporting the larger argument of the whole.[1] This structure, sometimes referred to as the *IMRAD format,* is described in numerous science writing guides (Katz 1985; Biddle and Bean 1987; Olsen and Huckin 1991; Day 1998) and is summarized by the Council of Biology Editors (1994) in Figure 3.1.

As the descriptions of these sections indicate, the IMRAD format consists of two sections in which the new study is actually described (methods and results), framed by two sections that place the new work in the context of previous knowledge (introduction and discussion). Both of these "framing" sections describe the current state of the field's knowledge. The introduction describes the current state at the start of the report—the state that created the need for the study—and the discussion describes the new state of the field's knowledge at the end of the report—now that the new results have been added to the knowledge pool.

The distinction between framing sections and describing sections is often signaled by verb tense: framing sections usually use present tense to describe the field's current knowledge ("Factors that limit the distribution of the cougar *are* not known entirely but *include* climatic features, availability of prey, and habitat features"), whereas describing sections typically use past tense to describe actions

---

[1]Notice that we use the term *argument* in the sense of "case building" here, as opposed to the everyday sense of "confrontation" or "debate." Thus, the question "What is your argument?" means "What is your line of reasoning?"

already taken and data already recorded ("We *searched* for cougar tracks between December and April each year"; "Summer and winter home ranges for individual females *overlapped* extensively").[2] In an extensive analysis of verb tense patterns in research reports, Hawes and Thomas (1997) found that writers frequently use past tense to describe particular examples in support of generalized trends, whereas present tense is used when stating the generalizations themselves. One of the effects of past tense, then, is to localize and limit findings to particular researchers or labs, whereas present tense identifies a claim or conclusion as part of the field's current understanding.[3]

## EXERCISE 3.1

Choose two or more full-length research reports from your field or from those included in this text. In each major section of the paper, circle the main verb in the first 10 or 12 sentences. Note whether and how verb tense shifts across sections and whether and how these verbs tend to localize findings or incorporate them into general knowledge claims.

# 3.2 The Logic(s) of Scientific Inquiry

We have included in this text three journal articles that follow the traditional IMRAD organizational pattern: Graham et al. (1992) and Chiba et al. (2002) in Chapter 10, both reporting clinical studies of treatments for the stomach bacterium *Helicobacter pylori*; and Mallin et al. (1995) in Chapter 11, reporting experimental tests of the relationship between the toxic dinoflagellate *Pfiesteria piscicida* and other estuarine predators. The IMRAD format is a natural choice for reporting results from these studies, for in each case the research involved posing a question, choosing and carrying out experimental procedures, interpreting the resulting data, and drawing some conclusions about the significance of the outcome. Thus the IMRAD format mirrors the basic logic of scientific method. Not surprisingly, this form is common in many scientific fields.

However, the emphasis on experimental methods and the presentation of new data in the IMRAD form are clearly unsuitable for papers reporting theoretical or historical research, in which the primary goal is to present new interpretations, theories, or models for understanding phenomena previously observed (Harmon 1992; Miller and Halloran 1993). Much of the research conducted in fields such as astrophysics, evolutionary biology, and geology, for example, is best described as theoretical or historical rather than experimental. Research in these areas may involve some data collection (e.g., core samples or satellite observa-

---

[2]Quoted sentences from Ross and Jalkotzy (1992, p 417, 418, 419). Italics added.

[3]You are already quite familiar with the effects of present and past tense verbs, for you use these distinctions in everyday conversation. Compare the statement "I mowed the lawn myself" with "I mow the lawn myself." The first sentence clearly refers to a specific occasion on which you mowed the lawn (a localized event), whereas the second refers to a general practice you've adopted (a recurring or general pattern).

tions), but frequently the primary goal of such research is not to test hypotheses but to formulate hypotheses—to propose theories or models that account for the field's observations to date.

The structure of papers in the historical and theoretical sciences therefore tends to be more variable than that of experimental papers. As a case in point, skim the section headings of the *Geology* paper by J. Z. de Boer and colleagues (2001) on pages 406–409 and the *Clinical Toxicology* paper by Spiller et al. (2002) on pages 410–417. Written by members of the same multidisciplinary research team and reporting on the same project, these papers nevertheless follow different organizational logics. Both begin by introducing the geographic location and history of the Delphic oracle. The researchers have compiled evidence that a gaseous vent was present in the ancient temple at Delphi and was responsible for intoxicating, and thus inspiring, the legendary priestesses who spoke prophecies there. After the introductory sections, the *Geology* article devotes separate sections to the tectonic setting, evidence of springs in the area, and analyses of gases present. Each of these sections incorporates both historical information and the results of the team's recent field survey and water sampling. In contrast to the IMRAD form, the survey and sampling methods are described briefly in each of the above sections; there is no separate section devoted to methodology, and few methodological details are included. This primary emphasis on observations or results over methods is common in geology reporting and consistent with the descriptive nature of this science. (See Figure 3.2 for a description of this pattern in another geology paper.)

## Reporting Observations in Sedimentary Geology

The article "Morphology and sedimentology of two contemporary fan deltas on the southeastern Baja California Peninsula, Mexico," by E. Nava-Sanchez, R. Cruz-Orozco, and D. S. Gorsline only partially follows the generic IMRAD structure. This article, as it appears in the journal *Sedimentary Geology* (98[1995]:45–61), is divided into eight general sections titled as follows: introduction, general physiographic setting, geologic setting, oceanographic setting, geomorphology, sedimentology, discussion, and conclusions. The purpose of this research is to describe the method of formation and the source of the formation sediments related to modern fan deltas in southeastern Baja California. From this information the authors concluded that a large section of this coastal area is dominated by similar active fan deltas.

The introduction of this article is consistent with the goals served by introductions in the IMRAD format. In this first section, the authors define the geologic structures that their research focuses on and the factors that influence these formations. The first several paragraphs state the nature and scope of the research as well as give a review of the pertinent information to be discussed throughout the remainder of the paper. This article introduction is definitely aimed at specialists and is composed almost entirely of jargon specific to sedimentary geology. Near the end of the introduction, the objective is clearly stated so as to create a fusion of the introduction and purpose into one segment. This is also characteristic of the general IMRAD format.

A section devoted to methods is absent from this research article, most likely due to the nature of the science of geology, which relies heavily on

*(continued on page 44)*

**FIGURE 3.2** Analysis of a research report from the field of sedimentary geology by David Brock. Student sample.

observations and descriptions and the subsequent correlation of this information with historic geologic episodes. In order to clarify this concept, the nature of geology as a science compared with other sciences should be considered. Geology is a speculative science that does not rely on cookbook experimentation due to the fact that geological processes occur on average over hundreds of thousands and millions of years. A short period of time to a geologist is 100,000 years.

In lieu of a methods section, this article contains three sections devoted to the description of the settings in which the observations occurred and two sections that describe the surface processes and sediment makeup in the area. The first section of setting description is labeled "General physiographic setting" and describes the location and weather patterns and provides a landscape evaluation of the study area. This type of regional description is valuable in geology because processes are highly dependent on weather patterns, climate, and existing topography. The second section of setting description is "Geological setting." This section focuses on when and how the Baja Peninsula formed and a definition of the topography in geologic terms. The regional geologic features are mentioned, such as faults, regional spreading, and subduction, which all characterize this area. This format is not in concordance with the general IMRAD formula; however, it substitutes sufficiently for a methods and materials section in a traditional research paper. Formatting the article in this manner is more suited to the field of geology, and particularly suitable in sedimentary geology, because this sort of information is crucial to understanding the remainder of the paper. The last section describing the setting is "Oceanographic setting." In these paragraphs the morphology of the underwater basins, specific formations, and the water circulation in the area are mentioned. This serves the same purpose as the previous sections on the setting, the main difference being that it concentrates more specifically on the facets of the landscape influenced by the ocean.

The remaining two sections before the discussion and conclusion become more specific and technical. In relation to the general IMRAD format, these two sections are most clearly related to a re-sults section. The fifth section, "Geomorphology," is a technical discussion of the surface processes responsible for alluvial fan formation in the area of research. This section becomes technical enough to make major generalizations about the study area as well as to present pertinent observations or data to support them. The sixth section, "Sedimentology," focuses on exactly what the paper is about. It describes the sediments that compose the modern fan deltas, the percentage makeup of these sediments, and their thicknesses. The sedimentology discussion is the final descriptive section before the discussion and conclusion. It finalizes the observations and narrows the scope of the paper to focus on the structures specifically mentioned in the title.

The final two segments of the paper are the discussion and conclusion. The discussion in this case classifies the two major fan deltas that were studied and supports this classification. This effectively fits the results into the context of the field by relating them to an already existing classification scheme. Overall, this section does not state how these observations contribute to the advancement of the field. It simply brings together the previous observations and classifies the fan deltas. Although this is not in accordance with the general IMRAD format, it is characteristic of the descriptive nature of geology. The last section, the conclusion, summarizes the five major findings by the authors. In general, the findings pertain to how the deltas were formed, the source of formation sediments, the modifications of these sediments, and the character of the resulting fan deltas. Again, this format is specific to the descriptive nature of geology. Rather than discuss the results of an experiment, the paper generalizes and classifies observations.

Throughout this research article in sedimentary geology, it is apparent that the main focus is on observation and description of landforms. Although the introduction and discussion sections tend to follow the general IMRAD format, the other sections of the article are geology specific. Despite this specificity, the article accomplishes the same general goals as other research reports. The majority of differences can be attributed to the difference in the scientific process between geologists and classical scientists.

**FIGURE 3.2**  *(continued)*

The second report of the Delphi research—Spiller, Hale, and De Boer's *Clinical Toxicology* article—is organized somewhat differently. Though this article reports basically the same information as the *Geology* report, the historical background is treated as evidence in this report and the authors explicitly state their intention to defend the gaseous vent theory, an ancient theory that had long been dismissed as myth. After the introduction and an overview of the controversy, the paper devotes separate sections to each of three types of evidence the authors have compiled: historical, geological, and chemical. The paper does not follow a predetermined form but is organized to highlight the critical components of this particular argument.

Notice that all the report structures discussed above provide explicit opportunities for authors to describe the assumptions and implications of their theories, in a sense the "methods" and "results" of theoretical work. It is also important to notice that these papers provide clearly announced introductory and concluding sections or paragraphs that contextualize the argument just as the introductory and concluding sections do in the standard IMRAD form. In stating the purpose of the study, the introduction lets readers know what kind of research is to be presented. The American Institute of Physics *Style Manual* explicitly reminds authors that the type and scope of the work, whether theoretical or experimental, should be clear from the introduction (AIP 1990).

In short, all research reports will include a framing introduction and discussion or conclusion (labeled or unlabeled), but the form of the body of the paper, the actual description of the work itself, will be determined by the type of work to be described. Because written texts communicate not only facts and observations but also the ideas and logic of their authors and their fields, texts will take different forms in different research communities. You will want to notice how reports in your field tend to be organized and how that organization compares with the organizational logic of the IMRAD form that we discuss here. As you read, pay close attention to the purpose and logic of each section. Notice how authors in your field convince readers of the need for the study and the significance of the outcome, as well as how they describe the methods, assumptions, and observations that serve as the evidence in their arguments.

In the next sections of this chapter, we describe the logic of each of the standard IMRAD elements in more detail.

## 3.3 Introducing the Research Problem

All research reports begin with an introduction of some sort, no matter what structure is followed in the rest of the paper. This is where you explain your research objectives, argue that the research is important, and place your study in the context of previous research. As Figure 3.1 demonstrates, editors and other readers expect the opening paragraphs of a journal article to describe the state of knowledge that motivated the research in the first place and to introduce the purpose of the study. We have referred to this opening section as a "framing" section above, to underscore its role in establishing a context or framework for interpreting the new research.

# Blood Flow Through the Human Arterial System in the Presence of a Steady Magnetic Field

A. Such studies of flow through single arteries, however, have somewhat limited practical applicability because the actual human arterial system is composed of a large number of interconnected vessels of different lengths and cross sections which cannot be treated as independent.

B. In the present work, the finite-element method (FEM) is used to analyse the effects of a magnetic field on blood flow through a model of the human arterial system.

C. In recent years some studies have been reported on the analysis of blood flow through single arteries in the presence of an externally applied magnetic field.

D. It is known from magnetohydrodynamics that when a stationary, transverse magnetic field is applied externally to a moving electrically conducting fluid, electric currents are induced in the fluid.

E. These include the work of Belousova (1965), Korchevskii and Marochunik (1965), Vardanyan (1973) and Sud *et al* (1974, 1978).

F. Quantitative results on the effects of field intensity and orientation on flow and arterial pressures are presented and discussed.

G. This could occur for instance in nature or in physics laboratories, during space travel, or in hospitals.

H. The interaction between these induced currents and the applied magnetic field produces a body force (known as the Lorentz Force) which tends to retard the movement of blood.

I. Human subjects could by accident or design be made to experience magnetic fields of moderate to high intensity.

**FIGURE 3.3**    Scrambled introduction to Sud and Sekhon (1989), from *Physics in Medicine and Biology*, p 795.

---

**EXERCISE 3.2**

Take a minute to read the scrambled introduction in Figure 3.3. How would you assemble those sentences into a focused and readable one-paragraph introduction? List the sentence letters in an appropriate sequence in the margin before reading on.

---

In Exercise 3.2, many of you probably chose to start the paragraph with sentence C, D, or I, presumably because you felt some background or announcement of the topic was warranted at the beginning, or perhaps because you were following the familiar "funnel" introduction pattern, which starts with a broad statement of the general topic and then narrows to the particular issue at hand. If you placed sentence B or F at the end of the paragraph, you chose to end by introducing the particular study to be reported in the paper, also in keeping with the funnel or inverted pyramid pattern. The actual introduction is reproduced in Figure 3.4.

Notice in Figure 3.4 that the authors Sud and Sekhon begin as we've just described, using sentence I to announce the general topic or issue (human beings occasionally encounter magnetic fields), briefly elaborated with sentence G (which tells us where this might happen). Next, sentences D and H summarize some basic

## Blood Flow Through the Human Arterial System in the Presence of a Steady Magnetic Field

(I) Human subjects could by accident or design be made to experience magnetic fields of moderate to high intensity. (G) This could occur for instance in nature or in physics laboratories, during space travel, or in hospitals. (D) It is known from magneto-hydrodynamics that when a stationary, transverse magnetic field is applied externally to a moving electrically conducting fluid, electric currents are induced in the fluid. (H) The interaction between these induced currents and the applied magnetic field produces a body force (known as the Lorentz Force) which tends to retard the movement of blood. (C) In recent years some studies have been reported on the analysis of blood flow through single arteries in the presence of an externally applied magnetic field. (E) These include the work of Belousova (1965), Korchevskii and Marochunik (1965), Vardanyan (1973) and Sud *et al* (1974, 1978). (A) Such studies of flow through single arteries, however, have somewhat limited practical applicability because the actual human arterial system is composed of a large number of interconnected vessels of different lengths and cross sections which cannot be treated as independent. (B) In the present work, the finite-element method (FEM) is used to analyse the effects of a magnetic field on blood flow through a model of the human arterial system. (F) Quantitative results on the effects of field intensity and orientation on flow and arterial pressures are presented and discussed.

**FIGURE 3.4**    Original (unscrambled) introduction to Sud and Sekhon (1989), from *Physics in Medicine and Biology*, p 795.

principles of the field to show readers why this phenomenon is of interest or concern (exposure to a magnetic field might have adverse health effects: it may retard the flow of blood). Sentences C and E then refer very briefly to the research that has been done on the topic so far (five studies, including two of Sud's own, which have looked at the effects of magnetic forces on blood flowing through single arteries).

This quick review of prior knowledge on the topic sets up the critical next move: in sentence A, Sud and Sekhon point out that the previous research is limited (because focusing on single arteries does not provide an accurate picture of effects on blood flow through the entire, interconnected arterial system). Hence the need for the new study reported in this paper. In the next move, sentence B announces the purpose and value of the new study (it presents a model for analyzing effects of magnetic fields on the whole arterial system, not just on single arteries). Finally, sentence F previews the types of information that will be reported in the paper to follow. Notice that both independent variables (field intensity and field orientation) and dependent variables (blood flow and arterial pressures) are announced in this final sentence. Brief as it is, it gives readers an overview of critical features of the research design. The introduction section in scientific journal reports is often quite compact, conveying a great deal of information in a relatively small space.

In asking you to try the "scrambled intro" exercise, we have borrowed an instructional technique from John Swales, a linguist who has studied the logic and form of scientific papers (Swales 1984). In a study of introductions from 48 scientific papers published in three research fields, Swales discovered a remarkable degree of consistency across fields and journals. He found, as we did in the Sud and Sekhon piece, that authors use introductions to demonstrate that their

## Common Moves in Research Article Introductions

| | |
|---|---|
| Move 1 | Establish topic and significance. |
| Move 2 | Summarize previous knowledge and research. |
| Move 3 | Prepare for present research<br>by indicating a gap in previous research and/or<br>by raising a question about previous research. |
| Move 4 | Introduce the present research<br>by stating the purpose and/or<br>by outlining the research. |

**FIGURE 3.5**    Adapted from Swales (1984, p 80).

research responds to a gap in the field's knowledge. They typically begin by announcing the topic at hand and then proceed to give an overview of recent research on the topic, in order to point out the gap or question that their study addresses. Swales thus identified four interconnected moves commonly found in journal article introductions, which we have summarized in Figure 3.5.

These four rhetorical moves can easily be seen even in Sud and Sekhon's brief introduction. (Take a minute to find them in Figure 3.4.) Notice that Moves 3 and 4 in particular are clearly signaled by Sud and Sekhon. Readers can't miss Move 3 in sentence A because it is signaled by the contrastive adverb *however:* "Such studies of flow through single arteries, *however,* have somewhat limited practical applicability because the actual human arterial system is composed of a large number of interconnected vessels of different lengths and cross sections which cannot be treated as independent." As we have noted, Sud and Sekhon must point out this limitation of previous studies in order to show why their new work is needed—to argue for its significance to the field. They proceed logically from Move 3 to introduce their work, Move 4, in sentence B. Move 4 is also clearly signaled, in this case with an introductory phrase: "*In the present work,* the finite-element method (FEM) is used to analyse the effects of a magnetic field on blood flow through a model of the human arterial system." There's no mistaking what the authors consider their primary contribution to be.

### EXERCISE 3.3

Read the introduction to the study by Graham et al. (1992) reprinted on page 254. Mark off the four basic moves outlined in Figure 3.5, circling any clear signals included in the text. Describe the line of reasoning presented in this introduction, as we did for the Sud and Sekhon introduction.

As you continue to read formal research reports in your field and others, you will come across many variations on these four moves. Sometimes the moves will be made in a different order, one or more moves may be only implied, moves may be made more than once, or moves may overlap. For example, authors often cite previous research *while* announcing the topic. Citations of prior studies may appear in any number of places in an introduction; in such cases it would be difficult to distinguish Move 2 from other moves. In fact, Swales (1990) later revised his model to underscore this variability. We believe the original model in Figure 3.5 is quite useful in identifying the basic gestures authors tend to make in introductions, as long as these gestures are understood as a set of flexible rhetorical moves that authors combine in diverse ways to achieve their rhetorical goals.

The Hendrick and Reynolds (2001) report in Chapter 12 offers a useful illustration (see page 394). Hendrick and Reynolds begin with a five-paragraph introduction. The first paragraph establishes the topic in the first two sentences (Move 1), then proceeds directly to Move 3 by indicating a limitation or gap in the research thus far ("However, from radio synchrotron observations, we learn only about electron energies in the range 1.0–10 GeV . . . "). This opening paragraph ends with a general statement of the purpose of the present research, Move 4 ("Understanding the cosmic-ray electrons should give us some information on the cosmic-ray ions, which contain most of the energy in the cosmic-ray spectrum"). Move 2, the review of previous knowledge, is made in the second, third, and fourth paragraphs. Paragraph 2 summarizes three factors theorized to affect electron energies. Paragraph 3 explains how maximum energies can be derived theoretically and shows how this explanation is consistent with previously published observations. Paragraph 4 provides a brief justification for the study's methodology by referring to a previous Reynolds study that the current investigation is modeled after, and by explaining why the supernova remnants (SNRs) in the Large Magellanic Cloud, the site of the current investigation, represent an appropriate sample for exploring questions about SNRs in general. After this extended review, the fifth and final paragraph utilizes a second common Move 4 strategy, outlining the organization of the report.

Despite variations in length, organization, headings, documentation format, and other text features, effective introductions in all fields include similar rhetorical moves because they share the same rhetorical goal: the authors want to convince readers that the topic is important and that their work on the topic will advance the field's knowledge. Thus the introduction serves several functions: It orients the reader to the research topic, reviews the current state of knowledge in the field, and establishes the need for the new work.[4]

The effectiveness of this argument for the significance of the new research rests largely on the authors' description of what the field knows now, which not only helps establish the authors' credibility in the field but in so doing creates common ground with readers and a shared context for understanding the new work. In an analysis of journal articles at early and later stages in the development of chaos theory, Paul and Charney (1995) found that when scientist authors were unable to use the conventional method of establishing shared context through research citation (i.e., when the research field was in its infancy and no prior research existed), they went to great lengths to create context through

---

[4]In fact, the structure of the introduction, in which the research gap or problem is made prominent, can be understood to create the psychological desire for a solution (see Olsen and Huckin 1991). As rhetorician Kenneth Burke puts it, "form is the creation of an appetite in the mind of the auditor, and the adequate satisfying of that appetite" (1968, p 31).

alternate means, for example, by introducing shared exemplars requiring extended elaboration. Once a research record developed, the same authors easily adopted the conventional research citation pattern that Swales observed. Paul and Charney aptly describe research citation as a "disciplinary shorthand" (p 427) for establishing shared context with other members of a research community. Conventional ways to summarize the state of knowledge in the field and to acknowledge the work of others are discussed at more length in Chapter 4.

## EXERCISE 3.4

Read the multiparagraph introduction to the study by Mallin et al. (1995) reprinted on pages 328–330. Mark off the four basic moves in this section, circling any clear signals included in the text. Write a paragraph in which you describe the authors' line of reasoning.

## EXERCISE 3.5

Choose a published research report from a journal in your field. Type up a scrambled version of the introduction to this report, as in Figure 3.3. (If you choose a multiparagraph introduction, break it into "packets" or clusters of sentences instead of individual sentences.) In class, have a partner read and unscramble the introduction, with the goal of reassembling the original line of reasoning.

## 3.4 Describing Methods

The second major component of the research report is a description of the methods and materials used in the research. The "materials and methods" section traditionally follows the introduction, though some journals have begun to move these details out of the main body of the article to the end of the paper and/or to set this section in smaller type. Berkenkotter and Huckin (1995) speculate that the trend toward deemphasizing the methods section in some journals indicates that readers rely heavily on peer reviewers' assessment of methodological details, saving their own reading time for inspection of the study's results. Diminished attention to methods may also reflect the maturation of a field or research area: as particular methods become more established and familiar, their description requires less and less detail (VandeKopple 1998).

The location of the methods section and the amount of attention it receives will also vary according to its role in the research argument being presented. In papers reporting theoretical as opposed to empirical results, the basic assumptions of the theory or model play a critical role in the unfolding research argument. If the paper follows the IMRAD form, these assumptions will be presented in a methodological section early on. For example, in the Hendrick and Reynolds (2001) paper described above (see pages 394–399), the authors apply theoretical models to archived satellite data to determine the maximum energies to which supernova remnants can accelerate electrons. Hendrick and Reynolds follow the basic IMRAD form, including a prominent second section labeled "Analysis Technique." This methods section is considerably expanded to allow the authors to describe the SNRs they've chosen to study and the satellite data they used, as well as to explain the assumptions and calculations their models entail. The length and location of this section in the article reinforce the impression that this methodological information is at least as important as the results that follow. Wherever it appears in the journals in your field, the methods section remains an important component of the overall argument of the research report, for it explains the framework within which the study's results and conclusions were generated.

As noted earlier, the concrete information in the methods section is usually presented in simple past tense, either active voice ("We *collected* water samples every three days") or passive ("Water samples *were collected* . . . "). Although the "scientific passive" has a long and venerable tradition, it is often easier and more direct to write in active voice, which is the mode preferred by many journal editors in the interests of brevity and clarity. The editors of the *Journal of Heredity*, for example, directly inform contributors that "first-person active voice is preferable to the impersonal passive voice" (Heredity 2002).

On the other hand, it is observed in the American Institute of Physics *Style Manual* (AIP 1990) that "the passive is often the most natural way to give prominence to the essential facts" (p 14). The AIP editors offer as illustration the sentence "Air was admitted to the chamber," in which it is not important to know who turned the valve (p 14–15). The AIP does recommend shifting to active voice where necessary to avoid confusion or awkwardness. You'll notice that active voice often leads naturally to the use of first person ("We collected . . . "), which is increasingly common in scientific prose, a trend also endorsed by the AIP (p 14) and other editorial panels. In addition to the use of active and passive voice for clarity and emphasis, scientists sometimes use these stylistic features strategically to highlight or minimize—to make "stylistic arguments." For example, note how Mallin et al. (1995) use passive voice in their introduction when describing their own research, but active voice when describing the new dinoflagellate they are investigating (Chapter 11, pages 328–330). The effect is to put the new dinoflagellate in the forefront, to call attention to its "active" existence and thus its toxicity, while downplaying the role of the researchers in discovering it.

### EXERCISE 3.6

Look for instances of active and passive voice in one or more of the research reports in Chapters 10 to 13 or in sample articles from your field. Do authors tend to use these modes consistently? Do you perceive any patterns or strategies in the use of active and passive voice? Compare the types of information that tend to be presented in each mode.

The content and organization of the methods section also will vary according to the type of research to be described. This variation is illustrated in Figure 3.6

**Example A**, from Chiba et al. (2002), *British Medical Journal* (reprinted in Chapter 10)
    Methods
        Selection of patients
        Randomisation and interventions
        Adherence to drugs
        Outcome measures (six subsections—see pages 281–282)
        Determination of sample size
        Statistical evaluation

**Example B,** from Lorimer et al. (1994), *Journal of Ecology*
    Study areas
    Methods
        Experimental design
        Plot measurements and analysis

**Example C,** from Lim et al. (1995), *Journal of Bacteriology*
    Materials and Methods
        Bacterial strains, plasmids, and culture conditions
        DNA manipulation and sequencing
        Data bank analyses
        Plasmid constructions
        Construction of *M. tuberculosis* genomic libraries
        Alkaline phosphatase assay
        Antibody preparations, SDS-PAGE, and immunoblots

**Example D,** from Hendrick and Reynolds (2001), *Astrophysical Journal* (reprinted in Chapter 12)
    2. Analysis Technique
        2.1 The Sample
        2.2 The X-Ray Data
        2.3 The Models
        2.4 The Spectral Fitting Methods

**Example E,** from Quigley and Slater (1994), *Southern Journal of Applied Forestry*
    Equipment
    Methods
        Field Procedure
        Laboratory Procedure

**FIGURE 3.6**     Subheadings from the methods sections of selected journal articles.

with a simple comparison of subheadings used by different authors to organize this section of their reports.

Although the logic of the methods section is dictated by the logic of the research and therefore varies widely across fields, notice in Figure 3.6 that there are some basic organizational similarities. Methods descriptions begin by identifying the subjects of study, whether they are bacterial strains, forests, supernova remnants, or human beings. It is conventional in many environmental science fields, for example, to include a separate "study area" section in which the ecosystem under study is identified and located and its distinctive features described (as in Example B). Papers reporting research involving human or animal subjects typically include a subsection labeled "subjects" or "patients" or "population characteristics" at the start of the methods section (Example A). In Example E, from a study that tests a new forest plot mapping technique, the technique itself is the object of study, so the authors inserted a separate section describing the mapping equipment before going on to describe how it was tested. Examples C and D similarly begin by identifying the bacterial strains and the sample of supernova remnants that are the focus of those studies.

Once the subjects of study have been identified, the materials and procedures used to study them can be described. Subheadings are often used to subdivide the methods section to highlight the various types of analyses that were performed, as in Examples A and C, or to distinguish groups of related procedures, as in Example E. It is often helpful and sometimes essential to use graphics in describing methods, such as maps of study sites or diagrams of experimental apparatus. In Section 3.5 we will present some guidelines for incorporating graphics into your text.

## EXERCISE 3.7

Examine the organization of the methods section in the research report by Graham et al. (1992) on pages 254–255 or by Mallin et al. (1995) on pages 330–333. Both of these papers follow conventional formats prescribed by the journals' editors; neither journal encourages the use of subheadings within sections. Suppose the editors had a change of heart and decided to allow authors to subdivide their methods section. What subheadings would you recommend, and where would they appear?

Although it is generally recognized that the methods section cannot describe every action of the experimenters, the description should be detailed and complete enough to enable knowledgeable colleagues to repeat the experiment, observation, or calculations successfully. Standard procedures that will be familiar to your in-field readers can be identified quickly without citations or further explanation. Procedures established in previous studies typically appear with citations acknowledging the precedent. And new procedures or substantial modifications will be explained and justified. Just as the introduction section argues that the study was needed, the methods section argues that the study was sensibly designed and carefully conducted. This is the occasion to defend the decisions you made in designing and carrying out the research. Often these explanations are simply offered in passing, as in, "We increased the volume to 30 ml for the copepod *to minimize containment effects*." At other times, an explicit justification is offered to explain a methodological choice: "The dinoflagellate's TFVCs require an unidentified substance in fresh fish excreta; *hence, it was necessary to maintain cultures using live fish*." (Both examples are from Mallin et al., 1995, emphasis added.)[5]

---

[5]Methods are sometimes described without such justification in student lab reports (because the primary audience, the instructor, has chosen the methods and does not need to be convinced of their validity), but in most other scientific contexts you will be describing methods *you* have chosen to readers who were not privy to your methodological decision-making. How you explain and justify these choices is crucial to your *ethos* as a professional scientist.

Matthew Barker noticed several levels of procedural explanation in a report published in the *Journal of Protein Chemistry* (see Figure 3.7). As Barker's analysis illustrates, the degree of justification needed for a procedure depends on the status of these procedures in the research community. We noted earlier that this status may well change over time as a research area matures. For example, in an analysis of *Physical Review* articles reporting research in spectroscopy over the course of the last century, VandeKopple (1998) found that authors publishing as the field first emerged in the 1890s included many more experimental details than authors reporting similar types of work in the 1980s. The early researchers also tended to adopt a more cautious tone in describing their instruments and materials, taking care to depict their methods "as applying to one particular time and place, rather than . . . as firmly established and widely generalizable" (p 190).

In sum, the methods section establishes the conditions under which the results of the study were generated, and therefore the context in which they must be interpreted. In this sense, the methods "frame" the results (and represent a frame within a frame). Notice that describing the methodological details of an experimental study is comparable to the articulation of assumptions required in theoretical papers. In describing their methods, researchers articulate the assumptions they made about the phenomenon under study: for example, that it will or will not vary over time or as a function of temperature or nutrient content or medical treatment or altitude or other factor; that the effects of these variables are observable with these instruments or observational techniques; that these techniques are reasonably free of bias; that these subjects are representative of the larger population to which the researchers wish to generalize; and so forth. The validity of the study design rests on the persuasiveness of the researchers' description of the methodological decisions they made.

## EXERCISE 3.8

In the article by Mallin et al. (1995), reprinted in Chapter 10, look for examples of each of the three levels of procedural explanation described above: (1) standard or routine procedures, (2) procedures established in previous studies, and (3) new procedures or substantial modifications. How much explanation or justification is provided in each case? Why?

## EXERCISE 3.9

Turn the methods section from one of your recent class lab reports into a methods section suitable for a journal audience in your field.[6] Where appropriate, state the rationale for your methodological choices. Target a specific journal, and use informative subheadings if editorial guidelines permit.

---

[6]We are indebted to Olsen and Huckin (1991) for this exercise.

## Analysis of Citation Use in a Research Report in the Field of Biochemistry

The article "Chemical Modification of Cationic Residues in Toxin α from King Cobra *(Ophiophagus hannah)* Venom" by Shinne-Ren Lin, Shu-Hwa Chi, Long-Sen Chang, Kou-Wha Kuo, and Chun-Chang Chang reflects the general IMRAD format quite closely with only a slight modification to the form. The article appears in *Journal of Protein Chemistry,* Vol. 15, No. 1, 1996, pages 95–101. The research report is divided into three main sections. The introduction is section one, the materials and methods is section two, and the results and discussion make up section three. In all these sections, the use of citation based on the knowledge of the audience can be seen.

In the first two sentences of the introduction the topic of research, snake toxins, is announced. Although these sentences use technical jargon that might be confusing to someone outside the field, the information they furnish is actually quite general and requires no citations. The authors then use eight extensively detailed sentences supported by ten citations to summarize the current knowledge and research in the field. In the first sentence of the second paragraph, the topic is more tightly focused on the toxins of the king cobra. The second sentence of the paragraph points out the low number of studies on the α-neurotoxins of the king cobra. The next three sentences list the research previously done in this subfield, mostly by the authors, and are supported by four citations. The fact that all of the current research cited was performed by members of the research team greatly enhances the credibility of this study. The final sentence of the introduction states the objective of the research.

The materials and methods section is divided into six subsections. The section opens with a paragraph describing the materials used. The rest of the section is broken down into five enumerated subsections as follows: 2.1 Modification of Arginine Residues with HPG, 2.2 Modification of Amino Groups with TNBS, 2.3 Localization of the Incorporated Groups, 2.4 Assay for Lethal Toxicity, and 2.5 nAChR-Binding Assay.

The materials paragraph describes the materials used in the experiment and gives the names and locations of the chemical and biological companies that supplied the materials. The paragraph also refers to a procedure used to isolate and purify cobra venom and supports it with two citations.

The methods subsections describe the two different modification procedures used and the two different assay procedures employed to assess the effects of the modifications. Of the ten citations within the materials and methods section, four of the research reports cited were authored by members of the research team, again indicating the expertise possessed by this group of scientists.

The methods sections themselves briefly describe the procedures and parameters under which the research was carried out. Citations are used where possible rather than writing summaries of the procedures. A procedure from 1938 is even cited. Familiarity with a procedure from so long ago demonstrates extensive knowledge of the field. It is possible that the procedure is actually fairly common knowledge in the field, but citing it suggests awareness of research and procedures discovered between then and the present.

As stated earlier, the results and discussion sections are consolidated into one section rather than separated. Even though that is the case, the section is divided into two subsections. The first, titled "3.1 Characterization of the Modified Derivatives," is actually a results section; and the second subsection, titled "3.2 Biological Activity of the Modified Toxins," discusses the results.

Subsection 3.1 presents the results and supports the data with four figures and three tables. The figures and tables are clearly labeled and complement the corresponding textual support for the data. The data from a polyacrylamide gel that was run was omitted and is so noted in the report. The reason for such an omission can only be guessed, but perhaps the reputation of the researchers has been so well established that they could forego visual support in this instance. It is also possible that the data can stand alone and did not necessitate displaying the gel, or the omission of such a figure is commonplace in the field. Aside from the one anomalous omission of a figure, the results subsection otherwise conforms to the IMRAD format.

*(continued on page 56)*

**FIGURE 3.7**  Analysis of a research report from the field of biochemistry by Matthew Barker. Student sample.

The discussion section is contained in subsection 3.1. Here the significance of the results and the conclusions of the research are displayed. Four other studies are cited as having carried out similar research, and this report is offered as an extension of the information that those previous studies first elucidated. Conclusions are further supported by citing another study that supports another aspect of the current study.

Unlike the citations found in the other sections of the report, none of the authors contributed to any of the citations found in the discussion section. Because all of the current research on the α-neurotoxins of the king cobra has been conducted by members of the team, it seems that those studies could easily have been chosen to lend support to the conclusions made in this report. Such support might be construed as being biased. By using outside support, however, the possibility of bias is avoided, and the conclusions gain credence. The researchers also successfully place their research within the context of other preexisting, accepted research in the field.

**FIGURE 3.7**   *(continued)*

## 3.5 Reporting Results

The presentation of results plays a critical role in the developing argument of the research report, for it is here that new evidence is presented to address the gap or question outlined in the introduction. At first glance, the content of this section of the IMRAD form seems obvious: this is where the data are presented. But presenting data is not a simple matter. Researchers cannot display all the information accumulated during the study, which may consist of pages and pages of lab or field notebooks or disk space. Journals do not allow room to report each day's striped bass capture rate in a three-month study period, for example, or pH levels from all soil samples in a 200-sample design.

Instead, the data are summarized. The first step is to *reduce* the data to a manageable size for presentation; in experimental studies this is often accomplished by converting raw data to means (averages) and by using figures and tables. Mean capture rates might be reported for each week or month, depending on how much variation was observed over those time periods; the results of soil tests might be grouped and averaged according to the location of the samples or the distance from a waste discharge point. Once the data have been reduced in these ways, comparisons can be made; for example, changes in capture rate can be examined over time, pH levels at different sampling sites can be compared. It is these observed changes or differences (or the lack of change or difference) that constitute the answers to a study's research questions.

Reducing the data thus enables authors to take a second step in presenting results, that of *generalizing* from the data. Researchers want to point out the trends they've noticed in the data so their readers can see why they drew the conclusions they did. If they are to convince readers that their conclusions are valid, they must first ensure that the patterns they saw in the data are readily apparent. Olsen and Huckin (1991) point out that both levels of information must therefore be presented: "(1) the major generalization(s) you are making about your data and (2), in compact form, the data supporting the generalization(s)" (p 363).

Compare the following sentences, both of which refer to the data contained in a table:

**Example A**

Bioassay trials showed that all 10 rotifers (*B. plicatilis*) exposed to nontoxic green algae survived; all 10 exposed to dinoflagellates survived; all 10 exposed to both survived; 3 of the unfed group survived (Table I).

**Example B**

Bioassay trials with the rotifer, *B. plicatilis,* showed no mortality among any of the three algal food treatments (Table I). In the unfed treatment, 7 of 10 replicate animals died during the 9 day experimental period, and survival of unfed rotifers was significantly lower than in the other treatments (Fisher's exact test, $P = 0.0015$).

Example B is from the report by Mallin et al. (1995), reprinted on pages 328–340. Example A is invented. If you look at Table I in the Mallin report, you'll see that Example A simply repeats data that can easily be read in the table itself. Example B, on the other hand, generalizes from the data, emphasizing the similarity among the three treatment groups and the difference between those groups and the control. Such generalizations are often stated in topic sentences at the beginning of paragraphs, as in Example B.

The relationship between data and generalizations is easily demonstrated by looking at how tables and figures are referred to in well-written reports, as in Example B above. Tables, graphs, and other visual aids are essential tools for reporting scientific results, but it is important to keep in mind that graphics only present the data; the generalizations needed to interpret those data must be provided in the text. We recommend two essential steps in integrating visual and verbal information:

- Refer readers to the graphic explicitly.
- Tell them what patterns to notice.

In Example B, the researchers refer readers to Table I by referencing it in parentheses at the end of the first statement describing data in that table. Other references to tables and figures are even more explicit, as in the following example from Chiba et al. (2002), reprinted in Chapter 10:

**Example C**

Table 3 shows the impact of eradication treatment on disease specific measures of quality of life. The difference in the change in scores from pretreatment to study end showed significantly greater improvement in three of the five domains for the eradication arm.

The first sentence points readers to the table, identifying the kind of information to be found there (turn to Table 3 on page 283). The authors then go on to highlight the trends to be noticed in this table, helping readers interpret the mean differences listed in the second column by calling attention to the significance levels (P values) reported in the fourth column. (Three of the five P values fall below the commonly accepted probability threshold of 0.05, indicating that these differences are unlikely to have occurred by chance and are therefore considered "real" or significant differences.) As in Example B, the authors of the excerpt in Example C do not simply repeat data from the table in the text itself; rather, they use the text to summarize and characterize the data, to help readers see what they see in this set of numbers.

## EXERCISE 3.10

Look for explicit references to figures and tables in sample articles in this text and from journals in your field.

1. Notice that the parenthetical reference to Table I in Example B is different from the direct reference to Table 3 in Example C. How common are these two "reporting formulae" in the articles you examined? Do these forms vary across fields? Across journals? Pay special attention to the ways in which authors in your field refer to figures and tables.
2. Find 5 to 10 sample parenthetical references to figures and rewrite them as direct references. You may find that you develop a formula of your own for transforming these structures.
3. Describe how the raw data have been transformed in the figures or tables you have found; that is, how have the raw data been reduced into a summary visual form?

## EXERCISE 3.11

In Figure 3.8 we have reprinted a table and a figure representing the same set of data. What do you notice about the two forms of visual presentation? List the types of information highlighted in each form, as well as the types of information that are subordinated or not available. What is gained or lost in the transformation from one representation to the other?

As you will have noticed in Exercise 3.11, different types of visuals serve different purposes. Tables are particularly useful for summarizing data when the exact values of the data are important or when there are no clear patterns that would lend themselves to graphical presentation (Monroe et al. 1977). Tables are the least "visual" of visuals insofar as they don't actually transform data into a pictorial form; however, the spatial arrangement of columns and rows enables researchers to display and categorize information, facilitating comparisons between groups or study sites or along other dimensions. Tables may be used to organize verbal as well as numerical data. In the paper by Spiller et al. (2002), reprinted in Chapter 13 (see page 410), Table 1 presents verbal data in two columns: a list of characteristic behaviors of the Delphic priestesses from the historical record is compared with a list of effects of ethylene inhalation documented in 20th century science. In another paper from this study, De Boer et al. (2001) use tables to present and compare numerical data: hydrocarbon gas concentrations at different study locations (see Tables 1 and 2, reprinted on page 408).

Table 4. Gastritis Scores of the 175 C. pyloridis Culture-Positive Patients Before and After Therapeutic Regimens

| | | Treatment | | | | | | | | | |
|---|---|---|---|---|---|---|---|---|---|---|---|
| | | CBS | | Amoxicillin | | CBS + amoxicillin | | Cimetidine | | Sucralfate | |
| | Culture | $m^a$ | $n^b$ | m | n | m | n | m | n | m | n |
| Before treatment | + | 5.9 | 67 | 5.5 | 22 | 6.2 | 20 | 5.5 | 53 | 5.8 | 13 |
| After treatment | + | 3.8 | 37 | 2.9 | 7 | 6.5 | 2 | 5.4 | 52 | 5.3 | 13 |
| | − | $1.2^d$ | 30 | 0.9 | 15 | $1.1^c$ | 18 | | | | |
| 1 mo after treatment | + | 5.6 | 55 | 4.2 | 17 | 5.0 | 12 | 5.8 | 51 | 5.3 | 13 |
| | − | $0.7^d$ | 12 | $0.7^d$ | 5 | $0.5^d$ | 8 | | | | |
| 3 mo after treatment | + | 5.8 | 54 | 5.1 | 16 | 5.3 | 12 | 5.5 | 52 | 5.7 | 13 |
| | − | $0.5^d$ | 10 | $0.2^d$ | 5 | $0.9^d$ | 8 | | | | |
| 6 mo after treatment | + | 5.7 | 54 | 5.2 | 16 | 6.0 | 12 | 5.5 | 48 | 5.6 | 13 |
| | − | $0^d$ | 10 | $0.6^d$ | 5 | $0.6^d$ | 8 | | | | |
| 12 mo after treatment | + | 6.6 | 52 | 5.5 | 16 | 5.5 | 13 | 5.3 | 51 | 5.9 | 13 |
| | − | $0^d$ | 10 | $0^d$ | 5 | $0^d$ | 7 | | | | |

Gastritis was assessed by scoring for four characteristic pathological parameters of chronic active gastritis (see Methods). +, positive; −, negative. [a] Mean score. [b] Number of patients. [c] $p < 0.05$ vs. culture-positive patients by Wilcoxon rank-sum test. [d] $p < 0.01$ vs. culture-positive patients by Wilcoxon rank-sum test.

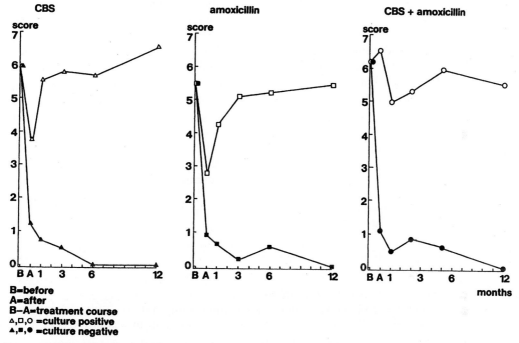

B=before
A=after
B−A=treatment course
△,□,○ =culture positive
▲,■,● =culture negative

Figure 2. Mean gastritis scores in patients with chronic active gastritis and positive culture for C. pyloridis, before (B) and after (A) treatment with CBS, amoxicillin, or a combination of CBS and amoxicillin. Note the ultimate complete disappearance of inflammatory changes of gastric mucosa (gastritis score = 0) after persistent eradication of C. pyloridis during follow-up of 1 yr. (See Table 4 for actual values.)

**FIGURE 3.8**   Two different representations of a data set. Table and figures from Rauws et al. (1988, p 54).

Other types of visuals are better able to highlight continuous trends and patterns. For example, line graphs illustrate chronological trends well, for lines as a visual form naturally illustrate movement and direction over time. Line graphs are included in the papers by Graham et al. (1992) in Chapter 10 (page 255), and Burkholder et al. (1992) in Chapter 11 (page 290). You'll notice in both that the line graph's x-axis represents time, measured in units such as weeks or days.

Compare these x-axes with the x-axis in the bar graph Burkholder and Rublee (1994) included in their proposal in Chapter 11 (Figure 5, page 321). In this bar graph, eight sampling sites—four control sites and four discharge sites—are arrayed along the x-axis. Each bar represents one observation (the number of zoospores counted at the site), rather than a sequence of observations over time. The bar graph enables readers to see at a glance the relative magnitude of the counts at each site. The bar graph presents data in discrete categories, much like a table, but the visual depiction dramatizes the comparisons: similarities among control sites and among the discharge sites are easily apparent, as is the striking difference between the two conditions.

Graphs are labeled as "figures" in most scientific texts. Other types of figures include diagrams, cross-sections, maps, photographs, and flowcharts, each of which serves a distinctive purpose and reflects the logic of the research design. For example, photographs and similar media are useful when observational detail of the object is important; Warren and Marshall use electron micrographs in their 1983 *Lancet* letters to show the shape and location of the *Campylobacter* bacteria they discovered (page 234). Diagrams are useful for highlighting structural features and relationships (Olsen and Huckin 1991), as in Burkholder and Rublee's Figure 2 (page 311), in which the authors illustrate the many life stages of the toxic dinoflagellate, the relationships among these stages in the life cycle, and interactions with other factors in the environment (finfish and phosphates). De Boer et al. (2001) use three different types of maps to show Delphi's proximity to two geologic faults (Figures 1, 2, 3; pages 406–407). In addition to their role in summarizing results, visuals such as maps, diagrams, and photographs are often used to illustrate methods or to locate study sites.

Some research fields are characterized by the use of particular types of nonverbal representations. For example, papers in organic chemistry rely heavily on diagrams of molecular models, whereas modeling studies in physics often include mathematical equations and formulae. When such visual or numerical information is part of a theoretical argument, it is typically inserted in the text itself rather than in a separate figure, as can be seen in the 1996 Reynolds paper reprinted in Chapter 12 (pages 385–389).

Guidelines for designing figures and tables can be found in style manuals (e.g., AIP Style Manual 1990; CBE 1994), writing guides (Monroe et al. 1977; Katz

## EXERCISE 3.12

Go back to the sample articles you examined in Exercise 3.10. List the types of graphics used in these articles, the kinds of data contained in them, and the patterns revealed. What claim in the text does each visual support? Why were these visual forms selected, and not others? Pay special attention to the purpose of the visuals and the information contained in table titles and figure captions.

1985; Pechenik 1987; Day 1998), and in the instructions to authors published by individual journals, particularly in fields that rely heavily on specialized types of visuals. For example, the Instructions for Contributors to the journal *Poultry Science* provide guidelines for preparing tables and figures and for submitting photographs of line drawings and graphs, computer-generated figures, and light and electron micrographs (Poultry Science 2002). These varied sources all agree on the following general guidelines for preparing visuals:

- Each table or figure must be independent, that is, self-explanatory. Readers should be able to understand what kind of information is presented in the table without having to find the description in the text.
- For this reason, table titles (which appear *above* the table) should be as informative as possible: not "Table 1. Vegetation in trial plots" but "Table 1. Percentage coverage by low vegetation (< 1.5 m high) in midsummer by treatment, year, and site" (Lorimer, Chapman, and Lambert 1994). Titles should identify the variables being compared in the table. All labels in the table (column headings, row headings, notes, etc.) should be clear and brief.
- Similarly, figure captions (which appear *below* the figure) should be as informative as possible: not "Figure 1. Study area" but "Figure 1. Location of Waccabassa Bay study area, sampling sites, and distribution of storm deposit" (Goodbred and Hine 1995). Symbols should be clearly identified in a key; axes on graphs should be clearly labeled.
- Despite their independent status, figures and tables must also be *interdependent* with the text. The visual and verbal presentations of results must work together. Visuals should be clearly referred to in the text, as discussed above, and they should appear in the text as soon after the textual reference as possible (visuals should never precede their text mention, as this can confuse readers).

Keep in mind that even though graphics are intended to be self-explanatory, readers will not know what *you* find interesting or noteworthy in the table or figure unless you tell them. Whether your data are presented graphically or discursively, that is, visually or verbally, be sure to highlight the trends you see and explain what they are based on.

It is conventional in many fields to subdivide the results section to highlight the different types of analyses reported. If subheadings are used in both the methods and results sections, they typically reflect the same principle of division, as in the Chiba et al. (2002) paper, reprinted in Chapter 10 (pages 280–286). The results section of this paper includes four subsections describing the four outcome measures introduced under methods (these parallels are not as clear as they might be because the order and labeling of these measures is not consistent across the two sections).

In sum, though researchers present their formal conclusions in later sections of the report, a great deal of interpretation goes on in results sections as well (Thompson 1993). Summarizing, reducing, and generalizing from the data are all highly interpretive processes. It should be clear by now that researchers do not let the data speak for themselves; in summarizing their results, they also interpret them for the reader. They must present their data clearly, in the text and in any accompanying tables and figures, to demonstrate to readers that their interpretations are warranted.

Read the introduction and methods sections (only) of the study by Graham et al. (1992) reprinted in Chapter 10 (pages 254–257).

1. Before reading the results section, study Table 1 in this paper. What trends do you see in these data? Draft a few sentences describing the patterns in this table as though you were writing up these results.
2. Then study Figure 1 in the same paper, and again draft some sentences about the patterns in the data. How would you describe the results presented in this figure?
3. Compare your descriptions with the authors' descriptions of this table and figure.

Choose a sample research question for which you can gather some actual data quickly from your classmates, friends, or fellow dorm residents. For example, you might be interested in who spends more time studying: women or men, seniors or first-year students, on-campus or off-campus residents. Or you might gather information about time spent writing term papers for different classes, or the number of visits to the library in a semester. To enable comparisons, be sure to include more than one group in your sample (e.g., freshmen versus seniors) or more than one variable (e.g., time spent writing versus grade on paper).

Present the data in a table or graph with an informative title or caption. Write a one-paragraph description of the purpose and research question (or hypothesis) of your mini-study. After you've reviewed the data yourself, exchange research questions and data sets with another student. Have that student describe the major trends he or she sees in the data and tell you the answer to your research question. If you've stated a formal hypothesis, have your reader tell you whether the hypothesis was supported.

## 3.6 Discussing Trends and Implications

As noted earlier, the discussion section of the IMRAD form completes the frame opened in the introduction by returning to the significance argument. The purposes of introduction and discussion sections are inversely related. Whereas the introduction introduces the research question and reviews the state of knowledge in the field that motivated the question, the discussion explains how the question has been answered (at least in part) by the new research and shows how the field's knowledge is changed with the addition of this new knowledge. Thus, the discussion describes a new state of knowledge in the field, one that in turn motivates further research questions, which are typically highlighted in this final section.

Structurally, the introduction and discussion are mirror images as well. Berkenkotter and Huckin (1995) describe the introduction as proceeding "from outside in" in that it begins by talking about the general topic and reviewing the existing research in the field, and then it narrows to focus on the particular work at hand. Conversely, the discussion proceeds "from inside out" by first summarizing major trends in the new results and then situating those results in the context of the field (p 41).

The new results are situated by comparing them with findings of previous studies and with the field's general understanding of the phenomenon under study. This is where the researchers offer their interpretation of their findings. They discuss how the new results extend, refine, or challenge previous findings or assumptions (Olsen and Huckin 1991). As the editors of the *British Medical Journal* (2002) explain, a given study may have

> asked and answered a new question (one whose relevance has only recently become clear); contradicted a belief, dogma, or previous evidence; provided a new perspective on something that is already known in general; [or] provided evidence of higher methodological quality for a message which is already known. (http://bmj.com/advice/sections.shtml)

This discussion of implications must be clearly tied to the current results and must demonstrate an awareness of the limits of those results. In sum, the discussion section is the place to

- Briefly summarize the major findings, including the magnitude and direction of the effects observed (compared with what others found or compared with what was predicted or might be expected).
- Acknowledge the advantages and limitations of the methods used in the research (and comment on how these features may have influenced the observed effects).
- Explain the implications of the findings for current practice or theory.
- Outline the research questions that remain.

Notice that these discussion issues are interdependent. In commenting on the magnitude and implications of the new findings and the research questions that remain, it is important that the researchers recognize and acknowledge the limitations of their study. Scientists must anticipate the questions other researchers might have about their methods, findings, and conclusions, and must answer these questions in the text before they are asked. It must be clear to readers that the authors have considered "rival hypotheses," that is, possible alternative explanations for their findings. Philosopher Stephen Toulmin, an expert on scientific argumentation, calls these answers to potential questions, objections, and counterarguments "rebuttals" (Toulmin et al. 1984). According to Toulmin, rebuttals are the recognition of the circumstances or exceptions that might undermine the argument of the research.

Rebuttals often consist of further qualifications or acknowledgments of limitations, exceptions, and the need for further study and research. Rebuttals therefore are a way for a writer to verbally close loopholes, seal leaks, and tie up loose ends in the logic of the report. This sentence from the third paragraph of the discussion section in Graham et al. (1992) is a good example (see page 257):

> Although some may argue that the lack of double-blinding introduced an important bias into our study, no objective data support such a contention, and we believe such a scenario extremely unlikely, especially considering the equipment now available for studying gastroduodenal mucosa.

All arguments, of course, are open to rebuttal; as we discussed in Chapter 1, scientific statements must be falsifiable to be accepted as valid. But in anticipating objections and answering them, scientists show they understand the limitations of their research, the nature of research problems in their field, and the expectations of their colleagues. Thus, the professional use of rebuttals is another way scientists enhance the credibility of their research as well as their *ethos* as researchers.

As is the case in other sections of the report, the emphasis in discussion sections varies somewhat across fields. It is conventional in journals in applied fields, such as the *Journal of Wildlife Management,* to include a separate section on management implications after the main discussion. Similarly, discussion sections in agricultural journals may include advice for producers on feeding practices or litter composition. Contributors to medical journals may offer recommendations for clinicians and/or for public health policy. In these and other applied areas, the discussion section is the place for researchers to show how their results pertain to actual practice or policy in the field.

In basic research areas the focus is not on practical applications but on the theoretical or methodological implications of the new results. This is certainly the case in reports of theoretical studies like the Hendrick and Reynolds (2001) piece included in Chapter 12 (pages 394–399). Notice that Hendrick and Reynolds use their discussion section to explore how their new analyses of older supernova remnants in the Large Magellanic Cloud extend the field's previous understanding, which was based on a Galactic remnant sample of younger SNRs. The implications of these findings are discussed in a separate conclusion section, in which the authors state very directly that their findings indicate "some revision may be needed" in the field's understanding of the origins of Galactic cosmic rays.

Thus, discussion and conclusions sections, whether combined or labeled individually, represent the conclusion to the broader argument of the research report. Research authors do not leave it up to readers to draw conclusions about the data. They tell readers *their* conclusions, the ones the researchers believe the data support, and they provide enough information about the data and the conditions under which the data were generated to enable readers to determine whether their conclusions are justified. The American Institute of Physics (1990) describes conclusions as "convictions based on evidence" (p 4). The discussion section is the place to state your convictions and demonstrate how they follow from the evidence you present in other sections of the report.

## EXERCISE 3.15

The Graham et al. (1992) report contains several rebuttals in the last two paragraphs of the discussion, leading up to the final recommendation supporting Marshall's ulcer therapy (see page 257). What rebuttals have the authors included? What counterarguments are these rebuttals to? Why do you think the authors felt they had to include these rebuttals in their report? It may be useful to know that Graham was a former opponent of Marshall's treatment.

1. Analyze the discussion and conclusion sections of three or more full-length research reports from journals in your field (include any separate sections describing recommendations or management implications). List the types of discussion issues raised in these sections. Use the categories listed in this section as a guide (summary of results, advantages and limitations of methods, implications for practice or theory, remaining questions), but be prepared to expand the list or create subcategories as necessary to best describe the kinds of issues raised in your field.

2. In groups of three or four students from different fields, compare lists and prepare to discuss similarities and differences across these research areas. How do these similarities and differences reflect the nature of research in these fields?

Plan a discussion section to contextualize the results of the informal study you conducted in Exercise 3.14. What kinds of previous research would you want to be familiar with if you were to conduct this study on a larger scale and prepare a formal report for a research audience? What implications can you draw from your results? What methodological limitations would you discuss? If your study were more formal, would you recommend changes in practice or policy on the basis of your results? What further research directions would you suggest? List, as specifically as you can, (1) the issues you would raise in such a discussion and (2) any further information you would need to complete this task.

We have focused thus far on the logic and structure of the arguments presented in research reports, but it is also important to notice distinctive features of the language in which these arguments are expressed. The persuasive and interpretive nature of scientific texts can be detected even in the language of individual sentences if we look carefully at how scientists phrase their claims and conclusions. Take a minute to fill in the blanks in the sentences listed in Figure 3.9 before reading further.

In a study of scientific claims, Penrose and Fennell (1993) asked experienced scientists from a variety of fields to complete the sentences in Figure 3.9. The verbs *suggest, indicate, show,* and *demonstrate* were the most common responses from these experts, regardless of field. The finding that responses from these individuals were fairly uniform indicates that these terms are conventional across disciplines and are not unique to writing in, say, geology or botany. Notice that we've used one of these conventional verbs, *indicates,* in the preceding sentence to signal our interpretation of Penrose and Fennell's results. The sentence tells you that from the responses of these individual scientists we are generalizing about scientific discourse at large.

**Directions**

The following selections are from a variety of research reports published in scientific journals. Read the selections carefully. When you come to a blank, fill in an appropriate word or phrase to complete the sentence. You may insert more than one word in the blank if necessary, and you may use the same word(s) in more than one sentence if you wish.

1. Eleven of the trials have shown the treatments to be ineffective, yielding an overall response rate of 4/278 (1.4%). . . . These data _____ that the minimal response rate of interest should be .15.
2. The above observations _____ that (1) fertilized soils tend to attain apparent equilibrium with orthophosphate solid phases and (2) soils with moderate to high P-fixing capacity tend to have limited movement of P when fertilized with inorganic P sources.
3. Statistical analysis _____ that corn yields were not influenced by N fertilizer rate in 1980, but were in 1981 (Table 1). The lack of influence of N fertilizer in 1980 was attributed to high levels of native N in the soil and climatic conditions unconducive to high corn yields (Fig. 2).
4. More recent studies of modern thickly sedimented convergent margins _____ that the Washington margin is anomalous. For example, the Makran (Platt et al. 1985) and Barbados (Westbrook 1982) convergent margins are thickly sedimented and have convergent rates similar to the Washington margin (about 5 cm/yr). However, only the Washington margin is dominated by landward-verging structures.
5. Results of this study _____ that significant genetic divergence has occurred among geographically separated groups of raccoons. The average differentiation among the 14 localities examined (37.4%) is similar to the value obtained among populations of pocket gophers (41.0%; Patton and Yang 1977).

**FIGURE 3.9**    Sample claims from a study of scientific discourse by Penrose and Fennell (1993). Sources: Sylvester (1988, p 833); Schwab and Kulyingyong (1989, p 180); Clay et al. (1989, p 321); Byrne and Hibbard (1987, p 1165); Hamilton and Kennedy (1987, p 270).

When we say that verbs like *suggest* and *indicate* are "conventional" in scientific discourse, we mean that they are commonly used constructions that carry particular, agreed-upon meanings in the community. These conventions enable scientists to make claims within the established parameters of knowledge-making in science. That is, these expressions enable a researcher to put forth a conclusion—to contribute to the advancement of knowledge in the field—while at the same time acknowledging that it *is* a conclusion—an interpretation of the facts and not a fact itself. Use of these tentative or hedging verbs signals to other scientist readers that this author is aware of the interpretive nature of scientific knowledge and that the claim is to be interpreted as "true" only within the boundaries of current knowledge and conditions (Hyland 1996). In examining research texts in psychology, Madigan et al. (1995) observed that hedged conclusions may be more convincing than more strongly worded claims because they convey "proper respect for the empirical process" (p 432). In addition to enhancing author credibility, there is some evidence to suggest that the inclusion of appropriate hedges helps readers comprehend the text (Crismore and VandeKopple 1988).

Scientists also use other kinds of hedges or qualifiers to convey the interpretive nature of their claims. Adverbs and adverbial phrases are often used to note limitations or special conditions; examples include *possibly, probably, very likely, necessarily, certainly, without doubt, presumably, in all probability, hypothetically, maybe, so far as the evidence suggests,* and *as far as we can determine* (Toulmin et al. 1984). Such qualifiers indicate the strength or extent of the claim being made, as in the following examples from the 1995 article by Mallin et al. (emphasis ours):

> *Pfiesteria*-like species *apparently* are widespread; recent investigations have demonstrated their presence at "sudden-death" estuarine fish kill sites from the mid-Atlantic . . .

> Since the fecundity of *B. plicatilis* in the two *P. piscicida* treatments was slightly elevated overall, relative to fecundity when given green algae alone, this dinoflagellate is *probably* a nutritious food source for the rotifer.

In addition to verbs and adverbs, a third type of qualifier is the modal auxiliary verb (Toulmin et al. 1984). Verbs like *may, might, would, could, should, must, can,* and *shall* are used to indicate qualifying conditions, as in the following examples:

> Given its broad temperature and salinity tolerance, and its stimulation by phosphate enrichment, this toxic phytoplankter *may* be a widespread but undetected source of fish mortality in nutrient-enriched estuaries. (Burkholder et al. 1992)

> The stomach *must* not be viewed as a sterile organ with no permanent flora. (Warren and Marshall 1983)

Hedging verbs, adverbs, and modal auxiliary verbs can and do occur anywhere in a scientific text where researchers need to qualify or limit their claims.[7] In a sense, then, qualifiers are related to rebuttals insofar as scientists use them to acknowledge the limitations of their work and to anticipate and head off questions and counterarguments that readers might pose. These linguistic devices thus reflect the interpretive and interactive nature of knowledge-building in science. As you join the community of scientists, you are acquiring knowledge of both the logic of scientific arguments and the language in which that logic is expressed.

---

### EXERCISE 3.18

1. Penrose and Fennell (1993) also asked students from different majors and levels of experience to complete the sentences in Figure 3.9. They found that first-year students and non-science majors were more likely to supply the word "prove" in these sentences than were science majors or expert scientists. Why is this usage less conventional among scientists? What is the difference between "suggest" and "prove" in these types of scientific claims?

2. Another unconventional choice some students made in this study was to use words like "conclude" or "hypothesize" in the sample sentences. What is unconventional about these choices? How does "conclude" differ from "suggest" or "indicate" in the context of these sentences?

---

[7]See Hyland (1996) for a detailed analysis of these and other forms of hedging.

Examine the discussion and conclusions section(s) of a research report from your field or one of the sample articles in this text, for example, Hendrick and Reynolds (2001), on pages 394–399, or Chiba et al. (2002), on pages 280–286. Look for instances of hedging in the text, particularly the three categories of qualifiers described above: verbs, adverbs, and modal auxiliary verbs. Look for other hedging devices as well. Would you propose any additional categories? What purposes do the hedges serve? How common are they in the sample articles you've chosen? Compare your findings with those of students examining other fields. If you're analyzing sample texts from this book, add one of the Delphic oracle studies from Chapter 13 to your survey to broaden the range of research types. Do you see any differences across fields in the use of hedges?

The original title of the Burkholder team's letter to *Nature* (1992) was "New 'phantom' dinoflagellate is implicated as the causative agent of major estuarine fish kills." This title was shortened during the publication process (see page 289), much to Dr. Burkholder's dismay. What is the effect of the title change, and why do you think Burkholder was dismayed?

## 3.7 The Research Report Abstract

The abstract is a summary of a document's major points, appearing at the start of each article in most journals. Along with the title of the report, the abstract helps readers decide whether the paper is pertinent to their research interests. The American Institute of Physics *Style Manual* (1990) advises authors to "[b]ear in mind that [the abstract] will appear, detached from the rest of the paper, in abstract journals and on-line information services. Therefore it must be complete and intelligible in itself; it should not be necessary to read the paper in order to understand the abstract" (p 5). The abstract therefore must represent the full argument contained in the report: the topic and purpose of the study, the methods used, the results obtained, and the conclusions drawn from those results. However, this argument must be outlined in extremely condensed form, usually only about 5 percent of the length of the full paper (AIP 1990).

Given its role in representing a larger document, the abstract can be described as a "contingent" genre: its content is contingent on the content and organizational logic of the document being abstracted. We will talk about research report abstracts in this chapter and will provide guidelines for the conference presentation abstract in Chapter 5 and the research proposal abstract in Chapter 6. Abstracts will vary considerably within these categories as well, however, reflecting variations in purpose and logic across research fields, journals, and articles. Some journals provide formal guidelines for the abstract that embody the standard research logic of their respective fields. *Annals of Internal Medicine*, for example, requires authors of research articles to follow a structured format for the abstract, as illustrated in the paper by Graham et al. (1992) in Chapter 10 (see page 254). This type of structure—in this case with separate sections for the study's objective, design, setting, participants, intervention, measurements, results, and conclusions[8]—is used in other journals reporting clinical trials as well. (See Chiba et al. [2002], on page 280, for an illustration of the *British Medical Journal*'s very similar form.)

Many journals, however, specify only a maximum length for the abstract (typically 100 to 250 words) and leave its structure up to the author. Because the abstract mimics the form and logic of the full report, the structure should be relatively straightforward. If the report follows the IMRAD form, the abstract is likely to include four basic moves representing the major parts of this pattern, and verb tense will shift accordingly (Olsen and Huckin 1991): the topic will be introduced in present tense, usually in a sentence or two; the background and/or need for the study will be outlined in another few sentences (but without references to individual prior studies [AIP 1990; CBE 1994]); methods and results will be briefly described in past tense; and the major conclusions and implications of the study will be stated in present tense. Abstracts for IMRAD articles can be easily generated using this four-part structure.

If the report follows a logic other than the IMRAD pattern, however, the structure of the abstract will vary accordingly. For example, the abstract for Spiller et al.'s (2002) *Clinical Toxicology* paper defending the gaseous vent theory (see page 410) begins with one sentence summarizing the historical claim that intoxicating gases inspired the oracle, followed by three sentences summarizing the geological and chemical evidence that supports this claim. The fifth and final sentence states the authors' conclusion that the priestess's trance was caused by ethylene inhalation. Thus, the abstract encapsulates the line of reasoning followed in this particular article.

The IMRAD and non-IMRAD samples mentioned above are all examples of *informative* abstracts, that is, abstracts that summarize the information in the larger text. This type is distinguished from the *descriptive* abstract often adopted for research reviews (Olsen and Huckin 1991) and conference proposals. Descriptive abstracts (also called *indicative* abstracts [Day 1998]) describe the kinds of information that will be contained in the paper, rather than providing a summary of that information. Two examples from the *Journal of Computational and Graphical Statistics* are presented in Figure 3.10. As these examples illustrate, the descriptive approach is particularly appropriate when describing research involving mathematical procedures or models, which cannot be presented in condensed form. Rather than attempting to abbreviate these models, the authors in Figure 3.10 preview the basic steps or components of their model, telling readers what will be demonstrated in the paper.

The descriptive abstract is less appropriate for reports of experimental research, for this approach provides little information about the study itself or its results and is therefore less useful to readers (Olsen and Huckin 1983). For example, compare the following hypothetical descriptive abstract with the informative abstract that accompanies Graham et al.'s (1992) paper in the *Annals of Internal Medicine*, contained in Chapter 10 (page 254):

---

[8]Since the Graham et al. paper was published in 1992, the *Annals* guidelines have added a ninth item, "Background," to the start of the list (see www.annals.org).

## Calculation of Posterior Bounds Given Convex Sets of Prior Probability Measures and Likelihood Functions (Cozman 1999)

### Abstract

This article presents alternatives and improvements to Lavine's algorithm, currently the most popular method for calculation of posterior expectation bounds induced by sets of probability measures. First, methods from probabilistic logic and Walley's and White-Snow's algorithms are reviewed and compared to Lavine's algorithm. Second, the calculation of posterior bounds is reduced to a fractional programming problem. From the unifying perspective of fractional programming, Lavine's algorithm is derived from Dinkelbach's algorithm, and the White-Snow algorithm is shown to be similar to the Charnes-Cooper transformation. From this analysis, a novel algorithm for expectation bounds is derived. This algorithm provides a complete solution for the calculation of expectation bounds from priors and likelihood functions specified as convex sets of measures. This novel algorithm is then extended to handle the situation where several independent identically distributed measurements are available. Examples are analyzed through a software package that performs robust inferences and that is publicly available.

## Reference Bands for Nonparametrically Estimated Link Functions (Ruckstuhl and Welsh 1999)

### Abstract

We explore the shape of the link function in a generalized linear model by estimating the link nonparametrically. We consider the problem of comparing the nonparametrically estimated link with a particular link function. Using reference bands that consist of pointwise confidence intervals for the nonparametrically estimated link centered at the hypothesized parametric link, simple graphical methods are obtained for comparing the two link functions. The resulting diagnostic plots are demonstrated with both artificial and real examples.

**FIGURE 3.10**    Sample descriptive abstracts from the *Journal of Computational and Graphical Statistics*.

The purpose of this report is to determine the effect of treating *Helicobacter pylori* infection on the recurrence of gastric and duodenal ulcer disease. Results of a clinical study of recent ulcer patients under two treatments are reported.

Though this hypothetical abstract does provide a general outline of Graham's study, it withholds critical information that the clinical audience of this journal would be interested in, namely the types of treatment tested and the outcomes of the tests. Given the type of research being reported, the extended informative abstract that this journal requires is much better suited to its readers' needs.

### EXERCISE 3.21

Read the abstracts for the papers by Mallin et al. (1995) in Chapter 11 (pages 328–340) and De Boer et al. (2001) in Chapter 13 (pages 406–409). Identify each as an informative or a descriptive abstract. Identify the major moves in the abstract and corresponding verb tense shifts.

Write a 100-word informative abstract for the research report by Brody and Pelton (1988), presented in Figure 3.11.

## Seasonal changes in digestion in black bears

ALLAN J. BRODY[1] AND MICHAEL R. PELTON

*Department of Forestry, Wildlife, and Fisheries, The University of Tennessee, P.O. Box 1071, Knoxville, TN 37901, U.S.A*

Received June 23, 1987

### Introduction

Winter dormancy in black bears (*Ursus americanus*) is likely an adaptation to predictable seasonal food shortages. Additionally, embryo development and parturition in bears are physiologically linked to hibernation (Herrero 1978). Fat is the major source of calories during hibernation and bears may lose as much as 25% of their body weight during the winter while maintaining lean body mass (Nelson et al. 1973). Thus, the ability to increase fat reserves before denning is necessary for winter survival and successful reproduction. Fat storage occurs during the fall (the "hyperphagia stage" of Nelson et al. (1983)), and many field studies have documented pronounced seasonal shifts in the diets of free-ranging bears (e.g., Tisch 1961; Hatler 1972; Beeman and Pelton 1980; Eagle and Pelton 1983; Grenfell and Brody 1983). Dietary shifts track plant phenology, and generally involve a transition from green forage and soft mast during spring and summer to hard mast in the fall.

Despite a strong dietary dependence on vegetable matter, bears exhibit only minor dental adaptation to herbivority and have retained the short, unspecialized gut of their carnivorous ancestors. Bunnell and Hamilton (1983) suggest that in grizzly bears (*Ursus arctos*) the evolution of a few morphological adaptations to herbivority combined with the conservation of physiological adaptations to carnivority make possible the rapid weight gains that occur before denning. They assume an evolutionary trade-off between the ability to digest food rapidly (a trait of carnivores) and the ability to digest low quality food efficiently (a trait of herbivores), and conclude that, in grizzlies, rapid processing at the expense of efficient digestion of fiber allows bears to take advantage of the large amounts of food available during the foraging period. Their experiments demonstrated that the digestive efficiencies of grizzlies are not markedly different from those of obligate carnivores. This note describes an effort to determine the digestive abilities of black bears, and to determine any seasonal differences in those abilities.

### Methods

Six adult bears, ranging from 91 to 178 kg (mean weight = 63.4 kg, SD = 28.6 kg) at the Ober-Gatlinburg Black Bear Habitat in Gatlinburg, Tennessee, were used in digestion trials in August and November 1983. The bears were housed in three enclosures, each holding one male and one female. The normal ration for the bears consisted of dry dog food (Tennessee Farmers Cooperative, Lavergne, TN 37086, U.S.A.), fed *ad libitum*, supplemented by produce discarded by local grocers. For the trials, all produce was withheld and the dog food ration was reduced to approximately 95% of *ad libitum* consumption for 7 days; portions of three feces in each enclosure were collected on the 7th day.

Fecal and ration samples were frozen until laboratory analysis was performed, at which time the samples from each enclosure in each trial were thawed and dried at 101°C. three subsamples were drawn from the fecal samples from each enclosure. Crude protein content was estimated by the macro-Kjeldahl technique (Association of Official Analytical Chemists 1970) (three replicates per subsample). Gross energy was estimated in an adiabatic oxygen bomb calorimeter (Parr Instrument Company, Moline, IL 61265, U.S.A.) (two replicates per subsample). Acid-insoluble ash content was estimated by the method of Van Soest (1966) (three replicates per subsample).

[1] Present address: Department of Ecology and Behavioral Biology, University of Minnesota, 318 Church Street SE, Minneapolis, MN 55455, U.S.A.

Printed in Canada / Imprimé au Canada

*(continued on p. 72)*

**FIGURE 3.11**    Text (without abstract) from Brody and Pelton (1988), from the *Canadian Journal of Zoology*.

TABLE 1. Composition of experimental diets and fecal samples from bears, on a dry-weight basis

|  | Crude protein (%) | Gross energy (kJ/g) | Acid-insoluble ash (%) |
|---|---|---|---|
| This study |  |  |  |
| Dog food ration |  |  |  |
| August trial | 36.7 | 19.292 | 1.12 |
| November trial | 30.9 | 19.212 | 1.22 |
| Fecal samples* |  |  |  |
| August mean ($n$ = 9) | 20.4 (0.60) | 17.403 (1.333) | 2.58 (0.002) |
| November mean ($n$ = 9) | 19.3 (0.76) | 15.038 (0.063) | 2.72 (0.001) |
| Grizzly rations used by |  |  |  |
| Bunnell and Hamilton (1983) |  |  |  |
| Basal ration | 36.3 | 5577 | 0.11 |
| Basal and beet pulp | 21.2 | 4729 | 1.42 |

*SE given in parentheses.

TABLE 2. Apparent digestibilities in bears

|  | Approximate consumption* | Apparent digestibility coefficients† | |
|---|---|---|---|
|  |  | Crude protein | Gross energy |
| August trial |  |  |  |
| Enclosure 1 | 4.4 (0.33) | 0.765 | 0.645 |
| Enclosure 2 | 3.3 (0.24) | 0.745 | 0.626 |
| Enclosure 3 | 2.5 (0.41) | 0.764 | 0.645 |
| August mean | 3.4 (0.70) | 0.758 (0.0113) | 0.638 (0.0110) |
| November trial |  |  |  |
| Enclosure 1 | 7.5 (0.45) | 0.729 | 0.651 |
| Enclosure 2 | 5.0 (0.34) | 0.708 | 0.640 |
| Enclosure 3 | 7.0 (0.21) | 0.723 | 0.656 |
| November mean | 6.6 (1.12) | 0.720 (0.0108) | 0.649 (0.0082) |
| Grizzlies on basal and beet ration (Bunnell and Hamilton 1983)‡ |  | 0.751 | 0.620 |

*Amount provided − orts. in kg/day for each enclosure. SD given in parentheses.
†SE given in parentheses.
‡Mean values for two bears.

Apparent digestibilities of crude protein and gross energy were estimated using the indicator method (McCarthy et al. 1974; Bunnell and Hamilton 1983), with acid-insoluble ash serving as the indicator:

$$\text{apparent digestion coefficient} = 1 - \frac{(\text{AIA in feed}) \times (Y \text{ in feces})}{(\text{AIA in feces}) \times (Y \text{ in feed})}$$

where AIA is the dry weight proportion of acid-insoluble ash and $Y$ is the dry weight energy content or proportion of crude nitrogen.

### Results and discussion

The dog food ration used in this study was substantially higher in crude protein than the natural food plants normally eaten by black bears, which typically contain 2−19% crude protein (Mealey 1975; Eagle and Pelton 1983). Composition of the ration used in this study (Table 1) was most similar to the "basal and beet pulp" ration used by Bunnell and Hamilton (1983) and the apparent digestibilities (Table 2) were correspondingly similar.

Digestibility of crude protein decreased (paired $t$-test, $t$ = 26.41, $P$ = 0.0014) while the digestibility of gross energy increased (paired $t$-test, $t$ = 4.58, $P$ = 0.0445) from August to November. Increased food consumption in November implies an increased transit rate which in turn could have caused the decrease in apparent protein digestion (Castle and Castle 1956; Rerat 1978). If transit rate were the only factor affecting seasonal changes in digestibility, however, a simultaneous decrease in apparent gross energy digestion would be expected. Instead, we found an increase in apparent energy digestion, indicating that bears were selectively digesting and (or) absorbing carbohydrates and fats at the expense of protein.

Inhibition of amino acid absorption by carbohydrates has been well documented in several monograstric animals (Alvarado 1971), but there is little evidence that the degree of inhibition is controlled by factors other than concentrations of specific substrates in the gut. This mechanism could operate in the wild, where summer foods are typically much higher in protein than fall foods (Eagle and Pelton 1983), but cannot explain our experimental results because rations in both trials were similar. We suggest that a systemically, possibly hormonally, mediated increase in carbohydrate and fat assimilation and decrease in protein assimilation occurs during the predenning hyperphagic period. If lean body growth ceases in the fall, as data from Nelson et al. (1983) imply, protein

**FIGURE 3.11** *(continued)*

requirements would be reduced and preferential assimilation of dietary substrates most efficiently converted to fat would appear to be adaptive.

This type of hormonal control of assimilation has yet to be documented, but would be consistent with the array of physiological adaptations, particularly those of nitrogen metabolism (Nelson et al. 1973, 1975, 1983), already described in bears.

### Acknowledgements

We wish to thank J. R. Carmichael and D. S. Carmichael for technical assistance. We benefitted from conversations with T. C. Eagle, R. L. Moser, R. A. Nelson, and W. D. Schmid. Comments from S. Innes and an anonymous reviewer improved an earlier draft of this paper.

ALVARADO, F. 1971. Interrelation of transport systems for sugars and amino acids in the small intestine. *In* Intestinal transport of electrolytes, amino acids, and sugars. *Edited by* W. M. Armstrong and A. S. Nunn. Charles C. Thomas, Springfield, IL. pp. 281–315.

ASSOCIATION OF OFFICIAL ANALYTICAL CHEMISTS. 1970. Official methods of analysis. 11th ed. Washington, DC.

BEEMAN, L. E., and PELTON, M. R. 1980. Seasonal foods and feeding ecology of black bears in the Smoky Mountains. Int. Conf. Bear Res. Manage. **4**: 141–147.

BUNNELL, F. L., and HAMILTON, T. 1983. Forage digestibility and fitness in grizzly bears. Int. Conf. Bear Res. Manage. **5**: 179–185.

CASTLE, E. J., and CASTLE, M. E. 1956. The rate of passage of food through the alimentary tract of pigs. J. Agric. Sci. **47**: 196–204.

EAGLE, T. C., and PELTON, M. R. 1983. Seasonal nutrition of black bears in the Great Smoky Mountains National Park. Int. Conf. Bear Res. Manage. **5**: 94–101.

GRENFELL, W. E., and BRODY, A. J. 1983. Seasonal foods of black bears in Tahoe National Forest, California. Calif. Fish Game, **69**: 132–150.

HATLER, D. F. 1972. Food habits of black bears in interior Alaska. Can. Field-Nat. **86**: 17–31.

HERRERO, S. 1978. A comparison of some features of the evolution, ecology, and behavior of black and grizzly/brown bears. Carnivore (Seattle), **1**: 1–17.

McCARTHY, J. F., AHERHE, F. X., and OKAI, D. B. 1974. Use of HCl insoluble ash as an index material for determining apparent digestibility with pigs. Can. J. Anim. Sci. **54**: 107–109.

MEALEY, S. P. 1975. The natural food habits of free ranging grizzly bears in Yellowstone National Park, 1973–1974. M.S. thesis, Montana State University, Bozeman.

NELSON, R. A., WAHNER, H. W., ELLEFSON, J. D., and ZOLLMAN, P. E. 1973. Metabolism of bears before, during, and after winter sleep. Am. J. Physiol. **224**: 491–496.

NELSON, R. A., JONES, J. D., WAHNER, H. W., McGILL, D. B., and CODE, C. F. 1975. Nitrogen metabolism in bears: urea metabolism in summer starvation and in winter sleep and role of urinary bladder in water and nitrogen conservation. Mayo Clin. Proc. **50**: 141–146.

NELSON, R. A., FOLK, G. E., JR., PFEIFFER, E. W., CRAIGHEAD, J. J., JONKEL, C. J., and STEIGER, D. L. 1983. Behavior, biochemistry, and hibernation in black, grizzly, and polar bears. Int. Conf. Bear Res. Manage. **5**: 284–290.

RERAT, A. 1978. Digestion and absorption of carbohydrates and nitrogenous matters in the hindgut of the omnivorous nonruminant animal. J. Anim. Sci. **46**: 1808–1837.

TISCH, E. L. 1961. Seasonal food habits of the black bear in the Whitefish Range of northwestern Montana. M.S. thesis, Montana State University, Bozeman.

VAN SOEST, P. J. 1966. Non-nutritive residues: a system of analysis for the replacement of crude fiber. J. Assoc. Off. Agric. Chem. **49**: 546–551.

**FIGURE 3.11**  *(continued)*

# 3.8 Brief Report Genres: Research Letters and Notes

In addition to the full-length research report, many journals publish brief reports in the form of *letters* or *notes*, which offer a range of forums for communicating urgent or preliminary findings. The Brody and Pelton (1988) piece in Figure 3.11 is classified as a research note. The journal *Physical Review Letters* is devoted exclusively to "short, original communications in active and rapidly developing areas of physics" (see Figure 2.1 on page 26), and many other journals solicit similar types of articles: *Nature Biotechnology* (2003) publishes letters and "brief communications"; *Analytical Chemistry* (2003) publishes "Correspondence" ("brief disclosure[s] of new analytical concepts of unusual significance") and "Technical Notes" ("brief descriptions of novel apparatus or techniques"). Most of these forms have evolved from earlier, less formal forms: *Physical Review Letters* essentially began as a "Letters to the Editor" section in the journal *Physical Review* (Blakeslee 1994).

In most cases, research letters undergo a separate, expedited review and publication process designed to enable researchers to communicate important findings to the research community as quickly as possible. The *Astrophysical Journal*, for example, publishes *Astrophysical Journal Letters* as Part 2 of the journal. The *Letters* is described as "a peer-reviewed express scientific journal" (Astrophysical Journal 2003). *Letters* articles first appear on a Rapid Release website sponsored by the journal's publisher, University of Chicago Press. They then appear with a complete electronic issue of the journal and only later in a print version along with Part 1, which is published three times a month. We've included in Chapter 12 a research letter by Stephen Reynolds that appeared in this journal (pages 385–389). Though limited to four pages in length, the letters are not governed by any special organizational constraints and may be similar in form to Part 1 articles, which may run up to 20 pages (note the similarities between the Reynolds letter and the full-length Hendrick and Reynolds piece from Part 1 of the journal, on pages 394–399). Letters are distinguished from full articles primarily by their urgency—their rhetorical exigence. In addition to satisfying the evaluative criteria established for Part 1 articles, letters must meet two further requirements, timeliness and brevity, which the editors describe as follows:

> **Timeliness**—A Letter should have a significant impact on the research of a number of other investigators or be of special current interest in astrophysics. Permanent, long-range value is less essential. A Letter can be more speculative and less rigorous than an article for Part 1 but should meet the same high standard of quality.
>
> **Brevity**—A Letter must be concise and to the point and require no more than 4 journal pages. . . . Within this space limitation, sufficient introductory material should be included, and the content of the paper should be such that it is generally understood by scientists who are not specialists in the particular field. (Editorial Criteria for Publishing in the ApJ: http://www.journals.uchicago.edu/ApJ/criteria.html)

Though the brief report category includes forms serving a variety of purposes, from presenting preliminary results, to introducing significant analytical concepts, to describing new techniques, research letters are typically used to present innovative research or to announce novel findings. In addition to the *Astrophysical Journal*, both *Nature* and *Physical Review Letters*, two of the most prestigious letter journals in publication, use the letter forum for this purpose. Watson and Crick's 1953 letter to *Nature* on the double helix structure of DNA is a case in point. We noted in Chapter 2 that the Burkholder team also first announced their discovery of the "phantom" dinoflagellate in a letter to *Nature* (Burkholder et al. 1992; see pages 289–292). Marshall and Warren initially reported the presence of bacteria in the stomach in joint letters to the *Lancet* (Warren and Marshall 1983; see pages 233–235). Notice that neither the dinoflagellate, *P. piscicida,* nor the "ulcer-causing" bacterium, *H. pylori,* had yet been named at the time the letters were published, indicative of the early stage at which these reports appeared. Both might be considered cases of "revolutionary" or extraordinary science (Toulmin et al. 1984; Kuhn 1996). That is, both research teams were reporting findings that, if further supported, could significantly alter basic knowledge paradigms in their respective fields: the Burkholder team described a new family, genus, and species of algae that evidenced predatory behavior and was found to be responsible for numerous previously unexplained fish kills; Marshall and Warren described a bacterium that not only was found to live in the supposedly sterile environment of the stomach but also was potentially linked to illnesses the field firmly believed to be caused by psychological rather than physical factors. In both cases, these early announcements were (and continue to be) followed by a wealth of further research in which these preliminary findings are tested and extended.

*Nature* limits letters to 1000 words and requires that they contain no subheadings and fewer than 30 references (contrasted with the 50 references allowed in full-length reports). The *Lancet* also publishes "preliminary reports and hypotheses" of up to 1500 words, but letters in the *Lancet* are even more abbreviated, with a 500-word limit (Lancet 1990, Information for Authors). Notice that the Warren and Marshall *Lancet* letters are formatted like letters to the editor, with the traditional salutation, *Sir,* at the start and the author's name at the end. The Burkholder team's letter to *Nature* looks quite different. No subheadings are used, but the piece includes all the standard components of the IMRAD format, although of course these moves are less elaborated, and they are adapted to reflect the more general purpose of the letter: Burkholder and her colleagues are not describing the methods and results of a single new study; though they do report some new data, they are also summarizing findings to date from a developing line of research.

In a study of the journal *Physical Review Letters,* Blakeslee (1994) points out that the brevity of these reports may make it difficult for reviewers to evaluate the quality of the science being reported, and indeed the letters forum allows little room for the types of evidence we expect to find in full-length research arguments. Nor does it allow for detailed explanations of unexpected findings or extensive rebuttals (as illustrated by the responses elicited by Fleischmann and Pons's letter [1989] in the *Journal of Electroanalytical Chemistry,* announcing the creation of cold fusion). The letters genre is well suited, however, for presenting quick summaries of new developments and recent observations, as the letters by Reynolds, Warren and Marshall, and the Burkholder team illustrate. When compared with the more formal and fully developed research report, letters often have something of a journalistic flavor, consistent with their function of presenting "cutting edge" research (Blakeslee 1994).

## EXERCISE 3.23

Look for the basic IMRAD elements in the Burkholder team's letter to *Nature* (Chapter 11). In what ways does this short report follow the IMRAD pattern? In what ways does it deviate from this pattern? Where are the basic research report moves made (introducing the problem, describing methods, reporting results, drawing conclusions)? Notice that table and figure captions are quite long in this report. What kind of information is contained in table and figure captions, and why is it there?

## 3.9 How Scientists Write Reports

The order in which report sections are written is rarely the same as the order in which they appear in the finished text. Your abstract is the first thing readers will see, but you won't be able to write it until you've at least sketched out the main body of your text. Perhaps you'll sketch out your methods section first, while specific details and decisions are still fresh; or you may begin by constructing the tables and figures you will use to present your results, filling them in as the data are generated and analyzed (Monroe et al. 1977). As noted earlier, your graphics represent the logic of your research design; the form of your tables and figures is likely to be determined long before you begin to write.

As is the case with many types of writing, the introduction to the research report is often revised throughout the drafting process; some authors even write it last. The focus and argument of the introduction depend in part on the outcome of the study. For example, a team of researchers may discover that their findings raise questions they did not expect to be important, and thus the study makes a somewhat different contribution to the field than the researchers had anticipated. The introduction needs to prepare readers for this contribution so that it is recognized and appreciated as important. Even though the purpose and design of the study have not changed, the researchers will need to adjust their review of background issues and research to establish a context for later commenting on these new questions. In short, you may not be able to finalize your introduction until you've thought about the implications you want to raise in the discussion section.

Writing a report is thus a recursive process in which authors continuously revisit and rethink their earlier arguments in light of their results and further thinking. In a study of the composing processes of nine eminent biochemists, Rymer (1988) found they employed a wide range of writing strategies and styles. Some subjects reported that they typically spend a great deal of time planning and outlining before drafting sections of the text, whereas others begin by writing a full "impressionistic" draft of the whole paper, refining it in later stages of the process. Most scientists use a combination of these approaches, adjusting their processes according to the type of research they're reporting, their familiarity with the prior research and the target journal audience, the amount of time they have available, and so forth.

The number of authors involved in a project also influences the structure of the writing process. In most scientific fields, collaborative projects are increasingly the norm. The average number of authors per article in the *New England Journal of Medicine,* for example, increased from one in 1925 to about six in 1995 (NAS 1995, p 13), by which time *NEJM* had instituted a cap of 12 authors per article (Kassirer and Angell 1991). In July 2002, *NEJM*'s editors lifted the cap, allowing the number of authors to continue to increase. The editors explain that as medical research has advanced, "investigators with a broader range of skills than were required in the past are often needed to take new ideas from the bench to the bedside and to conduct large clinical trials" (Drazen and Curfman 2002; http://content.nejm.org/content/vol347/issue1/index.shtml). This trend in multiple authorship has been a matter of some debate in science and in academia at large, for it raises issues of professional ethics and responsibility, as we will discuss in Chapter 9.

Research teams manage the writing process in a variety of ways. In some teams, one author drafts the entire paper and sends it around to others for their comments and revisions. In other working groups, different members of the team may draft different sections according to their role in the research; for example, those who carried out the procedures and collected the data will draft the methods and results sections pertaining to their parts of the project, while the project leader who initially conceptualized the study will write the initial draft of the introduction and discussion (Rymer 1988).

If you completed the interview assignment in Chapter 2, you no doubt discovered that scientists spend a great deal of time writing, revising, and polishing papers before submitting them for journal review. One of Rymer's biochemists described a "predrafting" stage lasting approximately three years, during which the study was conceived and carried out and the postdoctoral fellows on the project drafted figures and tables as well as the methods and results sections. Once the lab work and analysis were completed, the project leader drafted and revised the introduction and discussion sections over a period of two weeks and then spent another two months pulling the manuscript together, which included revising the methods and results sections provided by the postdocs and soliciting feedback on the draft from colleagues (Rymer 1988).

Clearly the form of the final written product tells us little about the process through which that text came to be. Unlike the conventions of the written report, writing processes seem to be governed not by field-specific convention but by the habits, skills, and preferences of the individuals who make up the research team.

## 3.10 How Scientists Read Reports

It should be clear from this chapter that the written reports themselves *are* governed by the conventions of their respective research fields and by the conventional practices of journals within fields. You'll recall that "conventional" practices are those that are customary in a community and thus familiar to all members. Readers in a community use their knowledge of these conventions to facilitate the reading and interpretation of the written texts those conventions govern. Thus, readers familiar with the IMRAD form will know where to find the results in a report quickly, where to look for the authors' interpretations of the results, where to find a description of methods, and so forth. They will not need to read the paper from start to finish in order to discover what kind of research was conducted and what was found.

In fact, although editors and reviewers may take the time to read a paper from start to finish, scientists reading for their own purposes are far less likely to do so (Burrough-Boenisch 1999). Few scientists have the time or the inclination to read journal articles in their entirety in the order in which they appear in print (Bazerman 1985; Berkenkotter and Huckin 1995). Readers rely heavily on the titles and abstracts of published reports, not only to help them decide whether to read the paper but also to help them quickly learn the key features of the study. They then can turn immediately to pertinent sections of the report for the information they are most interested in. From consultations with researchers in physics and biology, Berkenkotter and Huckin (1995) determined that readers typically began by scanning the title and abstract of an article and then looking for the data, focusing on the tables and graphs in which the data are summarized. Only after examining the data themselves did these scientists read the results section provided by the authors. The rest of the reading process was quite variable; readers selectively read or skimmed the introduction, methods, and discussion sections, depending on how much they already knew about the topic, the methodological approach used, and the results of prior studies, and depending on their familiarity with other work of the authors. In an earlier study of physicists, Bazerman (1985) found similar variation in reading strategies. On some occasions readers would read an article's introduction and conclusion to get a sense of the focus of the study for future reference, skipping over details of methods and results; but in other contexts, as when reading a paper on a very familiar topic, the same readers might invert this pattern, ignoring the contextual information in introduction and conclusion and concentrating instead on methodological details, calculations, and results.

Given these varied reading strategies, we can see why section headings and figure captions are important in research reports, and why the content of titles and abstracts is critical. The final form of the report may reflect neither the order in which it was written nor the order in which it will be read, but it must match the expectations of readers in the research community, who will expect to find particular kinds of information and particular parts of the overall argument in conventional locations. The paper must reflect the logic of presentation with which readers in the field have become familiar.

It's important to keep in mind that readers have expectations about the logic within sections as well. Paul and Charney's (1995) observations of scientists reading articles on chaos theory indicated that their readers were generally familiar with the four basic introduction moves described in Section 3.3 (though they may not be aware of this formula). When the authors had followed this pattern, these readers used this familiar structure to help them incorporate the new work into their current knowledge of the topic. In fact, the readers in this study—chaos researchers from the fields of physics, mathematics, ecology, engineering, and meteorology—all paid more attention to the context-setting information provided in the introduction than to the description of the new study itself. That is, readers' first concern was "whether they could relate the reading to their prior knowledge and to their own work" (p 427). In short, readers expect to find these connections in the introduction and authors must take care to provide them.

This discussion has implied that readers are active participants in the generation of knowledge through text. Scientists don't just gather facts when they read; they interpret those facts and carefully consider the interpretations proposed by the authors. They pay attention to how the authors designed and conducted the study, the conditions under which the work took place, and the operating assumptions underlying both the design of the project and the interpretation of its results. In other words, research readers are cautious consumers. Readers continually weigh the merits of the research and of the written argument as they read. The outcome of this evaluation process determines whether they accept the research as sound and important.

# 3.11 How Reviewers Evaluate Reports

In the manuscript review system, this process of weighing the merits of research as we read becomes more formal and explicit. Editors and reviewers must evaluate the merits of the research not just for their own edification but for other members of the field as well. They are making the decision of whether to read or not to read on behalf of the larger community; the outcome of their deliberations determines whether the wider community gets to see the paper at all. As we discussed in Chapter 1, the peer-review process is the mechanism by which the scientific community monitors the integrity of its work—in effect, "filters" the work of individuals. Though the system has its detractors (it has been criticized as a cumbersome process, subject to abuses such as favoritism and censorship), the peer-review system remains a formidable structure enabling the "volatile microcosm of individual scientists" to contribute to the "stable macrocosm of the scientific enterprise" (NAS 1989, p 10).

The logistics of the peer-review (or "peer-referee") system are relatively straightforward. Most journals publish their procedures and at least a tentative timetable in their instructions for authors. Authors submitting their work to *Nature*, for example, are told that the paper is first reviewed by a member of the editorial staff, who decides whether the research will be of significant interest to *Nature*'s interdisciplinary audience (Nature 2000). If the paper does not pass this initial assessment, the authors are notified immediately, usually within a week. Papers that do pass then qualify for peer review and are sent to two or more referees, who are chosen on the basis of their expertise in the area, their lack of connection to the authors, and their availability to complete the review in the specified time period, usually one week. (This turnaround period is relatively quick compared to other journals. The *Annals of Internal Medicine* and the *Astrophysical Journal*, for example, both allow 3 to 4 weeks at this stage.) The editor assigned to the paper summarizes the reviews when they come in, drafts a recommendation and a letter informing the authors, and has these materials reviewed by others on the editorial staff before sending the decision to the authors. The decision will be one of five options: acceptance without further changes, acceptance pending revisions suggested by reviewers, deferred decision (e.g., after further experiments), rejection with resubmission invited, or rejection with no resubmission invited. (See http://www.nature.com/nature/submit/get_published/index.html for further information about *Nature*'s review process.)

The vast majority of scientific papers undergo significant revision in response to peer review. *Nature*'s guidelines note that it is "extremely rare" for a paper to be accepted without revision. The *Astrophysical Journal* accepts "fewer than 3% of [submitted papers] . . . without significant revision" (http://www.journals.uchicago.edu/ApJ/guide-ed_text.html). Although peer review has come under extensive examination, especially as scientific organizations, journal editors, and researchers themselves consider the implications of e-journals, these figures from *Nature* and from the *Astrophysical Journal* demonstrate that the peer-review system is still a critical component of the scientific writing process, and will likely remain so (e.g., see Harnad 1990). As *Nature*'s editors explain, "Some of the papers of which *Nature* is most proud of publishing are those which were originally weak but intriguing, and have become really striking only by virtue of extensive attention from referees and editors" (http://www.nature.com/nature/submit/get_published/index.html).

There is a fair amount of consensus among journal editors and reviewers about what constitutes a publishable paper. Recall from Chapter 2 that journals clearly identify their target audience and the research areas they are interested in (review Figure 2.1 on page 26). It follows, then, that the minimum requirement for any paper is that it be of clear interest and importance to the readers of the journal to which it is submitted. Indeed, journal editors report that "suitability for the journal" is the primary concern when deciding whether to accept or reject a paper (Huler 1990). Lack of suitability is the most commonly cited reason for rejection of a manuscript (Davis 1985, cited in Olsen and Huckin 1991). The *Journal of Heredity*, for example, announces to contributors that "acceptance will depend on scientific merit and suitability for the journal" (Heredity 2002); as noted above, *Nature*'s editors make the suitability assessment in the first stage of review.

You cannot begin to write a research report, then, until you've decided which journal audience you would like to reach. You cannot develop a persuasive argument for the significance of your research without thorough knowledge of whom you're trying to persuade. In addition, the choice of journal dictates certain matters of presentation, from the purpose and organization of the article to, in some cases, details of style and formatting. Thus, prospective authors must understand not only the audience of the journal but also the nature of the communicative forum(s) the journal represents (Benson 1998) in order to write a paper that will be deemed appropriate for the journal and its audience. With the enormous number of papers competing for space in scientific journals (*Annals of Internal Medicine* published 15 percent of submissions in 2001 [Annals of Internal Medicine 2002]; *Nature*'s rate is 10 percent [Nature 2000]),[9] there is no room for papers that do not specifically address the interests of the journal's target audience and meet the expectations of its editors.

The following list of review criteria from the journal *Phytopathology* (http://www.apsnet.org/phyto/submit.asp) provides a good summary of the features that reviewers are asked to evaluate and thus a good checklist for writers of research reports:

- Importance of the research
- Originality of the work
- Analysis of previous literature
- Appropriateness of the approach and experimental design
- Adequacy of experimental techniques
- Soundness of conclusions and interpretations
- Relevance of discussion
- Clarity of presentation and organization of the article
- Demonstration of reproducibility

---

[9]Lest new researchers despair, we point out that acceptance rates vary a great deal. The *Astrophysical Journal* reports an overall acceptance rate of 78 percent, though all but 3 percent of articles undergo extensive revision (Kennicutt 2003).

# Activities and Assignments

1. Using the IMRAD form as the organizational framework for your analysis, analyze the organizational structure of a research report in your field, as illustrated in Figure 3.2. Does this report illustrate the generic features described in Sections 3.3 through 3.6, or does it follow an alternative form? How does the form of the report reflect the goals, objects, or methods of research in this particular field? Your goal is to *analyze,* not just describe, the form of this particular research report. You'll need to think about why it takes the form it does.

   *Note:* In this sort of analysis, it is conventional to refer to the authors in the third person and to describe their strategies in present tense (e.g., "Sud and Sekhon divide their presentation of results into three sections . . ."; "Their findings indicate that . . ."). To orient your reader, include the following identifying information in your opening paragraph: the authors' names, the title of the article, the name of the journal, and the purpose or main conclusions of the research. Be sure to turn in a copy of the article itself with your analysis.

2. In groups of four or five, read the analyses you wrote in Activity 1, looking for similarities and differences among the fields your articles represent. Prepare an oral report for the class in which you identify the major structural variables in this set of papers. You'll want to describe *how* the papers differ and *why* they differ in these ways. What do these formal features reveal about the goals and methods of these different research areas?

3. Either as part of Activity 1 or as a separate activity, analyze the use of explanation, justification, and citations in a research report in your field, as illustrated in Figure 3.7. When, and how much, do the authors explain and/or justify statements in each section, and why are those explanations included? When, and how, are citations used (e.g., to cite relevant studies, to support an argument, to indicate awareness of a controversial or undecided issue)? What is the relationship of these strategies to the logic and purpose of each section of the IMRAD form? What does your analysis tell you about the knowledge and expectations of the authors' audience? How does the authors' awareness of the knowledge and expectations of the audience enhance the authors' *ethos?*

4. Turn to Figure 1 in the Graham et al. (1992) study on page 255. From the data in this line graph, try to create a table, a bar graph, and a diagram. What data did you select for each visual, and what data did you exclude? Why? What do your visuals show? Which visual forms were most useful in representing the findings of the Graham team and which weren't as useful? Why?

5. The *British Medical Journal* asks authors to provide "a thumbnail sketch of what [their] paper adds to the literature, for readers who would like an overview without reading the whole paper" (http://bmj.com/advice/sections.shtml). Distinguished from the required abstract, this two-paragraph "What this paper adds" box appears at the end of the paper and essentially summarizes the motivation for the study and its major outcome. In other words, it condenses the arguments from the introduction and discussion sections. Authors are to write a two- to three-sentence paragraph explaining "what the state of

scientific knowledge was in this area before you did your study and why this study needed to be done," followed by a second paragraph that succinctly answers the question, "What do we now know as a result of this study that we did not know before?" See the Chiba et al. (2002) *BMJ* paper in Chapter 10 (pages 280–286) for an example, and see the URL above for further guidelines (scroll through the guidelines under "Original research").

Choose a research report from this text or from your field and write a "What this paper adds" box that meets the *BMJ* guidelines.

6. If you are currently involved in a research project, write a research letter or full-length report of your study. Your instructor may ask you to work alone or with a partner or group. You will need to begin by reviewing prior research on the topic in order to identify what the field already knows, how the question has been approached in the past (what methods have been used), and what questions remain for you to address. See Chapter 4 for advice on locating previous research.

Write this report as though you were preparing an article for submission to a research journal in your field. Choose an appropriate journal to target, and develop your paper with that journal's audience and editorial guidelines in mind. Select a sample report from your target journal to use as a model. Submit the guidelines and sample along with your report.

# CHAPTER 4

# Reviewing Prior Research

## 4.1 The Role of Prior Research in Scientific Argument

Scientific work is often conducted under solitary and isolated circumstances: late nights in the lab, long hours in the field collecting samples, visits to remote research sites. This work is neither planned nor interpreted in isolation, however. New studies are conducted because they promise to shed light on issues considered important by the research community; research questions are valid in that they reflect gaps or inconsistencies in the field's understanding; methodological decisions are made with an awareness of what has been done by others and how it has worked; and the outcomes of research are interpreted in light of the theories, questions, methods, and findings that have come before. Research becomes meaningful only when viewed in the context of the field's developing knowledge.

This social or "situated" nature of scientific research is clearly reflected in scientific texts, not only in the adherence to conventionalized forms and terminology but also more directly in the practice of citing previously published reports. Most scientific arguments rely heavily on references to previous research to support the claims and methods reported or proposed. The practice of citing prior research is critical in arguments written for research audiences, including the arguments contained in grant proposals, research reports, and many types of brief research notes and letters. The National Academy of Sciences describes this practice as follows:

> Citations serve many purposes in a scientific paper. They acknowledge the work of other scientists, direct the reader toward additional sources of information, acknowledge conflicts with other results, and provide support for the views expressed in the paper. More broadly, citations place a paper within its scientific context, relating it to the present state of scientific knowledge. (NAS 1995, p 12)

Berkenkotter and Huckin (1995) argue that this process of contextualizing one's work is essential, for "it is only when scientists place their laboratory findings within a framework of accepted knowledge that a claim to have made a scientific discovery, and thereby to have contributed to the field's body of knowledge, can be made" (p 47). The importance of situating one's work is dramatically illustrated in their case study of a biologist, June Davis, whose initial submissions of a manuscript were twice rejected by journal reviewers who were not convinced that the study's findings were significant enough to warrant publication. The manuscript was accepted only after Davis bolstered her argument by responding to reviewers' calls for more explicit connections to previous research in the introduction and discussion sections of her report (Berkenkotter and Huckin 1995). See Dong (1996) for a similar account of a doctoral student's struggle to situate his work in the context of related literature in order to complete his dissertation in genetics.

As described in Chapters 3 and 6, reports and proposals always begin by reviewing the current state of the field's knowledge of the phenomenon under study. Many proposal formats require a separate background section devoted to reviewing prior research, sometimes referred to as a "review of the literature." Research reports tend to include a more abbreviated review of research in the introduction section, but as the Davis case illustrates, this brief review serves a critical role in establishing the context for the study. Citations of prior research also appear in the discussion sections of research reports, where authors tend to make very specific references to previous studies in order to help readers interpret the scope, scale, and significance of the reported findings and to help them understand how the new research extends, refines, or challenges the state of knowledge in the field. Previous studies are frequently cited in methods and results sections as well, often as precedents for specific methodological decisions or data analysis procedures. As we discussed in Chapter 3, decisions about what statements need support or justification, and how much support or justification is appropriate, depend on the writer's awareness of what the audience knows and expects. Knowing when and to what degree to qualify, explain, justify, and cite is one of the ways scientists create a professional *ethos* (Herrington 1985).

In thus acknowledging the work of others, researchers demonstrate not only that their knowledge is up to date but also that they have taken advantage of the best of the field's expertise in designing, carrying out, and interpreting their own work. Researchers who fail to acknowledge the relevant prior research will appear either naïve (if readers are charitable) or arrogant (if they have no reason to be charitable). Pons and Fleischmann again represent a case in point. In their first paper in the *Journal of Electroanalytical Chemistry* (Fleischmann and Pons 1989), they claimed that the energy in their cold fusion experiment was produced by "an hitherto unknown nuclear process or processes" (p 301). In a critique of this paper, Huizenga (1992) criticized the researchers not only for neglecting to qualify their "risky assumption" but also for failing to "acknowledge the extensive literature on nuclear reactions acquired and the basic principles established over the last half century" (p 25). As described in Chapter 1, the general disapproval with which this research was met was due as much to the way in which Pons and Fleischmann presented and situated their work as it was to the quality of the research itself.

Choose one or more full-length research reports or grant proposals, either from this text or from your own field. In each major section of the paper, find two or three sentences that include citations of previous research. How are citations used in different sections of the paper? Look for citations that support qualifications, explanations, or justifications made in the text. What other functions do citations serve in the texts you've examined?

## 4.2 Reviewing as a Genre: The Review Article

Before we discuss general strategies for reviewing research in scientific reports or proposals, we will briefly examine the *review article,* a distinct genre in which these strategies are paramount. Many journals publish full-length review articles, often solicited from experts in the field on topics of particular interest to the journal's readers. Review articles tend to be written for a journal's broadest readership, including researchers in related fields (Day 1998), and thus are pitched at a somewhat more general level than research reports. As described in *Nature,* review articles "survey recent developments in a topical area of scientific research or, on occasion, can be more wide-ranging" (Nature 2003b). Such surveys serve an important function in their respective fields in that they offer a comprehensive synthesis of the results of a wide and complex set of studies. In so doing, the review may have a substantial influence on how readers perceive the nature and implications of recent developments in a field and thus may influence the direction of subsequent research (Myers 1991).

The review article we've included in Chapter 10 (Blaser 1987, on page 236) illustrates two primary traits implied above: comprehensiveness (it reviews 123 studies!) and recency or timeliness. Blaser's review, published in *Gastroenterology* in 1987, was occasioned by the renewed interest in "gastric bacteria" sparked by Marshall and Warren's work in the mid-1980s. Like other important reviews, this piece helped the field take stock of a rapidly developing research area, as the following excerpt from Blaser's introduction indicates:

> . . . Although the presence of gastric bacteria has long been known, their significance has been uncertain. The development of fiberoptic endoscopy, permitting collection of fresh clinical specimens, has ushered in a new era for the management and investigation of gastroduodenal inflammatory conditions. Gastric bacteria now are being observed with regularity (2–4), and recently, Marshall and Warren (5,6) were able to isolate a spiral bacterium that had never been cultivated before. This organism, which they called *Campylobacter pyloridis,* has since been isolated by many other investigators (7–9). The field has moved quickly, and a review of its current status is appropriate.

The goal of such status reports is to describe what the field has learned so far: what, if any, consensus is developing, and what questions remain to be answered? Review articles are especially useful to those who distribute resources, both financial and human, including, for example, program officers at funding agencies and administrators in research agencies such as the EPA. Similarly, reviews are useful to practitioners, including medical professionals, agricultural producers, and resource managers, who must make daily decisions about courses of treatment, feed composition, management practices, and so forth, and who want to base those decisions on the most current information available. Comparable to the discussion and implications sections contained in reports of individual studies, review articles discuss the implications of the complex set of findings under review. Blaser, for example, helped readers make sense of the many reported observations of gastric bacteria by pointing out trends and patterns across studies. His concluding paragraph emphasized directions for future research and implications for clinical practice:

> The rediscovery of spiral gastric organisms and the cultivation of *Campylobacter pylori* has opened a new era in gastric microbiology that has great clinical relevance. The next few years may provide answers to the perplexing problem of chronic idiopathic gastritis and possibly provide insights into the pathogenesis of peptic ulcer disease. Studies to clarify the role of GCLOs [gastric *Campylobacter*-like organisms] in the pathogenesis of these conditions should be a high priority. At the present, however, clinicians might wait for more definitive clinical studies before attempting to obtain cultures from affected patients or initiating specific antimicrobial treatment.

As noted in Chapter 1, as early as 1993 over a thousand *Campylobacter pylori* studies had appeared in the new era heralded by Blaser's review. The issues Blaser raised have not been fully resolved, but the field's understanding of the relationship between *C. pylori* and gastric illness has advanced to the point where much current research focuses on clinical techniques for detection and treatment.

In contrast to the trends toward multiple authorship in research reports and proposals, review articles are often single-authored. The review thus represents one expert reader's interpretation of the state of knowledge in the field. Given the important role that review articles play in shaping a field's understanding of the research base, journal editors are particularly careful to ensure the integrity of this interpretive process. To guard against potential bias, some journals have established stricter conflict-of-interest policies for authors of review articles. The *New England Journal of Medicine*, for example, will accept research reports from authors with financial ties to companies whose products are affected by their research, provided those relationships are disclosed. But such relationships are not allowed for authors of review articles and editorials (NEJM 2003).

Review authors must convince not only the journal's editors but also its readers that the selection and interpretation of prior studies are free of bias. In his review, Blaser based his advice to clinicians on his interpretation of the field's current understanding of the relationship between these bacteria and specific gastric conditions. For this interpretation to be accepted by readers in the field, it needed to be carefully and clearly supported, not only with citations of relevant studies but also with enough information about those studies to enable readers to see the trends the author had seen. We'll return to this issue later in this chapter.

Notice that research reviews typically present a synthesis of *findings* rather than a synthesis of *views*. Direct quotations are rarely found in research reviews because the primary focus is not on what previous authors have believed or said but on what their studies have demonstrated. Reviewers are interested in researchers' claims only insofar as they are supported by the empirical evidence they present. Even in reviews of theory, common in fields like geology, meteorology, and astrophysics, theories tend to be discussed in the context of the physical observations they seek to explain. In short, the goal of the research review is to help readers make sense of all the available evidence. The reviewer offers a description of what the field does and doesn't know on a given topic at a given point in time.

In keeping with this descriptive goal, scientists tend to adopt an objective and respectful tone when synthesizing the work of other researchers. In comparing and contrasting research findings, review authors often point out limitations in the scope or methods of individual studies, but by now it should be clear that *review* in this context means to synthesize or characterize a body of information, not simply to point out flaws (as is often the case, for example, in film reviews in the popular press). Though pointed and personal criticisms are somewhat more acceptable in the humanities literature where authorial presence is generally more prominent (Madigan et al. 1995; Hyland 1999), such tactics are inconsistent with the *ethos* of the objective scientist and the high value placed on consensus in the scientific community. In the sciences (Gilbert and Mulkay 1984) and social sciences (Madigan et al. 1995), public discussion tends to stay focused on the strengths and limitations of the work itself. Though there is clearly an evaluative component to the review, the overall goal is descriptive.

## EXERCISE 4.2

It will be useful to have some potential topics in mind as you continue to read this chapter about reviewing prior research. Researchers frequently discover topics for research while reading the work of others. Review articles are excellent sources of research topics (e.g., see Blaser's conclusion quoted above). As you saw in Chapter 3, research reports also tend to outline directions for further research in discussion or conclusion sections. Review the discussion sections of two or more research reports, either those contained in this textbook or, ideally, papers from your field. What kinds of further studies are suggested? List the potential research questions proposed or implied by these authors.

## 4.3 Locating the Literature

Whether writing stand-alone review articles or reviews embedded in other kinds of scientific texts, experienced researchers rarely begin their reviews of research from scratch. By the time they are ready to write the introduction to a research report or the background section of a grant proposal, they have become very familiar with previous work in the area, some of which may be their own. Thus, the scientist author already knows which previous studies are pertinent to the argument at hand. But new scientists, or researchers working in new areas, may need to do a more extensive search of the available literature in order to develop this sort of familiarity with the research base. Therefore, some search strategies are in order.

Research reviews focus on primary sources—original reports of individual studies published in professional research journals—as opposed to secondary sources such as textbooks or magazine articles written for nonexpert audiences. But secondary sources are excellent places to start in your search for the primary literature. Biddle and Bean (1987) recommend beginning a search with the sources you find at home, for example, your textbooks and lab manuals, both of which may include lists of works cited or suggestions for further reading. To this we would add as easily accessible starting points other course readings, class discussions, and conferences with your professors. If you have chosen a topic that was raised in class, your professor or lab instructor should be able to refer you to a recent paper on the subject. If your topic was suggested in the discussion section of another researcher's report, then you've already identified the first study to include in your review.

The easiest way to search for prior research on a topic you are investigating is to begin with such a reference in hand. This advice is more helpful than it sounds. First, the paper will contain at least a brief review of relevant research and a discussion of implications, providing you with a quick introduction to the topic. Second, if you have even one study in hand, you'll be able to locate the network of previous work by searching "backwards" through its references (and through the references in those references, and so forth). This will take some trial and error; you will undoubtedly come across studies that are not directly relevant to your specific topic. Streamline your search by using the authors' descriptions of these studies in the text to help you decide which are likely to be most pertinent to your particular research interest.

For example, turn to the first paragraph of the Mallin et al. article on *Pfiesteria* (page 328). This text is dense with citations, each selected to provide support for the statement in which it appears. If you want to learn more about how *Pfiesteria piscicida* was first discovered, the first sentence clearly identifies studies by Smith et al. (1988) and Noga et al. (1993) as appropriate sources. If instead you want to explore how the species preys on fish, the third sentence refers you to a 1993 paper by Burkholder and another paper by Burkholder and Glasgow, also in 1993. If you want to learn more about the life cycle of this organism, the next paragraph refers you to a later study by Burkholder and Glasgow (1995). Even brief reviews such as this one can provide important clues to promising sources.

Lastly, your in-hand source can help you navigate online indexes and abstract databases if you need to do some broader searching. Find your initial article in the database (perhaps by searching on the first author's name) and see how it is indexed—that is, see what keywords or identifiers are associated with the paper. Then use those keywords to help you design a search for related articles. If your original source is not contained in the database, check to make sure the database includes appropriate journals in your research area.

Electronic indexes and abstract services can also be useful if you are starting from scratch, but be prepared to spend some time experimenting with different keywords and combinations. If you are having difficulty finding or accessing databases online from home or the computer lab, the reference staff in your campus library is familiar with the databases accessible to students on your campus. (Just as libraries subscribe to journals for use by library patrons, they also subscribe to databases for access by library patrons.) The reference staff will be able to

***Astrophysical Journal* (Copyright 2002 American Astrophysical Society)**

The articles in this journal are indexed in the Science Citation Index, Philadelphia, PA.

***Gastroenterology* (Copyright 2002 American Gastroenterological Association)**

*Gastroenterology* is abstracted/indexed in Biological Abstracts, CABS, Chemical Abstracts, Current Contents, Excerpta Medica, Index Medicus, ISI, Nutrition Abstracts, and Science Citation Index.

***Behavioral Ecology* (Copyright 2003 Oxford University Press)**

*Behavioral Ecology* is covered by the following major indexing/abstracting services: Animal Behavior Abstracts; Current Contents/Agriculture; BIOSIS; Ecology Abstracts; Elsevier BIOBASE/Current Awareness in Biological Sciences; E-Psyche; Geo Abstracts; GEOBASE; Research Alert; SCISEARCH; Wildlife Review; Zoological Record; PsychInfo.

***Journal of Toxicology—Clinical Toxicology* (Copyright 2002 Marcel Dekker, Inc.)**

Articles published in *Clinical Toxicology* are selectively indexed or abstracted in: BioSciences Information Service of Biological Abstracts (BIOSIS); British Medical Journal; CAB Abstracts; Chemical Abstracts; Current Contents/Clinical Medicine; Current Contents/Life Sciences; Derwent Information, Ltd.; Elsevier BIOBASE/Current Awareness in Biological Sciences; EMBASE/Excerpta Medica; Index Medicus/MEDLINE; INIST-Pascal/CNRS; International Pharmaceutical Abstracts (IPA); Medical Documentation Service; NIOSHTIC; Pub-SCIENCE; Referativnyi Zhurnal/Russian Academy of Sciences; Reference Update; Research Alert; Safety and Health at Work, ILO-CIS Bulletin; Science Citation Index; SciSearch/SCI–Expanded; Toxicology Abstracts.

***The American Journal of Archaeology* (Copyright 2002 Archeological Institute of America)**

The *Journal* is indexed in the Humanities Index, the ABS International Guide to Classical Studies, Current Contents, the Book Review Index, the Avery Index to Architectural Periodicals, Anthropological Literature: An Index to Periodical Articles and Essays, and the Art Index.

**FIGURE 4.1**    Indexing information from the masthead pages of sample journals.

help you get started on keyword searches. They also can help you choose appropriate indexes and databases for topic searches in your field.

A broad variety of electronic databases are currently available in the sciences, ranging from general multidisciplinary services such as *Science Citation Index* to domain-specific sources such as *Agricola* (indexing journals and books in agriculture and related fields), *Medline* (biomedical sciences), and *GeoRef* (geology), to sources focusing on specific research areas within domains, for example, *Fish and Fisheries Worldwide, Textile Technology Digest,* and *Bacteriology Abstracts*. To identify the most useful indexes and abstract databases for your research topic, consult with your professors and/or check the masthead page of the primary journals in your field. This page, which provides information about the journal publisher, copyright notices, subscription procedures, and so forth, appears in print journals somewhere near the table of contents. The masthead page often lists the databases in which the journal is indexed. This information can often be found at the journal's website as well. Several sample indexing lists are reprinted in Figure 4.1.

**EXERCISE 4.3**

Either online or in print, find the masthead information for three to five major journals in your field. List the indexing and abstract services used by these journals. Put a star by those currently available through your university library.

**EXERCISE 4.4**

Working alone or with a partner, choose a topic in your field that you would like to know more about. It may be a topic you identified in Exercise 4.2; it may be an area you are currently exploring in another course or at work; or it may come from your own reading or browsing through journals in the library. Conduct a keyword or subject search on this topic in at least two different indexes appropriate for your field (print or electronic). Compare the outcomes of your searches. How useful was each database? Did the two systems produce similar sets of sources? What journals were cited in each search? Which database contained more relevant references on your topic? How easy was it to narrow the search in each system? Which provided greater flexibility in defining and combining keywords? Which search was more efficient? In what other ways did these indexes differ? List the advantages and disadvantages of each system you examined. Your instructor may ask you to compare these findings with those of your classmates.

# 4.4 Reading Previous Research

In most cases, you will be reviewing research as part of a research report or grant proposal. Thus the goal of your literature search is to determine the context for your research: What does the field already know about this topic? What kinds of studies have been done? What methods have been used, and how useful have they turned out to be? What has been found? What kind of information is still needed? Your answers to such questions will help you design a project that represents a reasonable next step for your field. When you write the review itself, you will aim to help your readers see the trends that you have seen in this literature.

In conducting your search, use paper titles and abstracts to help you sort through the sources you've located. As you begin to identify the studies that seem most relevant, read the introduction and discussion sections carefully. Your goal in reading each paper is to understand why the authors conducted this research, what questions they hoped to shed light on, and what conclusions they came to. What were they trying to find out, and how does this relate to what you're trying to find out?

As you begin to compare and contrast the studies to be included in your review, skim the methods and results sections as well to see what kinds of materials were used (or sites or subjects observed), what kinds of measurements were taken or observations made, and what kinds of analyses were performed. Locate the major findings of the study. Recall from our discussion in Chapter 3 that in well-structured results sections major findings are highlighted in the text. Look for the authors' generalizations about their results, often contained in topic sentences.

## 4.5 Identifying Trends and Patterns

Whether your review is a stand-alone document or part of a report or proposal, readers will expect you to have read widely in the research literature and to have selected the most significant and most relevant studies to include in your review. Your goal is to present an overview of what this research has demonstrated. That is, you will want to sift and synthesize, pointing out similarities and differences in the findings these researchers report and, if pertinent, in the methods they used and the focus of their experiments or observations. This synthesizing goal will lead you to talk about the studies in groups or clusters, rather than describing each in isolation. (Notice in the sample texts in Chapters 10 to 13 that studies are frequently cited in clusters of two or three.) The review should not be a list of individual article summaries but a discussion of the trends that you noticed across studies.

A useful way to identify trends is to construct a grid to help you record distinctive features of the studies as you read them. List the studies down the left-hand side of the page, and mark off several columns across the top. Early in your reading, the grid might include column headings such as "research question," "methods," and "principal results." As you become more familiar with the literature and the issues raised by these studies, you will want to develop more specific column headings.

For example, when he began reading the clinical research on *Campylobacter pylori*, Blaser (1987) might have used a column headed "conditions associated with *C. pylori*" to organize his notes. After a while, he would notice by reading down this column that some researchers documented the presence of *C. pylori* in patients with gastritis, others found associations with peptic ulcer, others with still other conditions. The grid would thus help him notice clusters or subgroups of studies that could usefully be discussed together in his review and may suggest subheadings he could use to organize the body of the paper. (Skim the subheadings in Blaser's review in Chapter 10 to see where the gastritis and peptic ulcer "clusters" were included.) Blaser could then use a similar grid to help him notice differences and similarities within each cluster. For example, he could compare the different types of methods used to study gastritis to see if results were consistent under different conditions. In fact, the table Blaser created to summarize the results of the gastritis studies looks very much like the sort of grid we have been describing (see Table 2 on page 239).

Whether you use a formal grid or some other note-taking system, your primary goal is to identify trends in this body of research. Are findings consistent across the set of studies? If the phenomenon was studied in different regions, at different times of year, with different methods, or under different conditions, were the findings similar or different? Is there theoretical consensus in the field, or have different interpretations been put forward? To help readers understand the current state of the field's knowledge on the topic, your review should highlight consistent patterns and points of agreement as well as inconsistencies and issues that are unresolved.

Two examples are presented in Figures 4.2 and 4.3. In Figure 4.2, Blaser (1987) synthesizes the findings from the gastritis studies he had listed in Table 2. Since the presence of *C. pylori* and the condition of gastritis were strongly associated in all but one of these studies, Blaser highlights this consistent trend in the topic sentence that opens the paragraph. Contrast this emphasis on consensus with the example in Figure 4.3. In this mini-review excerpted from a research report, astrophysicists Fulbright and Reynolds (1990) describe an unresolved issue in shock acceleration theory, the question of whether quasi-parallel or quasi-perpendicular shocks are more efficient in accelerating electrons. Their review emphasizes the lack of consensus in the field, a gap in the field's knowledge that their study will go on to address.

Notice in both examples that the authors offer conclusions or *generalizations* about the set of studies under review, and they cite the *specific studies* on which their conclusions are based. Notice also that, as discussed in Chapter 3, generalizations tend to be stated in present tense (because they describe the current state of knowledge), whereas past tense is used to describe the results of specific studies (which were conducted in the past).

Investigators on four continents have now identified GCLOs in gastric biopsy specimens and have shown an association between the presence of GCLOs and gastritis diagnosed by histology in adults (Table 2). Although methods employed in these studies to document the presence of GCLOs have varied, as have the definitions of gastritis used, it is notable that in all but one study the GCLO detection rate was significantly greater in patients with gastritis than in those without. The exception occurred in a small study in Australia in which nearly equal rates of GCLO detection were found in the two groups (44). Whether idiopathic antral gastritis in children is specifically associated with the presence of GCLOs is not yet settled (53–55). In attempts to answer the important question of whether GCLOs are present in healthy persons, three endoscopic studies of asymptomatic volunteers have been reported. Despite absence of symptoms or risk factors and the young age of the volunteers (mean ages 27–30 yr), 20.4% were found to have histologic gastritis; all of these subjects had GCLO present (Table 2). In contrast, no gastritis was found in 79.6% of the subjects and GCLOs were not detected in any of these cases. In total, the volunteer studies indicate that gastritis may be present in asymptomatic young adults, that this gastritis is associated with the presence of GCLOs, but that in the majority of subjects neither gastritis nor GCLOs are present.

**FIGURE 4.2**   Sample review paragraph from Blaser (1987, p 373): Gastric *Campylobacter*-like organisms, gastritis, and peptic ulcer disease. *Gastroenterology.* Full text included in Chapter 10 (pages 236–248).

The most obvious mechanism to produce a bipolar structure is the compression by a factor of 4 of magnetic field where it is perpendicular to the shock normal, compared to no amplification where it is parallel (van der Laan 1962; Whiteoak and Gardner 1968). However, this mechanism can produce only a limited amount of azimuthal modulation of intensity. Roger et al. (1988) and Leckband, Spangler, and Cairns (1989) point out that another possible mechanism for producing bipolar structure in shell remnants is a systematic dependence of the efficiency of shock acceleration on the obliquity angle $\theta_{Bn}$ between the shock normal and the external magnetic field, if the field is assumed to be fairly well ordered on the scale of the remnant diameter. Shock acceleration theorists are divided on whether quasi-parallel ($\theta_{Bn} \sim 0°$) or quasi-perpendicular ($\theta_{Bn} \sim 90°$)

shocks are more efficient in accelerating electrons. The quasi-parallel geometry seems more adapted to classical diffusive shock acceleration (see reviews such as Drury 1983 or Blandford and Eichler 1987), while quasi-perpendicular geometry allows the so-called shock drift mechanism (Pesses, Decker, and Armstrong 1982; Decker and Vlahos 1985, among others) in which electric fields along the shock front accelerate particles, a process which can be considerably more rapid (Jokipii 1987). Leckband, Spangler, and Cairns (1989) attempted to study this issue by examining the limb-to-center ratios of SNRs, inferring the direction of the external magnetic field for each remnant using a model of the galactic magnetic field, and comparing with model calculations for profiles of SNRs. They could not come to a definite conclusion.

**FIGURE 4.3**   Sample review paragraph from Fulbright and Reynolds (1990, p 592): Bipolar supernova remnants and the obliquity dependence of shock acceleration. *The Astrophysical Journal.*

## EXERCISE 4.5

Read the paragraph in Figure 4.4, from the introduction to Mallin et al. (1995). References to prior research have been deleted from this paragraph. Use asterisks to indicate where you think the authors would have included citations to previous research. Then compare your text with the original article, included in Chapter 11 (see page 329).

The ciliated protozoan *Stylonichia* cf. *putrina* has been observed to consume TFVCs of *P. piscicida* without apparent adverse toxic effects. During prolonged feeding events, however, remaining planozygote stages of the dinoflagellate can transform into large amoebae which in turn, engulf the ciliate as predator becomes prey. Interactions between toxic stages of *P. piscicida* and other potential predators

are currently unknown, such as whether mesozooplankton or microzooplankton aside from *Stylonichia* can consume or control it, or whether they are adversely affected by its toxin. Other toxic dinoflagellates are used as food resources by some zooplankters, but they reduce fecundity and survival in other zooplankton species or adversely affect higher trophic levels through toxin bioaccumulation.

**FIGURE 4.4**   Review paragraph, citations deleted, from Mallin et al. (1995, p 352): Response of two zooplankton grazers to an ichthyotoxic estuarine dinoflagellate. *Journal of Plankton Research.* Full text included in Chapter 11 (pages 328–340).

**EXERCISE 4.6**

Choose a sample paper or proposal in your field. Modify one paragraph of a re-search review section by stripping out research citations as we did in Figure 4.4. Type up the "research-free" review paragraph, and exchange with another member of the class. Read the paragraph you've been given, and note in the text where you think the authors would need to cite prior research and why. Compare your analysis with the original review.

# 4.6 Organizing the Review

There is no standard organizational format for the research review, for the scope and purpose of reviews vary widely. Whether you are writing a review article or the literature review section of a proposal or report, use basic principles of good writing as your guide:

- Introduce your discussion by establishing the significance of the topic. It is helpful to give a quick preview of the major trends or topics to be covered in the review. (For a good example of such a preview, see the opening paragraph of the background section in the research proposal written by Reynolds, Borkowski, and Blondin in Chapter 12; see page 363.)
- Organize the body of the review to reflect the clusters or subtopics you have identified, using headings if the review is lengthy.
- Use topic sentences at the start of paragraphs and sections to highlight similarities and differences and points of agreement and disagreement.
- Conclude with an overview of what is known and what is left to explore.

Though review articles originally tended to survey historical trends (Day 1998), today they are more likely to concentrate on recent history, as indicated by *Nature*'s (2003b) emphasis on "recent developments in a topical area." On some topics, however, the significant trends are chronological—that is, changes in the field's understanding or methodological approaches over time—in which case a chronological structure for the review may be warranted. The literature review in the introduction to Stephen Reynolds' (1996) research letter, for example, traces a succession of theoretical models proposed over time in response to an evolving body of observational data (see pages 385–386). Notice that Blaser (1987) includes a section entitled "Historical Developments" at the start of his review article on *C. pylori*. This section is particularly appropriate in Blaser's review because a notable feature of Marshall and Warren's discovery was the fact that gastric bacteria had been observed for decades but largely ignored. Now that the relationship between these bacteria and gastrointestinal conditions is becoming clearer, this "old" evidence is suddenly interesting. In this situation, a brief discussion of past history helps to contextualize the more recent advances discussed in the main body of the review. The main body of the review is then organized not by chronology, but around those advances.

Notice that the headings in Blaser's (1987) review identify the major subtopics covered in his article (e.g., "Microbiologic Characteristics of *Campylobacter pylori* and Related Organisms"; "Pathological Associations With Gastric *Campylobacter-like* Organism Infection"). These *topical* headings are quite different from the *functional* headings of the IMRAD form followed in research reports. The IMRAD headings—introduction, methods, results, and discussion—let readers know what function each section of the report serves: the introduction introduces, the methods section describes methods, and so forth. Functional headings are useful because they enable readers who are familiar with a standard format to quickly locate the kinds of information they expect the document to include. But unlike the research report, the review article is not subdivided by function and therefore follows no standard functional format. All sections of the review have the same goal or function: to review research. Each section does, however, describe a distinctive trend, feature, or area of that research, and headings can be used to highlight those subtopics. Thus, in a research review, headings signal shifts in the focus or content of the discussion.

A striking example can be found in the background section of the research proposal by Burkholder and Rublee (1994) in Chapter 11 (see pages 307–314). Burkholder and Rublee's first two headings summarize not just the major areas of research they will describe but the major claims they are basing on that research: that a linkage between fish kills and *Pfiesteria* has been established, and that highly specific molecular probes are needed to detect the pathogen. In this text, topical headings are used to outline the "plot" or argument of the review (Myers 1991).

## 4.7 Citing Sources in the Text

Two primary citation systems are used in scientific journals: the name-year system and the citation-sequence system (CBE 1994). Check the journals in your field to see which system they require authors to use. This information will appear in the "instructions to authors," typically posted along with editorial goals and guidelines at the journal's website and/or published annually in a hard-copy issue of the journal. Full descriptions of these systems can be found in style manuals, such as those published by the American Institute of Physics (AIP 1990) or the Council of Science Editors (formerly the Council of Biology Editors; CBE 1994). Some journals will simply refer you to the appropriate style manual for guidelines on handling citations. Others will include samples of their required citation and reference formats in their instructions to authors.

The two citation systems are illustrated in Figure 4.5. Under the **name-year system,** which we are using in this textbook, any sources referred to in the text are identified by the author's last name and the date of publication. These sources are then listed alphabetically in the list of references or works cited at the end of the text. This system is followed in such journals as the *Journal of Plankton Research,* the *Astrophysical Journal,* and *Geology* (see sample articles in Chapters 11–13).

In the **citation-sequence system,** cited sources are numbered in the order in which they are mentioned in the text. These numbers are used as identifiers for citations in the text, and the full references appear in this order in the list of works

## Name-Year System

### Example A
Koyama et al. (1995) argue convincingly that thermal bremsstrahlung models are very unlikely.

### Example B
Other toxic dinoflagellates are used as food resources by some zooplankters (Watras *et al.*, 1985; Turner and Tester, 1989), but they reduce fecundity and survival in other zooplankton species or adversely affect higher trophic levels through toxin bioaccumulation (White, 1980, 1981; Huntley *et al.*, 1986).

## Citation-Sequence System

### Example C
Gastric bacteria now are being observed with regularity (2-4), and recently, Marshall and Warren (5,6) were able to isolate a spiral bacterium that had never been cultivated before.

### Example D
Recent studies have suggested that the eradication of *Helicobacter pylori* infection affects the natural history of duodenal ulcer disease such that the rate of recurrence decreases markedly (2–6).

### Example E
A worldwide increase in toxic phytoplankton blooms over the past 20 years[1,2] has coincided with increasing reports of fish diseases and deaths of unknown cause.[3]

**FIGURE 4.5** Sample citations from the following sources included in Chapters 10 through 12: *A.* Reynolds (1996); *B.* Mallin et al. (1995); *C.* Blaser (1987); *D.* Graham et al. (1992); *E.* Burkholder et al. (1992). Sample reference list entries for A–D are presented in Figure 4.6.

cited. Numbers may be inserted in superscript (above the line) or in parentheses, depending on the capacities of your word-processing program. Of the journals represented in this textbook, the citation-sequence system is used in *Nature, Journal of Toxicology–Clinical Toxicology*, and in the four medical journals: *Annals of Internal Medicine*, the *Lancet, Gastroenterology*, and the *British Medical Journal* (see sample articles in Chapters 10, 11, and 13).

Some journals (e.g., the *Journal of Bacteriology, SIAM Journal of Scientific Computing*) use a less common numerical system that combines features of both systems described above: all cited sources are listed alphabetically at the end of the text and then numbered in this order. These numbers are then used as identifiers for citations in the text. You'll be able to recognize this variation of the numerical system quickly by checking the list of works cited to see whether the numbered sources are listed alphabetically.

## EXERCISE 4.7

Compare a research report or proposal using the name-year citation system with one using the citation-sequence system. What do you notice about the effects of these different systems on your reading? List the advantages and disadvantages of each system.

In addition to the basic form of the citation, the examples in Figure 4.5 illustrate a number of other variations in citation practices, as described below.

***Direct Versus Indirect Citations.***    Authors may be identified directly in the text, as in Examples A and C of Figure 4.5, or they may be cited indirectly, using parenthetical or numerical identifiers as in Examples B, D, and E. Swales (1984) notes that direct citation focuses discussion on the researcher(s) who did the work; in contrast, indirect citation features the research claim or finding as the subject of the sentence and thus focuses attention on the research itself. Reviewers use these different "reporting formulae" not only for variety but also to emphasize the contributions of individual researchers or to highlight trends in the research.

In fact, the use of reporting formulae can become a strategic dimension of a scientific argument. In a section entitled "Modern Controversies" in their *Clinical Toxicology* piece, Spiller, Hale, and De Boer (2002) discuss one researcher, Amandry, by name, but they use indirect citation to refer to similar conclusions by several other archaeologists (see page 411). Spiller et al. hold Amandry responsible for discouraging scientific interest in the Gaseous Vent Theory in the 1950s, a theory that their new evidence now supports. Using direct citation in this instance enables the authors to focus attention (and, in this case, blame) on this particular scientist's conclusions, underscoring their argument that ancient accounts of the oracle had been wrongly discounted. Highlighting the role of this individual researcher and downplaying the line of research he represents makes it easier to discredit this work, for it comes across as the mistaken assumption of one man.

The practice of direct citation is more often used to allocate credit than to place blame, however, as can be seen in the second paragraph of the letter by Marshall (Warren and Marshall 1983) contained in Chapter 10 (pages 233–235). Note how Marshall uses direct citation when highlighting the fact that earlier researchers had observed bacilli in the stomach lining of ulcer victims, a finding that supports Marshall and Warren's own claim. But Marshall uses indirect citation when reporting the fact that other researchers failed to confirm those observations or recognize their importance. The effect of indirect citation here is to downplay the negative findings of these other researchers. These two examples illustrate the potential impact of stylistic choices such as reporting formulae.[1]

As the Amandry example illustrates, it is conventional to refer to authors by last name only when using direct citation; first names and titles are not included. Once you have included the authors' names in the text, it is unnecessary (in fact redundant) to include them in the parenthetical citation. If you are using the name-year system, insert only the date of the source in parentheses immediately after any author mentions in your text, as in Example A. In the citation-sequence system, the numerical identifier remains the same in direct and indirect citations.

***Placement of Indirect Citations.***    Indirect citations appear most frequently at the end of a sentence, as in Examples B, D, and E of Figure 4.5, but they are also

---

[1]In reviewing and citing, scientists indicate what work they consider valuable as well as what work remains to be done. In so doing, they are engaging in what might be considered *"epidiectic"* argument, or the rhetoric of praise and blame (see Sullivan 1991), one of three types of classical argument that we will talk more about in Chapter 6. For further discussion of how scientists use citations strategically, see Latour (1987) and Paul (2000).

commonly found at the ends of clauses (Example B) or phrases (Example E). A parenthetical or numerical citation should be inserted immediately after the statement, word, or phrase to which it is directly relevant (CBE 1994), so that it is clear to readers which part of your claim or observation is based on that particular source.

**Citing Work by Multiple Authors.**    Whether you are citing directly or indirectly, always acknowledge all the authors of a work. If the cited work has two authors, include both names, as in Examples B and C of Figure 4.5. For works with more than two authors, use *et al.*, as in Examples A and B. An alternative form preferred by the Council of Biology Editors (1994) is to replace the Latin abbreviation with the English equivalent: "Koyama and others (1995)."

### EXERCISE 4.8

Obtain copies of the instructions to authors for three major journals in your field. What citation systems do these journals follow? What style guidelines do they offer? Which style manuals are recommended? Where can these manuals be found on your campus? Are they available online?

## 4.8 Preparing the List of Works Cited

Any sources cited in your text should be included in a reference list, alternatively called a works cited list, at the end of the paper. The general format for journal references is the same in the two systems, except for the placement of the year of publication. In the name-year system, the year of publication must appear directly after the authors' names, so that references can be easily recognized from their name-year citations in the text (CBE 1994). In the citation-sequence system, the year appears later in the reference. The basic format of the two types of references, as prescribed by the Council of Science Editors (CBE 1994, p 634), is as follows:

**Name-Year System [articles in print journals]**

Author(s). Year. Article title. Journal title volume number(issue number):inclusive pages.

**Citation-Sequence System [articles in print journals]**

Author(s). Article title. Journal title year month;volume number(issue number):inclusive pages.

Sample references are listed in Figure 4.6. You'll notice that though the general ordering of reference components is relatively consistent, some formatting details vary from journal to journal. For example, though the generic format includes the issue as well as the volume number, issue numbers tend to be omitted if pagination is continuous throughout a volume. In comparing the sample references in Figure 4.6, you'll find that some journals require first and last page numbers; others allow only the starting page number. Some put dates in parentheses; others do not. Some journals will limit the number of author names. (The *Lancet* and some other biomedical journals follow a common set of editorial requirements,

## Name-Year System

### Example A

Koyama, K., Petre, R., Gotthelf, E. V., Matsuura, M., Ozaki, M., & Holt, S. S. 1995, Nature, 378, 255.

### Example B

Watras, C.J., Garcon, V.C., Olson, R.J., Chisholm, S.W. and Anderson, D.M. (1985) The effect of zooplankton grazing on estuarine blooms of the toxic dinoflagellate *Gonyaulax tamarensis*. *J. Plankton Res.*, **7**, 891–908.

## Citation-Sequence System

### Example C

2. Steer HW. Ultrastructure of cell migration through the gastric epithelium and its relationship to bacteria. J Clin Pathol 1975:28:639–46.

### Example D

2. **Coghlan JG, Gilligan D, Humphries H, McKenna D, Dooley C, Sweeney E, et al.** Camplyobacter pylori and recurrence of duodenal ulcers—a 12-month follow-up study. Lancet. 1987;2:1109–11.

**FIGURE 4.6**   Sample references from the following sources included in Chapters 10 through 12: *A*. Reynolds (1996); *B*. Mallin et al. (1995); *C*. Blaser (1987); *D*. Graham et al. (1992). In-text citations are presented in Figure 4.5.

which allow only the first six authors' names to be listed. Others must be indicated with *et al.* as in Example D.) Some reference formats omit article titles altogether. Most journals do not use quotation marks around article titles in references (but some do!). Punctuation around authors' initials and journal title abbreviations varies across journals, as does capitalization and the use of italics and boldface. The format of in-text citations varies as well: some journals require a comma between name and year in parenthetical name-year citations, others do not.

Though these differences may seem arbitrary, in most cases the modifications were adopted in an effort to save space, enhance consistency, or make typesetting processes more efficient. Unfortunately, journal staffs have tended to experiment with these modifications independently, making a common set of rules difficult to maintain. For this reason, even if you're following the general formatting guidelines listed above, it is important to read the journal's instructions to authors and to use a sample reference list from the target journal as a model.

The generic format above covers basic references for articles in print journals only. For references to other types of print sources, such as conference proceedings, technical reports, monographs, and chapters in books, you'll need to consult the appropriate style manual in your field or find an example in the reference list you're using as a model. Style manuals will also provide guidelines for citing multiple articles by the same author, citing authors with the same last name, citing organizations as authors, citing unpublished work, and many other special cases.

The "special case" of the Internet encompasses a wide variety of source types and has set editors in all fields scrambling. The National Library of Medicine, whose *Recommended Formats for Bibliographic Citation* sets the standard for many scientific journals and style guides, recently published a 106-page supplement devoted solely to Internet sources (NLM 2001). Available at the NLM website (http://www.nlm.nih.gov/pubs/formats/internet.pdf), this guide provides formats for citing articles in electronic journals, as well as web pages, databases, and other electronic forms. The Council of Science Editors bases their recommended formats on this source. Internet citations differ from the formats specified above primarily in identifying the Internet as the medium and including the URL and copyright date, if available, as well as the date the site was accessed and cited. Because web materials can be continually updated (and even discontinued), the access date serves to identify which version of the source you read—in effect rendering a potentially transitory source permanent by fixing it in time. Some representative samples from the NLM guidelines are reprinted below.

**Sample web references: Homepages for government agencies, professional associations, and universities (NLM 2001, p 67).**[2]

Animal Welfare Information Center [Internet]. Beltsville (MD): National Agricultural Library (US); [updated 2001 Mar 1; cited 2001 Mar 2]. Available from: http://www.nal.usda.gov/awic/.

PsycPORT.com [Internet]. Washington: American Psychological Association; c2000 [cited 2000 Mar 7]. Available from: http://www.psycPORT.com/.

University of Maryland [Internet]. College Park (MD): The University; c2001 [updated 2001 Apr 28; cited 2001 May 1]. Available from: http://www.maryland.edu/.

**Sample web reference: article from electronic journal (NLM 2001, p 35).**[2]

Tong V, Abbott FS, Mbofana S, Walker MJ. In vitro investigation of the hepatic extraction of RSD1070, a novel antiarrhythmic compound. J Pharm Pharm Sci [Internet]. 2001 [cited 2001 May 3];4(1):15–23. Available from: http://www.ualberta.ca/~csps/JPPS4(1)/F.Abbott/RSD1070.pdf.

These special considerations apply only when the journal being cited is designed and published as an online serial. If your source is actually an article from a print journal that you have accessed through an electronic database or downloaded from the journal's website, then it should not be cited as an e-journal. Check the style guides in your field to see whether guidelines for citing sources obtained electronically have been established. Many librarians advise simply following the guidelines for print citations. A simple solution developed by editors of the American Psychological Association (http://www.apastyle.org/elecsource.html) is to add a retrieval statement at the end of the reference that provides the date you retrieved the source and the name of the database:

**Sample web reference: Print source obtained through electronic database (APA 2003)**[3]

Borman, W. C., Hanson, M. A., Oppler, S. H., Pulakos, E. D., & White, L. A. (1993). Role of early supervisory experience in supervisor performance. *Journal of Applied Psychology, 78,* 443–449. Retrieved October 23, 2000, from PsychARTICLES database.

---

[2]The NLM samples follow the principles of the citation-sequence system, with the various dates appearing late in the entry. This format can be modified for use under the name-year system by moving the publication date to appear after the name of the author or authoring organization. The date of modification or update and the date of citation remain in the order above. If the date of publication is not available, insert date of modification after the author, with the date of citation remaining in its usual position. If neither is available, use the date of citation. Note that dates of modification/update and of citation are written in full (including month and day) after the author and are enclosed within square brackets. This hierarchy of dates will be codified in the next edition of the CSE manual, which is in preparation as of this writing (K. Patrias, NLM editor, personal communication, 8-14-03).

[3]The APA uses a name-year system. To modify for use under the citation-sequence system, the year of publication should appear after the name of the journal.

# 4.9 The Research Review Abstract

Finally, if you are writing a review article—as opposed to a review section embedded in a research report or grant proposal—you will also need to prepare an abstract summarizing the focus and scope of the project. As we noted in Chapter 3, the abstract is a contingent genre; that is, its form is contingent upon the form of the paper it is intended to represent. Your review abstract may be either informative (summarizing the trends you observed in the literature and the conclusions of the review) or descriptive (identifying the topics you will cover but not what you found). Check the journals in your field to see which is preferred. In either case, the abstract should preview the major topics under which you have organized the review itself.

## EXERCISE 4.9

Read the abstract for Blaser's (1987) review article, included in Chapter 10 (see page 236). How do the content and structure of Blaser's abstract reflect the purpose and content of the research review genre? Is this an informative or a descriptive abstract? Write an alternative abstract in the other mode.

# Activities and Assignments

1. Choose a research report on a topic in your field. Read the introduction carefully. Find three to five of the works cited in this section. Write a one-paragraph summary of each cited paper and a one-sentence explanation of why it is cited in the review.
2. Read the first five paragraphs of the introduction to the article by Mallin et al. (1995) in Chapter 11 (pages 328–340). For the purposes of this exercise, consider this section as a stand-alone research review article. Write both an informative and a descriptive abstract for this review article.
3. Conduct an electronic search on a topic in your field or a topic assigned by your instructor. You may want to limit your search to the 10 to 15 most recent publications. Find and read as many of the sources as you can. Summarize the critical features of these studies in a grid that you could use to plan a literature review on the topic.
4. Use the grid you developed in Activity 3 to help you write a two- or three-page review of recent research on a topic in your field. Follow a citation format appropriate for a primary journal in your field.

# CHAPTER 5

# Preparing Conference Presentations

## 5.1 The Role of Research Conferences in the Sciences

Although this book is primarily concerned with writing, much if not most day-to-day communication in science takes place orally in informal one-on-one discussions, lab meetings, and participation in professional conferences. It is through these oral, more personal modes of interacting that much of the work of science gets done. Thus, the ability to interact and orally communicate with other scientists is essential for scientific progress. For an individual to participate in the knowledge-sharing and consensus-building business of science, publication is not enough (Reif-Lehrer 1990).

A scientist's personality and interpersonal skills come into play in oral modes of communication, perhaps more directly than in writing. We saw in the case of Barry Marshall in Chapter 1 that a scientist's personality and style are an important dimension of conference presentations and of scientific communication generally (see Chazin 1993; Monmaney 1993; SerVaas 1994). As Peter Raven, former Home Secretary of the National Academy of Sciences, states: "In contrast to published manuscripts, oral presentations establish personal contact with an audience. The favorable or unfavorable impression created through this contact often permanently shapes a scientist's reputation—a reputation that can either help or haunt the scientist for the rest of his or her career" (Anholt 1994, p ix). Of all the oral modes of communication, the conference presentation is the most formal and structured, and therefore it is more explicitly governed by generic conventions that can be studied.

You discovered in working through Chapter 2 how important professional conferences are in your field. Most scientists present their research at several conferences in their field per year. They attend conferences to share information, get feedback on new ideas or ongoing research, catch up with the latest developments in the field, and create professional relationships—to "network." Conferences provide a more immediate means for scientists to share new research or research in progress, often before it is even published.

Probably the best way to learn about conference presentations is to go to a few professional conferences yourself. If you have not already done so, make plans to attend a meeting in your field with your professors and/or peers. Observe the format and conventions of talks or poster sessions. Do presenters speak from notes, from visual aids, or from prepared texts? What kinds of visuals are common (tables, graphs, photographs), and what kinds of equipment are used (for example, are PowerPoint slides the norm?) What is the time limit for presentations, and is it strictly enforced? How are the presentations organized? How are they different from written research reports?

We will discuss these and other features of conference presentations in this chapter. We'll focus on two presentation modes common in the sciences: the oral presentation, often referred to as the conference "talk" or "paper"; and the research poster. The poster is an increasingly common form of presentation in many scientific fields. Posters offer several advantages over the traditional oral presentation, especially for conference organizers. Large organizations typically receive proposals for far more conference papers than can be accommodated in the time available, despite the practice of scheduling concurrent sessions. Because many posters can be displayed simultaneously (limited only by the size of the room or hall available), conference organizers are able to accept a much larger proportion of proposals than would be possible if each presenter needed a 10- or 20-minute presentation period. In 1990, this practice enabled the program committee for the Annual Meeting of the American Society for Microbiology to promise to include *all* acceptable proposals on the conference program, 75 percent of which were then scheduled as poster sessions (ASM 1990). In 2003, all presentations in the ASM General Meeting Program were poster sessions (ASM 2003).

As with oral presentations, preparing and presenting research posters require both visual and verbal skills, as well as detailed knowledge of the expectations of conference goers in your discipline. We offer here some basic principles that will help you investigate the conventions for presenting research in either mode at conferences in your field.

## 5.2 Writing Conference Proposal Abstracts

We discussed the writing of abstracts for research papers in Chapter 3. You'll recall that the generic abstract summarizes the major points of the research report in four moves: purpose, method, results, implications. The same is generally true for conference proposal abstracts. For example, under "Content" in Figure 5.1, "Abstract Submission Guidelines," the American Society for Microbiology requests that the one-paragraph abstract proposals contain background, methods, results, and conclusions; note that ASM's descriptions of the contents of these sections exactly mirror the IMRAD form. Also note the amount of attention and detailed instructions in Figure 5.1 under the heading "Key Submission Elements": formatting, style, graphs and tables, number of characters, contact information, subject category,

## Abstract Submission

Over 3000 abstracts are submitted each year for inclusion in the ASM General Meeting Program. Each properly completed submission is peer-reviewed for its scientific content and merit by the appropriate Division Officers. Accepted abstracts are grouped into poster sessions that have a common theme and are scheduled throughout the week of the Meeting. Since abstracts are grouped by theme, an abstract submitted for review in one Division may be scheduled with those of another. **Authors must be prepared to present their poster whenever it has been scheduled.**

### Preparing Your Abstract

The abstract is a short description of your work and should contain all the elements necessary to define your goals and results to the reader. It is not meant to be a complete and lengthy report of your work.

### Key Submission Elements

The use of an electronic submission process requires that each abstract submission includes the following discrete elements:

1. **Title**: The title should clearly identify the contents of the abstract. Capitalize the first letter of each word except prepositions, articles, and species names. Italicize scientific names of organisms (e.g., *Candida albicans*).
2. **Authors and Affiliations**: Each Author is entered separately. Each name is entered by completing fields for first (given), middle initial(s) and last (family) name. You should provide information on each author's institution/affiliation. **Do not give** department, division, branch, street address, etc. when completing the institution information. Enter the City and State/Province (if from the United States or Canada) and zip/postal code. If an author is from a country other than the United States, be sure to provide this information. Use abbreviations whenever possible (see table on page 10). You will be able to designate the presenting author, add additional affiliations, and modify the order of the authors using the appropriate tools in the submission area. Please note that it is a General Meeting policy to list authors using all initials and last name*. Therefore, a sample author/affiliation listing for presentation would be:

   G. E. F. Brown, Jr.[1], P. J. Smith[2], and **R. B. Jones**[1,2], [1]CDC, Atlanta, GA and [2]Bournemouth Univ., Poole, UNITED KINGDOM

3. **Abstract Text:** Because of the large number of abstracts presented at the meeting, the text of each abstract should consist of no more than 1850 characters (excluding spaces). **Do not include your title and author information as part of the abstract body**. The abstract text may be prepared in any standard word processing program. Insert sub or superscripts, boldface, italics, or other required symbols as necessary. Tables, if presented, should be as simple as possible. Please note they will be reduced in size in the final product to fit in a column approximately 2.5" (6.3 cm) in width. Graphics (graphs, pictures, etc.) **are not recommended** as they are very difficult to properly size in the space available. Files should be saved either as a standard word-processed file or as a Hyper Text Markup Language (HTML) file that is viewable on the internet.
4. **Other Information**: The following information must be provided with your submitted abstract.

---

*The submission system will insert affiliation superscripts and show the presentation author in boldface type.

**FIGURE 5.1**   Abstract submission guidelines, from Call for Abstracts, 103rd General Meeting of the American Society for Microbiology (2003).

keywords, student travel grants, and "Content." Further attention is paid to the deadlines for submission and to the submission procedures themselves, which as of 1999 became completely web-based (instructions and options for coding the

- **Contact Information**: Regardless of who submits the abstract, ASM will correspond with the presenting author. Therefore, complete mailing address information, telephone number, e-mail address and facsimile number must be given in the abstract submission. Please be sure that the address you provide is complete in order for either standard mail or e-mail to reach the presenter. (NOTE: be sure to include mail-stops or box numbers in the address.) **If the presenting author should relocate before the 103rd General Meeting, he/she should provide the ASM Meetings Department with the address change.** Please note that the database supporting the abstract submission system is not connected to the ASM Membership database. Therefore, changing this address in the Membership database does not cause a change in the abstract submission database. NOTE: The presenting author must be available at the General Meeting for the poster session in which the paper is scheduled. If the original presenting author becomes unavailable, a co-author must be ready to substitute and present the poster.
- **Subject Category**: From the listing provided on page 8, select a category in which the abstract should be reviewed.
- **Keywords**: Provide up to three keywords to aid others in searching for your abstract in the index or a searchable database. Try to select keywords in common usage, such as those used in *Medline* or *Index Medicus*. Each keyword should be able to stand alone in the index and should be typed in lowercase unless it is a proper noun or a genus name (e.g., Pseudomonas).
- **Student Travel Grant Applicant**: If the presenting author is both a student and a member of ASM, he/she is eligible to submit an application for a Student Travel Grant. Two types of student travel grants are available. The Corporate Activities Program Student Travel Grant is a $500 grant given to approximately 160 students who

are presenting an abstract. The Richard and Mary Finkelstein Travel Grant also is a $500 grant given to 6 students whose research is in the area of microbial pathogenesis. See the sidebars for further details on procedures and requirements.

### Content

**Abstracts must be based on results that have not been published or presented in any journal or at any public scientific conference, nationally or internationally, before May 22, 2003.**

1. The abstract must demonstrate scientific merit and significance.
2. While all abstracts may not exactly fit the following guidelines (e.g., a case study), abstracts should contain the following key points:
   - *Background*—the problem under investigation or a hypothesis
   - *Methods*—the experimental methods or protocols used to accomplish the research
   - *Results*—the key points derived from experiments. Data should be summarized and enough presented to allow the reviewers to judge the content. Generalizations such as "will be discussed" or "will be presented" are not acceptable.
   - *Conclusion*—a summary of your findings that are supported by the data presented
3. Be clear and concise. Avoid the use of jargon or catch phrases whenever possible.
4. If you and your team are submitting several abstracts for consideration, be certain that each abstract will stand on its own. If you submit several abstracts that vary only in the smallest details, most will be rejected.
5. Be sure to resolve all questions dealing with internal company or sponsor reviews as well as patent issues prior to submission.

**FIGURE 5.1**   *(continued)*

abstracts for web submission have not been included in the figure). All these requirements are essential in the electronic submission of an abstract to ASM; failure to adhere to these guidelines may result in your abstract being rejected (automatically, by the website!).

In Figure 5.2, "Criteria for Abstract Acceptance," ASM lists its scientific and presentation criteria for accepting abstracts, as well as reviewers' reasons for rejecting them. At the end of Figure 5.2, note that ASM states that there's no time for rejected abstracts to be appealed. Likewise, most conference organizers usually do not edit or proofread proposal abstracts, which are reproduced in conference programs exactly as they are submitted. Thus, the abstract you submit to a conference must be ready for publication. While not all calls for abstracts provide as much detail as the ASM, any guidelines that are provided should be adhered to with great care. Conference proposals are typically reviewed under a tight schedule. As ASM points out in Figure 5.2, abstracts that do not meet the criteria are excluded from the review pool altogether, with no time to appeal. So too are abstracts that exceed a stipulated length (see Figure 5.1).

## Criteria for Abstract Acceptance

The General Meeting Program Committee has established rules for the submission of abstracts. Each abstract will be reviewed for adherence to these rules and then the scientific content and presentation. The elected division officers conduct the peer-review process and have the authority to accept or reject any submission.

1. The main factors considered by the Program Committee (and the division reviewers) in qualifying abstracts for acceptance are:

   - The quality of the research
   - The importance/significance of the topic to the membership
   - The content of the abstract
   - Adherence to submission criteria

2. Abstracts will be rejected by the reviewers for the following reasons:

   - No Hypothesis—The abstract does not clearly indicate the reason for conducting the research and the question being tested by the experiments performed.
   - Inadequate Experimental Methods—The investigators failed to describe procedures used or neglected to include important or essential controls.
   - Insufficient Data Presented—The investigators failed to show the outcome(s) of their research. Insufficient data are presented to support the authors' conclusion(s).
   - No Summary of Essential Results—The investigators failed to concisely summarize the result(s) of their research.
   - No Conclusion—The investigators failed to describe the conclusions of their research with regard to the hypothesis being tested.
   - Duplicate Abstract—The abstract contents substantively overlap with contents of another submitted abstract by the same author or co-author.
   - Not Appropriate for the General Meeting—The content of the abstract is not relevant to any ASM Division.
   - Promotional in Nature—The abstract was written to promote a specific product or procedure on behalf of a specific company or organization.
   - Poorly written—Improper use of the English language renders the abstract incomprehensible.

Because of the time constraints associated with a meeting of this size, if an abstract is rejected, there is insufficient time to appeal the decision and resubmit the abstract for consideration. Therefore, it is crucial to submit an abstract that meets the criteria as outlined.

**FIGURE 5.2**    Criteria for abstract acceptance from Call for Abstracts, 103rd General Meeting of the American Society for Microbiology (2003).

We have included this section on conference abstracts at the beginning rather than the end of this chapter because, while abstracts for research reports are almost always written last, abstracts for conference papers are often written "*before* the paper is written, indeed sometimes even before the work itself is done" (Olsen and Huckin 1991, p 369). Olsen and Huckin describe this type of abstract as "promissory." In promissory abstracts, the research purpose and methodology are described, but results and implications may be projected or occasionally omitted entirely. In some cases, an overview of prior or preliminary results is provided.

In Figure 5.3 we have reprinted two sample promissory abstracts on cold fusion from the 2002 Annual Meeting of the American Physical Society. Notice that these abstracts contain sufficient detail for reviewers and readers to know what will be discussed, even though the research may be incomplete and ongoing. Instead of detailing the new findings (which at this point in the research may be at

### [W21.002] Progress on the SRI/ENEA Collaboration to Investigate Gaseous D_2/Pd Nuclear Effects

*Michael C.H. McKubre, Francis L. Tanzella (SRI International, Menlo Park, CA), Paolo Tripodi, Vittorio Violante (ENEA, 00044 Frascati, Rome, Italy)*

A collaborative effort has been established formally between SRI and ENEA researchers to test and demonstrate the cross-laboratory replicability of gas phase Pd/D_2 excess heat, helium and tritium observations. Similar facilities are being established in both countries to allow on-line determination of heat effects correlated with helium-4, and ultimately helium-3 measurements from so called "Case" experiments involving the application of modest temperatures and D_2 gas pressures to a packed bed of palladium on carbon catalyst and other finely divided palladium materials. The results of experiments performed under similar protocols will be examined and compared. A second facet of this collaboration is the joint attempt to replicate the production of tritium in an "Arata-Zhang" hollow, double-structured cathode. Two massive hollow palladium electrodes were manufactured at ENEA and sealed to contain palladium black within the enclosed void. These electrodes presently are being operated at SRI as electrolytic cathodes in LiOD electrolyte. On experiment termination these will be sectioned and the contents examined for helium-4, helium-3 and evidence of tritium.

### [W21.006] Thermal Measurement during Electrolysis of Pd-Ni Thin-film-Cathodes in Li2SO4/H2O Solution

*C.H. Castano, A.G. Lipson, Kim S-O., G.H. Miley (University of Illinois at Urbana-Champaign, Department of Nuclear, Plasma and Radiological Engineering, Urbana, IL 61801, USA)*

Using LENR - open type calorimeters, measurements of excess heat production were carried out during electrolysis in $Li_2SO_4/H_2O$ solution with a Pt-anode and Pd-Ni thin film cathodes (2000–8000 Åthick) sputtered on the different dielectric substrates. In order to accurately evaluate actual performance during electrolysis runs in the open-type calorimeter used, considering effects of heat convection, bubbling and possible $H_2+O_2$ recombination, smooth Pt sheets were used as cathodes. Pt provides a reference since it does not produce excess heat in the light water electrolyte. To increase the accuracy of measurements the water dissociation potential was determined for each cathode taking into account its individual over-voltage value. It is found that this design for the Pd-Ni cathodes resulted in the excess heat production of \sim 20–25% of input power, equivalent to \sim300 mW. In cases of the Pd/Ni-film fracture (or detachment from substrate) no excess heat was detected, providing an added reference point. These experiments plus use of optimized films will be presented.

**FIGURE 5.3**   Two promissory abstracts from Session W21—Cold Fusion, American Physical Society Meeting (March 2002).

least partially unknown), these promissory abstracts announce what types of information will be included in the presentation.

Though promissory moves are common in conference proposals, it is important to note that some organizations discourage this practice. In fact, ASM's guidelines explicitly prohibit the use of such predictive statements. ASM stipulates that "data should be summarized and enough presented to allow reviewers to judge the content," and states that "Generalizations such as 'will be discussed' or 'will be presented' are not acceptable" (Figure 5.1). In doing so, the ASM explicitly requires informative abstracts much like the report abstracts described in Chapter 3, and explicitly bans promissory moves as less useful to reviewers.

Unlike ASM, the American Physical Society (APS) does not prescribe the content of the abstracts, nor does it specify whether proposals may include promissory statements. APS contributors select a meeting they wish to attend and submit their abstract via the Internet. FAQs are detailed on the website, but the only content information relates to length (http://www.aps.org/meet/abstracts/abstract-faq.html). Note, however, that the sample abstracts in Figure 5.3 seem to follow the general structure of the IMRAD form; this form can help you write abstracts when the content for proposals is not delineated. Also note that the APS abstracts contain references and include explicit acknowledgments of lab or funding sources, neither of which is necessary in a research report abstract (because this information appears elsewhere in a printed article). Remember, the conference abstract is often the only written version of your paper that conference attendees will see; thus it is conventional in many organizations to include these critical acknowledgments. As the contrast between the ASM and APS expectations illustrates, however, conference organizers, like journal editors, establish their own guidelines. As a consequence, the critical features of conference proposals vary across and within fields. Always check the organization's call for proposals before submitting.

In proposal abstracts, as in all the work you do as a scientist, hypotheses, methods, results, and implications must be logical and reasonable to professionals in your field, as well as appropriate to the conference. Since the abstract is the document conference organizers use to select the presentations to be given, it goes without saying that these abstracts must be exceedingly well written. The guidelines from the American Society for Microbiology make this absolutely clear.

## EXERCISE 5.1

In Chapter 13 we have reprinted the abstract of a talk given by Hale and his colleagues at the 103rd Annual Meeting of the Archeological Institute of America (page 424). Analyze the structure of this abstract. Does it include the standard IMRAD components? If so, where? How much attention does each component receive? How does the structure and logic of this abstract compare to the structure and logic of the journal articles you've read from this team (pages 406–409 and 410–417)? Are there any "promissory" moves in this abstract?

**EXERCISE 5.2**

Return to the list of conferences you developed while exploring your field in Chapter 2 and select one you'd be interested in attending. Obtain the conference guidelines and copies of sample abstracts from a researcher in your field or from the World Wide Web. Read over the conference instructions, and look at the logic and structure of the sample abstracts. Do these abstracts tend to include "promissory" statements? Are the abstracts overall more informative or descriptive? Describe the distinctive features of these abstracts, such as length, format, use of references, and acknowledgments of funding sources.

## 5.3 Organizing the Research Talk

It is important to recognize that the conference presentation, like its written counterpart the journal article, is primarily an act of persuasion, not simply of fact dissemination. As we noted in Chapter 3, when presenting results to other scientists, researchers seek to convince their audiences that their research questions are important, their methods sensibly chosen and carefully carried out, and their interpretations of results sound—in short, that their work represents a valid contribution to the developing knowledge of the field. Like the author of a journal article, a scientist presenting his or her research at a conference must build a convincing case for the project. The primary difference is that it must be done quickly—often within 10 minutes—and without the benefit of a formal written document in the "reader's" hand.

Because of the strict time constraints, scientists carefully plan oral presentations to ensure that the essential parts of the research argument are included. As in the written report, you will need to report your research questions, methods, results, and conclusions; but the amount of attention devoted to each component is somewhat different in the oral presentation. You will not have time for an extensive review of literature or a detailed description of your experimental methods. Instead, you will want to devote as much of your brief time as possible to reporting your new results and discussing their implications.

Thus, the first two components of the IMRAD form are rather abbreviated in the conference presentation. State the purpose and rationale for your study quickly but clearly at the start. The standard four-part strategy described in Chapter 3 provides a useful framework for conference paper introductions as well: announce the topic, briefly overview trends in previous research, explain the gap or question raised by that research, and state the purpose of your study. Though the first three of these moves will be abridged in the interest of time, the final move is often expanded in an oral presentation to more fully outline for the listening audience the research to be presented or the major ideas to be discussed. Some written reports also provide this sort of preview for readers (see Figure 3.5 on page 48), but it is particularly important in the conference setting because listeners cannot "turn back" later to the first page to see what your main point is. Thus, in addition to announcing the purpose of the study, introductions to oral presentations have the added burden of orienting the audience to the organization and scope of the presentation itself.

Carl Blackman, a biophysicist at the EPA's National Health and Environmental Effects Research Laboratory, told us an interesting story that highlights the importance of the orientation function of the introduction. Early in his career, Blackman discovered while preparing a conference talk that his data would take less time to describe than he had expected, leaving his presentation rather short. To fill the time, he finished his introduction by elaborating on the structure of the presentation itself, telling his listeners in some detail what he was going to talk about. After talking about the data, he also summarized what he had said. To his surprise, the president of the society hosting the conference and others attending his session told him that this presentation was one of the clearest presentations they had heard, and one of his best. Taking time to orient the audience can be well worth the effort.

After your introductory remarks, briefly outline your research design, identifying who or what was studied under what conditions (or where); but unless your research focuses on methodological issues, save the details for later. A slide or overhead transparency outlining the essential features of your design can be very effective in communicating this basic information quickly. A conference audience is not going to replicate an experiment on the basis of a talk alone, so listeners do not need the comprehensive description of materials and methods required in the written report. Interested listeners will ask for further information during the question-and-answer period after the session.

Spend most of your allotted time describing the results of your study, illustrating your presentation with carefully prepared visuals that clearly highlight the major trends in your results. If you are presenting a series of findings, it may be helpful to spend some time discussing the implications of each finding while the visual is still on the screen, rather than saving all your discussion comments for later. In any case, you will want to prepare your concluding comments with great care, as this is the part of your talk that is most likely to be remembered. Attention typically wanders during the course of a presentation (distractions, fatigue, and cogitation can all affect the attention span of your audience), but it returns when listeners sense the speaker is wrapping up (Olsen and Huckin 1991). Since the introduction and concluding sections of your talk are when you have your audience's greatest attention, take extra care in preparing these sections, and deliver them slowly and clearly. Stressing the importance of these sections, Gurak (2000) recommends actually memorizing an outline of your introduction and conclusion; you may also wish to write out these sections. We will discuss presentation options next.

## 5.4 Methods of Oral Presentation

Scientists employ a variety of oral presentation techniques, depending not only on the customary practices of their fields but also on such variables as the type of topic and its reliance on visual as opposed to verbal explication, the length of time allowed for the presentation, and the speaker's own experience, comfort level, and familiarity with the material. Physicist Stephen Reynolds, some of whose work is contained in Chapter 12, reported to us that he never writes out a talk. Instead, he practices his talks several times beforehand, giving careful thought to how the presentation will begin and end, and then he works from his overhead transparencies when presenting.

Researchers use prepared text and notes to varying degrees when presenting. Carl Blackman, the EPA biophysicist, prepared scripts for his talks early in his career. We've included an excerpt from one of his scripts in Figure 5.4. Notice that Blackman writes out his introduction, indicates which slides he will present in

---

### Introduction

For a number of years, we have studied the influence of electromagnetic fields, on the efflux of calcium ions from brain tissue, in vitro. Our aim has been to define all the characteristics of the field, that are essential for the effect to occur.

Today, I'll report on additional responses we have observed, that provide more clues, but do not in themselves provide any clear indication of a mechanism of action.

### Background

[slide #1] Procedure (Expose vs. Sham Expose)
[slide #2] Crawford Cell with Function Generator.
Sample orientation & load producing EM field.
[slide #3] Overview-Freq vs. Int. (White dots vs. Blue dots)
Tested only a few of the many possible combinations.
Have extended the frequency response at 15 Vrms/m to 510 Hz.

### Frequency Data

[slide #4] Differences 1–510 Hz
features: 1. includes previous data 1–120 Hz
2. repeats at 165, 180, & 405 Hz
3. basically, no apparent pattern

At this point, I will digress from a formal data presentation to a more intuitive viewing of the data. I don't claim that this view is rigorously correct, I'm just using it to help develop some crude hypotheses. Specifically, in order to distill some pattern in this frequency-response data, I've replotted it using a function that reflects both the magnitude of the difference between the exposed and the sham tissues, and also the variance of each group.

[slide #5] Log P vs Freq. (Note lines drawn at $P = 0.05$ and $p = 0.01$)
features: 1. one group of data at $p<0.01$
2. other data at $p<0.05$, which includes 60 and 180 Hz and 405 Hz

Remember, we must exercise caution, because I've imposed an interpretation on the data that was not established before the experiments began. These interpretations should only be used for hypothesis generation. Any hypothesis still must be tested. In addition, the step size in frequency is 15 Hz. A smaller step size might reveal more fine structure or in some way change this picture.

---

**FIGURE 5.4**    Excerpt from the script of a presentation for the Bioelectromagnetics Society (BEMS) by Blackman, Benane, and House (1985).

which order, and scripts some of his commentary on the slides. He believes that such scripts might be especially useful for young scientists, particularly those who do not teach and thus do not regularly practice speaking to groups.

Scripting sections of your talk beforehand can help you stay within the time limit and ensure the accuracy of your remarks. Blackman suggests that if your text is written out beforehand and rehearsed, your presentation may be better organized and more accurate, subtle, and polished in wording and argument, even if you don't read from the written script when you present. He cautions against simply reading a paper to an audience, however. When notes and texts are used, they serve as an aid or a prompt for the presenter, not as a substitute for careful practice and preparation. Blackman now prepares his talks by drafting detailed outlines, which he then condenses and incorporates into overheads or slides, which then serve as notes for a more extemporaneous presentation.

Title:
"The interaction of static and time-varying magnetic fields with biological systems"

I. Introduction
   A. Objective
      Identify field conditions responsible for changes in biological endpoints, NOT to examine physiological significance of changes, nor whether there are health implications.
   B. Question
      Can EMF cause changes in biological systems that were not due to heating? If so, how did this happen?

II. Research History
   A. Identification of Non-Thermal <u>Radiofrequency</u> Field Effects
      1. Biological preparation—excised chick hemisphere in test tube
      2. Frequency tuning curve, "window"—Bawin 1975
      3. Intensity window—Blackman 1979
      4. Predictions, different carrier frequencies—Blackman & Joines 1980
   B. Influence of Various Exposure Parameters. To simplify emphasize ELF.
      1. Intensity
         a. One intensity window, but few points—Bawin 1976
         b. Two intensity windows—Blackman 1982
         c. <u>RF</u> shows multiple intensity windows—Blackman 1989
         d. Low intensity ELF—Blackman 1989
      2. Frequency
         a. Frequency series (1—510 Hz)—Blackman 1988
         b. Islands of intensity and frequency—Blackman 1985a
      3. Static (DC) Magnetic Field
         a. Bdc assigns effective frequency—Blackman 1985b
         b. Orientation dependence of Bac/Bdc—Blackman 1990

**FIGURE 5.5**    Outline of talk prepared by biophysicist Carl Blackman for a conference on Scientific and Technical Foundations for Therapeutic MRI, sponsored by the Ohlendorf Foundation (1994).

## EXERCISE 5.3

In Figures 5.5 and 5.6 we have reprinted an outline, as well as the opening set of slides, Carl Blackman prepared for a 1994 presentation at a meeting sponsored by the Ohlendorf Foundation, an organization that supports biomedical research on the use of electromagnetic fields. Compare Blackman's outline with the overhead transparencies from which he presented. How has material in the outline been reworked into slides (that is, what has been changed, reworded, added)? Which version has more detail? Why? How difficult would it be to deliver the presentation without the slides? What is the value of preparing an outline to begin with?

4. Other Salient Factors
    a. Prior embryo exposure to E field affects tissue test—Blackman 1988
    b. Temperature history of tissue—Blackman 1991

III. Hypotheses—Mechanistic Basis for Response
  1. Initial transduction of EM signal into physical chemical change
    a. NMR/EPR-like signature in data (405; 15–315 Hz)
    b. Cyclotron resonance also possible (60, 90 & 180 Hz)
      —Liboff developed ICR for frequency dependence
      —Lednev added concept of intensity dependence
      —Models incomplete
  2. Amplification of initial effect by biological system
    a. Biological systems poised at instability point, like high diver
  3. Expression
    a. Observable—perhaps far removed from initial transduction

IV. New Biological System
  1. Requirements
    a. more direct physiological relevance
    b. standard, well-established and widely used assay
  2. Essential features
    a. NGF ->NO
  3. Results
    a. Dose response to B—Blackman 1993
    b. Induced E field not involved—Blackman 1993
    c. Frequency dependence—just like chick brain (in review)
    d. DC magnetic field influences results (unpublished)
  4. Consequences—collaboration w/JPB

**FIGURE 5.5**    *(continued)*

(1)

Interaction of Static and Time-Varying
Magnetic Fields with Biological Systems

Carl F. Blackman
Health Effects Research Laboratory
US Environmental Protection Agency

Presented at

"Scientific and Technical Foundations for Therapeutic MRI"
27-29 January 1994
Margarita Island, Venezuela

(2)

Opinions Voiced

during this talk are my own

and do not necessarily represent those of the

US Environmental Protection Agency

(3)

**Outline**

Objectives of Research

Results with Model System

Theoretical Considerations

New Biological System

(4)

Objective of Research

**Identify & Characterize**
EMF field conditions that cause biological change

**NOT**
to determine physiological significance, or
examine health implications

(5)

Question 1

Is there a non-thermally based
biological response to RFR exposure?

(6)

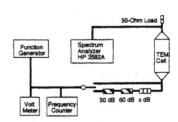

**FIGURE 5.6**     Slides prepared by Carl Blackman for the Ohlendorf Foundation meeting (1994).

(7)

## Biological Preparation
### Chick Brain

Label    Forebraín, separated at midline;
        37 °C, 30 min; Ca ions, sugars

Rinse    to remove loosely associated label

Treat    same salt solution w/o label; 20 min

Assay    aliquot counted

Analyze   compare exposed to sham-treated

(8)

## Early Experimental Results

• Frequency Tuning Curve - Bawin 1975

• Intensity Windows - Blackman 1979

• Carrier Frequency Influence - Joines/Blackman 1980

(9)

## Question 1
• Is there a non-thermally based biological
response to RFR exposure?

## Answer 1
• Yes, a reproducible response exists.

## Question 2
• How does this happen?

(10)

## Radiobiological Approach
### Identify Critical Field Parameters

DOSE RESPONSE
Kinetics -> Mechanism of Action

FREQUENCY
Action Spectrum -> Site of Action

(11)

## Dose Response

• One intensity window - Bawin/Blackman

• Two intensity windows in ELF

• RF shows multiple intensity windows

• Low intensity ELF

(12)

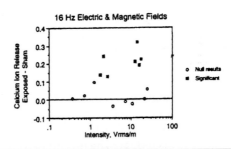

**FIGURE 5.6** *(continued)*

Both Blackman and Reynolds, then, work out their presentations ahead of time. Blackman develops a written outline; Reynolds rehearses from his slides. And both present from overheads or slides. The degree to which a presenter uses these strategies depends on his or her personality and confidence, according to Blackman, as well as on the audience, purpose, and conventions of the discipline. However, Blackman adds, "Students should not feel something is lacking if they can't do a presentation from overheads or slides; it is a skill that needs to be developed." Relying on prepared scripts can be an effective interim measure.

In actual practice, conference presenters usually combine three presentation methods: memorizing, reading, and speaking extemporaneously (Ehninger et al. 1988).

**Memorizing** can be useful for *short* presentations for which time limits are severe (Reif-Lehrer 1990): if a speaker sticks to the script and does not ad lib, memorizing will allow him or her to gauge the length of the talk fairly accurately; memorizing also may allow for continuous eye contact with the audience when presenting. Memorizing takes time, however, and a memorized delivery is often stilted and stiff; there also is the danger that if a speaker forgets one part, the rest of the presentation may unravel as well. Presenters rarely work entirely from memory, using it instead to help deliver short pieces of their talk or to make transitions from one part of a presentation to another.

**Reading** from a full text is rare at scientific conferences, as noted earlier, but speakers may prepare and read particularly critical or intricate parts of their presentation to facilitate clarity, coherence, and timing. The danger with the reading strategy is that it may hinder the speaker's ability to look at the audience. There is also the danger when reading a prepared text that the speaker's delivery will be flat and that the presentation will lack energy and variety. Scientists reading parts of their presentation must be very aware of these tendencies and make a special effort to maintain eye contact with the audience. Experienced speakers look up between sentences or paragraphs, as well as at important or strategic points that they want to emphasize in the text. Good speakers, even when reading, also vary their rate of speed, pitch, and intonation pattern to enliven their speech and hold listeners' attention.

**Speaking extemporaneously** from notes or slides is the most frequently used method of presenting at scientific conferences. Speaking extemporaneously is not the same as speaking impromptu or "off the cuff," which is often required after a formal presentation during the question-and-answer period. To "extemporize" means to elaborate, usually on written notes or graphics. As the examples of both Blackman and Reynolds demonstrate, professional scientists spend time thinking about and preparing visuals and notes to elaborate on during their presentations. (In fact, Blackman goes so far as to prepare extra slides to handle questions that may arise after his talk. Blackman anticipates his audience's reaction, considers his answers and arguments, and creates "rebuttal graphics" that will provide backup illustrations and evidence; he does not present this material unless he is asked, at which time he can extemporize rather than speak impromptu because he is better prepared.)

As we have seen, extemporaneous presenters may choose to prepare a formal introduction and conclusion, but they will usually speak only from slides or overhead transparencies or from an outline of the main points and supporting sub-points. Notecards, outlines, text, or graphics can be used as "cue cards" to organize a talk and keep the speaker from rambling or straying too far from the subject. Thus, the extemporaneous presentation is a combination of reading, speaking impromptu, and perhaps to some extent memorizing, and it shares some of their advantages and disadvantages. In an extemporaneous presentation important parts may be read so that they can be worded carefully; the discussion of the visuals may be part impromptu and part memorized, while the general structure and movement of the presentation may be both in outline form and in memory. The result is a presentation that appears spontaneous but is actually based on a careful structure worked out beforehand. Extemporaneous talks thus have all the energy of a spontaneous talk but are clearer, are better organized, and fit within the time allotted. To speak extemporaneously, Blackman adds, "you have to reach a comfort zone; to get there, you might need notes, crib sheets, etc."

No matter what method of presentation is used, both Blackman and Reynolds advise that speakers time themselves. Practice your entire presentation several times at the speed you will deliver it.

## 5.5 Delivering Conference Presentations

Although much about presenting orally is innate and physical and is influenced by factors such as the speaker's personality, voice, and physical appearance, presentation skills can be developed and improved through practice. In a well-known book on oral presentations, zoologist Robert Anholt asserts that a scientific presentation "stands or falls with delivery. . . . Delivery is important in establishing the impact a scientific presentation makes on an audience, and speaking skills can be determining factors in scientific careers" (1994, p 152). In addition to the content of the presentation, delivery speed, body language, eye contact, voice, gestures, and the clothes you wear all influence how effectively you communicate and how professional an *ethos* you create (see Gurak 2000).

Because papers are presented orally to listeners who do not have a text and are more distant than they would be in conversation, speakers need to deliver their presentation more slowly than they would normally speak if it is to be comprehensible. This is especially true if parts of the presentation are read. The normal rate of speech in conversation is probably between 200 and 250 words per minute; public speaking experts recommend that presentations to large audiences average 120 to 150 words per minute (Ehninger et al. 1988). Thus, for example, a 250-word typed page should take about two minutes to read. (To establish a good presentation speed, you can practice this yourself.) It is important to try to observe time limits: it is unfair and infuriating to other speakers to have someone use up a part of their time. (Imagine trying to adjust your carefully planned 20-minute presentation to fit the 12 minutes left to you by the previous speaker[s].)

Remember, presenting orally is a different experience from reading to yourself. Whether using notes, visuals, or text, it is much easier for you to lose your place when addressing a live audience. In addition to any nervousness you may experience, you may lose your place as you extemporize or look up from reading. To make your notes, visuals, or text easier to "perform," double- or even triple-space them (as newscasters do). Pick a readable typeface. A sans-serif typeface such as New Courier is easier to read than more ornamented fonts such as Times Roman. Underline words in your notes or text that you want to emphasize, and "underline" them vocally with a change of volume or intonation when you deliver your presentation.

Unlike written presentations, oral presentations are very physical—very much rooted in the body. It is the person (not a piece of paper) who presents, the person who is the focus of attention. Thus, in delivering a conference paper, the speaker becomes a part of the presentation; and except for situations involving broadcast, the audience is live and present too. A good speaker is aware of the reactions and needs of his or her audience throughout the presentation.

One of the most important ways of doing this is to maintain eye contact with the audience. Blackman notes that, in his experience, eye contact does not occur naturally but rather comes with verbal flexibility, must be cultivated, and requires practice. There are at least four reasons to maintain eye contact with the audience during a presentation (especially when reading from notes, visuals, or text). The first is to monitor the audience: Can they hear you? Can they see the graphics? Are they following you? Do they agree with you? Do they need a point explained? The second reason is more subtle: when speakers periodically make direct eye contact with individuals, audience members are less likely to let their attention wander; psychologically, they know the speaker is watching and so will want to pay closer attention. Third, by looking at people directly, speakers establish a good rapport with them—a personal relationship of sorts, which is very important in scientific oral presentations (Anholt 1994). Finally, eye contact helps establish the speaker's sincerity and belief in what he or she is saying (Can the speaker "look them in the eye" and say it?). In other words, eye contact helps establish the speaker's credibility, an important component of the scientist's *ethos* or professional character.

Of course, in oral presentations the voice is the medium of communication. Even if you use visual aids, the bulk of the presentation will be carried by your voice. Just as the typeface has to be dark enough and the manuscript free of errors to render the text as readable as possible, so your voice has to be loud enough and clear to render your words audible. There are several variables of voice that can be improved with practice: volume, rate, and pitch (Ehninger et al. 1988; Anholt 1994).

**Volume** is crucial even with electronic amplification, for a microphone can pick up only what's coming out of the speaker's mouth. The worst presentation is a mumbled one, essentially because there is no presentation; no one can hear what is being said, even if it's brilliant. You must speak loudly enough so that those at the very back of the room can hear you. You can do a sound check before you begin your presentation by asking people in the back of the room if they can hear you.

Voice is physical, and so it is difficult to change, but it can be improved. If you are someone who naturally speaks softly, try opening up your entire vocal passage and work on pushing air up from your diaphragm as you speak; think of your whole body as a hollow instrument, and let your voice resonate in it. Sometimes people speak softly or choke up because they are nervous; we all have noticed how people get louder as a presentation goes on and nerves subside and the speaker gathers more confidence. Breathing helps. So does a glass of water. Try to channel and use nervous energy to increase your volume and your presence, rather than letting it stiffen or choke you up. If you are nervous, look into the eyes of audience members in different parts of the room. No one wants you to fail, for they are more or less a captive audience and will have to suffer with you.

In addition to volume, the **rate** or speed of delivery is important in oral presentations. Again, since a speaker's voice is the text, the rate at which a talk is delivered should not be so fast that listeners will not be able to follow or keep up with it. Nor should it be so slow that listeners will fall asleep. An intelligible, but lively pace (150 words per minute), with some variation, is best. Some speakers tend to speed up when they're nervous. Taking a deep breath before every sentence will help you both slow down and calm down—and will lend a clarity and dignity to your presentation that will do much for your confidence if you are nervous!

**Pitch,** the musical tone of your voice, also is important in an oral presentation (Anholt 1994). Sometimes people start speaking at a higher pitch than they normally talk because they are nervous, and then find they are uncomfortable with that pitch but are too embarrassed to change it; find a starting pitch that is natural or comfortable for you. Also pay particular attention to your intonation pattern, the musical scale, of your voice. Variety of pitch (as opposed to a droning monotone) keeps your presentation pleasant, lively, interesting, and intelligible. Intonation can also be used to highlight your key points. In fact, just as volume can be used to underline key words, phrases, or concepts, so too pitch can be used for emphasis.

**EXERCISE 5.4**

For this exercise choose something that you have written, or a sample text from this book. For example, the Huyghe (1993) article in Chapter 11 (pages 293–298) would work well.

1. Read the first few paragraphs to yourself, listening to the volume, rate of delivery, and intonation pattern in your mind. Now read the paragraph out loud. Do the volume, speed of delivery, and intonation pattern best capture and express the nuances of meaning in this paragraph? Try delivering the paragraph again, varying the volume, the rate of delivery, and intonation pattern as needed, until your delivery conveys most accurately the nuances of meaning. How does the voice help communicate meaning? Deliver this paragraph out loud a few times to practice volume, rate, and pitch.

2. Now listen to someone else read the same passage, and compare your presentations. Where did he or she alter pitch, volume, or intonation? What different decisions did you make in presenting the same passage?

Just as the voice is a verbal text, the body is a visual element. Body stance and movement, facial expressions, and hand gestures, whether intentional or not, all become part of the message being communicated and in fact often reveal a speaker's feelings toward his or her subject and audience (Ehninger et al. 1988).

How do you feel about someone when they can't look you in the eye? If they hang their head and hunch their shoulders as they speak? If they keep jingling change in their pockets? If they keep twiddling their thumbs? If they have a pained expression on their face as they talk? In addition to being distracting, this kind of body language reveals much about a speaker's attitude and so becomes a part of his or her message—part of the speaker's *ethos*.

However, body language, like voice, can also be used to subtly communicate parts of the talk (Ehninger et al. 1988). Purposeful body movements can visually communicate a change in the focus or direction of a discussion, as when a speaker shifts weight or changes position as he or she moves to a new topic. Hand gestures too can help a speaker emphasize key points in a presentation. While hand gestures are generally toned down in scientific presentations, they can be useful, in moderation, to reinforce important points.

Your body is a natural visual, and it is always with you. Your body is part of your professional character. There's no getting around it: an oral presentation is a self-presentation.

# 5.6 The Use of Graphics in Oral Presentations

It has long been generally assumed that "two channels" working together can be more effective than one channel in getting a message across and retained (e.g., Hsia 1977). Therefore most speakers try to complement the spoken word with graphics. According to Anholt (1994), graphics are the heart of a scientific presentation. In Chapter 3 we discussed some of the more conventional graphics used in research papers; in Chapter 8 we will talk about those used by scientists to communicate with general audiences. In this section we will focus on conventions for using graphics in conference presentations.

The kinds of graphics used to represent trends or summarize results in written reports are also used in conference presentations. The difference between the two lies mainly in the amount of detail that can be accommodated in each medium. While many scientists simply reproduce the visuals from their text for use on an overhead or slide projector at a conference, visuals that are too "busy" are distracting and at worst confusing (Day 1998). If you believe this might be the case with one of your graphics, consider simplifying or redesigning it explicitly for the purpose and audience of your oral presentation. For example, turn to Figure 1 in the research report by Mallin et al. (1995) in Chapter 11 (page 334). Readers of this report can distinguish among the four curves on the graph by examining the figure closely, but the curves may not be as easily distinguishable when projected on a screen. If Mallin were preparing a slide from this figure, he could use a number of devices to clarify this figure for an oral presentation. He could eliminate the standard-error bars or de-emphasize them by putting them in a fainter color. He could draw the curves of the graph in different colors to make the distinctions among groups more obvious. He could indicate group means with symbols that are more easily distinguished from each other at a distance than are those in the printed graph.

In Chapter 3, we also discussed how graphics must be both independent (clearly labeled) and interdependent (referenced in the text). The same applies to a graphic in an oral presentation. A graphic without a label will be less comprehensible, especially if someone in the audience misses your verbal reference to it. Labeling all your graphics and making the letters large enough to be readable from the back of the room can solve this problem.

At the same time, be sure your visuals and your talk are interdependent. Even a labeled graphic should not be displayed without explanation. The time and attention the audience spends trying to figure out how the visual relates to what you're saying is time and attention taken away from your presentation. Take advantage of the opportunity to interact with your visuals, an opportunity unique to the oral presentation mode. Present your visual at the appropriate time in your discussion, refer to it directly ("Here are the results from the two trials"), and as you proceed, point to significant features ("As you can see in the final column, temperatures in the second trial were substantially higher than in the first"). Take care not to block the projection screen when you work with your visuals. If you are not able to highlight the image electronically, stand clear of the screen and use a pointer or a lightpen, always trying to face the audience.

The effective use of graphics is an adventure with many hidden pitfalls. But like conference presentations themselves, graphics can be improved by an awareness of the pitfalls and with practice. If they are designed well and are appropriate, graphics can make complex information visually comprehensible at a glance.

## 5.7 Preparing Research Posters

As noted at the start of this chapter, the research poster is an increasingly popular mode of presentation in the sciences. In addition to the logistical advantages for conference organizers, poster sessions offer presenters a chance to meet and talk one-on-one with scientists who are specifically interested in their research topic. In a poster session, presenters can field questions and provide clarification for their audience directly and immediately in a less formal setting than in the oral presentation format. The size of the audience and the effectiveness of poster sessions depend to a large extent on how the sessions are designed by conference organizers. If posters are displayed in an out-of-the-way room or hallway, attendance can be sparse, but if they are prominently displayed and easily accessible between speaking sessions, poster sessions can offer broad exposure. Steve Reynolds reports that at his physics and astronomy conferences posters are often displayed for one or two days in a central location; presenters thus are able to reach a much larger proportion of the attendees at a given meeting than are likely to be reached through a 10-minute talk in a small, remote meeting room.

Poster titles are listed in conference programs, and authors typically stand by their posters to be available for questions and discussion with conference attendees who have come to find out about their work. ASM (2003) requires that presenters be present with their posters on an assigned day as a condition of acceptance. At some conferences it is conventional to provide a written abstract or summary for "listeners" to take away with them, often including important figures and tables and always including contact information to facilitate further dialogue with the research team.

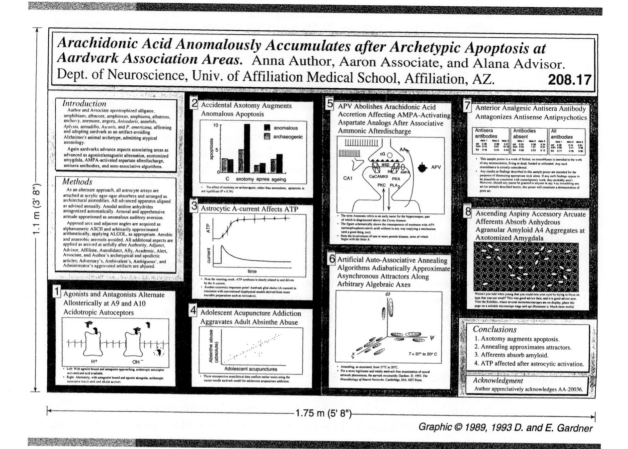

**FIGURE 5.7** Model poster developed by D. and E. Gardner for the Society for Neuroscience (1994) as a guide for presenters.

The size and format of posters vary across conferences and are constrained by the space and facilities available at the conference site. Always check for guidelines before you begin designing your poster. Most organizations provide easels or bulletin-board space and specify dimensions for the poster itself. Steve Reynolds reports that in the astronomical community each presenter receives about 1 square meter of bulletin-board space. The Society for Neuroscience allows a bit more room, specifying that materials be mounted on a poster board that is 5 feet, 8 inches (1.75 m) wide and 3 feet, 8 inches (1.1 m) high (Neuroscience 1994). In Figure 5.7 we have reprinted the sample poster that Daniel Gardner of Cornell Medical College and Esther P. Gardner of NYU School of Medicine developed for the Society's 1994 annual meeting.

In addition to guidelines provided by conference organizers, you'll want to adhere to general principles of effective page design when preparing your poster. As with any type of presentation, an aesthetically pleasing poster is more likely to communicate effectively (and to attract an audience in the first place) than one that is put together without regard to readers' needs. The sample poster in Figure 5.7 illustrates several important design principles:

■ Arrange the presentation in columns within the poster or allotted bulletin-board space. A columnar arrangement is congruent with natural reading and viewing patterns, which proceed from top to bottom and from left to right (Dondis 1973). Use ample space between "blocks" and be sure the vertical and horizontal margins clearly indicate the path readers are intended to take (Day 1998). The Society for Neuroscience recommends placing an introduction in the top-left position and a conclusion in the bottom-right portion of the poster, taking further advantage of natural reading preferences.

■ Use a large, readable typeface for titles and headings. Day (1998) suggests the title should be readable from a distance of 10 feet; the Society for Neuroscience (1994) requires letters at least 1-inch high for the main title, authors' names, and authors' affiliations. Highly visible titles ensure that conference goers can easily discern whether the project is relevant to their own research interests and thus worth a closer look.

■ Clearly label figures, tables, and other graphics. Figures should, of course, be uncluttered and easy to read, but they will not need to be simplified for the poster as much as is needed in the oral presentation, because poster audiences can take as much time as they need to examine figures closely.

■ Avoid large blocks of uninterrupted text. Use bulleted lists where appropriate, for example, in listing major conclusions.

■ If previous research is cited on the poster, include a list of works cited at the end.

■ Finally, color and variety are appropriate and desirable in posters (Day 1998), but don't get carried away. The poster should be readable and informative, not purely decorative. Substance and clarity impress professional readers more than glitz. The Society for Neuroscience guidelines recommend the use of background colors or shades as organizing elements to underscore connections between closely related portions of the presentation and to distinguish among major sections.

In Figure 5.8, we have reproduced in black and white a poster by Burkholder et al., which illustrates these principles of poster design. Although you cannot see their effective use of color, note the easy-to-read arrangement of the poster in columns, the use of smaller blocks of text within the columns, the clearly labeled sections and graphics, and the placement of the list of references, to create a visually balanced and appealing poster that will attract viewers from a distance and invite them to read as they get closer.

As with all forms of presentation, you will want to notice the distinctive features of research posters in your field. Look around your department for sample posters that faculty and graduate students may have on display in offices or hallways. What types of visuals are typical? What kinds of information are included in captions? How much color is used? How much variation is there in poster size and shape? What is the ratio of visual to verbal information?

The poster presentation is in effect a hybrid form of communication, combining features of both oral and written modes. The poster session offers the immediacy of one-to-one interaction between the researcher and his or her audience—as listeners—but also provides a carefully structured formal presentation that the same audience—as readers—can peruse at their own pace.

# THE LIFE CYCLES OF TOXIC *PFIESTERIA* SPECIES AND OTHER ESTUARINE DINOFLAGELLATES: TOWARD VERIFICATION OF PFIESTER'S HYPOTHESIS

J.M. Burkholder, Ph.D; H.B. Glasgow, Ph.D; J. Springer, M.S.; and M.W. Parrow, B.S.

Center for Applied Aquatic Ecology, North Carolina State University, Raleigh, NC 27606

## ABSTRACT

After working many years to contribute among the most elegant research published on dinoflagellates, the late Dr. Lois Pfiester hypothesized that "many if not all dinoflagellates will be found to have amoeboid stages." Research previously confirmed (1916- early 1970s) dinoflagellates with amoeboid stages in marine waters. Given that dinoflagellates with amoeboid stages were known from freshwater and marine habitats, we expected that species with amoeboid stages also would be found in estuaries. During the early 1990s we described transformations from isolated cells and isolated populations (clones) of *P. piscicida* from zoospores to distinct filose, lobose, and rhizopodial amoebae (each of which can be maintained in culture for weeks to months – thus, these are not merely transitional forms; cross-confirmed by 3 other laboratories with SEM and molecular probes) in response to changing prey availability, temperature, salinity, and other factors. Since the discovery of *P. piscicida* as the first estuarine dinoflagellate with a complex life cycle and amoeboid stages (length 5-750 μm), two other *Pfiesteria* species have been described with amoeboid stages (Landsberg et al. 1995, Glasgow et al. 2000), at least 1 of which is toxic (*P. shumwayae* sp. nov.); and several species of thus-far-benign *Cryptoperidiniopsis* gen. nov. have been documented to have amoeboid stages (Marshall et al. accepted, Steidinger and coworkers pers. comm.). Thus, increasing evidence over a relatively short time provides support for Dr. Pfiester's hypothesis, and support for the hypothesis that many other estuarine dinoflagellates will be found with complex life cycles and amoeboid stages.

## MATERIALS AND METHODS

**Clonal cultures** of amoebae and zoospores (routinely isolated and cleaned using flow cytometric procedures) originated from estuarine locations in mid-Atlantic and southeastern U.S. estuaries, mostly the Albemarle-Pamlico Estuarine System in NC and Chesapeake Bay in MD. Cultures were maintained on a diet of *Cryptomonas* sp. in an environmental incubator (15 ppt, 21 °C, 14:10 L:D). Light microscopic observations were made using an Olympus AX-70 research microscope equipped with water immersion optics under Nomarski DIC. SEM was completed with a JEOL 505T at 15kV, following the protocols of Burkholder & Glasgow (1995) and Glasgow et al. (2000).

An 18s rDNA-based PCR protocol was used to identify whether cultured isolates were *P. piscicida, P. shumwayae,* or *Cryptoperidiniopsis brodyi.* Alexa Fluor labelled *in-situ* hybridization (FISH) probes based on the 18s rDNA PCR primer pairs were applied to enable fluorescent identification of the above species. Epifluorescence (light microscope equipped with the appropriate band-specific filters -- 350, 488, 532 nm) was used to image labeled cells.

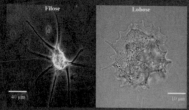

**Figure 1.** Representative amoeboid stages commonly observed under light microscopy in clonal cultures of estuarine dinoflagellates. The amoeboid stages that we have reported have also been separately maintained for at least 4 wk. in clonal culture; that is, they are not merely transitional forms.

## ACKNOWLEDGMENTS

We thank the National Science Foundation, the U.S. EPA, the Z. Smith Reynolds Foundation, and an anonymous foundation for funding support. We would also like to thank members of the Center for Applied Aquatic Ecology for the provision of technical assistance.

**Figure 2.** Scanning electron micrograph of a *P. piscicida* lobose amoeba adjacent to *P. piscicida* zoospores (Neuse clone).

**Table 1.** Examples of dinoflagellate species with multiphasic life cycles including amoeboid stage(s).

| Species | Habitat | Citation |
|---|---|---|
| *Cystodinium bataviense* (*Dinococcus oedogonii*) | Peat Bogs, Clay Pools (Central Europe, Java, USA) | Popovsky and Pfiester (1990) |
| *Dinamoebidinium coloradense* | Freshwater (alpine; amoeboid stage resembles *Amoeba verrucosa*) | Bursa (1970a) |
| *D. hyperboreum* | Marine (Arctic, coastal) | Bursa (1970b) |
| *Filodinium hovassei* | Marine (parasite of Appendicularians) | Cachon (1968) |
| *Gymnodinium austriacum* | Freshwater (oligotrophic lakes in Europe and N. America) | Popovsky & Pfiester (1990) |
| *Oodinium* spp | Predominantly marine (parasites of fish) | Popovsky & Pfiester (1990) |
| *Pfiesteria piscicida* | Estuarine (U.S. Atlantic, Gulf) | Burkholder et al. (1992) |
| *P. shumwayae* | Estuarine (similar distribution) | Glasgow et al. (2000) |
| *Piscinodinium*-like sp. | Freshwater (parasite of stickleback, *Gasterosteus* sp.) | Buckland-Nicks et al. (1990) |
| *Stylodinium sphaera* (= *Vampyrella pendula*) | Freshwater (parasitic on *Oedogonium* spp.) | Timpano & Pfiester (1986) |

## SELECTED REFERENCES

Buckland-Nicks, J.A., Reimchen, T.E. & F.J.R. Taylor (1990). A novel association between an endemic stickleback and a parasitic dinoflagellate. 2. Morphology and life cycle. *Journal of Phycology* 26: 539-548.

Burkholder, J.M. & H.B. Glasgow (1995) Interactions of a toxic estuarine dinoflagellate with microbial predators and prey. *Archives für Protistenkunde* 145: 177-188.

Bursa, A.S. (1970a). *Dinamoebidium coloradense* sp. nov. and *Katodinium auratum* sp. nov. in Como creek, Boulder County, Colorado. *Arctic and Alpine Research* 2: 145-151.

Bursa, A.S. (1970b). *Dinamoebidium hyperboreum* sp. nov. in coastal plankton off Ellesmere Island, N.W.T., Canada. *Arctic and Alpine Research* 2: 152-154.

Cachon, J.M. (1968). *Filodinium hovassei* gen. nov. sp. nov. *Protistologica* T. IV, facs.

Glasgow, H.B., Burkholder, J.M., Morton, S.L. and J. Springer (2000) A second species of ichthyotoxic *Pfiesteria* (Dinamoebales, Pyrrhophyta). *Phycologia.*

Popovsky, J. & L.A. Pfiester (1990) *Süβwasserflora von Mitteleuropa,* Eds., H. Ettl, J. Gerloff, H. Heynig, and D. Mollenhauer, Gustav Fischer Verlag, Stuttgart.

Steidinger, K.A., J.M. Burkholder, H.B. Glasgow, C.W. Hobbs, F. Truby, J. Garrett, E.J. Noga & S.A. Smith (1996) *Pfiesteria piscicida* gen. et sp. nov. (Pfiesteriaceae fam. nov.), a new toxic dinoflagellate genus and species with a complex life cycle and behavior. *J. Phycol.* 32: 157-164.

Timpano, P. & L.A. Pfiester (1986) Observations on *Vampyrella pendula-Stylodinium sphaera* and the ultrastructure of the reproductive cyst. *Am. J. Bot.* 73: 1341-1350.

## DISCUSSION

There is a general paucity of information on the distribution and ecology of amoebae in estuarine ecosystems. Various authors have hypothesized that amoebae, which are common among estuarine microfauna, play more important ecological roles in these ecosystems then previously conceived.

A number of freshwater and marine dinoflagellate species have been observed to have amoeboid stages. Dinoflagellates with amoeboid stages have been described from such diverse habitats as peat bogs, cold-temperate and sub-arctic coastal waters, and alpine streams. However, until the discovery of *Pfiesteria,* there was no information on estuarine dinoflagellates with amoeboid stages. *Pfiesteria* spp. were also the first toxic dinoflagellates with documented amoeboid stages.

An intensive sampling effort has yielded dinoflagellate clones with amoeboid stages, including several 'cryptoperidiniopsoid' species, scrippsielloids, and gymnodinioids in addition to the two *Pfiesteria* species. These isolates were obtained from the Indian River Inland Bay of Delaware; eight tributaries of Chesapeake Bay in Maryland and Virginia; the Neuse, Pamlico, and New River Estuaries in North Carolina; and four major estuaries of Florida (east and west coasts). Common morphotypes of amoebae observed in these clonal cultures have included filose, heliozoan, lobose, and rhizopodial forms. These amoeboid stages commonly co-occur in clonal cultures with motile, zoospore stages of the dinoflagellate species.

The development of primers specific to the 18s rDNA of *P. piscicida, P. shumwayae,* and *Cryptoperidiniopsis brodyi* sp. nov. (Steidinger & coworkers) have enabled molecular diagnosis of these amoeboid stages. The data from PCR and fluorescent *in-situ* hybridization testing support the observations from clonal cultures (e.g., Burkholder & Glasgow 1995, Marshall et al. in press), and the transformations described in single cells by Steidinger et al. (1996, J. Phycol.) that three estuarine dinoflagellates each have amoeboid stages in their complex life cycle.

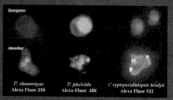

**Figure 3.** Application of fluorescent *in-situ* hybridization (FISH) probes toward the molecular identification of free-living amoeboid stages in clonal cultures of estuarine dinoflagellates.

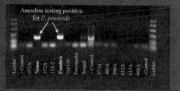

**Figure 4.** Gel electrophoresis of PCR products amplified using a 18s rDNA primer developed for the detection of *P. piscicida.* Assayed cultures included clonal isolates of amoebae collected from estuarine sites in Maryland and North Carolina.

**FIGURE 5.8**   Sample poster. Burkholder et al. The life cycles of toxic *Pfiesteria* species and other estuarine dinoflagellates: Toward verification of Pfiester's hypothesis. The CDC National Conference on Pfiesteria: From Biology to Public Health, 2000.

# 5.8 Conference Proceedings

As mentioned earlier in this chapter, conference proposal abstracts are often reproduced in print or electronic form prior to the meeting at which they will be presented. After the meeting, many organizations also publish selected "proceedings," which consist of extended summaries or even the complete texts of some or all of the presentations themselves. Check the conference guidelines to understand how papers will be selected and what the final form of the proceedings will be. Will you need to write up your presentation in the form of a conference paper after it is delivered? In effect, like posters, some conference proceedings are a hybrid document, for the paper intended to be delivered orally may wind up as a print or electronic text document. Often revision, as well as a full documentation of sources, is necessary for the publication or posting of a previously delivered paper in conference proceedings.

In Chapter 12 we have reprinted a proceedings article by the Reynolds team: "Maximum Energies of Shock-Accelerated Electrons in Supernova Remnants in the Large Magellanic Cloud" (Hendrick, Reynolds, & Borkowski 2001; see pages 390–393). Notice that this article generally follows the IMRAD form, but that like the oral presentation, less attention is paid to methodology; thus the methods section is abbreviated. Comparing this proceedings article to the journal article by the same title in the *Astrophysical Journal* (pages 394–399), you'll note that the journal article actually contains, in addition to the introduction, results, and discussion sections, a rather developed methods section under the title "2. Analysis Technique," and that this is divided into "2.1 The Sample," "2.2 The X-Ray Data," "2.3 The Models," and "2.4 The Spectral Fitting Methods." Note that the conference proceeding contains only three sections of the IMRAD form: "Introduction," "Data," and "Discussion." Although the same information about methods is described under "Data," this information is mentioned in passing, with a reference to the paper in *Astrophysical Journal* for a fuller discussion of the model the authors used and an equation at the end of the section. The "Data" section focuses mainly on results, which are also quickly summarized at the beginning of the discussion section by a second reference to Table 1. In fact, given the time constraints (of a presentation) and space constraints (of published proceedings), every section of the proceedings is abbreviated considerably.

**EXERCISE 5.5**

Compare each section of the published proceedings by Hendrick et al. with the paper in the *Astrophysical Journal*. In the proceedings, what information has been left out of the introduction, methods, results, and discussion sections? How were the authors able to edit these details out of the talk and the proceedings in a way that made sense; that is, what details do they focus on to retain or recreate in the proceedings the logic of their longer argument in the journal article.

# Activities and Assignments

1. Imagine it is 1987. You have been working as a research assistant for Allan Brody and Michael Pelton at the University of Tennessee. These two authors were at that time engaged in the research on black bears that they later reported in the *Canadian Journal of Zoology* (Brody and Pelton 1988). You've been asked to write a conference proposal abstract for this research, to be submitted to the International Conference on Bear Research and Management. Brody and Pelton's 1988 journal article is reprinted in Figure 3.11 (pages 71–73). Though the article would not yet have been published at the time of your employment with them, use it to help you understand the goals, methods, and outcomes of that research. Write a promissory abstract for the project.

2. A. Prepare and deliver to the class an oral presentation of a set of research findings in your field. You may choose to present the results of a study you have conducted in another class or at work. Or your instructor may suggest you practice by developing a presentation based on published research conducted by another research team. (In this case, identify the original researchers, and present yourself as a "representative" of that team.) Choose an appropriate professional conference for your presentation, and identify the target audience at the start of your presentation. Use either the extemporaneous or the reading method of presentation, whichever you are more comfortable with. Supplement and illustrate your presentation with carefully prepared visuals, using media available to you.

   B. As part of this assignment, you may be asked to serve as a respondent for one of the scheduled talks in your area of expertise. Your assigned speaker will give you a copy of his or her conference paper ahead of time so you can read it and prepare two or three follow-up questions to ask during a question-and-answer period after the presentation.

3. Use the criteria in Figure 5.9 to evaluate the effectiveness of in-class presentations. Even if you do not follow all the technical detail in presentations outside your research area, you should be able to tell from the discussion in this chapter which presentations were effective and why.

4. Prepare a poster presentation of a set of research findings in your field, following the topic guidelines outlined in Activity 2. As in that activity, identify an appropriate conference and audience for the poster session.

5. Develop a list of evaluative criteria to use in assessing the effectiveness of the poster presentations in your class. Using Figure 5.9 as a starting point, add and delete criteria as needed in order to prepare an effective and usable evaluation instrument. Your instructor may ask you to do this project in pairs or groups.

6. Use the criteria in Figure 5.9 and the poster criteria you developed in Activity 5 to evaluate presentations given at a professional conference that you attend. Which speakers gave the best presentations? Which posters were most readable? Which generated the most interest? Why were these presentations more effective than others you observed?

Rank your response to each question from 1 (low) to 5 (high). Beside each, write a brief statement explaining the reason for your ranking and/or the effect of that element on the presentation.

1. Was the type of presentation appropriate and effective? __
2. Was the length of the presentation appropriate? Were time limits adhered to? __
3. Was the rate (speed) of delivery effective? __
4. Did the introduction establish an adequate context? __
5. Was the presentation well organized? __
6. Was the conclusion effective in emphasizing the major point of the presentation? __
7. Was there sufficient (and direct) eye contact? __
8. Did the speaker:

   - Speak loudly enough? __
   - Vary pitch (intonation)? __
   - Use voice to emphasize important points? __
9. Was body language purposeful and appropriate? __
10. Were hand gestures meaningful and appropriate? __
11. Were visuals:

    - Appropriate (to the content, purpose, and method of presentation)? __
    - Independent (adequately labeled)? __
    - Interdependent (referenced and used in the presentation)? __
    - Well designed and professional in appearance? __
    - Informative? __
12. Was the medium for the visuals (PowerPoint slides, etc.) used effectively? __

**FIGURE 5.9** Criteria for evaluating conference presentations.

CHAPTER 6

# Writing Research Proposals

## 6.1 The Role of the Proposal in Science

In this book we have taken the position that scientific texts are not just informative but also persuasive documents. This persuasive dimension of scientific discourse is nowhere more obvious than in the research proposal. When writing reports and journal articles, researchers present and defend a particular interpretation of the prior research, of their new findings, and of the relationship between the two; the research report aims to convince readers that the work is valid and important. When submitting a grant proposal to a potential funding agency, however, the researchers must go a step further. Now they must convince readers not only that the work will be valid and important but also that the readers should pay for it!

In a classical rhetorical framework, the proposal may be classified primarily as a *deliberative argument*. Aristotle distinguished three types of argument: forensic, epideictic, and deliberative, which can be described as arguments of fact, of value, and of policy, respectively. Most types of writing contain elements of more than one argument type. In research reports the overriding goal is forensic: the authors try to convince readers to accept a set of "facts"—the results of their study. However, because the introduction and discussion sections assess the current state of knowledge, and argue at least implicitly about the value and limitations of that knowledge, the report has an epideictic dimension as well. And the discussion section's recommendations for future research or policy are deliberative. Similarly, the research review serves a forensic function in that it presents a summary of previous research, but its implicit goal of identifying those studies that are most pertinent or valuable to the field's understanding entails an epideictic dimension. The proposal is distinguished from both these genres in that, although it too contains forensic and epideictic elements, its purpose is primarily deliberative: the proposal argues for a specific research plan as well as a general research direction.

The stakes become clear when we think about the number of scientific questions that could potentially be explored; the amount of time, effort, and money that would be needed to support all these explorations; and the resources actually available for scientific research. The National Science Foundation funds roughly one-third of the 30,000 proposals it receives each year (NSF 2002). The success rate is similar at the National Institutes of Health, which funded 32 percent of the 26,000 proposals received in 1999 (NIH 2000). The proposal document plays a critical role in how these scientific resources are allocated.

Research proposals are written for a variety of audiences and purposes in science: they are submitted to funding agencies to solicit financial support for new research, to academic departments to request approval of dissertation projects, to research facilities to gain access to equipment and resources, and to other parties whose approval must be secured for research to proceed. For example, in Chapter 13 we have included the brief proposal that Jelle de Boer and John Hale sent to the Greek government to request permission to remove rock samples from the ancient site at Delphi (pages 402–403). Though the present chapter will focus primarily on proposals for research funding, the goal in addressing any of these proposal audiences is to convince the relevant gatekeepers that a particular problem is significant enough to justify the costs and consequences of exploring it, whether those costs be in terms of money, time, or risks to participants or resources. To succeed, a research proposal must present a well-reasoned and carefully documented argument that persuades its decision-making audience that the potential benefits of the research outweigh the costs.

## EXERCISE 6.1

Read De Boer and Hale's proposal to the Greek government to remove samples of stones from Delphi, and the reply from the Greek Ministry of Culture (pages 402–405). From the government's standpoint, what are the costs of allowing this research to proceed? What are the potential benefits? Where and how do De Boer and Hale address these costs and benefits in their short proposal? In what ways does the permit from the Ministry of Culture reflect its concern?

The U.S. Department of Energy's grant application guidelines summarize the content of the proposal as follows:

> The application should present the objectives and scientific significance of the proposed work; the rationale for selecting the proposed approach to achieve the objectives; qualifications of the principal investigator and the applicant organization; and amount of funding required. Since the application will compete with others on related topics . . . it should present the scientific merit of the proposed project clearly and

convincingly and should be prepared with the care and thoroughness of a paper submitted for publication. (DOE 2002)

This definition highlights the two primary persuasive goals of the research proposal:

- To convince your scientific audience that the problem you propose to investigate is important and worth exploring
- To convince them you will explore the problem in a sensible way

Thus your proposal must convince readers not only that the research problem is significant but that your research approach is likely to succeed. Proposal reviewers will need to ascertain whether the methods you propose to use represent the most efficient and most worthwhile use of their agency's resources. They also will need to determine whether you are well qualified for the job. Funding agencies will request a description of your track record as part of the grant application, usually in the form of a *curriculum vitae*, an "academic résumé" that lists your research training and experience. But this ancillary material comes later in the application package; it is not the primary focus of readers' attention as they read and evaluate your project description. The text of the proposal itself also must demonstrate that you know what you're doing.

Recall that a text creates a professional *ethos*; it projects a character. Readers will form an impression of you based on the competence revealed in your proposal: Does your review of research demonstrate a thorough knowledge of the field? Have you exercised good judgment in the design of the study and the choice of materials and procedures (and are you therefore likely to exercise good judgment in other phases of the study, such as in recording and interpreting data)? Does your description of methods demonstrate the technical competence needed to carry out the project effectively? Thus, while your proposal contains an explicit argument for the importance and validity of the study and its design, it also contains an implicit argument for your own research competence (Myers 1985). It is important to note that even superficial features of your presentation may enhance or detract from the professional *ethos* created in the proposal. The NIH cautions that "[y]our presentation can also make or break your application. Though reviewers assess science, they are also influenced by the writing and appearance of your application. If there are lots of typos and internal inconsistencies in the document, your score can suffer" (NIAID 2003b; http://www.niaid.nih.gov/ncn/grants/basics/basics_b4.htm).

**EXERCISE 6.2**

In Chapter 12 we have included a proposal that Reynolds, Borkowski, and Blondin (1994) submitted to NASA's Astrophysics Theory Program (pages 359–375). Read this proposal, paying special attention to the *ethos* it projects. What clues are there to the authors' experience and expertise? What have you learned about the researchers themselves from reading their proposal?

# 6.2 Multiple Audiences of the Proposal

Most granting agencies use peer review mechanisms that are similar in principle to, though typically more elaborate than, the peer review systems used by journal editors. Because most agencies fund research in many different areas and must allocate the agency's resources among these areas, the typical proposal document is reviewed by in-house "generalists" as well as outside specialists and thus must be comprehensible and persuasive to a broader range of readers than is typical for journal articles. In-house reviewers are program officers and other agency staff who read proposals from a number of related topic areas; though educated scientists, readers at this level are not likely to be specialists in the particular topic area of a given proposal. The in-house readers will solicit reviews from specialists in the scientific community, as many as three to ten in the case of the National Science Foundation (NSF 1995). The number of outside reviews requested may depend on the topic of the proposal and the expertise of the program staff, as NASA explains in their 1999 instructions for proposers (Figure 6.1).

The logistics of the proposal review process vary from agency to agency and sometimes within agencies as well, as Figure 6.1 illustrates. The NIH uses a two-stage review system (NIH 2003a) in which grant applications are first assigned to an appropriate Scientific Review Group (SRG) consisting of 18 to 20 outside experts. Some groups are standing committees, which meet three times a year to review proposals on a range of topics within a general area of interest to NIH; others are ad hoc Special Emphasis Panels selected for their expertise on specific topics (NIH 2001). In this first stage of review, panels are charged with evaluating proposals for scientific merit. In the second review stage, the SRG's evaluations are summarized and sent to the appropriate institute's advisory council, composed of about eight scientists and four lay people (Seiken 1992) who evaluate proposals in light of the institute's program goals and priorities. NIH's standing review groups include representation from a fairly wide range of medical fields,

## Evaluation Techniques

Selection decisions will be made following peer and/or scientific review of the proposals. Several evaluation techniques are regularly used within NASA. In all cases proposals are subject to scientific review by discipline specialists in the area of the proposal. Some proposals are reviewed entirely in-house, others are evaluated by a combination of in-house and selected external reviewers, while yet others are subject to the full external peer review technique (with due regard for conflict-of-interest and protection of proposal information), such as by mail or through assembled panels. The final decisions are made by a NASA selecting official. A proposal which is scientifically and programmatically meritorious, but not selected for award during its initial review, may be included in subsequent reviews unless the proposer requests otherwise.

**FIGURE 6.1**   Summary of NASA's proposal evaluation techniques. From Instructions for Responding to NASA Research Announcements for Solicited Basic Research Proposals (NASA 1999; NRA 99-OSS-04, Appendix B).

which further illustrates the importance of pitching a proposal at a more general level than is necessary for journal articles and other technical reports. As a case in point, the Allergy and Immunology SRG includes members representing departments of molecular immunology, medicine and rheumatology, pathology, microbiology, biochemistry, immunogenetics, and other areas (NIH 2003a). Given the size of these groups and the breadth of expertise represented, the National Institute of Allergy and Infectious Diseases (NIAID) urges researchers to keep in mind that their application "has two audiences: the majority of reviewers who will probably not be familiar with your techniques or field and a smaller number who are" (NIAID 2003a; http://www.niaid.nih.gov/ncn/grants/write/write_d1.htm).

In sum, research proposals will be read and evaluated by readers with varying types and degrees of experience and expertise, some of whom will know more about your research topic than will others. This multiplicity of audiences clearly complicates the writing process, for the proposal must address all these readers at once. Given the variety of audiences to be addressed and the multiple agendas to be accomplished, writing the research proposal is one of the most challenging—and most important—tasks scientists engage in.

---

**EXERCISE 6.3**

Choose a research topic in your field and identify a funding agency likely to support research in this area. (Review Exercise 4.2, page 87, on identifying topics.) Obtain a copy of the agency's proposal guidelines. In most cases, guidelines and calls for proposals will be posted online. For example, visit the NSF home page at http://www.nsf.gov/, the NASA funding site at http://research.hq.nasa.gov/research.cfm, or the Department of Energy at http://www.science.doe.gov/grants/. Once you've located the guidelines, find the section(s) that describe the review or evaluation process. Based on this information, write a paragraph describing the mixed audience for a proposal on your topic and your readers' likely areas of expertise. Follow this with a paragraph describing ways in which you might adapt your proposal for those readers.

---

## 6.3 Logic and Organization in the Research Proposal

Whether you are responding to a specific request for proposals or submitting an unsolicited proposal to a general program, your proposal must meet the specific guidelines established by your target funding agency. In addition to examining the main argument or project description, funding agencies will request an abstract or summary, a table of contents, a complete budget, biographical information about the investigator(s), and other supporting material pertinent to the type of research being proposed, for example, information about the treatment of laboratory animals, the protection of human subjects' rights and confidentiality, the handling of hazardous materials and other worker and environmental safety issues, or provisions for sharing research data. The scope, form, and ordering of these ancillary materials will vary across research fields and funding agencies.

The components of the main argument in a proposal are, however, fairly consistent. All proposals are designed to do the following:

- Introduce the purpose, significance, and specific objectives of the proposed research.
- Explain the background and rationale for the project by surveying previous research, summarizing the current state of the field's knowledge on the topic, and showing how the proposed project will further that knowledge.
- Describe the methodology to be used in the proposed study, and explain the rationale behind these methodological choices.

*Where* these goals are accomplished in the proposal document will differ from one proposal to another. The arrangement and labeling of sections and subsections may depend as much on the nature of your research topic or on your own writing preferences as it does on the funder's guidelines. For example, NSF and NASA use essentially the same language in their instructions for writing the main body of the proposal (Figure 6.2). But in responding to these same instructions in the early 1990s, the Burkholder team and the Reynolds team developed very different organizational plans. Figure 6.3 contains the table of contents for a proposal by Burkholder and Lewitus, written in response to the NSF instructions. Chapter 12 contains a proposal by Reynolds, Borkowski, and Blondin, written in response to the NASA guidelines.

**Instructions for Project Description, NASA 1999**

The main body of the proposal shall be a detailed statement of the work to be undertaken and should include objectives and expected significance, relation to the present state of knowledge in the field, and relation to previous work done on the project and to related work in progress elsewhere. The statement should outline the general plan of work, including the broad design of experiments to be undertaken and an adequate description of experimental methods and procedures. The project description should be prepared in a manner that addresses the evaluation factors in these instructions and any additional specific factors in the NRA [NASA Research Announcement].

**Instructions for Project Description, NSF 2002**

The main body of the proposal should be a detailed statement of the work to be undertaken and should include: objectives for the period of the proposed work and expected significance; relation to longer-term goals of the PI's project; and relation to the present state of knowledge in the field, to work in progress by the PI under other support, and to work in progress elsewhere.

The Project Description should outline the general plan of work, including the broad design of activities to be undertaken, and, where appropriate, provide a clear description of experimental methods and procedures and plans for preservation, documentation, and sharing of data, samples, physical collections, curriculum materials and other related research and education products.

**FIGURE 6.2** Instructions for project description from NASA (1999; NRA-99-OSS-04, Appendix B(c.4)) and NSF (2002; NSF 03-2, section II.C.2.d.i.).

# TABLE OF CONTENTS

**FIGURE 6.3**   Table of contents from proposal submitted to the NSF by JoAnn Burkholder and Alan Lewitus: Trophic interactions of ambush predator dinoflagellates in estuarine microbial food webs (Burkholder and Lewitus 1994).

## EXERCISE 6.4

Compare the table of contents (TOC) for the Burkholder and Lewitus (1994) NSF proposal, reproduced in Figure 6.3, with that for the NASA proposal by Reynolds, Borkowski, and Blondin (1994), included in Chapter 12 (pages 359–375). Aside from the obvious differences in level of detail (the TOC for the Reynolds et al. proposal omits subheadings; you'll have to find these in the text itself), how do these two organizational plans differ? Identify the distinctive features of each proposal in a list or brief paragraph. What do these distinctive features tell us about the nature of research in each area?

# 6.4 Introducing the Research Problem and Objectives

In some respects, the proposal introduction is similar in structure and content to the introductions found in most research reports. The four moves Swales (1984) observed in report introductions provide a useful heuristic for proposal introductions as well (see Figure 3.5, page 48). In both the report and the proposal, your goals are to introduce the research topic; summarize the current state of the field's knowledge on the topic; identify the gap, question, or problem that motivates the study; and announce the purpose of the study.

But within this general framework, emphasis and development may differ considerably in the two genres, due to the different audiences these texts address and the purposes for which they are written. In the research report, the introduction serves to quickly establish the context for the study and announce its purpose. Unless you are writing for a multidisciplinary journal such as *Nature* or *Science,* the journal audience consists of specialists in your field, and the stance is one of expert to expert; you are reminding fellow marine biologists or physicists or pathologists of the state of the field's knowledge on your topic, showing them which specific issues or previous findings you consider most pertinent to your study, and highlighting the gap your study responds to. In other words, you are guiding them through a review of information that is at least generally familiar to them, refocusing the discussion to situate the new results you are reporting. Because most of your report audience shares your specialized knowledge to some extent, this review can be accomplished rather quickly, sometimes in a single paragraph.

The proposal, however, carries a greater burden of proof and must therefore provide a more fully elaborated introduction to the project at hand. We noted in Section 6.2 that proposal readers are a diverse group, including specialists in your field and "generalists" with other areas of expertise. The significance of your line of research will not be obvious to all the members of this group; therefore you will need to make a stronger case than would be necessary in a journal article. Some of your proposal readers will also need to be educated about the general topic before they can make an informed judgment about the merits of your project. Lastly, keep in mind that you are asking more of your proposal readers than of a journal audience—you're asking them not just to entertain your ideas but to invest in them.

Thus, one of the primary goals of the proposal is to convince this audience of the significance of the proposed work. It will not be enough to assert that a problem exists, or that a question has not been answered. The researcher must convince proposal reviewers that the problem is important enough to spend money on, that answering the question will be worthwhile. What readers consider "worthwhile" will be largely a function of the goals and priorities of the agencies they represent. Just as research journals differ with respect to the topics and types of research they are interested in publishing (see Chapter 3), funding agencies also have distinctive research agendas and preferences. The first step in writing a successful proposal is to choose an appropriate agency to submit it to.

A funding agency's general priorities are well known to experienced researchers in the field, but they are also explicitly stated in the agency's proposal guidelines or requests for proposals (RFPs). Most funders accept proposals in their areas of interest on an ongoing basis but also issue RFPs to solicit proposals on particular topics for which they have set aside special funds. (The NIH uses the term *RFA*: request for applications; NASA uses *NRA*: NASA Research Announcement.) RFPs typically announce a new initiative in a carefully delimited research area. Researchers responding to RFPs must demonstrate that their proposed research fits within these parameters and will significantly further the announced goals. For example, the Reynolds team's proposal in Chapter 12 was submitted in response to an NRA soliciting proposals for "Theory in Space Astrophysics." The NRA was intended to encourage basic theoretical work needed for NASA's Space Astrophysics Program. Accordingly, the program description stipulates that "the proposed studies should facilitate the interpretation of existing data from Space Astrophysics missions or should lead to predictions which can be tested with Space Astrophysics observations" (NASA 1993, p A-1). This goal is implied throughout the Reynolds team's proposal and is directly addressed in a summary paragraph toward the end of the introduction:

> The proposed research is timely—it is new observations of SNRs with the ASCA satellite, and to a lesser degree with the ROSAT satellite, which have been the primary cause of the current renaissance of SNR studies. We bring to this field the only tool capable of addressing the spectral imaging data being collected by the ASCA satellite, and one well able to meet the theoretical challenge of the AXAF mission. (Reynolds, Borkowski, and Blondin 1994; page 362)

## EXERCISE 6.5

Read the Reynolds team's (1994) proposal to NASA and the accompanying program description, both reprinted in Chapter 12 (pages 358–375). Examine the other criteria mentioned in the program description. How and where are each of these issues addressed in the proposal?

As illustrated in the example just quoted, successful proposal authors leave nothing to chance. In discussing the significance of their research, they carefully highlight the specific ways in which the proposed work will further the goals and interests of their target funding agency. In so doing, the researchers are appealing to the agency's *values*. The appeal to values is one type of classical rhetorical argument. Aristotle described three types of rhetorical appeals: those based on the logic of the subject matter (*logos*), those based on the character of the speaker (*ethos*), and those based on the emotions and values of the audience (*pathos*). Most arguments contain all three types of appeal, and each can clearly be seen in the proposal genre: successful proposals present a logical, well-supported line of reasoning; project a professional *ethos* of competence and knowledgeableness; and clearly address the values and concerns of the funding agency.

These values are both articulated and implied in the RFP or proposal guide. Each agency has a fairly well-defined research domain that gives it a distinctive character or identity. For instance, the NSF was initially established in 1950 "to promote the progress of science; [and] to advance the national health, prosperity, and welfare by supporting research and education in all fields of science and engineering" (NSF 2002, p 2). This concern with the nation's science infrastructure and education is revealed in a number of critical places in the NSF Grant Proposal Guide, most notably in the discussion of review criteria. NSF's two primary review criteria, approved in 1997, place as much emphasis on the development of infrastructure and dissemination of results as on the intellectual merit of the proposed research (see Figure 6.4).

## NSF Merit Review Criteria

### What is the intellectual merit of the proposed activity?

How important is the proposed activity to advancing knowledge and understanding within its own field or across different fields? How well qualified is the proposer (individual or team) to conduct the project? . . . To what extent does the proposed activity suggest and explore creative and original concepts? How well conceived and organized is the proposed activity? Is there sufficient access to resources?

### What are the broader impacts of the proposed activity?

How well does the activity advance discovery and understanding while promoting teaching, training, and learning? How well does the proposed activity broaden the participation of underrepresented groups (e.g., gender, ethnicity, disability, geographic, etc.)? To what extent will it enhance the infrastructure for research and education, such as facilities, instrumentation, networks, and partnerships? Will the results be disseminated broadly to enhance scientific and technological understanding? What may be the benefits of the proposed activity to society?

**FIGURE 6.4**    Merit review criteria, NSF Grant Proposal Guide (NSF 2002; NSF 03-2, section III.A.).

An examination of NSF's description of who is eligible to submit proposals (Figure 6.5) also clearly underscores the agency's stated mission. The categories themselves, the ordering of the categories, and the descriptions within them all seem to point to the NSF's overarching concern with the infrastructure of American science and education. According to this list, proposals that do not appear to further these goals (e.g., research at foreign institutions or proposals from professional societies *not* "directly associated with educational or research activities") are rarely funded. Funding agencies spend time, money, and effort carefully wording their proposal guides in order to clearly indicate the kinds of research they will and will not consider. These guidelines can be an invaluable resource in selecting an appropriate agency for your proposal and in developing the significance argument for that audience.

### EXERCISE 6.6

Read the RFP excerpts from the National Sea Grant College Program (Chapter 11, pages 300–302) and from NASA (Chapter 12, page 358). Describe the values and goals of each agency as revealed in these documents. Now compare and contrast the values and goals of these two programs. What do you learn about the identity and priorities of these agencies by reading their proposal guidelines?

The introduction does not have to build the case for significance all on its own, of course; this is the purpose of the full proposal. In most cases the introduction serves as a preview of issues to be discussed further in other sections of the

## Who May Submit Proposals

Scientists, engineers and educators usually initiate proposals that are officially submitted by their employing organization. Before formal submission, the proposal may be discussed with appropriate NSF program staff. Graduate students are not encouraged to submit research proposals, but should arrange to serve as research assistants to faculty members. Some NSF divisions accept proposals for Doctoral Dissertation Research Grants when submitted by a faculty member on behalf of the graduate student. The Foundation also provides support specifically for women and minority scientists and engineers, scientists and engineers with disabilities, and faculty at primarily undergraduate academic institutions.

### Categories of Proposers

Except where a program solicitation establishes more restrictive eligibility criteria, individuals and organizations in the following categories may submit proposals:

1. **Universities and colleges**—U.S. universities and two- and four-year colleges (including community colleges) acting on behalf of their faculty members.

2. **Non-profit, non-academic organizations**—Independent museums, observatories, research laboratories, professional societies and similar organizations in the U.S. that are directly associated with educational or research activities.

3. **For-profit organizations**—U.S. commercial organizations, especially small businesses with strong capabilities in scientific or engineering research or education. An unsolicited proposal from a commercial organization may be funded when the project is of special concern from a national point of view, special resources are available for the work, or the proposed project is especially meritorious. NSF is interested in supporting projects that couple industrial research resources and perspectives with those of universities; therefore, it especially welcomes proposals for cooperative projects involving both universities and the private commercial sector.

4. **State and Local Governments**—State educational offices or organizations and local school districts may submit proposals intended to broaden the impact, accelerate the pace, and increase the effectiveness of improvements in science, mathematics and engineering education in both K-12 and post-secondary levels.

**FIGURE 6.5**     Categories of proposers, NSF Grant Proposal Guide (NSF 2002; NSF 03-2, section I.C.).

document (Pechenik 1987). The introduction identifies the problem, but specific research objectives are often elaborated elsewhere, for example, in the background or methods sections or in a separate section altogether, as in the two Burkholder proposals. Similarly, the introduction overviews previous research and the questions that motivated the study, but a more extensive discussion of prior research and explanation of the research problem are generally presented in a background section. The introduction also identifies the methodological approach to be used in the study, but a detailed description of methods is saved for the methods section. In some cases, the significance argument may be elaborated in a separate section as well, as in the proposals by Burkholder and Rublee (labeled "Expected Results"; see page 322) and Burkholder and Lewitus (Figure 6.3). These sections, appearing in both cases at the end of the document, provide a neat conclusion to the proposal argument as a whole, leaving readers with a strong sense of the importance and potential value of the proposed work.

In short, the introduction orients readers to the topic, purpose, and significance of the research, providing a framework or scaffold for other sections of the proposal to build on.

5. **Unaffiliated Individuals**—Scientists, engineers or educators in the U.S. and U.S. citizens may be eligible for support, provided that the individual is not employed by, or affiliated with, an organization, and:

- the proposed project is sufficiently meritorious and otherwise complies with the conditions of any applicable proposal-generating document;
- the proposer has demonstrated the capability and has access to any necessary facilities to carry out the project; and
- the proposer agrees to fiscal arrangements that, in the opinion of the NSF Division of Grants & Agreements, ensure responsible management of Federal funds.

Unaffiliated individuals should contact the appropriate program before preparing a proposal for submission.

6. **Foreign organizations**—NSF rarely provides support to foreign organizations. NSF will consider proposals for cooperative projects involving U.S. and foreign organizations, provided support is requested only for the U.S. portion of the collaborative effort.

7. **Other Federal agencies**—NSF does not normally support research or education activities by scientists, engineers or educators employed by Federal agencies or Federally Funded Research and Development Centers (FFRDCs). A scientist, engineer or educator, however, who has a joint appointment with a university and a Federal agency (such as a Veterans Administration Hospital, or with a university and a FFRDC) may submit proposals through the university and may receive support if he/she is a bona fide faculty member of the university, although part of his/her salary may be provided by the Federal agency. Under unusual circumstances, other Federal agencies and FFRDCs may submit proposals directly to NSF. Preliminary inquiry should be made to the appropriate program before preparing a proposal for submission.

**FIGURE 6.5**   *(continued)*

**EXERCISE 6.7**

The Reynolds team's proposal (pages 359–375) does not contain a special section or subheading for research objectives. Where are the specific goals of this study introduced? In one or two paragraphs, identify the objectives of the study and explain the rationale behind them.

**EXERCISE 6.8**

The Burkholder and Rublee proposal (pages 303–327) combines the introduction and background sections. As you read this section, look for the four standard introductory moves (announce topic, review prior research, identify gap, introduce new research). Are these moves recognizable in this elaborate introduction, and, if so, where is each move made? In one or two paragraphs, describe the structure and logic of this section.

# 6.5 Providing Background

Most proposals contain a separate background section or sections where the authors can present a more extensive explanation of the research problem, grounded in a thorough review of previous research. This section serves different purposes for the two segments of your audience. Your discussion should bring program officers and other generalists up to speed on the nature of the problem and the reasoning behind your specialized project. At the same time, it provides an opportunity for in-field readers to judge how familiar you are with the current state of knowledge in the field and how well you understand the issues and constraints involved in conducting research of this sort. This second purpose is clearly evident in the advice that one NIH institute offers on reviewing prior research:

> Citations show reviewers your breadth of knowledge of your field. Research proposals do not fare well when applicants fail to reference relevant published research, particularly if it indicates that the proposed research has already been attempted or the methods were found to be inappropriate for answering the questions you've posed. (NIAID 2003d; http://www.niaid.nih.gov/ncn/grants/write/write_p1.htm)

As this quotation indicates, the way in which you review the literature in the background section, as in the introduction, significantly influences the professional *ethos* projected by your text. Readers will want to know not only that you understand what other researchers have done but also that you appreciate the contributions their studies have made to the developing knowledge in your research area. Situating your work in this context is essentially a cooperative gesture as opposed to a competitive one. While oversights and methodological limitations must be taken into account in interpreting the results of individual studies, the primary goal in a review is to highlight what has been learned by the field so far (see Chapter 4). In the proposal, the review of research shows how far the previous research has gone and where it still needs to go. Once this groundwork is established, you will be in a position to explain how your proposed study will take the field forward.

Chapter 4 provides some general guidelines for structuring research reviews and citing sources. You'll see in the sample proposals in Chapters 11 and 12 that subheadings are a useful device for organizing the background section and are usually needed because of the length and complexity of these discussions. Notice that, as in the research report, the major headings in the proposal are *functional* headings (introduction, background, methods). But subheadings within sections are *topical;* that is, they identify the topic to be discussed in the section rather than simply announcing the function the section serves. Reynolds, Borkowski, and Blondin's (1994) background section contains three subdivisions: "X-Ray Emission of SNRs," "Radio Emission of SNRs," and "Dynamics of SNRs." Topical subheadings provide important cues to help readers navigate your background argument, and they are particularly effective when the organizational scheme is introduced at the start of the section, as in the Reynolds et al. proposal (page 363).

As you develop the background section, keep in mind the overall purpose of your proposal. You are reviewing research in order to introduce your study and show how it will further the field's knowledge and the agency's goals. You'll want to be sure to tie the background research clearly to the proposed research, especially as you bring this section to a close. Reynolds, Borkowski, and Blondin finish the background section with a review of the broad need their study addresses: "We shall work to correct this major gap between theory and observations, so that the observations can in fact serve as a useful check on the theoretical work" (see page 367). This statement brings the discussion back to the proposed research, clearly recalling NASA's emphasis in the program description on developing theory to help interpret satellite data.

Another way to connect the background argument to the proposed study is to follow this section with a list of specific research objectives (Pechenik 1987), as in the Burkholder and Rublee (1994) proposal in Chapter 11. Once readers have read the background section, they are well prepared to understand and appreciate these specific goals of your study. Another advantage of placing the specific objectives here (rather than earlier in the proposal) is that they can serve as a preview of the methods section, enhancing the coherence of the overall proposal argument.

**EXERCISE 6.9**

Choose a sample research topic in your field, perhaps one suggested in the discussion section of a research report you've read recently. List the research areas that would need to be reviewed in the background section of a proposal for this project. Design sample subheadings for this section.

## 6.6 Describing Proposed Methods

The methods section of a research proposal is distinguished from that of a journal article in two respects: the proposal generally contains fewer details but more explanation of rationale. Fewer details are to be expected, given that the research being proposed has not yet been conducted, but the need for rationale is more a function of the deliberative purpose of the proposal. This section must do more than describe how the study will be carried out; it must explain why this approach, as opposed to others, was chosen. The researcher must articulate and defend the methodological decisions he or she has made in such a way that the diverse readers in the proposal audience will be able to understand and appreciate those decisions. Remember that your target agency is being asked to pay for these activities. Its program boards must be convinced that this approach represents the best possible use of their limited funds.

As a consequence, the description of methods in the proposal tends to be heavily documented. In effect, it is an extension of the background discussion and serves a similar dual purpose. This extended explanation of materials and procedures helps your generalist readers understand what's needed to accomplish this research and allows your in-field readers to determine whether *you* understand what's needed for this research. The NIH cautions:

> While you may safely assume the reviewers are experts in the field and familiar with current methodology, they will not make the same assumption about you. . . .
>
> Since the reviewers are experienced research scientists, they will undoubtedly be aware of possible problem areas, even if you don't include them in your research plan. But they have no way of knowing that you too have considered these problem areas unless you fully discuss any potential pitfalls and alternative approaches. (NIH 1993, p 6)

**EXERCISE 6.10**

In either the Burkholder and Rublee (1994) proposal (Chapter 11) or the Reynolds team's (1994) proposal (Chapter 12), look for places where the authors include a brief or extended rationale for a methodological choice. What kinds of decisions have they chosen to explain? Why do these issues warrant further explanation? What role do citations play in these explanations? Look for examples of each of the three levels of procedural explanation described in Chapter 3: routine procedures, procedures established in previous studies, and new procedures or substantial modifications (see Section 3.4, page 50).

Of course you will describe your proposed methods in future rather than past tense, but otherwise the basic guidelines presented in Chapter 3 (Section 3.4) can be followed in developing this section. See Section 3.5 for advice on incorporating figures and tables. As in the background section, subheadings are common in this section of the proposal. Headings and subheadings help readers keep track of the basic components of your methodology. The Burkholder and Reynolds teams both use topical subheadings within their methods sections. Notice that Reynolds et al. begin their methods with an overview of topics to be covered, as they did in the background section. In some proposals, the research objectives introduced earlier become subheadings for organizing the methods section. In the Burkholder and Lewitus (1994) proposal outlined in Figure 6.3, for example, the research hypotheses, first introduced in Section B of "Project Description," reappear as subheadings under Section D, "Research Design," creating a clear link between the goals and methods of the study.

In sum, the proposal must convince readers that the project is significant and that it will be conducted expertly. Each section of the proposal document plays a role in building this case.

**EXERCISE 6.11**

Examine De Boer and Hale's proposal to the Greek government for permission to remove rock samples from the Delphi site (Chapter 13, pages 402–403). The researchers' proposed methods are mentioned in several places in this brief proposal, particularly under the "Request" and "Laboratory" headings. In contrast to proposals to funding agencies, in which methods must be described to withstand scrutiny by expert scientists, De Boer and Hale's plans were also evaluated by officials of the Greek government concerned with cultural preservation. What methodological details have the authors included to reassure this audience that their resources will not be abused and that the tests have a high probability of success?

# 6.7 The Research Proposal Abstract

The proposal abstract or summary is likely to serve a number of purposes and reach a number of audiences: it may be used at the start of the review process to help program officers sort proposals and select appropriate reviewers; it will be used by reviewers during the process as a preview of the larger document; and it may be used in reporting or publicizing an agency's funding decisions after those decisions are made. Thus, the NSF's proposal guidelines specify that the proposal summary should be "suitable for publication" and "informative to other persons working in the same or related fields and, insofar as possible, understandable to a scientifically or technically literate lay reader" (NSF 2002; NSF 03-2, section II.C.2.b).

The general shape of the proposal abstract reflects the shape of the proposal itself and thus includes a synopsis of the research problem, goals, and methods. Specifications for abstracts vary. A limit of 200 to 300 words is common, but some agencies will accept more elaborate summaries, as illustrated in the Burkholder and Rublee (1994) proposal, pages 304–305. Proposal summaries for NSF cannot be longer than one page. The summary Burkholder and Lewitus wrote for their NSF proposal is presented in Figure 6.6.

## Trophic Interactions of Ambush Predator Dinoflagellates in Estuarine Microbial Food Webs

### I. Project Summary

The diverse heterotrophic dinoflagellates include free-living estuarine species that demonstrate "ambush predator" behavior toward algal, protozoan, or fish prey. This behavioral pattern is widespread; thus far, it has been reported from the Mediterranean Sea, the Gulf of Mexico, and the western Atlantic. Within the past two years, we have discovered a new genus of ambush predator dinoflagellate, *Pfiesteria* (nov. gen.), with known species having a complex life cycle that includes at least 15 stages. Among these stages are persistent amoeboid heterotrophs and ephemeral "phantom-like" flagellates that are highly toxic to fish. *Pfiesteria* (nov. gen. et sp.) and its close relatives are common in eutrophic mid-Atlantic and Gulf Coast estuaries, and likely are widespread throughout warm temperate and subtropical waters. The ubiquitous occurrence and abundance of these stages in the water column and sediments, their voracious phagotrophy on bacterial, algal, and microfaunal prey, and their lethality to fish point to a major role of ambush predator dinoflagellates in the structure and function of estuarine food webs. In the proposed research, we plan to experimentally examine the nutritional ecology and trophic interactions of abundant stages of *Pfiesteria piscimorte* (nov. gen. et sp.) as a representative toxic ambush predator, toward determining the role of these organisms—in their many forms—within estuarine food webs. In a three-pronged approach, we will (1) examine saprotrophic nutrition of the biflagellate and amoeboid forms on stimulatory secretions of finfish prey; (2) characterize phagotrophic interactions between *Pfiesteria* (nov. gen.) and bacterial, algal and microfaunal prey; and (3) determine the influence of the biflagellate and amoeboid stages on representative microbial predators. The insights gained from this research will alter general paradigms about the role of dinoflagellates in the structure and function of food webs in eutrophic warm temperate estuaries.

**FIGURE 6.6**    Project summary from Burkholder and Lewitus (1994) proposal to the NSF. (The full proposal is outlined in Figure 6.3.)

---

**EXERCISE 6.12**

Read the project summary in Figure 6.6, from the Burkholder and Lewitus (1994) proposal to the NSF. Outline the "major moves" in this summary. Compare this summary with the table of contents for this proposal (Figure 6.3). What components of the proposal are highlighted in the summary? In what ways is this summary "understandable to a scientifically or technically literate lay reader"?

---

Whether brief or elaborated, the overview of methods in the abstract or summary must, of course, be "promissory" in that it describes what you *will* do. But the discussion of the research problem and goals is expected to be informative rather than descriptive (see Section 3.7); it must provide a clear, concise summary of what the project is about. Most, if not all, of your readers will read the abstract; some will read only the abstract (Olsen and Huckin 1991). The NIAID guidelines explain that although all proposals receive a careful reading from assigned primary reviewers who will represent the project in the review committee discussion, most of the 18 to 20 members of the Scientific Review Group "will likely scan your application, reading only your abstract, significance, and specific aims" (NIAID 2003c; http://www.niaid.nih.gov/ncn/grants/basics/basics_c2.htm). Given the vast number of proposals to be reviewed and the real-life constraints under which reviewers work, the abstract may well be the most important section of the proposal document.

---

**EXERCISE 6.13**

The research announcement for NASA's Astrophysics Theory Program asks authors to include a 200- to 300-word abstract "describing the objective of the proposed effort and the method of approach" (NASA 1993, p B-4). Write such an abstract for the Reynolds team's proposal, reprinted in Chapter 12 (pages 359–375).

---

**EXERCISE 6.14**

Write a 200-word abstract for the Burkholder and Rublee (1994) proposal, found in Chapter 11 (pages 303–327), using their one-page project summary as a starting point. What parts of the summary did you use, and what parts didn't you use? Why?

---

## 6.8 How Scientists Write Research Proposals

As discussed in Chapter 3, research scientists, like all writers, follow a wide range of writing processes and preferences. This variation applies equally to the processes of writing and revising research proposals. Individual scientists may prefer one drafting method over another; research teams may develop preferred patterns of collaboration that enable them to produce and revise documents efficiently. Authors may vary their methods of composing and collaborating from one occasion to the next. The one constant in all this variation is an overriding concern with audience in developing the proposal. A critical dimension of the proposal writing process is the assessment of the potential funding agency's interests, values, and goals.

The goals of most funding agencies are widely known and are articulated in RFPs, but many researchers also actively seek out new information about their target agency and establish a dialogue with program staff well before they submit proposals. Many funding agencies encourage this sort of early contact and information exchange. Program staff are readily available by phone and email for such consultations. It is in the agency's interests to provide this guidance in advance to ensure that the proposals it receives fall within its program guidelines and include all the information needed in the review process. In a study of the funding process, Mehlenbacher (1994) found this give-and-take among researchers and program staff to be quite common. The prominent researchers Mehlenbacher interviewed consistently described the proposal process as a long-term, interactive process.

If a proposal is not accepted on its first submission, the dialogue generally continues. Most agencies will send copies of reviewers' comments to the proposer and will continue to consult with the researcher as he or she revises the text for resubmission. In an extensive case study of this revision process, Myers (1985) found that proposal arguments changed significantly as researchers evaluated and responded to reviewers' concerns. Changes were effected on several levels, from the shape of the argument, to the amount of explanation, to the tone and *ethos* established in the text. But significantly, basic content remained relatively unchanged.

In sum, the proposal writing process is a complex, long-term endeavor that involves the participation of many. Given the high degree of interaction among researchers, program staff, and reviewers—and the continuing interaction among members of a research team (some of whom may be geographically separated)—it is no wonder that researchers in Mehlenbacher's (1994) study cited management and organizational skills as essential components of the research process.

## 6.9 How Reviewers Evaluate Research Proposals

The proposal review process exerts a powerful influence on the direction of research in scientific fields. In issuing RFPs on specific topics, funding agencies encourage research in some areas and not others; proposal guidelines ensure that researchers who work in these areas address the agency's goals and priorities; and in responding to reviewers' comments in the revision process, researchers may further tailor their research to fit within the parameters established by the targeted research program.

This degree of control makes some scientists nervous. As with the journal article review system, there are always worries about potential abuses such as favoritism, censorship, breaches of confidentiality, or misappropriation of ideas—concerns that may be enhanced by the political context in which these organizations operate; federal funding agencies such as NIH and NSF depend on congressional approval of their budgets and thus may be subject to the economic tug of political priorities. An additional concern in the granting process is the role that nonspecialists play in the review process, as, for example, in NIH advisory councils (Cohen 1996). In response to such concerns the NSF and other agencies have initiated changes in the review process designed to give proposers more "say" in the choice of reviewers and to provide authors of unsuccessful proposals more complete information about the reasons for the rejection (NSF 1995). Like the peer review process used by journal editors, the proposal review process may perhaps best be described as an imperfect system that nevertheless provides careful scrutiny of thousands of proposals every year, roughly a third of which are funded.

Proposal guidelines and RFPs routinely include a list of the criteria reviewers will use in evaluating new applications. The review criteria for the Sea Grant program to which Burkholder and Rublee (1994) submitted their proposal are presented in Figure 6.7. NASA's review criteria, in effect for the Reynolds team's (1994) proposal, are presented in Figure 6.8. Notice that these criteria clearly reflect the different goals of the two programs as represented in their respective RFPs (pages 299 and 358).

## Criteria for Evaluation of Proposals

The following criteria will be used to evaluate the proposals.

1. *Rationale*—the degree to which the proposed activity addresses an important issue, problem, or opportunity in marine biotechnology, and how the results will contribute to the solution of the problem.
2. *Scientific Merit*—the degree to which that activity will advance the state of the science or discipline through use and extension of state-of-the-art methods.
3. *User Relationship*—the degree to which users or potential users of the results of the proposed activity have been brought into the execution of the activity, will be brought into the execution of the activity, or will be kept apprised of progress and results.
4. *Innovativeness*—the degree to which new approaches (including biotechnological ones) to solving problems and exploiting opportunities will be employed or, alternatively, the degree to which the activity will focus on new types of important or potentially important issues.
5. *Programmatic Justification*—the degree to which the proposed activity will contribute an essential or complementary unit to other projects, or the degree to which it addresses the needs of important state, regional, or national issues.
6. *Relationship to Priorities*—the degree to which the proposed activity relates to guidance priorities in this document.
7. *Qualifications and Past Record of Investigators*—the degree to which investigators are qualified by education, training, and/or experience to execute the proposed activity and past record of achievement with previous funding.

**FIGURE 6.7**  Review criteria, National Sea Grant College Program (1994). "Statement of Opportunity for Funding: Marine Biotechnology." No page number. Reviewers of the Burkholder and Rublee proposal in Chapter 11 were guided by these criteria.

## Evaluation Factors

a. Unless otherwise specified in the NRA, the principal elements (of approximately equal weight) considered in evaluating a proposal are its relevance to NASA's objectives, intrinsic merit, and cost.

b. Evaluation of a proposal's relevance to NASA's objectives includes the consideration of the potential contribution of the effort to NASA's mission.

c. Evaluation of its intrinsic merit includes the consideration of the following factors, none of which is more important than any other:

   (1) Overall scientific or technical merit of the proposal or unique and innovative methods, approaches, or concepts demonstrated by the proposal;

   (2) The offeror's capabilities, related experience, facilities, techniques, or unique contributions of these which are integral factors for achieving the proposal objectives;

   (3) The qualifications, capabilities, and experience of the proposed principal investigator, team leader, or key personnel who are critical in achieving the proposal's objectives;

   (4) Overall standing among similar proposals available for evaluation and/or evaluation against the known state-of-the-art.

d. Evaluation of the cost of a proposed effort includes the consideration of the realism and reasonableness of the proposed cost and the relationship of the proposed cost to available funds.

**FIGURE 6.8**   Review criteria, NASA Astrophysics Theory Program (1993). p B-7. Reviewers of the Reynolds et al. proposal in Chapter 12 were guided by these criteria.

A comparison of Figures 6.7 and 6.8 also will reveal some common concerns that reflect the generic goals of the research proposal. Both evaluation guides highlight the relevance of the study to the agency's goals or mission, the scientific merit of the proposed work, and the capabilities of the investigators. This chapter has argued that while there are specific places in the proposal document to establish the relevance of the study and the scientific merit of the design, the professional competence of the researcher is indirectly demonstrated throughout. As these review criteria indicate, the researcher's professional *ethos* plays a critical role in the evaluation of his or her work.

## EXERCISE 6.15

The National Institute of Allergy and Infectious Diseases (NIAID), one of the Institutes of Health, has compiled a list of the most common reasons given by reviewers for rejecting a proposal. We have reprinted these in Figure 6.9. Consider this list carefully.

   A. Which of these reasons are related to the researcher's projected competence or professional *ethos?*

   B. How and where in the proposal argument should each of these potential concerns be anticipated and addressed?

## Common Problems Cited by NIAID Peer Reviewers

- Problem not important enough.
- Study not likely to produce useful information.
- Studies based on a shaky hypothesis or data.
- Alternative hypotheses not considered.
- Methods unsuited to the objective.
- Problem more complex than investigator appears to realize.
- Not significant to health-related research.
- Too little detail in the research plan to convince reviewers the investigator knows what he or she is doing (no recognition of potential problems and pitfalls).
- Issue is scientifically premature.
- Over-ambitious research plan with an unrealistically large amount of work.
- Direction or sense of priority not clearly defined, i.e., the experiments do not follow from one another, and lack a clear starting or finishing point.
- Lack of original or new ideas.
- Investigator too inexperienced with the proposed techniques.
- Proposed project a fishing expedition lacking solid scientific basis (i.e., no basic scientific question being addressed).
- Proposal driven by technology (i.e., a method in search of a problem).
- Rationale for experiments not provided (why important, or how relevant to the hypothesis).
- Experiments too dependent on success of an initial proposed experiment. Lack of alternative methods in case the primary approach does not work out.
- Proposed model system not appropriate to address the proposed questions.
- Relevant controls not included.
- Proposal lacking enough preliminary data or preliminary data do not support project's feasibility.
- Insufficient consideration of statistical needs.
- Not clear which data were obtained by the investigator and which reported by others.

**FIGURE 6.9**   Most common reasons cited by NIAID reviewers for rejection of proposals (NIAID 2003e; http://www.niaid.nih.gov/ncn/grants/write/write_d6.htm).

## 6.10 Accountability in the Research Process

Public trust in science rests on the assumption that all parties involved in research—from the researchers themselves to the reviewers and program officers who pass judgment on their work—are competent and trustworthy. This basic assumption shapes the nature of the relationship between researchers and granting institutions, as exemplified in the NASA policy statement reproduced in Figure 6.10.

In this statement, prominently displayed in the introduction to their *Guidebook for Proposers* (2003), NASA clearly identifies itself as an agent of the American citizenry, charged with facilitating work that is in the public's interest and thus worthy of public funds. The professional conduct of that research is thus NASA's responsibility, yet as the policy statement indicates, the agency exercises minimal oversight, placing their trust, as all funding agencies do, in the integrity of the research participants. We'll explore the ethics of research reporting at greater length in Chapter 9, but it is important at this point to acknowledge the mechanisms through which such trust is maintained in the granting process.

In addition to the checks and balances inherent in the peer review systems used to evaluate research proposals, funding agencies also rely on internal con-

## NASA's Partnership with the Research and Education Communities

Funding for NASA-related research and development projects is a privilege accorded to qualified science, engineering, and educational personnel by NASA acting on behalf of the citizens of the United States through Congressional action. NASA's proposal and selection procedures work only because the various research communities and NASA Program Offices together maintain the highest level of integrity at all stages of the processes. As a general rule, recipients of NASA research awards largely manage their own research projects with minimal oversight by the Agency. Throughout the entire process—starting with the identification of program objectives, the preparation and peer review of submitted proposals, the conduct of the research itself, and, finally, the exposition of new knowledge through publications, public outreach, and education—NASA sees itself as a partner with the scientific, engineering, and educational communities in making its programs relevant and productive.

**FIGURE 6.10**    NASA general policy statement (NASA 2003, p vi).

trols at the researcher's home institution. Most agencies require that proposals be accompanied by verification of approval by an institutional review board (IRB) at the home institution. Such boards are typically composed of researchers from a variety of departments on campus who review the proposed study design to ensure compliance with good research practice, as well as with federal regulations on such issues as the ethical treatment of human subjects and laboratory animals and the use of materials that may be hazardous to researchers, research subjects, or the surrounding community and environment. As noted in Section 6.3, funding agencies also may require special declarations on such issues in addition to documentation of IRB approval.

Once a grant is awarded, researchers are expected to keep the agency informed of their activities, usually through annual reports summarizing their progress to date. For multi-year projects such as that proposed by Reynolds, Borkowski, and Blondin to NASA's Astrophysics Theory Program, NASA "requires that a brief progress report be submitted to the Program Officer 60 days before the anniversary date of the award, in order to allow for the timely recommendation for continuation of funding" (NASA 2003, p F-3). At the end of the grant period, a final report is required, including "substantive results from the work, as well as references to all published materials from the work" (NASA 2003, p F-3). The first-, second-, and third-year (final) reports from the Reynolds project are included in Chapter 12 (pages 376–384).

Unlike journal reports, progress reports do not contain methods sections, for this audience has already evaluated and approved the design and will interpret the report in that context. The focus in progress reports is on showing how the research team is accomplishing the goals that the researchers and funder agreed upon through the proposal approval process. In keeping with the mixed audience of the proposal, progress reports are written at a fairly general level. Program officers are not scrutinizing results in detail; they are looking for a general overview of where the project stands and where it is headed. Note the growing bibliography generated over the course of the Reynolds team's three reports, testifying to the knowledge created and disseminated through this research partnership.

Lastly, notice that the Reynolds team's reports are organized using categories and headings established in the original proposal. Just as the proposal document explicitly addresses the funding agency's stated goals in arguing for the significance of the research, the resulting reports should tie the team's accomplishments directly to those goals and promises. The logic of the initial proposal thus provides the scaffolding not only for understanding the goals of the project but also for assessing its progress and outcomes. Decisions made in the proposal writing stage have consequences far beyond the success of that initial document.

## EXERCISE 6.16

NASA's policy statement on partnering with the research and education communities includes as the final stage of this partnership "the exposition of new knowledge through publications, public outreach, and education" (Figure 6.10). This provision conveys an interest in the nation's scientific infrastructure similar to that articulated by NSF (Figure 6.4) and other federal agencies. Read the three progress reports from the Reynolds, Borkowski, and Blondin project (pages 376–384). In what ways has this team furthered this broader agenda? How do these authors demonstrate that their work will have consequences beyond their immediate results?

# Activities and Assignments

1. In a two- or three-page essay, discuss the significance argument developed by Burkholder and Rublee in their proposal to the National Sea Grant College Program. After reading the proposal and excerpts from Sea Grant's RFP, both reprinted in Chapter 11, describe the primary goals of the marine biotechnology initiative, the primary goals of Burkholder and Rublee's proposed project, and the relationship between them. Explain where and how in the proposal document Burkholder and Rublee demonstrate that their research meets the initiative's goals and fits the specified parameters for the program.

2. Obtain a copy of an RFP from a funding agency in your field. Imagine that you and a colleague at another institution plan to submit a proposal to this agency for a collaborative project you have in mind. Based on the information contained in the RFP, write a letter to your colleague describing what you think are the most important goals and constraints the proposal will have to meet. Identify the intended audiences and their interests, needs, and knowledge. Summarize the program goals, the evaluation criteria that will be used, and any special considerations you will have to keep in mind in preparing the proposal.

3. Obtain a copy of a successful research proposal on a topic in your field, along with the appropriate RFP and/or proposal guidelines. Write a descriptive analysis in which you begin by examining the organization of the proposal as

revealed in the headings and subheadings of sections. Comment on any distinctive features of this organizational plan that reflect distinctive characteristics of research in this area (as in Exercise 6.4). Then examine the program goals and criteria listed in the RFP. Describe how and where each of these issues is addressed in the proposal.

4. Write a two- or three-page analysis of the rhetorical appeals in a research proposal from this text or one from your field. In your analysis, examine how logical, ethical, and emotional appeals (*logos, ethos, pathos*) are made or implied in the text. That is, you should describe the shape of the logical argument being presented (the major claims made in each section and the kind of evidence offered in support of those claims); describe the character or *ethos* projected by the author(s) and the ways in which that *ethos* is created in the text; and describe any appeals to readers' interests, needs, and broader social concerns. In which section(s) of the report is each type of appeal found?

5. We noted in Section 6.1 that most texts contain elements of all three types of classical argument: arguments of fact (forensic), of value (epideictic), and of policy (deliberative). Choose a proposal from this text or from your field and look for instances of each type of argument. In which sections of the proposal does each type appear? In which sections does one or the other type predominate? Write a two- or three-page analysis outlining the role each type of argument plays in the proposal.

6. Write a proposal for a new research project on a topic in your field. The project should build on, and represent a logical "next step" in a current line of research reported in the literature. (Review Chapter 4 on identifying topics and sources.) Target a specific funding agency that is likely to support work in this area, and obtain the agency's proposal guidelines and/or an RFP. Structure your text in accordance with the generic form described in this chapter (introduction, background, proposed methods), unless your guidelines include more specific instructions for organizing the project description. Use an appropriate documentation format in citing sources and in preparing your reference list. You may be asked to include a table of contents, a budget, biographies of investigators, and other ancillary information pertinent to your specific research focus.

7. Find a description of your university's requirements for internal review of research, that is, the guidelines for submitting proposals to an institutional review board for their approval. Following the IRB's form and/or guidelines, prepare a description of the study you designed in Activity 6.

8. In this activity you and your classmates will be organized into groups. Your group will role-play a peer review committee for a research proposal written by a team in your class. The team will provide copies of the RFP, proposal guidelines, or other agency material that they used to write their proposal, as well as the proposal itself. First, evaluate the proposal on your own: Is it well thought out, well argued, and well written? Then, use the criteria discussed in this chapter to help you evaluate the merits and limitations of the proposal: Does the proposal adhere to the agency's guidelines for organizing and formatting proposals (or follow the generic form discussed in this chapter if the agency does not specify one)? Have the authors avoided the problems listed in Figure 6.9? Does the introduction demonstrate the merits of the

project based on the goals, priorities, and values of the funding agency? Does the review of literature seem to demonstrate a knowledge of the subject and the field? Do the design of the study and choice of methods seem to reflect good judgment? Does the description of methods seem to demonstrate the technical competence needed to carry out the project? Is the proposal free of surface-level errors that would detract from the professional *ethos* of the writer? In short, are you persuaded that this project is worth funding?

After you have made your own assessment, discuss the merits and limitations of the proposal with your fellow reviewers, and arrive at a consensus concerning which of the following actions should be taken on this proposal.

a. Recommend this proposal for funding
b. Reconsider with revision (provide suggestions for revision)
c. Reject

Write a statement to the research team explaining the reasons for your decision. Your instructor may ask review teams to review their findings with the proposers.

# 7

# Documenting Procedures and Guidelines

## 7.1 Audiences and Purposes in Industry and Government

In prior chapters we have focused on the major genres of communication within academic research communities: the research report, literature review, conference presentation, and grant proposal. We have stressed that the conventions of these communication forms are largely determined by the needs and purposes of the scientific community. For example, the IMRAD form is well adapted to the tasks of conveying research results to fellow scientists and of arguing for the significance of those results to the field; this form provides the basic information journal reviewers and readers need to evaluate the quality and importance of the new work. We've also investigated some of the variability within these common research genres, but overall we've examined a somewhat narrow range of research texts thus far. When we look outside the immediate research community, however, we find that the audiences and purposes for scientific research vary widely, as do the written forms used to communicate scientific knowledge.

In the applied settings of industry and government, audiences for scientific texts might include policy specialists, marketing directors, field technicians, clinicians, resource managers, legal staff, transportation personnel, construction monitors, safety inspectors, legislators, public health workers, and the general public, to list just a few. The forms of communication used among these groups are equally varied, ranging from internal memos, manuals, and quality assurance guidelines used by employees of an organization; to executive summaries and reports for administrators, regulators, and other gatekeepers and decision-makers; to product inserts, health alerts, press releases, and other texts intended for private citizens and consumers. In this chapter, we'll explore some of this variation by focusing on one category of written text that has wide application across contexts in industry and government: procedural documents.

By procedural documents we simply mean texts that document actions or prescribe them for others. One way to understand the nature, purpose, and function of procedural documents is in relation to the traditional methods section of the IMRAD form. In prior chapters we talked about methods in the context of research reports and proposals, where the methods section is part of the record or projected record of the actions of the authors. In the research report, scientists document what they have done so that their work can be interpreted and evaluated; in the research proposal, they discuss what they will do so that the feasibility of their proposed research can be determined and its potential value assessed.

But in neither case do scientists expect their *readers* to act on or follow the procedures they've described. In fact, as we discussed in Chapter 3, while methods sections theoretically provide enough detail for scientists to replicate an experiment, the primary purpose of these sections in research reports or proposals is to establish the parameters within which the study's results were or are to be generated. The methods section plays a central role in the argument of the research report or proposal. Because the researcher is accountable to the scientific community, the description of methods serves to demonstrate compliance with the field's standard practices, as well as to show that the researcher's knowledge and techniques are current and thus that the methods are sound and the study's conclusions valid.

This evidentiary function of methods is rather unusual, however, when we consider that procedures and instructions in most contexts are written to be used rather than evaluated. When we move outside the research community and into applied settings of industry and government, procedural descriptions are often created by people who are not making research arguments at all, but rather are providing guidance to coworkers or other audiences who need to perform certain actions to achieve certain ends. In simple sets of instructions, for example, the "method" or procedure itself is not in question; it has already been established. In applied settings, the primary goal is to use the method or procedure consistently in order to observe required practices, to meet prescribed standards, or to achieve agreed-upon goals. In applied settings the purpose of procedural documents is to establish uniformity and consistency in performing a task so that trust can be placed in the resulting outcome or product. The use of standard procedures makes it possible for different people at different times and in different places to achieve the same desired goal or outcome. As we will see, such trust is based not only on the validity and efficacy of the methods themselves, but also on the documents that communicate those procedures clearly, effectively, and persuasively.

## 7.2 General Principles for Writing Basic Instructions

Many procedural documents provide basic instructions for completing a task. The structure of basic instructions is governed by a well-known set of general design principles outlined in most standard technical writing textbooks (e.g., Houp et al. 1998). In fact, the writing of instructions has a long and venerable history in the field of technical communication and is usually one of the basic genres taught in technical communication courses. Though instructions in research contexts and applied scientific settings have not been well studied, they figure prominently in standard operating procedures (SOPs) in laboratories, regulatory guidelines, and other common types of texts. As you will see below, even complex procedural documents in the sciences to some extent share a basic grammar or logic.

Thus, we will begin by briefly discussing the general principles for writing basic instructions before turning to more structurally and rhetorically complex forms. All these principles have one goal: to create a *usable* document. Like the methods section of the IMRAD report, instructions usually begin by assembling necessary materials and equipment. But basic instructions are written for people who are not merely reading, but also using the document: these users have to carry out actions, often *while* they are reading! A good set of basic instructions gives users direct access to the steps of the procedure, making the document as easy to use as possible. All the general design principles are geared toward providing this access to users. The following list summarizes and illustrates these principles:

### A. Purpose Statement

1. State the intended use of the procedures.
2. State the scope of application of the procedures.

### B. Materials and Equipment

1. Summarize all materials to be used.
2. List all equipment necessary to complete the steps described.

### C. Procedures

1. Write instructions as a formatted sequence of steps.

2. List all steps in their chronological sequence.

3. Number the steps.

4. Don't overlook important steps that may have become second nature to you.

5. Write in active voice, using the imperative mood (the "implied you").

6. Describe all steps in parallel form for easy comprehension.

7. Use white space for readability.

   **CAUTION: To avoid injury or damage, cautions and warnings should be placed before the step, not after. Cautions are especially important when the step is particularly tricky or important for the success of the procedure; warnings are absolutely essential when the step is dangerous.**

8. Put cautions and warnings before steps, not after!

9. Divide long lists of steps into logical sections that reflect separate stages of the actions.

10. For each step, separate all explanations, discussions, and examples from the step, so that both the step and the explanation are clearly visible.

11. Design the page for *usability*.

Remember: People use instructions because they need to accomplish some task, and they are often reading the document while doing the task. Keep this in mind when designing the document. For instance, don't break a sequence of steps in a place where the user will find it hard (or dangerous) to turn a page: Visualize what you are asking the user to do! For example, don't require the user to turn the page while pouring a hazardous substance!

These are some of the major principles of designing basic instructions. However, as you will see in this chapter, procedures may have rhetorical, commercial, and even political purposes beyond their basic instructional function or value. For example, one important dimension of written procedural documents is the control over practice they represent, that is, the authority they exert over users. Some of this authority is created by the appearance of procedures themselves: their numbered or bulleted lists of steps create an *ethos* of certainty and completeness (Gilbert and Mulkay 1984; Bell et al. 2000). But authority also comes from frequent use. When procedures become the standard method of accomplishing a task or the common practice accepted by a community, they are not generally questioned. If researchers are accountable to the scientific community for their methods, those who use procedures, such as lab workers, are held accountable to the procedures themselves. The sense of authority invested in written procedures is underscored by the fact that many, perhaps most, procedural documents in applied contexts are "authored" by organizations rather than by individuals.

Procedures and guidelines have many audiences; these audiences need and use scientific procedures for different purposes, and the difference in purpose undoubtedly changes the form of the written document. Variations in form might include, for example, (1) the use of unformatted rather than formatted lists of steps; or (2) the presentation of guidelines without numbered steps at all. But procedural documents can and do take on different and sometimes surprising functions that go beyond simply providing directions. In the remaining sections of this chapter, we will examine some of the other forms and functions of procedural documents, and the industry and governmental contexts in which they are applied.

## EXERCISE 7.1

In Chapter 11 we've included a set of instructions for handling fish pathology samples (pages 348–349). Posted by the North Carolina Department of Environment and Natural Resources in the wake of repeated *Pfiesteria* outbreaks in the state, these instructions constitute a set of standard procedures to be followed in studying fish with lesions suggestive of *Pfiesteria* attack. Aside from a summary list of steps at the end, these procedures are presented in paragraph form, which may be difficult to use as a guide in the field. Using the general guidelines presented above, rewrite these instructions to create a formatted list of sequenced steps that could be more easily followed by field technicians gathering samples from North Carolina's coastal waters.

# 7.3 Documenting Procedures for Quality Assurance

Perhaps the simplest forms of scientific instructions are the basic operating procedures developed by research organizations to govern their lab and field work. Standard Operating Procedures, or SOPs, are internal documents intended for use by scientists and technicians working within an organization, as well as contractors or others whose work the organization uses or monitors. The U.S. Environmental Protection Agency (EPA), for example, develops Good Laboratory Practice (GLP) Standards for industries that they regulate, such as the pesticide industry.

Because standard procedures allow many people to perform the same task and achieve the same goal, they play an important role in ensuring consistency in lab and field work, and thus the reliability of that work, enabling researchers to trust their data even if collected by multiple lab technicians in different locations and/or over extended periods of time. Such consistency ultimately forms the basis for trust in the reported findings. Simply put, our respect for the work of agencies such as EPA, NIH, and the Department of Energy (DOE) is predicated on the assumption that these research organizations have high standards, consistently applied. Indeed, federal law mandates that standard practices be established in federal research labs. We've included an excerpt from an EPA rule to this effect in Figure 7.1. Notice the clearly stated purpose of SOPs: "to ensure the quality and integrity of the data generated in the course of a study."

## EPA Proposed Rule on Consolidation of Good Laboratory Practice Standards

Sec. 806.81 Standard Operating Procedures.

(a) A testing facility shall have standard operating procedures in writing setting forth study methods that management is satisfied are adequate to ensure the quality and integrity of the data generated in the course of a study. All deviations in a study from standard operating procedures shall be authorized by the study director and shall be documented in the raw data. Significant changes in established standard operating procedures shall be properly authorized in writing by management.

(b) Standard operating procedures shall be established for, but not limited to, the following:

(1) Test system area preparation.
(2) Test system care.
(3) Receipt, identification, storage, handling, mixing, and method of sampling of the test, control, and reference substances.
(4) Test system observations.
(5) Laboratory or other tests.
(6) Handling of test systems found moribund or dead during study.
(7) Necropsy of test systems or post-mortem examination of test systems.
(8) Collection and identification of specimens.
(9) Histopathology.
(10) Data handling, storage, and retrieval.
(11) Maintenance and calibration of equipment.
(12) Transfer, proper placement, and identification of test systems.

(c) Each laboratory or other study area shall have immediately available manuals and standard operating procedures relative to the laboratory or field procedures being performed. Published literature may be used as a supplement to standard operating procedures.

(d) A historical file of standard operating procedures, and all revisions thereof, including the dates of such revisions, shall be maintained.

**FIGURE 7.1**    From EPA Proposed Rule on Consolidation of Good Laboratory Practice Standards (EPA 1999).

The EPA's rule places emphasis on what to us may seem self-evident: the fact that SOPs are to be in writing. The goal of quality assurance requires that such standards not only be established but recorded and readily available to those engaged in the regulated practices (see section (c) in Figure 7.1). The written text also serves as demonstration that appropriate standards are in place. Sections (a) and (d) of the rule articulate these accountability goals in requiring that all deviations are recorded and that an historical record is kept. This tangible record documents the procedures in place at given points in time and demonstrates that all necessary authorizations were secured. Thus the purpose of standard procedures extends beyond the practical goal of guiding the institution's work to the broader purpose of establishing the validity of that work by inscribing the organization's standards of performance.

EPA's mandate that all its labs develop Operating Procedures (OPs) is part of the agency's broader quality control requirements. In response to this mandate, the quality assurance staff of EPA's National Health and Environmental Effects Research Laboratory (NHEERL) developed a set of guidelines for writing OPs, distributed to all departments in the lab (NHEERL 1997). (The general term "operating procedure" includes both SOPs and ROPs, Recommended Operating Procedures, which are less prescribed and under development.) NHEERL's guidelines define the SOP as "a written document that details an operation, analysis, or action where procedures are thoroughly prescribed and appropriately validated, and which is commonly accepted as the method for performing certain routine or repetitive activities" (p 7). NHEERL's recommended components of OPs are listed in Figure 7.2., and a sample SOP is reproduced in Figure 7.3.

Following an organizational logic similar to that of methods sections in research reports, OPs begin by first establishing the scope or purpose of the procedure (which is also established in the introduction section of reports prior to the description of methods). The OP then describes the prerequisite materials and

National Health and Environmental Effects Research Laboratory
Office of Research and Development
U.S. Environmental Protection Agency
Research Triangle Park, NC

**Recommended Contents of Operating Procedures (OPs)**

- Scope of application (limitations)
- Prerequisites (equipment, supplies, training)
- Cautionary notes or special considerations
- Procedure
- Quality control
- References

**FIGURE 7.2**    NHEERL recommended contents of Operating Procedures (NHEERL 1997, p 12).

SOP No. NHEERL-H/ECD/GCTB/ADK/00/22/000
Page 1 of 1
Date: June 21, 2000

### STANDARD OPERATING PROCEDURE:

### Staining with Acridine Orange and Micronucleus Analysis

1.    **Scope of Application**

To prepare for and analyze rodent lymphocytes stained with acridine orange for micronuclei

2    **Prerequisites**

2.1    **Equipment and Supplies**

Prepared slides with lymphocytes exposed to cytochalasin B during culture (See SOP No.00/23 for details)
Pasteur pipets
Sorenson's buffer  (See SOP No. 00/29 for details of preparation)
40 uM Acridine orange staining solution (See SOP No. 00/00/28 for details of preparation)
Coverslips, 24 x 60 mm

2.2    **Training Requirements**

Laboratory safety training

3    **Cautionary Notes or Special Considerations**

**WARNING**

**Acridine orange  is a hazardous agent.  Review the appropriate safety protocol before use. Wear the required protective laboratory equipment and follow all safely procedures.**

4    **Procedure**

1.    Place slides in a coplin jar with 40 uM Acridine Orange for 30 - 60 seconds.
2.    Wash slides in a coplin jar with distilled water.
3.    Place a drop of Sorenson's buffer on the slide and cover with a cover slip.
4.    Place the slide onto the microscope and scan immediately.
5.    Score 2000 binucleated nuclei with good differential staining for the presence or absence of micronuclei.
6.    Cells containing micronuclei are also scored for the number of micronuclei per cell.

SOP NO. NHEERL-H/ECD/GCTB/ADK/00/22/000

GCTB Approval:_____Date:_____

QA Coordinator:_____Date:_____

Author Approval:_____Date:_____

Page 1 of 1
Date:
Revision No.: 0
Biennial Review Date:_____

**FIGURE 7.3**    Sample SOP from EPA's NHEERL lab, Research Triangle Park, NC (NHEERL 2000).

equipment before describing the procedure itself. In contrast to research reports, however, OPs are written not to document what has been done but to instruct those who will use the procedures in the future. Because these instructions are intended to be followed by others, writers of OPs are reminded to include a section describing any special precautions to be observed. Notice that these cautionary notes are to appear before the procedures are described, in keeping with the principles outlined in Section 7.2, above. In research reports, special precautions, if they are mentioned at all, are likely to be embedded in the description of procedures, for example, appearing in the form of justifications for particular methodological decisions, as in the mention by Mallin et al. (1995) of the safety equipment researchers used and the precautions taken against tank contamination (see page 330).

In addition to prescribing these content components, the NHEERL guidelines provide detailed advice for researchers preparing OPs, reiterating many of the basic stylistic principles outlined in Section 7.2. In particular, the procedures section is to consist of sequentially numbered steps "for ease in referencing and for use as a 'check-off' device during performance of activities" (NHEERL 1997, p 13). As the sample SOP in Figure 7.3 illustrates, procedural steps are listed in parallel form and stated in second person imperative mood ("Place slides in a coplin jar. . ."). By design, SOPs describe a single task or operation and frequently refer to other SOPs for sub-tasks necessary to complete the operation, as illustrated in the Equipment and Supplies section of the sample. The level of detail in the resulting network of SOPs "should be adequate to provide a suitably trained person with all the information necessary to perform the procedure consistently and correctly" (p 11). Thus, SOPs are not intended for the general public but for specified internal audiences with certain, perhaps minimal, levels of technical expertise.

The NHEERL guidelines state that OPs are to be "consistent with sound scientific principles and good research practices [and] adequate to establish traceability of standards, instrumentation, samples, and data" (NHEERL 1997, p 11), criteria that clearly echo the EPA's concerns with quality and accountability. Beyond these basic goals, these internal documents serve a number of important purposes within an organization, providing a basis for training, for maintaining continuity when personnel change, and for internal assessments of research procedures and facilities (p 8). Thus, although the immediate purpose of the SOP is the effective performance of a specific task, the effectiveness of the research organization as a whole rests to a large extent on this network of self-regulation.

**EXERCISE 7.2**

Examine the methods section of an article in your field or one in this text. If you were supervising the technicians carrying out these procedures, for what parts of the process would you need to have established SOPs? List the SOPs that would need to be developed in order to train new technicians.

# 7.4 Documenting Procedures for Regulatory Purposes

As described above, internal SOPs are established by organizations to maintain uniformity in practice within a lab, enabling the organization to "regulate" itself and ensuring consistent implementation of appropriate research standards. But written procedures play an equally important role in meeting standards set by outside regulators. The role of procedural documents in this process cannot be underestimated, for in government-regulated industries, the standards established by oversight agencies such as the EPA, the Occupational Safety and Health Administration (OSHA), the U.S. Department of Agriculture (USDA), or the Food and Drug Administration (FDA) usually have the force of law. Companies that write procedural documents for these regulatory agencies often need to have the procedures approved by these agencies to conduct their work. In addition to providing instructions and/or documenting internal processes for quality assurance, then, one of the primary purposes of procedures in the regulatory process is to assure the regulatory agency that the company will adhere to the agency's scientific and ethical standards for quality and safety.

One example of a procedural document used in a regulatory process is the clinical protocol, written by clinical researchers and medical writers within pharmaceutical corporations for approval by the FDA. A complex procedure for conducting a proposed clinical study in accordance with the FDA's "Good Clinical Practice" (CFR 1999), the clinical protocol must demonstrate to the FDA that every proposed study will meet the FDA's high standards for the "efficacy and safety" of a drug (CDER and CBER 1995, p 1). A protocol must be submitted and approved by the FDA prior to all clinical trials in every phase of the drug development process. The flow chart reprinted in Figure 7.4 illustrates the many stages of this long and complicated process.

Through its approval of procedures (as well as the results of the studies based on those procedures), the FDA monitors and regulates the development of every drug in the United States, and acts as an external check on quality control in the pharmaceutical industry.[1] Once a clinical trial has been conducted, the clinical protocol is appended to a Clinical Study Report (ICH 1996); the FDA often needs to send these back for revision, which in turn costs a company a lot of time and money. Eventually, if and when all the drug trials have been proven safe and effective, every clinical protocol and study report generated during the drug development process, as well as all other required documentation, are incorporated into the NDA (New Drug Application) for FDA approval before a drug can be marketed to the public. Thus, it is the job of the FDA to ensure that the procedures proposed for developing and testing drugs are carefully and consistently applied as a way of ensuring the reliability of drugs released into the marketplace.[2]

---

[1]Although there are no worldwide legal requirements for drug development and manufacturing, efforts to standardize regulations and practice of pharmaceutical agencies in the European Union, Japan, and the United States are made by the International Conference on Harmonization of Technical Requirements for Registration of Pharmaceuticals for Human Use (ICH), which issues nonbinding guidances that "represent the Agency's current thinking" (1996, p 1) for researchers and manufacturers in the industry.

[2]The FDA's requirements for submitting applications and investigating new drugs are primarily contained in the Code of Federal Regulations (CFR), a codification of rules published in the Federal Register (see Title 21 CFR §312 Investigational New Drug Application, and Part §314, New Drug Application).

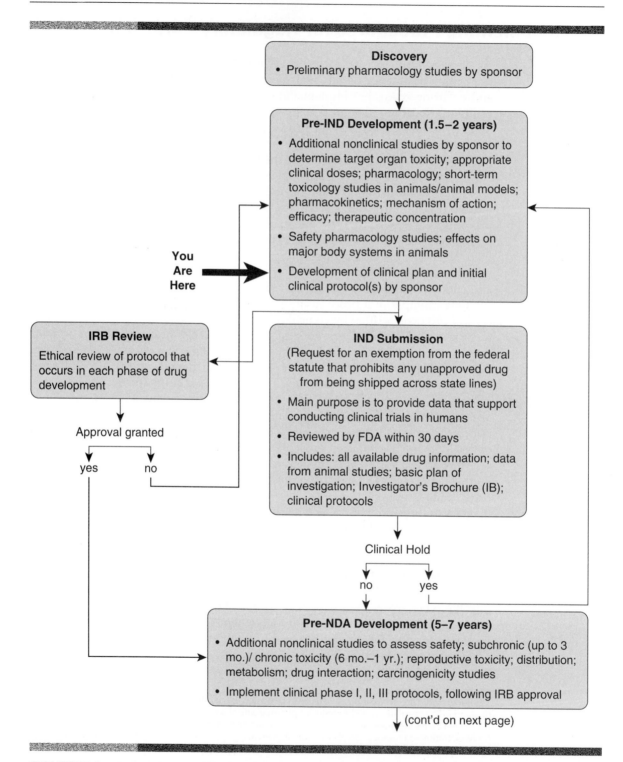

**FIGURE 7.4** The traditional drug development process in the United States (Bell et al. 2000, p 254). IND = Investigational New Drug Application; NDA = New Drug Application.

**Phase I Studies**

- Initial introduction into humans; closely monitored; conducted in patients or healthy volunteers

- Designed to determine: drug's metabolism; pharmacologic actions; side effects; early efficacy

(21 C.F.R. § 312.21 [a])

**Phase II Studies**

- Well-controlled; closely monitored; conducted in a small number of patients (i.e., several 100)

- Designed to evaluate: drug's effectiveness for a particular indication; short-term side effects; risks associated with the drug

(21 C.F.R. § 312.21 [b])

**Phase III Studies**

- Expanded; controlled and uncontrolled; conducted after drug is proven effective; several 100 to several 1000 patients

- Designed to: gather additional information about effectiveness, safety; evaluate overall benefits/risks; provide basis for product label

(21 C.F.R. § 312.21 [c])

**NDA Submission**
(Application to market new drug; reviewed by FDA from 6 months to 2 years)

- All information collected about drug

- Summaries of efficacy and safety results

- Marketing label

**FIGURE 7.4** *(continued)*

From this discussion it is obvious that the clinical protocol is a central and critical document in the drug development process (Bonk 1998): it presents the procedures for FDA review and approval, and subsequently guides the studies themselves. Based on their analysis of the clinical protocol in the drug development process, Bell et al. (2000) state:

> [T]he protocol defines the objectives of the study in clinical terms, outlines both the conduct of the trial and the ultimate analyses, and ties the objectives to the statistical hypotheses that are to be tested. But most importantly, the clinical protocol must demonstrate that the study represents a valid approach to assessing the safety and efficacy of the drug (p 252).

Unfortunately for clinical researchers and medical writers, no complete or adequate instructions are available for writing a clinical protocol (Bell et al. 2000). But in addition to the FDA's general rules and "guidances," there are a few books that provide some outlines of its form and content (e.g., Bonk 1998; Spilker 1984). According to medical writer Robert Bonk, every clinical protocol should have:

1) Clearly defined objectives (what specifically is the trial seeking to find out?)

2) Criteria for inclusion and exclusion (which patients are being studied?)

3) Drug dose and regimen (how is the drug administered, and at what strength?)

4) Clinical vs. pharmacological end points (how are results measured?)

5) Definitions of success and failure (which results are considered meaningful?)

6) Statistical design and analysis (how can data be reliably analyzed?)

(Bonk 1998, p 11–12)

We see clearly from this outline that the clinical protocol is a set of procedures that define and guide every part of the drug trial.

Because of their highly proprietary and confidential nature, sample clinical protocols are not available to reproduce, but in Figure 7.5 we provide a typical if abbreviated table of contents. As you may be able to infer from this table of contents, the clinical protocol is not simply a series of "steps," but rather an extended explanation and rationale for every part of the research design. As the functional headings in Figure 7.5 illustrate, sections range from a description of drug storage procedures to a discussion of confidentiality of patient records. In fact, the clinical protocol more closely resembles a research proposal than it does a set of instructions. Note in Figure 7.5 that the clinical protocol contains introduction sections (2–3) and proposed methods sections (4–10); like the proposal, it doesn't offer results but does allow for the acknowledgment of possible limitations ("Adverse Reactions," for example) and special conditions ("Patient Discontinuation").

As you can see, the clinical protocol is not only a procedural document submitted to a regulatory agency for approval, but also a set of directives for carrying out the scientific study, and a framework for reporting the clinical results. What you can't see from this table of contents is the level of specificity that the protocol as a proposed procedure contains. For example, because human patients are involved as subjects in experimental drug studies, and the collection of patient data and the maintenance of the drug regimen often occur at off-site clinics or in patients' homes, the FDA scrutinizes every detail of a proposed procedure. Thus the clinical protocol may contain information on how the company will monitor the external environment of patients, including the patients' activity and exercise, diet, availability of food, and between-meal snacks; fluids consumed; conversation; visual,

## SAMPLE TABLE OF CONTENTS

Page                                    Page

**FIGURE 7.5** Sample table of contents for a clinical protocol (Spilker 1984, p 6).

auditory, and tactile stimuli; stress; contact with other patients and the number of patients in a room; and the number and experience of personnel conducting the study (Spilker 1984, p 159). For shipping biological samples from clinical sites to a laboratory, the protocol might contain information on the proposed wall thickness of styrofoam containers, the ratio of packing material to volume, the number of kilograms of dry ice used in the packing, the preferred carrier, and the exact time windows allotted for shipping and receiving (Spilker 1984, p 160). To meet FDA standards for "adequate and well-controlled studies" (CFR 1999, §314.126), clinical protocols must be well-planned procedures indeed!

Parts of a clinical study may be contracted out, depending on the size and organization of the pharmaceutical company. Thus the clinical protocol has many readers in addition to external regulators. Like the SOP, the clinical protocol will guide the work of different people at different times—beyond as well as within the company. The protocol is therefore a complex procedural document because it not only has multiple purposes, but also multiple audiences. While the FDA, the first and final regulatory authority, is the primary audience for the clinical protocol, there are other external audiences who may need and use the document in different ways. For example, IRB reviewers must examine and approve the ethics of the procedures; clinical investigators follow the protocol in conducting their part of the study; medical writers use it as a guide in pulling together all the arguments and documents for clinical study reports. The problem of multiple users is further exacerbated by the fact that these different audiences may have different levels of knowledge: clinical investigators may be physicians, whereas IRB members may represent a range of disciplines.

As a procedural document guiding a scientific study in a regulated industry, the clinical protocol is rhetorically as well as structurally difficult to write. But we should not lose sight of the fact that it is a proposed procedure for a scientific study. In addition to understanding the principles of basic instructions, the heuristics we have discussed for writing introduction and methods sections under the standard IMRAD form may be useful in writing clinical protocols (Bell et al. 2000). Although the protocol's users vary even more widely than audiences of the research proposal, applying the principles discussed in other parts of this book may help authors in clinical settings write protocols that are better adapted to their many uses.

## 7.5 Describing Procedures for Commercial Purposes

In this section, we shift our attention to medical documents used for commercial purposes. Once a drug is approved, it must be marketed to medical professionals who will use or prescribe it. An example of such a text is the documentation accompanying the "Pylori-Chek Breath Test Kit," produced by the research and development firm Alimenterics, Inc. for detecting *H. pylori* infection. This document, included in Chapter 10 (pages 264–279), contains many types of information, including sections on the intended use of the test, warnings and precautions, how to prepare the patient, how to calculate the test results, and performance characteristics of the test.

But there are actually two types of procedures here, intended for two different types of "users" within the medical office: physicians and clinical staff. A study protocol, appearing late in the document (Section XIII.2.A.), summarizes the methods used in clinical trials of the breath test and is clearly intended for physicians; in conjunction with a summary of results (XIII.2.B.), it provides physicians with information about how the test has been evaluated and how it works compared to other methods. The step-by-step instructions presented primarily in Sections IV through IX, however, seem intended for use by clinical workers.

Although distributed throughout this rather lengthy document, the procedures for using the breath test kit follow the general guidelines for writing instructions outlined in Section 7.2: in the documentation for the "Pylori-Chek Breath Test Kit," the purpose is stated in Section I; warnings and precautions are listed in Section IV and also inserted as needed before individual steps (Section IX); required materials are listed before the procedures (VII–VIII); and the procedures themselves are described in a formatted sequence of complete, simple, and parallel steps, numbered and listed in chronological order, written in active voice using imperative mood, divided into logical subsections that reflect separate stages of the actions, and formatted with white space for easy visibility. Notes also are broken from the steps and put in boldface and italics.

One question that might come to mind at this point is why procedures for medical personnel would be broken into such basic steps. In Section 7.2, we discussed how procedures as mandated practice have authority over those who are supposed to use them. In the case of the breath test, responsibility for administering the procedure is most likely to fall to technicians in the medical office. It is clearly important that medical technicians or other clinical workers follow the prescribed instructions if the product is to work as intended. Since these procedures have been developed and tested to ensure the safety and reliability of the product, deviations from the procedures could lead to a failure of the product. In effect, the procedures confer expertise on the technicians who will ultimately administer the test without the aid of instructions, but the written document still exerts the force of authority over the procedure. While physicians have authority over how the test results are interpreted, there is little judgment allowed in how the test is administered; technicians must stick to the script.

Scholars studying medical discourse have shown that documents in the healthcare industry reveal established lines of authority and power; to various extents, medical documents have the social hierarchies of the healthcare practice and profession built right in to them.[3] Clearly, a certain level of proficiency and knowledge is required to administer this test. The Alimenterics document, like the breath test kit itself, is not intended to be used by patients in their homes, but rather by skilled healthcare workers in doctors' offices. However, the difference in audience, purpose, and structure in the different sections of the document perhaps also reflects the social hierarchy in the profession. The dual audiences for this document are made perfectly clear in the last two sentences of the first section: "For use by healthcare professionals. To be administered under a physician's supervision."

Beyond the issue of these "politics of procedures," this document does recognize its multiple audiences. It provides straightforward instructions for clinical staff administering the test to patients and offers extensive background information needed by physicians to evaluate the test's efficacy, to decide whether it is appropriate to use for particular patients, and to interpret its results. While the text does not state which sections are intended for which users or group sections intended for particular user groups in any clearly identifiable way, like many technical reports addressed to multiple audiences, in this sectioned document the topical headings themselves more or less indicate which audience is addressed in each section. Sections III ("Principles of the Test") and XIII ("Performance Characteristics"), for example, may be of more interest to physicians than to the staff; Sections VI ("Patient Preparation") and IX ("Procedure") are especially pertinent to the staff.

In meeting the needs of these various users, the document accomplishes several purposes. First, of course, is the simple purpose of making the product possible to use (this clearly applies to the basic instructions for administering the test, intended for the clinical staff). But what distinguishes this text from others we examine in this chapter is its commercial purpose. In showing how to use the product, the document also "argues" that the product is easy to use and a good one to choose. The need for this document to be persuasive is attested to by the fact that

---

[3]For example, Schryer (1994) demonstrates how, because of the research processes embedded in it, the IMRAD form used in the lab is held by some scientists to be a "higher," more advanced form than the applied problem-solution report used in the clinic. Pettinari (1988) has closely analyzed the privilege physicians exercise in their exclusive power to sign operating and patient reports. And Dautermann (1994) has examined the failed attempt by nurses in a Canadian hospital to redesign their documentation and communication procedures as a way of renegotiating their position and lack of authority in the hospital. For a discussion of authorship and power in the field of technical communication, see Kynell-Hunt and Savage (2003).

other *H. pylori* breath tests came on market at about the same time (see Fallone et al. 2000). As noted in Section 7.4, clinical trial information such as that provided in Section XIII is proprietary and usually closely guarded. However, in this context that information becomes important evidence for persuading physicians to choose this test over others. In the document by Alimenterics we see that clinical procedures are embedded not only in a larger set of scientific procedures intended for physicians, but also in a rather blatant but normal commercial context. Procedures on how to do something and how a study was conducted are embedded as a rationale for the purchase and use of the final product. The procedures thus not only provide data on the use and interpretation of the product, but also create a persuasive marketing appeal. The procedures become marketing tools. It is clear that the purpose of *all* the sections of the Alimenterics report is not only to report the findings of the breath test kit and provide a set of procedures for administering it, but also to persuade physicians of the value and usefulness of the kit.

Ultimately, the purpose of the document—the intent of all procedural documents—is to obtain compliance, in this case with directions for the prescribed use of this product. This is important for the manufacturer, not only to ensure the product performs as promised, but also to ensure customer satisfaction, which improves success rate and minimizes error and unintended effects (and lawsuits). Customer satisfaction and improved success rate, in turn, lead to continued sales and a good reputation—a corporate *ethos* built not only on the efficacy of the product, but also on the efficacy of the accompanying documentation.

### EXERCISE 7.3

Read the Alimenterics "Pylori-Chek Breath Test Kit" for *H. pylori* in Chapter 10 (pages 264–279). As you read, make a list of selling points. Which appear as part of a procedure? Which appear as scientific or experimental fact? (Be sure to pay attention to sales pitches created through text conventions and stylistic features.) How are these procedures and scientific/experimental facts used to sell the product? What do your findings tell you about the use of procedures for commercial purposes?

## 7.6 Procedures as Public Policy

Thus far we have examined procedures and instructions written for audiences with particular kinds and levels of expertise: lab technicians, federal regulators, physicians and their staffs. But procedures are frequently written for nonspecialist audiences as well: members of the general public who use instructions in everyday situations such as operating a fire extinguisher, taking medication, sorting recyclables, or fertilizing the lawn. In such situations, where no prior knowledge or training can be assumed, the clarity and usability of procedural documents are paramount. Much attention has been paid to how scientific knowledge is communicated to the general public. That attention typically focuses on public understanding of research *results* rather than of procedures. We'll defer the discussion of how the public comes to understand research until Chapter 8 and keep our focus here on procedural documents developed for public use.

Beyond their responsibility to regulate their own research and the industries they monitor, many federal, state, and local government agencies also have more direct responsibilities to the citizenry. Whether they conduct research themselves or review research generated elsewhere, government agencies are expected to maintain state-of-the-art knowledge of the science pertaining to issues under their jurisdiction and to use that knowledge to act in the public interest, for example, to develop procedures and policies to protect human health and environmental resources. Agencies such as EPA, FDA, and the Centers for Disease Control and Prevention (CDC) have a responsibility not only to keep the public informed but also to protect the public, and to set standards of good, safe practice for the public to follow.

The *Pfiesteria piscicida* threat is a case in point. The extensive research on this "killer algae" documents the environmental dangers of *Pfiesteria* and the potential threat to human health. Given this knowledge, it is the responsibility of state and federal agencies to regulate behavior in and around the water in threatened areas. The North Carolina Department of Health and Human Services (NCDHHS), for example, developed guidelines for state workers, fishermen, and others in affected areas in an effort to minimize the threat to human health and safety. We've included this text in Chapter 11 (see pages 341–344). Like many public documents, these guidelines were made available online, via posting to the North Carolina Department of Environment and Natural Resources (NCDENR) website (http://www.esb.enr.state.nc.us/Fishkill/persprotect.html). In this public safety document, precautions alluded to in the earlier research literature (see Mallin et al., page 330) and in Burkholder and Rublee's grant proposal (page 316) are elaborated and formalized as a set of directions for state workers and others who may be at risk. In effect, safety procedures developed for individual labs have in this document become institutionalized as public policy.

Like most government policies, NCDHHS's "Guidelines for Safety and Personal Protection During Fish Kills" were developed in a political context. The extent and severity of the *Pfiesteria* threat had been the subject of contentious debate within the local scientific community by the time these guidelines were posted in 1998, with some in scientific and legislative circles urging the state to act to inform and protect the public and others in the same communities arguing that the danger was exaggerated and insufficiently supported by the data (Burkholder and Glasgow 1999; Griffith 1999). Thus, posting safety guidelines was not a simple matter of identifying and describing appropriate procedures but of determining whether procedures were warranted at all—and of considering the political fallout of either course of action, posting or not posting. Scientists working in the realm of public policy need to be aware that even procedural documents may take on a political dimension.

In the first paragraph of the document we see how NCDHHS resolved this dilemma. The text begins by stating that the extent of the threat to human health is unknown, but follows this neutral claim with a characterization of the research that implies that the evidence is not convincing: "Some anecdotal reports . . . suggest that environmental health risk is plausible. . . . However, a careful review of the evidence to date also suggests that . . . impacts are not widespread, are minor, and are of short duration if they have occurred at all" (see page 341). But the author acknowledges that "definitive results, if found" will not be found soon and concludes that in the meantime precautions must be taken to manage the potential risk. This opening paragraph ends with a clear statement of the document's purpose: ". . . to make recommendations for work practices and the use of personal protective equipment to reduce potential exposure to toxins generated by *Pfiesteria* and MROs [morphologically related organisms]." Thus, despite the scientific debate raging on the topic, NCDHHS decides in favor of protecting public health against a potential, if to some officials unproven, threat by issuing safety guidelines.

The text adopts a more objective tone in describing the recommended practices. The guidelines do not offer a sequence of steps to be followed but a set of safe practices to be observed in the field. Thus they are organized under topical headings ("Protective Clothing," "Respiratory Protection") rather than a numbered list of actions. In contrast to the conventional use of second person "you" in instructions, these guidelines are written in third person and make frequent use of the modal auxiliary verb *should* to make strong recommendations: "Workers should not enter risky waters unless there is a critical need" (page 341); "Personal protective equipment (PPE) should be used only when exposure to risky water cannot be avoided" (page 342). This use of third person provides a clue to the intended primary audience of this document. By referring to workers in the third person instead of addressing them directly, the text seems to target the supervisor who will be held accountable for making sure precautions are followed, rather than the persons who will actually need to follow them. The heavy use of the modal *should* underscores the status of this document as advisory[4] (perhaps in keeping with the agency's skepticism about the danger *Pfiesteria* poses). Putting these elaborate precautions in place essentially allows NCDHHS to meet its obligations and satisfy those who feel the threat is real, without holding up the department's work in the field.

Thus, although these guidelines serve a straightforward informational purpose, it seems clear that the document also functions as an accountability instrument. Like the SOP, the text itself serves as a written record of the state's attention to these matters. The posting of the text to a public website further serves as demonstration that the state is meeting its responsibility to the public welfare. Despite its skepticism, the NCDHHS is taking appropriate precautions to protect state workers and to keep the public apprised of potential threats to their health and safety.[5] Thus, although the primary audience may be the supervisors who are in a position to act on these recommendations, the document is also available to the general public, who may or may not be involved in implementing these directives. The message conveyed by the act of posting the document may be just as critical as the information conveyed in the text itself.

---

[4]In Chapter 3 we discussed modal auxiliary verbs as qualifiers that indicate the conditions of a claim. Here the modal *should* is being used to indicate the strength of the recommendation. In fact, in the Code of Federal Regulations, the FDA use modals such as *should* to express many of its legally binding rules (Bell et al. 2000).

[5]For an extended discussion of the need to balance scientific and legal requirements in the development of public policy and procedures, see Cranor (1993).

Choose a section of the guidelines for "Safety and Personal Protection During Fish Kills" (pages 341–344) that could usefully be transformed into a set of instructions either for workers or for supervisors (choose one of these audiences). For example, using the information provided under the "Heat Stress" heading, develop a set of instructions for supervisors to follow in scheduling work and making the job site safe. Rewrite these third person guidelines, putting them in second person imperative mood instead, and follow other principles in Section 7.2 as applicable.

# 7.7 Procedures as Public Ethos

As just illustrated, an important function of procedures in governmental or civic contexts is to assure the public that procedures exist—that the responsible government agency is aware of an issue of public concern, has investigated it, and is acting appropriately. In the case of the North Carolina safety guidelines, assuring the public is one of several functions the procedural document is intended to serve. We observed that the text also serves a more direct instructional purpose: it conveys real information that readers—particularly state supervisors but potentially others who work or recreate near coastal waters—can learn from and act on. The text provides guidance to those who need it and at the same time serves as tangible evidence that NCDHHS is doing its job. The obligation of governmental agencies to provide such evidence is increasingly important, given the increase in the number and scope of potential hazards, the breadth of governmental responsibilities in democratic systems, and the need for public accountability. Certainly it's important for the public to know what the government agencies they've funded are doing. Thus it may come as no surprise that in some procedural documents this evidentiary purpose supersedes all others. In fact, some written procedures are not intended to be followed by readers at all.

We conclude this chapter with a look at just such a text, a set of guidelines developed by the state of Maryland for closing rivers affected by *Pfiesteria* or related organisms. The document is included in Chapter 11 (pages 345–347). Posted at Maryland's Department of Natural Resources (MDNR) homepage,[6] which provides information for fishermen, recreational users, educators, and other public audiences, these guidelines "generally describe the conditions considered for evaluating *Pfiesteria* or *Pfiesteria*-like events, as well as procedures followed for issuing closures" (page 345). A quick glance shows that the procedures described in this document are not intended to govern readers' behavior but to explain and justify the behavior of the state of Maryland. For example, in contrast to the safety guidelines discussed in Section 7.6, these guidelines are framed in future tense: "A closure will be recommended when. . . ." This text describes not what should

---

[6]http://www.dnr.state.md.us/bay/cblife/algae/dino/pfiesteria/river_closure.html

be done by readers, but what will be done by the author (MDNR), and what conditions will set these actions in motion.

Notice the similarity between this rhetorical situation and that of a researcher submitting a grant proposal. The grant proposal describes experimental procedures in future tense and also does not intend them to be followed by readers; in a proposal, research methods serve as documentation or evidence of the researchers' plans. The intended readers (proposal reviewers) evaluate those plans not only to understand the project and assess its feasibility but also to evaluate the research team's collective competence and expertise. As we argued in Chapter 6, the success of a proposal depends heavily on the professional *ethos* projected by the authors. Similarly, the public *ethos* created by texts such as the Maryland guidelines plays a critical role in shaping public perception of the agency and thus its successful interaction with its constituents.

Like the North Carolina safety guidelines, the Maryland river closing document begins with a quick overview of the exigence or the need for communication: these guidelines were developed in response to the public health risks that *Pfiesteria* poses. Notice that, in contrast to the NCDHHS document, this opening does not question the potential risk but takes it for granted. The overview acknowledges limitations in detection methods and explains that government officials will use their good judgment and the best available knowledge to make these decisions; there is no indication here that MDNR has any concerns about the quality of the research. Later, the concluding "Comments" section adopts a similar perspective. The statements in this section take the toxic nature of *Pfiesteria* as a given; the focus is on the nature and activity of those toxins, not on the validity of that basic claim. The effect is to create a much more reassuring *ethos* than the NCDHHS document does. Though the political nature of this document may not be apparent to those who are unaware of the research controversy, we can see in these opening and closing sections that Maryland's Department of Natural Resources has in fact taken a position in that controversy.

You will remember our discussion of the strategic use of active and passive voice in Chapter 3, where scientists intentionally or unintentionally use active voice to emphasize the researcher as agent and passive voice to de-emphasize the researcher's role. The body of the Maryland document contains an interesting and perhaps meaningful mix of active and passive voice in describing the river closing conditions and procedures. Passive voice seems to dominate in describing the conditions under which action will be taken: "closure will be recommended when . . ."; "the decision to close a river will be based on . . ."; "reopening will be recommended when. . . ." But under the Procedures subheads, active voice is used to identify specifically who in the government will take action: ". . . the Secretary of the Department of Natural Resources (DNR) will recommend closure of the affected area to the Secretary of the Department of Health and Mental Hygiene . . ."; "Natural Resources Police will regularly patrol the area . . ."; ". . . and the Governor may issue a public caution." The effect is to emphasize the active role the government will take in protecting its citizens.

These clear explanations of the chain of command and the individuals who are responsible for acting all underscore the public accountability function of this document. Such instructions could be, and probably are, also conveyed by internal documents, since they will be carried out by specified individuals in specified departments. The posting of this internal information for the public indicates that MDNR also has a broader goal of demonstrating that the agency is meeting its responsibilities. Though some may argue that a good public image is an end in itself and helps the government agency perpetuate itself, we prefer to presume that the ultimate goal of inspiring confidence and trust in an institution's preparedness is to lessen anxiety and help the population cope with potentially threatening situations. Interestingly, in Maryland the decision to close a river due to environmental threat is not ultimately made by the Secretary of the Department of Natural Resources but by the Secretary of the Department of Health and Mental Hygiene, suggesting that the decision will take into account (we are speculating) potential psychological as well as physical effects of both the threat and the act of closing the river.

Lastly, we should not overlook the information value of this document. The public does need to understand under what conditions the river will be closed, not only so they can rest assured that the state is doing its job but also so they understand the severity of the conditions that may lead the state to act, particularly since actions such as closing or patrolling a river may negatively impact individuals' livelihoods or recreational activities. A well-informed public also will minimize confusion and complaints should a river closing be required. Thus, this seemingly simple description of procedures serves a variety of purposes in maintaining the complex relationship between a government and the public it serves.

## EXERCISE 7.5

As we write this book, American culture abounds with examples of "accountability" documents, as institutions in private and public domains disseminate descriptions of security procedures they have instituted to comply with mandates issued by the newly created U.S. Department of Homeland Security in the aftermath of the September 11, 2001 terrorist attacks. Though such plans sometimes contain concrete instructions for employees or the public to act on, they are frequently designed to inform those groups of actions the organization has already taken or will take in case of emergency. Look for documents of this sort posted by your university, a local or state agency, or another organization you are interested in. Analyze the procedures described in the document. What proportion of the material is intended for readers, or some subset of readers, to act on? Who are those readers? What proportion of the material serves a reporting or evidentiary function instead? Compare your findings with others in class who have analyzed texts from different organizations. Are there any "rules" governing security documents? Is this an emerging genre? Do different types of institutions (e.g., profit/nonprofit; local/state/federal; large/small) create different types of procedural documents in response to the Homeland Security mandate, or are there some commonalities across these varied contexts?

In this chapter we have sought to illustrate some of the diverse purposes and audiences for scientific communication beyond the boundaries of the academic research community. The authors of the procedural texts sampled here must anticipate the needs of readers with widely varying types and levels of expertise and must envision the real-world contexts in which their documents are to be read and, more importantly, used. Each of these documents has been shaped by its purpose and rhetorical context, from the strictly regulated SOP and clinical protocol to the more loosely defined forms of the product information text and public safety guidelines. Further investigation of these latter categories may well reveal that they too are governed by more predictable conventions than are evident here, for persistent audience needs and recurring rhetorical purposes exert a powerful influence on the development of written forms.[7] Just as the content and structure of the research report and proposal are shaped by the needs and expectations of the research community, so are these procedural texts shaped by the immediate needs of those who will use them and the goals and standards of those responsible for monitoring and approving the organization's work, whether that be a governmental body monitoring an industry or the general public monitoring its government.

As we conclude this discussion we note that the concept of *ethos* has figured prominently in these analyses. We've argued throughout this book that the *ethos* projected by the authors of research reports and proposals plays an important role in determining acceptance for publication and the distribution of research funds. The success or failure of these research texts has consequences not only for the continuation of a particular research agenda but also for the development of a field's knowledge. When we move into the applied settings explored in this chapter, the stakes become more concrete and immediate. These documents must work well—to ensure that EPA's research is reliable, that a promising drug receives regulatory approval, that Alimenterics' new product works as promised, that North Carolinians are protected when they work or play near hazardous waters, that public anxieties over *Pfiesteria* are assuaged in Maryland. An agency's or organization's failure to meet these goals has "real" costs, for example, to public health and safety or to the financial well-being of the corporation, to say nothing of the lost opportunities to educate the public. The organization's reputation is on the line in all these documents, and in the realms of government and commerce, reputation matters enormously. The careful description and communication of scientific knowledge in these contexts reflect on the "character" of the authoring organization, enhancing or diminishing public trust in it.

# Activities and Assignments

1. Write an SOP for a common procedure you've conducted in a lab, following principles outlined in Sections 7.2 and 7.3. Then transform these instructions into a description of the procedure appropriate for inclusion in the methods section of a lab report or journal article. Aside from the shifts in person, tense, and mood, what kinds of details did you include in the SOP for an internal audience but exclude from the methods section for an external audience?
2. The principles for giving instructions outlined in Section 7.2 facilitate basic cognitive processes in reading and are applicable to any kind of procedural text. For example, they are equally applicable when developing instructions

---

[7]See, for example, genre theorists Bitzer (1968), Campbell and Jamieson (1978), Miller (1984).

for consumer products when the user is a nonexpert, not a medical professional. Look at the directions accompanying an over-the-counter medical product, such as a contact lens solution. Directions are typically included in the FDA-regulated product label or package insert.

a. Analyze these directions using the principles outlined in Section 7.2. Do the directions follow the basic principles for written instructions? If not, rewrite them.

b. In addition to examining the stylistic principles, compare these instructions to the structural components typically found in SOPs (Figure 7.2). For example, do the product directions explain the scope of application, include precautions or special considerations, or provide quality assurance information? In comparing your observations with those of other students analyzing other products, determine how common these various structural components are in consumer product documentation. Are these or other types of information routinely included in or excluded from product information inserts? What generic conventions can you discern from examining a range of such documents?

3. Look for samples of instructions or procedures at your state's department of natural resources or public health website. Write a rhetorical analysis of the document in which you explain who the target audience(s) is/are, what the occasion or exigence is (why this information is needed now), and the purpose(s) for posting the material. Research the motivating event or issue in the local press to develop a comprehensive understanding of the context in which the agency has chosen to post the material. Check the site to see when the material was posted. How often is it updated? Does the site offer further information or connect to recent research? Do the authors intend these instructions to be followed by those who visit the site? By some but not by others? If they are not to be followed by some or by any readers, what purpose(s) do they serve for those audiences?

4. Research the controversy over the effects of *Pfiesteria* on human health. As starting points, see Griffith (1999); Burkholder and Glasgow (1999); Samet et al. (2001), and the popular nonfiction account *And the Waters Turned to Blood* by Rodney Barker (1997). Write an analysis of this research case that explores the relationship between science and public policy. How have the standard scientific mechanisms of peer review, publication, and competitive grant funding figured into the development of knowledge on this topic? How has this knowledge been brought to bear on governmental policy and behavior? What factors have influenced and are influencing the flow of information and the interpretation of the science? It also may be useful to compare how state governments in North Carolina and in Maryland have responded to the emerging research on *Pfiesteria*. How have the processes of science and the processes of politics interacted in these cases? Have these combined processes served the public interest? If so, show how these processes have worked to ensure accurate interpretation and appropriate application of scientific knowledge. If not, explain where and how the system seems to break down and consider whether such problems can be avoided. What is the scientist's responsibility in resolving such controversies? What is the government's responsibility?

# 8

# Communicating with Public Audiences

## 8.1 Why Do Scientists Communicate with Public Audiences?

In Chapter 1 we noted that scientific disciplines are not only communities in themselves but also parts of the larger society in which these communities are situated and in which scientists create, produce, and live. This relationship between scientists and society is of great interest to several national science organizations, including the Committee on Assessing Integrity in Research Environments (CAIRE). In a joint report issued in 2002, CAIRE et al. state:

> The pursuit and dissemination of knowledge enjoy a place of distinction in American culture, and the public expects to reap considerable benefit from the creative and innovative contributions of scientists. As science becomes increasingly intertwined with major social, philosophical, economic, and political issues, scientists become more accountable to the larger society of which they are a part. (p 33)

In previous chapters we have explored genres and conventions that professional scientists use to communicate with each other, and some applied forms used in the realms of industry and government. In this chapter we will explore conventions used to communicate knowledge of science and scientific research to public audiences in the larger society.

The term *public audience* is a rather loose one and is meant here to imply a wide range of listeners and readers with a variety of interests, needs, and educational backgrounds. It may be a group or a professional person (even with a PhD in another field) in need of information about a particular scientific topic, or a general reader or listener simply curious about science. Public audiences for science thus can include, but are not limited to, those who regularly use the results of scientific research in the course of their daily work (such as agricultural producers, fish and game managers, medical professionals); administrators, local government agencies, and other public officials who need scientific information to make decisions about issues such as waste management, industrial and environmental regulations, and road construction; clubs, classes, and other educational or special-interest groups that want to learn about science; private citizens who use natural resources for hunting, fishing, hiking, and other recreational purposes; and the public at large, which has a vested interest in science insofar as they support it financially through government funding and must live with its consequences.

Thus, public audiences are not monolithic, but rather quite diverse, and so no single, "rationalistic" formula will suffice to define them or their interests (see Locke 2001). In addition to addressing more general audiences *within* the scientific community (e.g., readers of journals such as *Science* and *Nature,* or the mixed audience of the research proposal), scientists may wish to write articles for the much broader audiences that read science-oriented journals such as *Scientific American* and *Discover,* general-interest publications such as *Time* and *Newsweek,* or the feature section of a newspaper. They also may make presentations to public audiences, participate in question-and-answer sessions, take part in public policy debates affected by developments in their research areas, and give press releases and interviews on important discoveries or issues in their field.

## EXERCISE 8.1

Choose a topic in your field (perhaps derived from a research report you have read), and identify some specific public audiences you might address on this topic. Who outside your field might read or listen to what you as an expert have to say about this topic? Why would they be reading or listening? How do the needs and interests of these groups differ from those of experts in your field? What do they want or need from your presentation, and how much would they already know? (Are there any public audiences with whom experts in your field might communicate but currently don't?)

There are three major reasons scientists communicate with the general public: moral, economic, and political. The National Academy of Sciences asserts that scientists have an ethical responsibility to understand and explain the effect of the work that they do on the society in which they live:

> The occurrence and consequences of discoveries in basic research are virtually impossible to foresee. Nevertheless, the scientific community must recognize the potential for such discoveries and be prepared to address the questions that they raise. If scientists do find that their discoveries have implications for some important aspect of public affairs, they have a responsibility to call attention to the public issues involved. (NAS 1995, p 20)

Genetic research—including genetically modified foods (GMFs), stem cell research, and cloning—and nuclear power are two obvious cases in which scientific discoveries are morally controversial or have complicated, long-term implications for society (Associated Press 1996). Other controversial areas of research include environmental protection and wildlife preservation, the use of laboratory animals in medical research, the development of drugs, the application of medical technology to prolong life, and biological studies of race and gender.

A second reason scientists communicate with the general public is economic and hinges on the practical question: Who funds science? In addition to private labs, corporations, and universities, it is the government, and therefore the public, that funds science through tax dollars and so indirectly chooses which projects to support. Public financial support of science takes two forms: the funding of governmental agencies that conduct scientific research and the funding of government grant programs that support research by scientists at other institutions. The federal government's proposed budget for 2004 included $15,469,000,000 for research and development at NASA (NASA 2003), $74,000,000 for the Department of Agriculture (USDA 2003), and $27,893,000,000 for NIH (NIH 2003b). In 2002, $32,000,000 in federal and matching funds were awarded to the National Sea Grant Office (National Sea Grant 2003). And the President signed an authorization act increasing the National Science Foundation's (NSF) budget from $4,790,000,000 in 2002 to $9,840,000,000 in 2007 (NSF 2002b). Increases in some areas of scientific research and education were undoubtedly spurred by the terrorist attack on the United States on September 11, 2001, notably research on biochemical and biological agents. But other government-run programs have either been scaled back because of budget cuts (e.g., NASA's Mars expedition) or eliminated altogether (the Superconducting Super Collider). "Big science" projects in which the federal government plays a major role (such as the Hubble Telescope, the International Space Station, and the Human Genome Project) require costly equipment and the coordination of efforts of scientists around the world and can be prohibitively expensive. In a time of renewed budget deficits and economic belt tightening, scientists must be able to convince not only their peers but also the public and its official representatives in government of the worthiness of scientific projects. Responsibility for garnering public understanding of, enthusiasm for, and goodwill toward science ultimately rests with scientists.

The third reason scientists should learn to communicate with the general public is related to the politics of a democracy. A democratic society requires that its citizens (both electorate and elected) be informed about the issues that confront it. Since science is a major cultural force in our democracy, many of the policy decisions we make are about or based on science. As the National Academy of Sciences states:

> [S]cience and technology have become such integral parts of society that scientists can no longer isolate themselves from societal concerns. Nearly half of the bills that come before Congress have a significant scientific or technological component. Scientists are increasingly called upon to contribute to public policy and to the public understanding of science. They play an important role in educating nonscientists about the content and processes of science. (NAS 1995, p 21)

Thus, a democracy such as ours needs a scientifically informed citizenry to arrive at good decisions about what research to support and how to interpret and apply its results (see O'Keefe 2001). In a 1996 survey of the American public, the

National Science Foundation found only 25 percent performed well on a basic test in science and economics (Associated Press 1996). That same survey, however, found that 72 percent of the participants considered scientific research valuable. A later study by NSF found that "[M]ost Americans have a positive attitude about science and technology" (CAIRE et al. 2002, p 16), but also argues that scientists themselves have an important role to play in maintaining that positive relationship with the public. Scientists can help citizens understand and continue to appreciate science by becoming aware of the various genres through which the public gets its information and by learning to use the conventions of those genres to effectively communicate with the public. According to John Wilkes (1990), director of the Science Communication Program at the University of California–Santa Cruz,

> U.S. citizens get up to 90% of their information about science from newspapers, magazines, and, to a lesser extent, television. As producers of scientific knowledge, scientists are in the best position to use the media to teach the public what it wants—and needs—to know about developments in medicine, science, and technology. (p 15)

While most scientists are not professionally trained in speaking or writing to general audiences, many scientists have recognized that the general public is an important audience to reach. As Bazerman demonstrates in *The Languages of Edison's Light* (1999), Thomas Edison was a master of public relations; it would not be farfetched to say that the adoption of electric power and light was the result of his ability to employ various media (including the lightbulb itself in spectacular displays of electric power) to persuasively communicate with a diversity of audiences: investors, businessmen, the U.S. Patent Office, international governments, the press, and the public.[1] Books and articles by such famous scientists as physicist Stephen Hawking, paleontologist Stephen Jay Gould, marine biologist Rachel Carson, research physician Lewis Thomas, and anthropologist Richard Leakey—just to name a few—as well as the current popularity of television programs about science (*National Geographic Explores, NATURE,* and *NOVA*) suggest that rather than heading to ivory-tower labs and leaving the communication of science to journalists, scientists are taking to public pages and airwaves to explain their work and pique an interest in their science.

As a medium, the Internet also is increasingly being used as a means for scientists and scientific government agencies to communicate directly with public audiences. As discussed in Chapter 7, for instance, the World Wide Web provides a convenient medium for conveying safety information to a broad range of audiences, including technicians working in the field; swimmers, boaters, and fishermen; and the public at large.

With open and rapid dissemination of information and the potential anonymity of sources on the Web, one problem that has emerged in the public communication of science online is the authenticity and quality of information. It is a commonplace that because of the frequent lack of quality control mechanisms, not all the information found on the Web is reliable. This is an issue that every user

---

[1]Lievrouw (1990) has studied the Pons and Fleischmann debate as a modern-day example of the way "scientists strategically use popular media to make knowledge accessible to the public at large and to make themselves known." (p 1)

looking for information on the Internet faces, of course. On what basis can and does the public judge the validity of work (scientific or otherwise) displayed on the screen? How can the public know which information is reliable?

---

**EXERCISE 8.2**

On the World Wide Web, search for information on a subject that you are perhaps interested in but don't know anything about. After reading the material, speculate on whether the information is reliable—whether you can trust this website. How do you know? Make a list of specific features of the author(s), the organization(s), the website design and graphics, the writing, the content, and anything else that you are using to assess the validity and truthfulness of the information the site contains. What are some of the ways the public evaluates online information? What questions/issues about the validity and truthfulness of the information remain unanswered for you? What else do you need to know before you can trust this site? How can online communication with the public be improved?

---

One solution to the inconsistency of online communication may be the direct interaction of scientists with the public online. The interactive capabilities of the World Wide Web make the Internet a potential meeting place where scientists can directly interact not only with each other (Smith 1997), but also with other publics. In fact, the online journal, *Issues in Science and Technology Online*, "a forum for discussion of public policy related to science, engineering, and medicine" (2002, http://www.nap.edu/isssues/about.html) envisions itself as a public meeting place, focusing on social policy as well as policies designed to enhance specific research fields—and not only the communication of social policy, but also the interactive discussion of it. The journal's editors state their mission on their homepage:

> Although *Issues* is published by the scientific and technical communities, it is not just a platform for these communities to present their views to Congress and the public. Rather, it is a place where researchers, government officials, business leaders, and others with a stake in public policy can share ideas and offer specific suggestions.

> Unlike a popular magazine, in which journalists report on the work of experts, or a professional journal, in which experts communicate with colleagues, Issues offers authorities an opportunity to share their insights directly with a broad audience. And the expertise of the boardroom, the statehouse, and the federal agency is as important as that of the laboratory and the university.

The public also may learn to distinguish between reliable and less reliable information simply by becoming more familiar with websites. Earlier studies of the impact of technologies on audiences suggest that technologies do change "the range of experiences and skills that audiences bring to media" (Nightingale 1986, p 31). Improvement in the skills of the public also to some extent depends on continuing innovations and improvements in the accessibility, quality, and presentation of online information. For instance, computer companies learned in the 1980s that they could not simply dump printed information online, but rather had to adapt that information to the new medium, as well as develop the new medium in ways that would facilitate that adaptation. An understanding of the principles and techniques of audience adaptation, and of general audiences themselves, is essential for successfully communicating with the public in any medium.

# 8.2 Understanding "General" Audiences

Some research indicates that human reasoning, even in sciences, is "proverbial" —based on human experience and common sense (e.g., Shapin 2001). But of course, human experience varies. Public audiences can be as diverse as the general population in their knowledge, interests, and needs. Listeners and readers in these audiences will possess varying types and degrees of scientific knowledge. What these general audiences have in common is a presumed lack of knowledge about *your* topic. Just as scientists must understand the conventions governing communication with their peers to be successful professionals, so too must they understand the conventions of communicating with public audiences in order to do so successfully. In this book we cannot describe in detail the many forms public communication may take. But we can discuss some general strategies for adapting scientific information to meet the needs and interests of nonscientists in a wide range of situations. These strategies can then be applied to specific audiences, genres, and mediums as the occasion demands.

**EXERCISE 8.3**

In Figure 8.1 we have reproduced the introductions to five different articles on the same topic. You'll see that they are clearly intended for different audiences. As you read, think about what kind of publication these pieces would have appeared in. (This exercise has been adapted from Bradford and Whitburn [1982].)

A. Read the five introductions and categorize them according to the level of specialized knowledge assumed on the part of the audience. Use a scale of 1 (general audience) to 5 (most specialized audience). Speculate about where each piece might have been published.

B. After you've categorized the texts, reflect on what criteria you used to do so. In what ways do these introductions vary? What made you decide that one article is intended for a more general audience than another? Be sure to consider all dimensions of the text, including such features as content and organization, terminology and phrasing, formatting and visual presentation, tone and point of view. List the many ways in which these texts vary. Illustrate each of these features with a pair of contrasting examples from the passages.

C. Your instructor may ask you to work in small groups to develop a consensus ordering of the five texts, a consensus list of the features on which they vary, and a set of contrasting examples to illustrate each feature.

## Introduction A

RECENT studies have provided reasons to postulate that the primary timer for long-cycle biological rhythms that are closely similar in period to the natural geophysical ones and that persist in so-called constant conditions is, in fact, one of organismic response to subtle geophysical fluctuations which pervade ordinary constant conditions in the laboratory (Brown, 1959, 1960). In such constant laboratory conditions a wide variety of organisms have been demonstrated to display, nearly equally conspicuously, metabolic periodicities of both solar-day and lunar-day frequencies, with their interference derivative, the 29.5-day synodic month, and in some instances even the year. These metabolic cycles exhibit day-by-day irregularities and distortions which have been established to be highly significantly correlated with aperiodic meteorological and other geophysical changes. These correlations provide strong evidence for the exogenous origin of these biological periodisms themselves, since cycles exist in these meteorological and geophysical factors.

In addition to possessing these basic metabolic periodisms, many organisms exhibit also overt periodisms of numerous phenomena which in the laboratory in artificially controlled conditions of constancy of illumination and temperature may depart from a natural period. The literature contains many accounts, for a wide spectrum of kinds of plants and animals, of regular rhythmic periods ranging from about 20 to about 30 hours. The extent of the departure from 24 hours is generally a function of the level of the illumination and temperature.

It has been commonly assumed, without any direct supporting evidence, that the phase- and frequency-labile periodisms persisting in constant conditions reflect inherited periods of fully autonomous internal oscillations. However, the relationships of period-length to the ambient illumination and temperature levels suggest that it is not the period-length itself which is inherited but rather the characteristics of some response mechanism which participates in the derivation of the periods in a reaction with the environment (Webb and Brown, 1959).

It has recently been alternatively postulated that the timing mechanism responsible for the periods of rhythms differing from a natural one involves, jointly, use of both the exogenous natural periodisms and a phenomenon of regular resetting, or "autophasing," of the phase-labile, 24-hour cycles in reaction of the rhythmic organisms to the ambient light and temperature (Brown, Shriner and Ralph, 1956; Webb and Brown, 1959; Brown, 1959). It is thus postulated that the exogenous metabolic periodisms function critically as temporal frames of reference for biological rhythms of approximately the same frequencies.

**FIGURE 8.1** Five introductions. Bradford and Whitburn. "Analysis of the same subject in diverse periodicals." Technical Writing Teacher 9 (Winter 1982).[2]

The basic principle of audience adaptation is that we build a discussion or argument on the knowledge, goals, values, and experience of the audience. This principle is the basis of *all* successful communication and teaching, including that between professional scientists. It is also essential for the scientist communicating with public audiences. In adapting scientific information for nonspecialists, the writer or speaker introduces new knowledge by trying to relate it to what the audience knows or values; this new knowledge, grounded in what the audience already knows, then becomes the foundation for more new knowledge and so forth.

When you write for other scientists in your area of specialization, as in the research report, you can assume your readers have some degree of familiarity with the topic to begin with, as well as some degree of interest. When you write for public audiences, however, you need to be more cautious. Instead of assuming knowledge on the part of your readers, you must help them develop that knowledge; instead of assuming interest in the topic, you must generate interest. The way to begin is by assessing your audience's knowledge, needs, and goals.

---

[2]Original formatting and typeface are reproduced as closely as possible. Sources for these introductions are listed in the Works Cited list and identified in the Instructor's Manual.

## Introduction B

One of the greatest riddles of the universe is the uncanny ability of living things to carry out their normal activities with clocklike precision at a particular time of the day, month and year. Why do oysters plucked from a Connecticut bay and shipped to a Midwest laboratory continue to time their lives to ocean tides 800 miles away? How do potatoes in hermetically sealed containers predict atmospheric pressure trends two days in advance? What effects do the lunar and solar rhythms have on the life habits of man? Living things clearly possess powerful adaptive capacities—but the explanation of whatever strange and permeative forces are concerned continues to challenge science. Let us consider the phenomena more closely.

Over the course of millions of years living organisms have evolved under complex environmental conditions, some obvious and some so subtle that we are only now beginning to understand their influence. One important factor of the environment is its rhythmicality. Contributing to this rhythmicality are movements of the earth relative to the sun and moon.

The earth's rotation relative to the sun gives us our 24-hour day; relative to the moon this rotation, together with the moon's revolution about the earth, gives us our lunar day of 24 hours and 50 minutes. The lunar day is the time from moonrise to moonrise.

The moon's arrival every 29.5 days at the same relative position between the earth and the sun marks what is called the synodical month. The earth with its tilted axis revolves about the sun every 365 days, 5 hours and 48 minutes, yielding the year and its seasons.

The daily and annual rhythms related to the sun are associated with the changes in light and temperature. The 24.8-hour lunar day and the 29.5-day synodical month are associated most obviously with the moon-dominated ocean tides and with changes in nighttime illumination. But all four types of rhythms include changes in forces such as gravity, barometric pressure, high energy radiation, and magnetic and electrical fields.

Considering the rhythmic daily changes in light and temperature, it is not surprising that living creatures display daily patterns in their activities. Cockroaches, earthworms and owls are nocturnal; songbirds and butterflies are diurnal; and still other creatures are crepuscular, like the crowing cock at daybreak and the serenading frogs on a springtime evening. Many plants show daily sleep movements of their leaves and flowers. Man himself exhibits daily rhythms in degrees of wakefulness, body temperature and blood-sugar level.

We take for granted the annual rhythms of growth and reproduction of animals and plants, and we now know that the migration periods of birds and the flowering periods of plants are determined by the seasonal changes in the lengths of day and night.

In a similar fashion creatures living on the seashore exhibit a rhythmic behavior corresponding to the lunar day. Oyster and clams open their shells for feeding only after the rising tide has covered them. Fiddler crabs and shore birds scour the beach for food exposed at ebb tide and retreat to rest at high tide.

*(continued on page 184)*

**FIGURE 8.1** *(continued)*

# Introduction C

Familiar to all are the rhythmic changes in innumerable processes of animals and plants in nature. Examples of phenomena geared to the 24-hour solar day produced by rotation of the earth relative to the sun are sleep movements of plant leaves and petals, spontaneous activity in numerous animals, emergence of flies from their pupal cases, color changes of the skin in crabs, and wakefulness in man. Sample patterns of daily fluctuations, each interpretable as adaptive for the species, are illustrated in Fig. 1. Rhythmic phenomena linked to the 24-hour and 50-minute lunar-day period of rotation of the earth relative to the moon are most conspicuous among intertidal organisms whose lives are dominated by the ebb and flow of the ocean tides. Fiddler crabs forage on the beaches exposed at low tide; oysters feed when covered by water. "Noons" of sun- and moon-related days come into synchrony with an average interval of 29½ days, the synodic month; quite precisely of this average interval are such diverse phenomena as the menstrual cycle of the human being and the breeding rhythms of numerous marine organisms, the latter timed to specific phases of the moon and critical for assuring union of reproductive elements. Examples of annual biological rhythms, whose 365¼-day periods are produced by the orbiting about the sun of the earth with its tilted axis, are so well known as scarcely to require mention. These periodisms of animals and plants, which adapt them so nicely to their geophysical environment with its rhythmic fluctuations in light, temperature, and ocean tides, appear at first glance to be exclusively simple responses of the organisms to these physical factors. However, it is now known that rhythms of all these natural frequencies may persist in living things even after the organisms have been sealed in under conditions constant with respect to every factor biologists have conceded to be of influence. The presence of such persistent rhythms clearly indicates that organisms possess some means of timing these periods which does not depend directly upon the obvious environmental physical rhythms. The means has come to be termed "living clocks."

## Autonomous-Clock Hypothesis

From the earliest intensive studies of solar-day rhythmicality during the first decade of this century by Pfeffer (*1*), with bean seedlings, certain very interesting properties of this rhythm became clearly evident. Pfeffer found that when his plants were reared from the seed in continuous darkness, they displayed no daily sleep movements of their leaves. He could easily induce such a movement, however, by exposing the plants to a brief period of illumination. Returned to darkness, the plants possessed a persisting daily sleep rhythm. The time of day when the leaves were elevated in the daily rhythm was set by the time of day when the single experimental light period commenced. It was apparent that the daily rhythmic mechanism possessed the capacity for synchronization with the outside daylight cycles while having its cyclic phases experimentally altered by appropriate light changes made to occur at any desired time of day. These alterations would then persist under constant conditions. Since Pfeffer's time, this property has been abundantly confirmed for numerous other plants and animals. The daily rhythms, therefore, exhibit the capacity for synchronization with external, physical cycles while having freely labile phase relations.

A second discovery, also made by Pfeffer, was that the daily recurring changes under constant conditions could occur earlier, or later, day by day, to yield regular periods deviating a little from the natural solar-day ones. Periods have now been reported ranging from about 19 to 29 hours. The occurrence of persisting rhythmic changes under constant conditions, with regular periods of other than precisely 24 hours, clearly indicated that these observed rhythmic periods could not be a simple direct consequence of any known or unknown geophysical fluctuation of the organism's physical environment.

A third fundamental contribution to the properties of the daily rhythms was made by Kleinhoonte (*2*). While confirming, in essentials, all of Pfeffer's findings, she discovered that the daily sleep movements of plants could be induced to "follow" artificial cycles of alternating light and dark ranging from about 18-hour "days" to about 30-hour "days." When the "days" deviated further than these limits from the natural solar-day period the plants "broke away" to reveal their normal daily periodicity, despite the continuing unnatural light cycles. This observation clearly emphasized the very deep-seated character of the organismic daily rhythm.

**FIGURE 8.1**   *(continued)*

## Introduction D

A deep-seated, persistent, rhythmic nature, with periods identical with or close to the major natural geophysical ones, appears increasingly to be a universal biological property. Striking published correlations of activity of hermetically sealed organisms with unpredictable weather-associated atmospheric temperature and pressure changes, and with day to day irregularities in the variations in primary cosmic and general background radiations, compel the conclusion that some, normally uncontrolled, subtle pervasive forces must be effective for living systems. The earth's natural electrostatic field may be one contributing factor.

A number of reports have been published over the years advancing evidence that organisms are sensitive to electrostatic fields and their fluctuations. More recently Edwards (1960) has found that activity of flies was reduced by sudden exposures to experimental atmospheric gradients of 10 to 62 volts/cm., and that prolonged activity reduction resulted from gradient alternation with a five-minute period. In 1961, Edwards reported a small delay in moth development in a constant vertical field of 180 volts/cm., but less delay when the field was alternated. The moths tended to deposit eggs outside the experimental field, whether constant or alternating, in contrast to egg distribution of controls. Maw (1961), studying rate of oviposition in hymenopterans, found significantly higher rates in the insects shielded from the natural field fluctuations, whether or not provided instead with a constant 1.2 volts/cm. gradient, than were found in either the natural fluctuating field, or in a field shielded from the natural one and subjected to simulated weather-system passages in the form of a fluctuating field of 0.8 volts/cm.

A study in our laboratory early in 1959 (unpublished) by the late Kenneth R. Penhale on the rate of locomotion in *Dugesin* suggested strongly that the rate was influenced by the difference in charge of expansive copper plates placed horizontally in the air about six inches above and closely below a long horizontal glass tube of water containing the worms. Locomotory rates in fields of 15 volts/cm. (+ beneath the worms) were compared with those in fields between equipotential plates. The fields were obtained with a Kepco Laboratories, voltage-regulated power supply. A comparable study with the marine snail, *Nassarius,* by Webb, Brown and Brett (1959), employing a Packard Instrument Co., high-voltage power supply, confirmed the occurrence of such responsiveness to vertical fields of 15 to 45 volts/cm., and advanced evidence that the response of the snails displayed a daily rhythm.

*(continued on page 186)*

**FIGURE 8.1** *(continued)*

In *The Rhetorical Act,* Karlyn Kohrs Campbell (1982) gives the following advice on assessing audience (p 149–150). Consider your subject *from the point of view of your readers or listeners.* What expectations does your audience have about the subject? About you? How is your audience likely to see and/or understand the topic or issue? Are there conflicting beliefs or concepts that will have to be dealt with, and how will you deal with them? Are familiar explanations trite or boring? Does the audience have firsthand experiences that you can draw on to illustrate points in your discussion?

To answer these questions, you will need to find out as much as possible about the background, areas of expertise, probable beliefs, values, and general interests of members of your audience. One way to learn about your audience is to ask the organization sponsoring your talk about who will be in attendance and why. Or read the magazine you are writing for and look at the kinds of text features you noted in Exercise 8.3; these features can often provide hints about how to interpret and adapt to a particular audience.

## Introduction E

Everyone knows that there are individuals who are able to awaken morning after morning at the same time to within a few minutes. Are they awakened by sensory cues received unconsciously, or is there some "biological clock" that keeps accurate account of the passage of time? Students of the behavior of animals in relation to their environment have long been interested in the biological clock question.

Most animals show a rhythmic behavior pattern of one sort or another. For instance, many animals that live along the ocean shores have behavior cycles which are repeated with the ebb and flow of the tides, each cycle averaging about 12½ hours in length. Intertidal animals, particularly those that live so far up on the beaches that they are usually submerged only by the very high semimonthly tides when the moon's pull upon the ocean waters is reinforced by the sun's, have cycles of behavior timed to those 15-day intervals. Great numbers of lower animals living in the seas have semilunar or lunar breeding cycles. As a result, all the members of a species within any given region carry on their breeding activities synchronously; this insures a high likelihood

of fertilization of eggs and maintenance of the species. The Atlantic fireworm offers a very good example of how precise this timing can be. Each month during the summer for three or four evenings at a particular phase of the moon these luminescing animals swarm in the waters about Bermuda a few minutes after the official time of sunset. After an hour or two only occasional stragglers are in evidence. Perhaps even more spectacular is the case of the small surface fish, the grunion, of the U.S. Pacific coast. On the nights of the highest semilunar tides the male and female grunion swarm in from the sea just as the tide has reached its highest point. They are tossed by the waves onto the sandy beaches, quickly deposit their reproductive cells in the sand and then flip back into the water and are off to sea again. The fertilized eggs develop in the moist sand. At the time of the next high tide when the spot is again submerged by waves, the young leave the nest for the open sea.

Almost every species of animal is dependent upon an ability to carry out some activity at precisely the correct moment. One way to test whether these activities are set off

by an internal biological clock, rather than by factors or signals in the environment, is to find out whether the organisms can anticipate the environmental events. The first well-controlled experimental evidence on the question was furnished by the Polish biologist J. S. Szymanski. In experiments conducted from 1914 to 1918 he found that animals exhibited a 24-hour activity cycle even when all external factors known to influence them, such as light and temperature, were kept constant. During the succeeding 20 years various investigators, especially Orlando Park of Northwestern University, J. H. Welsh of Harvard University and Maynard Johnson (currently in the U.S. Navy), demonstrated that comparable rhythmic processes persisted in many insects, in crustaceans and in mice. Persistent daily rhythmicity has been found in animals ranging from one-celled protozoa to mammals. And the Austrian biologist Carl von Frisch, using a slightly different approach, discovered that bees could be trained to come to a feeding station at the same time on successive days but not at different times—a finding which suggested that bees have an internal daily cycle.

**FIGURE 8.1** *(continued)*

Take a moment to consider the rhetorical situation described in Exercise 8.4. To extend a distinction drawn by Flower (1993), Jane was "speaker-based" rather than "listener-based": She was more concerned with what she as the speaker was interested in and had to say about lasers than in what Mike as the listener was interested in and may have wanted to hear about lasers. The story of Jane and Mike is a microcosm of large- and small-scale breakdowns in communication that occur every day in our society. In fact, communication breakdown has been cited as a major contributing factor in a number of serious technological accidents. In the case of the Three Mile Island nuclear reactor meltdown in 1979, for example,

**EXERCISE 8.4**

Consider this scenario. Jane, a premed student working with lasers, wanted to show her friend Mike, a zoology major, how a new laser in her lab worked: "Come on over to my lab and I'll give you a demonstration." Mike had never studied or worked with lasers, but from what he had heard, they seemed fascinating; so one day he took Jane up on her offer. When Mike got to the lab, Jane escorted him through a maze of machines to the lab table where her laser was set up, and she proceeded to take the laser apart and explain each major component. Mike quickly lost interest and wandered away to look at the other machines, while Jane continued to discuss at length the technical details she had learned about lasers, oblivious to the fact that Mike was no longer listening. What do you think happened?

A. Using the information given about Mike and your own common sense and empathy (both necessary in audience adaptation), what do you think Mike expected when he walked into Jane's lab? What would Mike like to see and learn about lasers?

B. Now, if Mike were an administrator of the lab—say in charge of personnel and finance—why would he be interested in the laser? What would he want to see and learn?

C. If Mike were a parent whose child's school was about to purchase a laser for use in science classes, what would he want to know about it?

faulty assumptions about how to communicate with the public hampered the efforts of officials to find the best way to inform them and to control the emergency and led to widespread panic and social disarray (Farrell and Goodnight 1981). The inability of engineers and managers to understand each others' values has been directly implicated as a major cause of the Challenger shuttle explosion (Herndl et al. 1991), and is more than likely an important factor in the Columbia accident as well (see Columbia Accident Investigation Board 2003, esp. Chapter 7). And the inability of experts and government officials to consider the values and emotions of public audiences has been a factor in unsuccessful attempts to site low-level radioactive waste facilities all around the country (Katz and Miller 1996).

In all these situations, communicators in one area of expertise seriously misunderstood the audience outside their fields, with dire consequences. Unlike the expert colleague, who has both knowledge of the subject and an intrinsic interest in it, public audiences have different perspectives on and interests in the subject, and thus different expectations and needs that must be appealed to. While experts are interested in theory and technical details, in methods and results, public audiences are generally interested in what things "do" and their effect on public safety, health, and welfare.

In addition to the three general modes of appeal (*logos, pathos, ethos*) discussed in Chapter 6, two special appeals often come into play in the accommodation of scientific knowledge to public audiences (Fahnestock 1986). The first is the *wonder appeal,* which emphasizes the sense of surprise and joy and awe people (both generalists and specialists!) often feel when confronted with an exciting scientific discovery. The second is the *application appeal,* which emphasizes the practical benefits of a scientific concept or discovery for a particular audience, a society at large, or humankind. This appeal is especially effective with administrators and public officials.[3]

Most of us are fascinated by the accomplishments and spectacle of science, and interested in what things do from the point of view of common experience or daily life. Practical application is also important to the general public. Figure 8.2 contains a blurb by John O'Neil in the *New York Times* reporting results on the use of an H-pylori breath test that are similar to Chiba et al.'s (2002) published in the *British Medical Journal* (contained in Chapter 10, pages 280–286). Comparing the blurb to the Chiba et al. article, note the amount of scientific detail that is not in the blurb. Also note the appeals to "wonder" in the title and the graphic of the blurb that is not in the scientific article in *BMJ,* and the two direct appeals to "application," which are merely referenced in the *BMJ* article, in the last paragraph of the blurb: the breath test (which is as good as endoscopy) is less invasive than endoscopy, and is cheaper than endoscopy.

Both the wonder appeal and the application appeal work well with general audiences. In the scenario described above, Mike may have secretly wanted to ask: "Can we see the laser burn a hole in something?" (wonder). "How can lasers be used to shoot down missiles in outer space and also perform delicate eye surgery?" (wonder and application). "How could I use lasers in my zoology major?" (application). He never got a chance.

---

**EXERCISE 8.5**

Assume you're a member of Stephen Reynolds's research team. You've been invited to speak to a local amateur astronomy club on the topic of supernova remnants. Using the audience analysis questions on page 185 as your starting point, write a paragraph describing your audience. Write a second paragraph describing what kinds of appeals you would use with this audience.

---

In the remainder of this chapter, we will explore several different strategies writers can use to adapt scientific and technical discussions for general audiences.

---

[3]The administrator is interested in what things do from the point of view of cost, production, public health, environmental safety, and resource management: How can this new piece of equipment be used in the lab? How much does it cost to purchase and maintain? What products (or discoveries) are likely to arise from it, and will they be profitable for the lab or company? Is the piece of equipment or product cost-effective? Safe? Efficient?

# Easy Way to Find Ulcers: Breathe

Every year thousands of patients with digestive problems undergo endoscopy, the examination of the gastric tract through a tube lowered down the throat, to look for signs of ulcers.

But a study published on Saturday in The British Medical Journal finds that a breath test for the presence of the bacteria that cause ulcers is just as good as the invasive endoscopy at making the diagnosis.

Researchers from the Western Infirmary in Glasgow randomly assigned 708 patients younger than 55 who had digestive problems to have endoscopy or a breath test for the presence of the ulcer-causing bacteria, Helicobacter pylori.

If either test was positive, the patients were given antibiotics to kill the bacteria.

A year later, the two groups reported similar reductions in symptoms, and no signs were found that any significant illness requiring other treatment had been missed in the group given the breath test instead of the endoscopy.

The researchers said their findings were good news for patients.

"The nonendoscopic strategy has two potential benefits," they wrote in the medical journal. "The first is that patients find the procedure of noninvasive H. pylori testing less uncomfortable and distressing than endoscopy. The second is that noninvasive H. pylori testing is substantially cheaper than endoscopy."

**FIGURE 8.2**    Blurb on H-pylori breath test (O'Neil 2002, p D6).

## 8.3 Adapting Through Narration

Narration is a powerful means of audience adaptation. Research has shown that stories are basic to the formation of human identity and knowledge—even scientific knowledge (Fisher 1987; Polkinghorne 1988). By creating a human story through which an audience can identify with a scientific subject, narration can make science accessible and acceptable to general audiences (Katz 1992a). Jorgensen-Earp and Jorgensen (2002) have examined the way Alexander Fleming, unlike his less remembered rival, Howard Florey, used narratives from the British wartime press to make his penicillin research salient to a victorious World War II public. Unlike the research report—with its hypothesis, methods, results, and conclusion sections—science, when presented to the general public, often takes the form of a dramatic story, with characters, plot, conflict, and resolution.[4] When the *New Yorker* reported the discovery of *H. pylori,* for example, its story was titled "Marshall's Hunch" (Monmaney 1993) and focused on Marshall himself, tracking his rise to prominence from relative obscurity and emphasizing his iconoclastic style. Marshall the rebel, not *H. pylori,* was the hero of this story for a general audience. (Haller [1998] analyzes a similar characterization of researchers studying chronic fatigue syndrome.)

Parts of the "story of science" often take the form of a *history,* a brief chronology of events. It may be a recounting of steps leading to a discovery or the development of a phenomenon, concept, project, or field. Historical narrative can be especially effective if the story can be accommodated to the audience's interest and/or general (cultural) experience. Such narratives are usually given dramatic flair, as in the following "scene" in Huyghe's description of the discovery of the killer algae (contained in Chapter 11): "The story properly begins one night early in 1988, with a massacre in Edward Noga's laboratory at North Carolina State University in Raleigh" (1993, p 72).

## EXERCISE 8.6

Read the Huyghe (1993) article (Chapter 11, pages 293–298) from *Discover* magazine about the Burkholder team's research on toxic dinoflagellates. Look for instances of narrative storytelling and character development. What parts of this project are dramatized in this article written for a lay audience? Who are the main "characters"? What is the "plot"? What is the "conflict"? How is it "resolved"?

## EXERCISE 8.7

Watch an episode of a science program such as *NATURE* or *NOVA* on television. Is the scientific information adapted to a general viewing audience through the use of narrative storytelling? What parts of the science are being dramatized? Who are the main "characters"? What is the "plot"? What is the "conflict"? How is it "resolved," and how does the resolution support scientific research?

---

[4]As you will see in this chapter, the application of what appears to be "literary technique"—not only narrative, but also metaphor, analogy, and even imagery—can be very effective in making abstract or complex science tangible and accessible to the public (see Arroliga [2002]). Because of this ability to make abstractions palpable, literary technique is increasingly being used to develop and teach empathy in medical education: Martin Blaser, who in 2000 became Chairman of the Department of Medicine and Professor of Microbiology at New York University and whose work is represented in Chapter 10, believes so much in the value of the study of writing and literature for scientific and medical education (http:// www.blreview.org/staff.htm) that he cofounded the *Bellevue Literary Review,* a prestigious literary journal published by Bellevue Hospital that features poetry and fiction "that touch upon relationships to the human body, illness, health and healing" (http://www.blreview.org/index.htm).

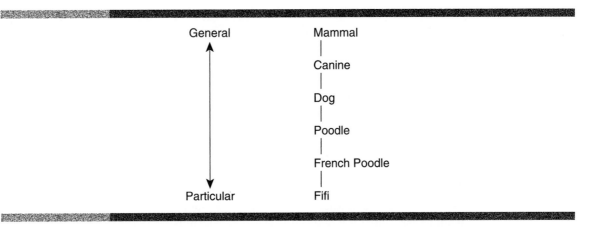

**FIGURE 8.3**    The use of examples: Movement from general classes to particular instances.

## 8.4 Adapting Through Examples

Examples play a prominent role in audience adaptation and are essential for comprehension. As shown in Figure 8.3, examples provide a more particular instance of a general concept or class. Thus, they tend to be more concrete, and so they tend to make the general concept or class easier to grasp. Red tide, which many coastal visitors have seen firsthand, is an example of the class of biological phenomena called *algal blooms.* In this book we have used examples to illustrate our discussion of various genres and conventions. Imagine this book without them! Examples also provide an effective way to relate unfamiliar concepts to what readers or listeners already know, value, and are interested in—if they are drawn from the reader's or listener's experience. As you discovered in Exercise 8.4, different audiences are interested in different issues. Figure 8.4 outlines three sets of issues that might be useful in explaining lasers to three different audiences. In

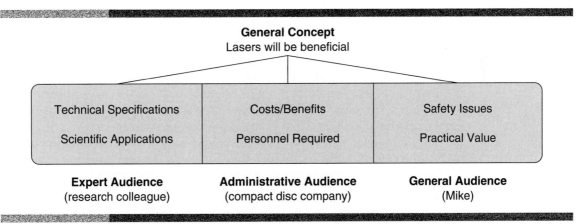

**FIGURE 8.4**    Topics for different audiences.

speaking to or writing for one of these audiences, you would want to draw examples from that particular audience's area of interest.

Thus, there are two general rules concerning the use of examples as a strategy for audience adaptation: (1) examples must *logically* support your point by representing the general class of items they're meant to illustrate; (2) examples should *rhetorically* support your point by being drawn from the knowledge and experience of your audience.

## 8.5 Adapting Through Definition

Several definitional strategies can be used to "unpack," or "translate," scientific terms for a general audience. Before you consider any of these strategies, you will want to ask yourself whether the particular term is needed by the audience at all. Think about the knowledge, needs, and interests of your audience. Is the term absolutely essential for the audience to learn? If you decide it will bog down the reader, omit the term. For example, when you read our descriptions of the appeals to wonder and application in Section 8.2, did you need to know that the technical term for the appeal to application is the *teleological appeal* and that the appeal to wonder is the *deontological appeal* (Fahnestock 1986)? What good does it do you to know this information now? What would have been the effect if we had used these terms instead of the general terms in the earlier discussion? Are there circumstances in which this technical terminology might be appropriate?

If you decide a term is important enough to take the time to explain it to your audience, then you can use one of the following definitional strategies to define it. These strategies are not just different ways of explaining terms but also different ways of thinking about concepts. Each definitional strategy gives you (and thus the audience) a different "angle" on a term, a different perspective on a concept. Thus, your choice of definitional strategy will depend on (1) the term or concept; (2) the audience's knowledge, values, interests, and needs; and (3) what you want to say to this audience (your purpose).

When you define using the *classical* (*Aristotelian*) *definition*, you place a term or concept in a general class of things to which it is similar and then delineate how the term or concept is different from other members of this class. This genus-differentia definition is the one we find most often in dictionaries. It depends on classification as the primary intellectual strategy to explain a concept. For example, the term *nova* (used as a matter of course for an expert audience in Reynolds's paper on supernova remnants in Chapter 12) is defined in an ordinary dictionary as "a star that suddenly becomes thousands of times brighter and then gradually fades" (Random House). The *genus* is "a star," and everything that follows is the *differentia*, that part of the formal definition that distinguishes novas from other stars.

The definitional strategy *etymology* explains a term by examining the history of the word(s). Notice how in the following etymological definition Huyghe (1993) employs the narration strategy to enhance audience appeal as well:

> They have proposed calling [the dinoflagellate] *Pfiesteria piscimorte* (the genus name was picked to honor the late dino specialist Lois Pfiester of the University of Oklahoma and also because Burkholder liked the name's echo of both *feast* and *cafeteria;* the species name means 'fish killer'). (p 75)

## 8.6 Adapting Through Analysis

*Analysis* breaks a whole into constituent elements. Thus, in analysis, division, not classification, is the primary strategy used to unpack a concept. Analysis can be used to discuss the structure of a molecule or the stages of a supernova. Huyghe (1993) uses analysis to divide the life cycle of dinoflagellates into stages to emphasize the scientific oddity of these creatures and their devastating effect:

> [The new dinoflagellate's] life cycle consists of more than 15 stages. . . . The dino lives most of its life as an amoeba; in one toxic "giant" amoeboid stage it grows to be nearly 20 times the size of the flagellated toxic cells. The stage when the dino leaves its cyst to attack fish is actually quite ephemeral, appearing only when fish are present. The researchers also found that this stage is the only time the creature sexually reproduces. (p 75)

## 8.7 Adapting Through Comparison

Comparison is a particularly useful method for explaining concepts to nonspecialists. Using this strategy, the scientist indicates how a phenomenon is similar to or different from other phenomena the audience is more familiar with: "Dinoflagellates are twilight-zone creatures: half-plant, half-animal. They produce chlorophyll, but they also move about, using their two flagella, or whiplike tails, to swim rapidly through the water" (Huyghe 1993, p 72).

The use of synonyms is one type of comparison strategy. *Synonym*, substituting one or more words for another, can be used to define a term in a context the audience is more familiar with, as in the example above: "flagella, or whiplike tails." This sort of definition in context is also known as *parenthetical explanation* (if in parentheses), and *apposition*, if inserted without parentheses.

It is because no two words have exactly the same meaning or connotation that synonym falls under the category of comparison: for the writer or speaker, the use of synonyms really involves a comparison of terms and concepts rather than exact substitution—a search for a similar word that the audience will be more likely to understand and be able to relate to. Given that technical terms are part of a precise vocabulary of a field, the scientist using synonym usually has to trade off some accuracy for the sake of gaining the audience's comprehension.

You've all learned that *simile* is a comparison that uses *like* or *as*. Simile can be used to drive home a point figuratively, as in the following example: "Efforts to describe [the dinoflagellate] have evoked the strangest comparisons: like grass feeding on sheep, said one scientist. And that's not stretching things much" (Huyghe 1993, p 72). Unlike simile, *metaphor* does not use *like* or *as*. It is thus a less obvious, but a much more pervasive, method of comparison than we might think. By omitting *like* or *as*, metaphors not only conceal the comparison but also imply an identification of the things compared, as in the following example: "The creature is in fact a tiny plant with a Jekyll and Hyde personality, one that preys on animal life a million times its size" (Huyghe 1993, p 72). Notice that the attribution of a specific behavior to the alga through the personality metaphor of "Jekyll and Hyde" is grounded in the audience's cultural knowledge.

Some researchers believe that metaphors actually structure, and to some extent determine, the way we conceptualize the world (e.g., Lakoff and Johnson 1980). It has been argued that metaphors are implicit in scientific models (Turbayne 1970; Kuhn 1979; Leary 1990; Boyd 1993); Neils Bohr's early model of the atom as a solar system immediately comes to mind, but there are many others. In any case, common sense tells us that the metaphors we choose both color and reflect the way we think about phenomena. Metaphors are used not only in communicating with general audiences (e.g., "biological clock" [Brown 1959]; "killer algae" [Huyghe 1993]), but also by scientists themselves in exploring new phenomena (e.g., "genetic code" [Watson and Crick 1953]; "phantom dinoflagellate" [Burkholder et al. 1992]).[5]

In creating metaphors for nonexpert audiences, it is important to remember two things: (1) Find the metaphors in the language, knowledge, and experience of your audience; do not merely switch to other technical metaphors embedded in your field that the audience will not understand. (2) Use metaphors consistently; mixing metaphors confuses readers and may introduce inconsistencies into your discussion.

## EXERCISE 8.8

An excerpt from Monmaney's (1993) profile of Barry Marshall in the *New Yorker* is presented in Figure 8.5. Identify and circle the metaphors Monmaney uses to discuss the *H. pylori* bacteria. What are the bacteria being compared to? Why? Are the metaphors related to each other (i.e., consistent)? What do the metaphors communicate? Compare the description of this bacterium in the *New Yorker* with its description in the Marshall and Warren letters (1983) in Chapter 10. Are there any metaphors in the technical letters? What conclusions can you draw about the use of metaphors for scientific and nonscientific audiences? What is the scientific basis for the metaphors in the *New Yorker* excerpt? What is the popular basis of the metaphor?

## EXERCISE 8.9

Identify and compare the metaphors on the first page of Huyghe's essay in *Discover* (1993) and the first page of the Burkholder team's (1992) letter in *Nature* (1992), both contained in Chapter 11. You will see that Huyghe and the Burkholder team use similar metaphors to describe microorganisms to specialist and nonspecialist readers. What does this tell you about the role of metaphors in science?

---

[5]Quotation marks are often used around "popular" terminology by scientists writing to general and specialist audiences to distance themselves from it, reflecting an abiding concern for their professional credibility.

It takes a kind of cunning for *Helicobacter pylori* to fill its niche, for the adult human stomach is one of nature's most hostile habitats. Each day, the stomach normally produces about half a gallon of gastric juice, whose strong hydrochloric acid and digestive enzymes readily tear meat and microbes apart. Gastric juice is like a binary chemical weapon—so destructive that it's constituted only on the way to the target. As cells in the stomach lining secrete the raw ingredients of gastric juice into the mucus that coats the stomach lining, the ingredients mix into an even more caustic brew, which then oozes into the cavity. There the gastric juice breaks pabulum down chemically while muscles in the stomach wall act to crush it. The viscoelastic mucus, as thick as axle grease, keeps the stomach from digesting itself.

Once *Helicobacter* reaches the stomach, it probably does not linger out in the open cavity—a tossing sea of toxic chemicals. It heads for cover. The bacterium's helical shape seems to have been designed for speedy travel in a dense medium. *Helicobacter* is living torque; a microscopic Roto-Rooter, it corkscrews through the mucus. Then, instead of penetrating the cells of the stomach lining, it settles in the mucus just beyond the lining. More often than not, it settles in the pylorus. No one knows why. Under the microscope, a *Helicobacter* infection looks like a satellite image of an armada gathered off a ragged shore. At one end of the bacterium is a cluster of long, wispy, curving flagella, which may serve as anchors. It's a graceful menace.

*Helicobacter* possesses a vital defense against stomach acid, and this adaptation, too, is a marvel of evolutionary design. Its coat is studded with enzymes that convert urea—a waste product, virtually unlimited supplies of which can be found in the stomach—directly into carbon dioxide and also into ammonia, a strong alkali. Thus *Helicobacter* ensconces itself in an acid-neutralizing mist. In like fashion, it generates another antacid—bicarbonate, as in Alka-Seltzer.

A *Helicobacter* infection that establishes itself succeeds largely because the immune system can't reach it. In response to a *Helicobacter* invasion, immune-system cells in the bone marrow produce white blood cells, killer cells, and other microbe destroyers, and those float through the bloodstream to the very edge of the stomach lining—and go no farther, because the lining holds them back. The *Helicobacter,* hovering in the mucus, are out of range. And yet the immune system sends reinforcements. Killer cells pile up, gorging the stomach lining; permanently alerted, seldom engaged, the killers become trigger-happy. Some die, fall apart, and spill their microbe-fighting compounds into the host tissue. Friendly fire begets friendly fire. Metabolic hell breaks loose. The lining is now inflamed: acute gastritis. Micronutrients are pumped from the bloodstream to the front lines, to feed the killers, but loads of them seep out of the stomach lining and into the mucus. Offshore, the *Helicobacter* feast; having drawn the immune system into battle, the bacteria now loot the provisions. "I propose that inflammation is good for *Helicobacter,*" Blaser says. "That's what it wants."

Chronic gastritis, a standoff between the bacteria and the host's immune system, may persist for years—decades, according to some estimates. As it happens, whatever serious damage is done to stomach or intestinal tissue is apparently done not by the bacteria themselves but by the inflammatory response they provoke. That the host plays such a large role in his own pathology may help explain why *Helicobacter* infection affects different people differently. (Also, scientists recently discovered that there are at least two strains of *Helicobacter pylori,* and that one of them is far more likely to cause a peptic ulcer than any other.) Inevitably, though, for *Helicobacter* to be really successful it has to meet a parasite's final challenge: to start another colony before the host dies.

**FIGURE 8.5**  Metaphor identification: Excerpt from "Marshall's Hunch" (Monmaney 1993).

*Analogy* is an extended comparison that may include similes and metaphors. In fact, an analogy might comprise a series of related metaphors running throughout a text. Like other methods of comparison, the advantage of an analogy is that the writer can begin with what the audience knows and then move back and forth between the known and the unknown as the concept is explained or new terms are further defined. For example, in the Huyghe piece, the metaphor "killer algae," announced in the title of the article, "surfaces" throughout the text.

---

**EXERCISE 8.10**

In the Huyghe (1993) piece, look at definitions, similes, metaphors, and other places where the metaphor "killer algae" is extended in the discussion through analogy. Are there places where the analogy breaks down or is violated? Why? What is the effect? Do you see any similiarity between the analogy in the Huyghe piece on dinoflagellates and the analogy in the excerpt from Monmaney's piece on *H. pylori?* What do these analogies tell you about common attitudes toward microorganisms?

---

You should note that *analogy* **is not** *example.* While examples are drawn from the general class of concepts they are a part of, the "examples" in analogies are drawn from similar or parallel classes of concepts; that is, they are based on comparison. Scientists can draw on family and social relationships as well as causal ones in comparing unfamiliar concepts with concepts more familiar to their general audiences (Young et al. 1970). In the example in Figure 8.6, from Zukav's (1979) *The Dancing Wu Li Masters: An Overview of the New Physics,* the population of a city is used not as an *example* of the concept of discontinuity in New Physics; people are not members of the class of particles. Rather, the parallel between people and particles is being used to explain the concept of discontinuity in New Physics by *comparison* with something the audience is more familiar with: population statistics.

Usually several strategies are used together to unpack a scientific concept; they even may be embedded in each other, as in the following excerpt: "Under the right conditions, for example, sodium reacts to chlorine (by forming sodium chloride—salt), iron reacts to oxygen (by forming iron oxide—rust), and so on, just as humans react to food when they are hungry and to affection when they are lonely" (Zukav 1979, p 71). In this passage, we see that in *comparing* chemical and

---

What had [Max] Planck discovered that disturbed him so much? Planck had discovered that the basic structure of nature is granular, or, as physicists like to say, discontinuous.

What is meant by "discontinuous"?

If we talk about the population of a city, it is evident that it can fluctuate only by a whole number of people. The least the population of a city can increase or decrease is by one person. It cannot increase by .7 of a person. It can increase or decrease by fifteen people, but not by 15.27 people. In the dialect of physics, a population can change only in discrete increments, or discontinuously. It can get larger or smaller only in jumps, and the smallest jump that it can make is a whole person. In general, this is what Planck discovered about the processes of nature.

---

**FIGURE 8.6**   Excerpt from *The Dancing Wu Li Masters: An Overview of the New Physics* (Zukav 1979, p 73).

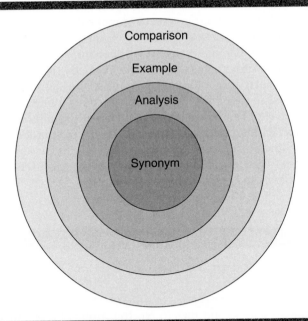

**FIGURE 8.7**   Orbits of explanation in Zukav's comparison.

human reactions the author uses *examples;* these examples are *analyses* of chemical reactions; the results of the reactions are clarified by *synonyms* drawn from common experience. Thus, several layers of explanation revolve around each other, as illustrated in Figure 8.7.

## 8.8 Adapting Through Graphics

Graphics are visual representations of phenomena. Like metaphor, they also have the power to persuasively portray abstract *concepts* as physical phenomena (see Gilbert and Mulkay 1984). As discussed in previous chapters, visual aids such as tables, line graphs, diagrams, and site maps are conventionally used to present information to scientific audiences in research reports and conference presentations. But graphics can also be used to explain concepts to general audiences. Note the use of graphics to explain audience adaptation strategies in this chapter. Like verbal examples, visual examples should be based on your audience's knowledge, interests, and values.

**EXERCISE 8.11**

If you did Activity 1 at the end of Chapter 1, look at the three different ways you represented the information contained in that exercise. What interests, values, and needs does each (graphic) representation appeal to (or assume) on the part of your audience?

For general audiences, photographs, slides, maps, drawings, bar graphs, pie charts, and simple diagrams serve to make the verbally abstract visually concrete. One advantage of these types of pictorial representation is that they can be made colorful, attractive, and thus very appealing. They also can be discussed at any level of generality that is appropriate. A catchy or attractive photograph can even be used to introduce a topic. For example, the opening page spread of the "Killer Algae" article (Huyghe 1993) catches readers' attention by juxtaposing a "super-sized" image of the toxic dinoflagellate with smaller pictures of the fish that become its victims, effectively dramatizing the magnitude of the threat posed by the microscopic algae. In the original piece Martin Blaser (1996) wrote for *Scientific American* (Chapter 10), a colorful diagram of a stomach lining inhabited by *H. pylori* decorates the first two pages of the text.

## EXERCISE 8.12

Imagine that you are an editor working for *Discover* or *Scientific American*. You have assigned one of your staff writers to write an article describing Stephen Reynolds's recent research on supernova remnants. While your writer reviews Reynolds's papers and flies to North Carolina State University to conduct an interview, you call in your design staff to start planning the visual presentation. After reviewing Reynolds's work yourself (see Chapter 12), think about what kind of visual effect you'd like to create. Describe (or better yet, sketch) the first page spread for the article. Think of an effective title for the piece, and incorporate it into your design.

The medium, size, shape, and color of a visual representation also can be used to communicate symbolically. Line graphs use colors (e.g., black and red) as well as the direction of the line (up and down) to symbolically (and dramatically) show increase and decrease *visually*. Thus, a pie chart showing percentages of food consumed by Americans during the day versus the night, for instance, might color the day part of the graph yellow and the night part blue, communicating basic information at a glance through color as well as labels. A graph showing annual deforestation by fire might have bars colored red or orange (known as "hot colors," as opposed to blue or green, known as "cool colors") or even shaped like flames! The use of size, shape, and color in graphics is especially important for nonexpert audiences when knowledge of the appearance of phenomena cannot be assumed but can be communicated quickly through the visual medium, or when dramatic effect is desired to make a point.

When using graphics to communicate to a general audience, consider all the possibilities the visual medium has to offer you. Recall from Chapter 5, though, that one of the dangers of pictorial representation is that the visual aid may contain too much detail, especially if it is not drawn specifically for the purposes of your presentation. Visual aids, like text, should be focused and uncluttered when communicating with nonspecialists.

**EXERCISE 8.13**

Burkholder and Rublee diagram the life cycle of the toxic dinoflagellate in Figure 2 of their 1994 Sea Grant Proposal (see page 311). Patrick Huyghe presents a brief verbal description of the life cycle on the last page of his *Discover* piece (1993, page 298). Assume that you are working with Huyghe on a revision of this piece. Your task is to adapt Burkholder and Rublee's technical diagram for the general readership of *Discover*. Redraw this diagram, using both the original diagram and Huyghe's verbal description as your guide. You may add to or revise Huyghe's verbal description to accommodate your graphic if you wish.

# 8.9 Logic and Organization in Writing for Public Audiences

Unlike the research report and proposal, there is no standard structure for articles written for general audiences. Thus, you will want to consult the publication you are writing for to get a sense of stylistic and formatting options. Major topics or points are often indicated by headings, as in Blaser's (1996) *Scientific American* article on *H. pylori*, contained in Chapter 10. When headings are used, they are topical rather than functional; that is, they provide clues to the content rather than the structure of the text. Functional headings (such as "Introduction," "Methods," "Results," "Discussion") are useful in genres governed by a standard structure, for they enable readers familiar with that structure to quickly locate the types of information they're most interested in. But such headings are of no use when no particular pattern of organization is expected by readers.

Topical headings help readers see at a glance what major topics or issues will be raised in each section of an article. Thus this type of heading serves to introduce the unique pattern of organization used in a given essay. To do so, they must be clearly comprehensible to the general reader. Topical headings also provide an opportunity to catch readers' attention and pique their interest. Thus, authors writing for general audiences aim to create titles and headings that are not only informative but intriguing.

**EXERCISE 8.14**

Compare the headings in Blaser's (1996) *Scientific American* piece (pages 258–263) with those of a standard research report contained in Part Two of this book, such as those by Marshall and Warren (1984) or Graham et al. (1992). Then compare the title and headings of the *Scientific American* piece with the review article Blaser (1987) wrote for *Gastroenterology* (pages 236–248). (These readings are all contained in Chapter 10.) Write a brief (two-page) analysis in which you use these sample articles to illustrate the differences between topical and functional headings and the differences between specialized and general audiences.

Articles in general-reader publications may end with a list of suggested readings but usually do not include a formal reference list. These articles rarely report the results of individual experiments, focusing instead on the general outcomes of a body of research. They therefore do not typically include citations of individual research reports. Such reports are occasionally included in a list of further readings (see Blaser [1996], Chapter 10), but it is not common practice to include sources written for research journal audiences when writing for newspapers or general-interest magazines. It is more helpful to those readers to list resources that are specifically designed for general readers and are readily available in local bookstores or public libraries.

Instead of drawing on prior research reports to support their claims, authors often use interview data when writing for general audiences. Direct quotations are rarely used in other genres of scientific discourse (see Chapter 4), but they are legitimate and popular forms of evidence in general-reader publications such as newspapers and magazines. In fact, *quoting* can be understood to constitute another major strategy for adapting science to a general audience. Quotations from "characters" in a scientific story can enliven it, and thus are frequently used when the narrative strategy is employed. The Huyghe (1993) article in *Discover* includes numerous examples of this strategy. Another common use for quotations is to provide support for or elaboration of a claim made in an essay, as when the author quotes a prominent authority's opinion of a new finding or its implications. In classical argumentation terms, quotations thus represent either *appeals to experience* (testimonials) or *appeals to authority*. Both types of testimony enhance the credibility and authenticity of the scientific story being told.

## EXERCISE 8.15

Blaser (1996) does not include direct quotations in his *Scientific American* piece (Chapter 10), most likely because he himself is one of the researchers involved in this research and thus provides the authenticity that quotations would contribute. However, if you were to write on his topic for a general readership, what kinds of quotations would you want to obtain to supplement your article? In a paragraph or two, choose a target publication and audience for your piece, and explain whom you would like to interview, what kind of information you'd hope to obtain from those interviewed, and how you would use this material in your article.

## EXERCISE 8.16

Chapter 11 includes a "fact sheet" prepared by EPA titled "What You Should Know About *Pfiesteria piscicida*" (pages 350–355). Also posted on the web (http://www.epa.gov/owow/estuaries/pfiesteria/fact.html), this document utilizes a range of strategies for communicating complex scientific knowledge to nonspecialists. Analyze this text, describing the adaptation strategies it employs and speculating on the effectiveness of each. Don't limit your analysis to the strategies discussed in this chapter. Overall, does this seem an effective document? What public needs, interests, or concerns does it seem intended to address? Does it do a good job?

# Activities and Assignments

1. In this chapter we have discussed the use of visuals and the use of subtopics in Blaser's *Scientific American* essay (1996). Read this essay, included in Chapter 10 (page 258). Identify other audience adaptation strategies being used, and describe how they work together.

2. Watch a local weather forecast on television. In a two- or three-page paper, identify and describe the audience adaptation strategies being used, and assess their effect on you as a member of the audience.

3. A. Choose a paragraph from an expert text in your field (e.g., from a textbook or journal). Select and analyze a general audience for which this topic might be appropriate. Adapt the paragraph for that audience using the audience adaptation strategies you have studied in this chapter. Be sure to specify the intended audience (by publication, age, education, background, interests, goals, etc.). Compare your adaptation with the original, and explain the similarities and differences.
   B. Now select and analyze an appropriate administrative audience, and adapt the same piece for them (e.g., suppose you are working in a lab and need to explain a piece of equipment or an experiment to the lab administrators so that they will continue funding your project). Note what changes you had to make in content and strategy from the piece written for a general audience. Compare this piece written for an administrative audience with the original text written for an expert audience; explain any similarities and differences between these texts.

4. Write a scientific essay for a popular magazine or newspaper that traditionally covers science (*Time, National Geographic, Scientific American, Discover,* the *New York Times,* etc.). Choose an essay from that magazine as a model. Analyze the audience for the magazine using the questions in Section 8.2 as a guide, and examine your model for the audience adaptation strategies we have discussed. Use these strategies where necessary and appropriate in your essay. Create visuals designed for this journal and audience to include with your text. Be sure to include your model with your assignment.

5. Choose a particular public audience and occasion for an oral presentation on a topic in your field. For example, you might address a club, a junior high classroom, a congressional committee, the press. Give an oral presentation appropriate to this audience. The talk may be based on a previous topic you wrote about in this or other classes for an expert audience, or for Activity 4, but be sure to adapt it fully for an oral presentation to your new audience, using the principles and strategies in this chapter. Remember to eliminate any terms, concepts, and details that don't fit your general audience's needs and interests; make sure other terms are appropriately explained. Design visuals for this audience. Before you present this talk in class, inform your listeners of the intended purpose, audience, and occasion for the presentation—who and where they are supposed to be as they role-play the target audience. Because they most likely represent a diverse public audience for your topic, they will be able to give you direct feedback about what they understood.

6. Write an online "fact sheet" for the public on an issue in your field. EPA's *Pfiesteria* fact sheet provides a good model (pages 350–355). View the online

version at http://www.epa.gov/owow/estuaries/pfiesteria/fact.html. Consider where your fact sheet should be posted: Where would interested publics look for it or be likely to come across it? What government agency or private organization should provide such a fact sheet? As you plan your text, review the strategies you identified as most effective in the EPA document (Exercise 8.16). Design a fact sheet that is clearly adapted to the needs of the general public and that uses the Internet medium to full advantage. What text features does the Internet enable you to include that would not be available in a hard copy text?

# 9

# Considering Ethics in Scientific Communication

> Society's confidence in and support of research rest in large part on public trust in the
> integrity of individual researchers and their supporting institutions. . . . It is therefore
> incumbent on all scientists and scientific institutions to create and nurture a research
> environment that promotes high ethical standards, contributes to ongoing profes-
> sional development, and preserves public confidence in the scientific enterprise.
> (Committee on Assessing Integrity in Research Environments [CAIRE] et al., 2002,
> p 33)

## 9.1 Scientific and Social Ethics

As members of society, scientists are influenced by the dominant values of the
culture in which they live: truth, justice, freedom, progress, happiness. As societies in themselves, scientific disciplines also are governed by conventions, both
written and unwritten, which come to constitute standards and principles of professional conduct and practice. That is, the conventions of your field constitute (or
imply) a system of ethics. In exploring communication practices in your field, you
have been learning about some of the conventions that shape your discipline. As
you become a member of that discipline, it also is important that you understand
the ethical dimension of these conventions and some of their implications for you
as a scientist.

Scientists usually learn professional ethics informally, by observing how other
scientists behave. But as the National Academy of Sciences remarks, "science has
become so complex and so closely intertwined with society's needs that a more formal introduction to research ethics and the responsibilities that these commitments
imply is also needed . . ." (1995, Preface). While informal observation is still the primary venue for learning ethics, individual efforts to formally integrate the teaching
of ethics into science curricula across the country began in the early 1980s (see

Benditt 1995; Sweeting 1999;[1] DuMez 2000; Macrina 2000; CAIRE et al. 2002). On the national level, a collaborative effort has been launched by the U.S. Department of Health and Human Services' Office of Research Integrity (ORI), in association with the nonprofit National Academies (Institute of Medicine, the National Academy of Sciences, the National Academy of Engineering, and the National Research Council). At the request of and with funding from the Department of Health and Human Services (DHHS), the National Academies' Committee on Assessing Integrity in Research Environments, in conjunction with other organizations in the National Academies, began by articulating core values and behaviors associated with scientific integrity, listed in Figure 9.1. As you examine this list, notice how many of these core values pertain to issues of communication. Because the conduct of science includes the ways in which scientists share their results and interact with others in and beyond their research communities, communication practices are a central focus of efforts to examine and foster scientific integrity. CAIRE et al. has set out not only to develop a rigorous program of defining, teaching, and promoting scientific integrity, but also to develop more formal procedures of collecting, measuring, and evaluating data on integrity in science education and research (CAIRE et al. 2002). Their report, "Integrity in Scientific Research: Creating an Environment that Promotes Responsible Conduct" (2002), represents an early step in the effort to formally identify, track, and study scientific integrity for the National Academies.

NSF as well, through its Office of Inspector General, continues to be concerned with scientific ethics and the integrity of scientific communication. Like ORI, NSF has invested in research on these issues, awarding grants for projects such as "Openness, Secrecy, Authorship, and Intellectual Property" in 1991; "Sharing Research Data: An Examination of Practices" in 1993; "Dynamic Issues in Scientific Integrity: Collaborative Research" in 1994; and "Historical Perspectives on Scientific Authorship" in 1996 (CAIRE et al. 2002, p 24). We will touch on many of these ethical issues in this chapter as we conclude Part One of this book and reflect on the ethical dimension of the communication conventions we have discussed, as well as on the ethical implications of the socialization process you are now engaged in as a new member of the scientific community.

Since scientific disciplines are participants in the larger society in which they are situated, it stands to reason that scientists would face some of the same ethical issues in communication faced by society at large. These might be organized under two categories: misconduct in publication (misrepresenting data or authorship) and corruption and bias in the gatekeeping institutions that oversee communication (prejudice, underrepresentation).

Occasionally we read about incidents of misconduct in scientific communication. For example, we read about Dr. Roger Poisson, who falsified and misrepresented data in a seminal report on breast cancer treatment on which major research and health recommendations had been based (Taubes 1995), and about W. Bezwoda who fabricated data involving cancer treatment in South Africa (Hagmann 2000); about allegations that Dr. Robert Gallo "made a misstatement in a science publication" and erroneously claimed sole credit for the discovery of the

---

[1]Dr. Linda Sweeting (2000) and her students have compiled a comprehensive and annotated bibliography of ethics in science, available at: http://www.towson.edu/~sweeting/ethics/ethicbib.html. We are indebted to this bibliography for several of the referenced sources in this chapter.

# Integrity in Research

### Individual Level

For the individual scientist, integrity embodies above all a commitment to intellectual honesty and personal responsibility for one's actions and to a range of practices that characterize the responsible conduct of research, including

- intellectual honesty in proposing, performing, and reporting research;
- accuracy in representing contributions to research proposals and reports;
- fairness in peer review;
- collegiality in scientific interactions, including communications and sharing of resources;
- transparency in conflicts of interest or potential conflicts of interest;
- protection of human subjects in the conduct of research;
- humane care of animals in the conduct of research; and
- adherence to the mutual responsibilities between investigators and their research teams.

### Institutional Level

Institutions seeking to create an environment that promotes responsible conduct by individual scientists and that fosters integrity must establish and continuously monitor structures, processes, policies, and procedures that

- provide leadership in support of responsible conduct of research;
- encourage respect for everyone involved in the research enterprise;
- promote productive interactions between trainees and mentors;
- advocate adherence to the rules regarding all aspects of the conduct of research, especially research involving human participants and animals;
- anticipate, reveal, and manage individual and institutional conflicts of interest;
- arrange timely and thorough inquiries and investigations of allegations of scientific misconduct and apply appropriate administrative sanctions;
- offer educational opportunities pertaining to integrity in the conduct of research; and
- monitor and evaluate the institutional environment supporting integrity in the conduct of research and use this knowledge for continuous quality improvement.

**FIGURE 9.1**    "Integrity in research" from CAIRE et al. (2002, p 5).

HIV virus (Associated Press 1992, p A7); about charges by Sarvamangala Devi, an assistant researcher who discovered a vaccine that could be used in the treatment of AIDS patients, that she was being denied credit for her discovery by senior scientists at the National Institutes of Health because of her gender and ethnic background (Marshall 1995); and about the scandal in the Department of Oncology at the University of Freiburg Medical Center in Germany, where hematologist Frieldhelm Herrmann and colleagues were accused of falsifying or manipulating the data of numerous research reports, prompting *Nature* to state that there are serious flaws in the infrastructure of German science (AAAS 2000).[2]

---

[2]For continuously updated cases of investigations by the ORI into scientific misconduct, see http://ori.dhhs.gov/html/misconduct/casesummaries.asp.

We also read about charges of bias in the gatekeeping and educational institutions of science. We read about how NIH stopped its funding for a controversial conference examining the relationship between genetics and crime after protest from black leaders (Associated Press 1992); about the possible theft of data by peer reviewers who worked at Immunex Corporation from a competing drug company, Cistron Biotechnology (Marshall 1996a), which ended with an out-of-court settlement that established neither guilt nor the legality of the confidentiality of peer review (Marshall 1996b); about charges of elitism and bias against "low income students, women, minorities and the disabled" in science education (Cole 1990, EDUC 18); and about continuing problems of underrepresentation, prejudiced evaluation, and suppression in science and industry based on gender (NAS 1989; Horn 1998; Jacobs and Storck 2000; Science 2000a)—even in the National Academy of Sciences itself (Angier 1992). Questions regarding the representation and tracking of foreign scientists and science students were debated even before the terrorist attacks on the U.S. mainland on September 11, 2001 (see Heylin 2000; Lawler 2000a, 2000b). Given issues and charges like these, U.S. science would appear to be a mere microcosm of American society. As science ethicist Francis Macrina puts it,

> Both the DHHS and the NSF definitions clearly forbid "fabrication, falsification, and plagiarism." This is commonly referred to as the FFP core. Fabrication means making up results; falsification means tampering with results; and plagiarism means passing off another's ideas as your own. In other words, in science as in life, it is wrong to lie, cheat, or steal. (2000, p 11)

Indeed, today the scientific community is a microcosm of the entire world. The reaction of scientific communities around the globe to unethical incidents in individual countries not only reflects a concern for the ethical infrastructures of nations; it also reflects the internationalization of ethics in the community of science. The accessibility of research and interaction of researchers through electronic communication make the internationalization of science even more possible, and thus make ethics to guide global science even more important. As science becomes more international, the actions of one scientist affect scientists everywhere.

For example, the internationalization of ethical concern in science is reflected in the establishment of the Scholars at Risk Network to help persecuted scientists outside the United States (AAAS 2000). It also is reflected in the call by the French national agency of scientific research, L'Institut de Recherche pour le Développement, for an international code of ethics (AAAS 2000). Because the Internet has outgrown national boundaries, there even has been a call for the revision of international treaties to address such ethical and legal issues as ownership and copyright (Liedes 1997).[3] Just as political economies are globally interdependent, in today's world

---

[3]The ethical and legal issue of ownership and copyright (as opposed to authorship) is a tricky one. Contrary to popular belief, most authors do not legally own their work once it is published; when a manuscript is accepted for publication, the author usually is required to sell various "rights" to the journal; these rights determine legal ownership and thus distribution and reprinting of the work (see the Instructions to Authors sections of journals in your field). These rights might be First American Rights, for instance, although they generally include all International Rights. The problem is that international copyright is currently governed and determined by a seeming hodge-podge of national copyright laws, and not generally addressed in international treaties. The need to address issues of international copyright at the level of treaties between countries has become even more urgent not only with the advent of the "information age," but also with the explosion of electronic publication. Many issues regarding republication and distribution worldwide are yet to be resolved.

there is an increasing awareness that the validity and quality of work done by scientists in one country affect the perception and quality of science done everywhere.

As this quick overview has indicated, scientists working in research environments are subject to pressure from a variety of social, economic, and political forces. By all accounts, the pressure is mounting: The very success and "affluence" of science (Hoshiko 1991, p 11)—and the enormous pressures to compete for dwindling funds, publication, prizes, promotions, prestigious labs and equipment, research assistants, students, and media attention—seem to be changing the ethical environment in which science is conducted (Cohen 1995; Macrina 2000; CAIRE et al. 2002). In Figure 9.2 we have reprinted CAIRE et al.'s graphic depiction of these "environmental influences." For institutions as well as individual scientists, maintaining scientific integrity is a matter of negotiating these influences, including the pressure to publish in order to maintain employment and enhance professional reputation; the need to tailor research agendas to the interests of funding agencies; and the constraints imposed by journal editors, professional societies, and governmental regulation on what, and how, research is presented to the scientific community.

Because of alleged and reported cases of misconduct in the United States, there is strong political pressure for congressional oversight of science—a pressure that scientists naturally resist, insisting that they can monitor themselves (Koshland 1989; Hoshiko 1991; CAIRE et al. 2002). Although most scientists "have experienced or know first-hand some cases of misconduct" (Hoshiko 1991, p 11), misconduct in

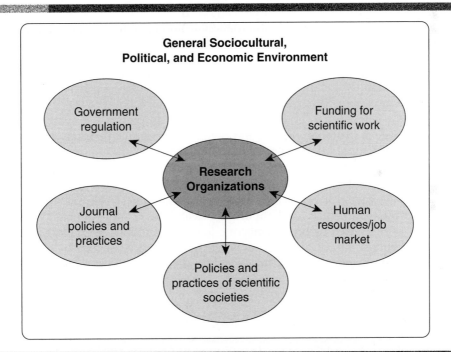

**FIGURE 9.2**    External factors influencing the research environment. CAIRE et al. (2002, p 8).

science is in fact relatively rare. In 1989, *Science* editor Daniel Koshland Jr. estimated that while three million research articles had been published over the prior ten years, only about a dozen major cases of fraud had been reported. In 2000, an evaluation involving two hundred scientists by the Department of Health and Human Services' Office of Research Integrity reported only one case of fraud out of ten thousand studies per year (ORI 2000). There are perhaps several reasons for the relatively low number of incidents involving misconduct in science.

As already discussed, several U.S. government organizations—including the National Science Foundation and the Department of Health and Human Services—continue to articulate new and better standards and policies for ethical conduct in every facet of scientific research and communication (AAAS 2000; Macrina 2000; CAIRE et al. 2002). Not only does the rapid evolution of scientific knowledge and techniques frequently outstrip current ethical guidelines and laws, but the enormous impacts of these developments on society and humanity generally, as in the case of cloning, demand more up-to-date and complete ethical guidelines and laws.

In addition to this institutional oversight, scientific journals have had in place the system of checks and balances we discussed in Chapter 1, based on peer review and credit by publication, to guard against bias and fraud in science. This process also is governed by strict ethical guidelines, for the validity of the peer review system depends on the integrity of editors and reviewers. Readers must be able to trust that the work they see in print has been carefully scrutinized and that all high-quality work in the field will have the opportunity to be published. For authors to be willing to submit their work to a peer-reviewed journal, they must trust that their work will be evaluated fairly and their authorship will be recognized. This system thus places a high priority on the values of skepticism and fairness. Editors are expected to assign reviewers impartially, to keep all information about submitted research confidential, and to recuse themselves from decisions pertaining to manuscripts in which they are coauthors or have a vested interest. Peer reviewers are similarly obligated to review without bias, maintain confidentiality, and reveal potential conflicts of interest when agreeing to review a manuscript.[4]

Though it is generally assumed that all parties involved in the peer review system—authors, editors, and reviewers—conduct themselves ethically, some scientists doubt that the above principles can be reliably sustained, given the sheer number of people involved and the many sources of pressure that might potentially undermine ethical behavior. The efficacy and fairness of these internal gatekeeping mechanisms thus have been questioned (Hoshiko 1991; Roy 1993; Dagani 1995), particularly in light of the possibilities of more open and interactive publication raised by proponents of e-journals (Odlyzko 1995; Harnad 1997). But even proponents of e-journals acknowledge the importance of the checks and balances that since the 17th century have been designed to protect the rights of individual scientists, maintain the integrity of research, enhance the reputation of scientific

---

[4]For samples of editorial statements on these and other ethical issues, see the guidelines published by the International Committee of Medical Journal Editors (ICMJE 2001) at http://www.icmje.org or those developed by the American Society of Civil Engineers (ASCE 2000) at http://www.pubs.asce.org/authors/index.html#ethic.

institutions, and preserve the public trust in science (NAS 1995; Harnad 1997; Day 1999).[5] There also has been a concerted effort to reform the peer review systems of major funding agencies such as NIH and NSF (Agnew 1999), and revised and expanded guidelines for many scientific organizations are now readily available online (see Macrina 2000, p 71–72).

The ethical standards articulated by journal editors and funding agencies reflect the fundamental values of scientific research itself—"honesty, skepticism, fairness, collegiality, openness" (NAS 1995, p 21), or what Jacob Bronowski (1965) called the "Habit of Truth." Scientists' respect for these shared values is perhaps the ultimate safeguard against misconduct in science. Because scientific communities are themselves social groups, the behavior of scientists also is strongly influenced by the opinion of their peers, with shame being a major deterrent of misconduct (Cohen 1995). Even an allegation of misconduct in science can be damaging to a scientist's career (Macrina 2000). As Hoshiko notes, "A scientist's most valuable possession is his or her reputation, which, once lost—whether deservedly or not—may be virtually impossible to regain" (1991, p 11).

---

**EXERCISE 9.1**

Obtain a copy of the research policies or guidelines for your department, lab, university, or professional association, or look at the special concerns addressed in a funding agency's RFP or the editorial guidelines for a journal in your field. What kinds of ethical issues are raised in these guidelines? Categorize these issues (e.g., humane treatment of subjects, misrepresenting data, authorship, sexual harassment). What do these categories tell you about the focus of ethical concern in your field?

---

## 9.2 The Ethics of Authorship

These general values of honesty, skepticism, fairness, collegiality, and openness imply distinct responsibilities for authors as well, which journal editors often articulate quite explicitly. The editors of *Gastroenterology*, for example, begin their Instructions to Authors with statements on ethics and authorship, reprinted in Figure 9.3. As these guidelines illustrate, misconduct in scientific communication includes not only falsification but also questions about authorship, including plagiarism and the allocation of credit. The Ethics section of the guidelines instantiates the values of honesty and collegiality by requiring authors to certify that the submitted manuscript does not contain intentional error or plagiarize the work of others.

---

[5]Some of the arguments for the wider acceptance of electronic publication in science have to do with reforming the peer review system of journals (e.g., Smith 1997). A few scientists have called for the replacement of the traditional peer review system with one based on trust in individual scientific authors, the number of papers previously published (Dagani 1995), and/or, as Ginsparg believes, the opinion of scientific readers, who can and will determine what is good themselves (Cornell News 2002). The latter most likely depends on the high degree of familiarity and trust that develops in a smaller field, such as exists in high-energy physics (Fry 2003).

### Ethics

GASTROENTEROLOGY strongly discourages the submission of
more than one article dealing with related aspects of the
same study. In almost all cases, a single study is best
reported in a single paper.

The journal editors consider research/publication misconduct
to be a serious breach of ethics and will take action as
necessary to address such misconduct, which includes submission or publication of information that:

(1) Is intentionally erroneous,

(2) Has been published elsewhere by a different author without acknowledgment (plagiarism),

(3) Has been published elsewhere by the same author without
acknowledgment (duplicate publication), or

(4) Is subsequently published elsewhere by the same author
without acknowledgment, attribution, or permission from
the AGA, as holder of the copyright, to reprint or adapt
the material.

Breaches in these standards may result in proscribed submission for all authors of the concerned manuscript and, when
appropriate, notification of the authors' institutions. All
authors are fully responsible for the content of the manuscript.

The publication of abstracts is not considered duplicate publication but should be disclosed in the cover letter accompanying the manuscript submission.

### Authorship

Each author must have participated sufficiently in the work to
take public responsibility for the content of the paper and
must approve of the final version of the manuscript. Authorship should be based on substantive contributions to
each of the following:

(1) conception and design of the study;

(2) generation, collection, assembly, analysis and/or interpretation of data;

(3) drafting or revision of the manuscript;

(4) approval of the final version of the manuscript.

**FIGURE 9.3**   Guidelines on ethics and authorship from *Gastroenterology* Instructions to Authors
(Gastroenterology 2003).

*Plagiarism*—the appropriation of other scientists' ideas, research, or language
without proper attribution—is a serious offense in science since proper attribution
is essential to the trust necessary among scientists for the free exchange of information. A scientist's reputation can be permanently damaged by acts of plagia-

rism. But as with other ethical questions, it is sometimes difficult to determine what constitutes plagiarism and what to do about it. As suggested in Chapter 3, the issue of plagiarism is complicated by the fact that decisions about what to cite depend to some extent on the writer's knowledge of the field: when describing methods, for example, procedures used in other studies may not need to be cited if they represent common practice in the field. Thus, the question of plagiarizing is to some degree a function of a scientist's socialization into and awareness of his or her field (see Latour and Woolgar 1979; Winsor 1993).

## EXERCISE 9.2

Read the following case study, "A Case of Plagiarism" (NAS 1995, p 18), about a student who finds she has committed plagiarism; then consider the questions that the National Academy of Sciences asks. In addition to answering these questions, think about why NAS asks these particular questions; that is, what ethical questions does NAS want you to focus on in the case? Are there some mitigating circumstances here, and if so, what are they?

> May is a second-year graduate student preparing the written portion of her qualifying exam. She incorporates whole sentences and paragraphs verbatim from several published papers. She does not use quotation marks, but the sources are suggested by statements like "(see . . . for more details)." The faculty on the qualifying exam committee note inconsistencies in the writing styles of different paragraphs of the text and check the sources, uncovering May's plagiarism.
>
> After discussion with the faculty, May's plagiarism is brought to the attention of the dean of the graduate school, whose responsibility it is to review such incidents. The graduate school regulations state that "plagiarism, that is, the failure in a dissertation, essay, or other written exercise to acknowledge ideas, research or language taken from others" is specifically prohibited. The dean expels May from the program with the stipulation that she can reapply for the next academic year.
>
> 1. Is plagiarism like this a common practice?
> 2. Are there circumstances that should have led to May's being forgiven for plagiarizing?
> 3. Should May be allowed to reapply to the program?

Related to the concern about acknowledging the work of others are scientists' concerns about allocating credit appropriately among members of the research team itself. As illustrated in earlier chapters, scientists pay close attention to the allocation of credit. In a standard research report, review, or grant proposal, credit is allocated in three places: the list of authors, the acknowledgments, and the citations. Each of these indicates a particular relationship of scientists to the report, review, or proposal. Citations, obviously, refer to published (and occasionally unpublished) work outside the present study that create the context for the new work. Acknowledgments are expressions of appreciation for funding, support, equipment, or other contributions to the research, such as technical assistance (the help of a statistician, for example). However, the question of who should be listed as an author (to say nothing of in what order) can be complicated.

We've noted elsewhere the trend toward increasing numbers of authors in some fields. Papers in the life sciences have listed more than 100 authors (Regalado 1995). In physics, the number of papers with over 50 authors rose dramatically between 1985 and 1994; over 150 papers had more than 100 authors, and about 30 papers had more than 500 authors (McDonald 1995)! In fact, the problem of the soaring number of authors has become so vast that a number of scientists have decried this "author inflation" and have called for reforms in the system (see Garfield 1995; McDonald 1995; Schwitters 1996). The Council of Science Editors (formerly CBE) created a task force to investigate the problem and recommend possible solutions (CBE Task Force on Authorship 2000). Many journal publishers have or are being urged to develop written policies concerning authorship and the allocation of credit (Hoshiko 1991; AAAS 2000; Scheetz 2000), and a push is on to make these policies more complete and uniform for publications featuring interdisciplinary collaboration, as well as for e-journals (DuMez 2000). This concern is especially important given the degree of collaboration that occurs in the sciences.

*Gastroenterology*, like many other journals, addresses this concern directly in their statement on authorship (Figure 9.3), which specifies that all listed authors must make "substantive contributions" to all stages of the research, including the initial study design, the collection and interpretation of data, and all versions of the manuscript. Generally, credit for authorship is given to the primary and all secondary authors, including assistants, postdocs, and students (Cohen 1995, p 1706)—but only to those who have played a direct or important role in the research (NAS 1995). According to Macrina (2000), "Authorship encompasses two fundamental principles: contribution and responsibility. An author must make a significant intellectual or practical contribution to the work reported in the paper. With such authorship goes the responsibility for the content of the paper" (p 59–60). The editors of the *Annals of Internal Medicine* make this point succinctly: "Authorship means accountability" (Annals 2002).

In addition to certifying their role in the research, authors also are required to certify that they have followed appropriate standards for the ethical conduct of the research itself, as discussed in Chapter 6, and they must declare any financial relationship they may have with companies whose products are involved in or affected by the research. All coauthors are held accountable to these ethical standards. *Gastroenterology*'s ethics guidelines in Figure 9.3 also make it clear that authors are expected to respect the journal's right to exclusive publication and copyright, thus preserving the journal's as well as the authors' claim of ownership. The prohibition against duplicate publication reflects the community's concern that all published research be original. In sum, the integrity of the journal publication process clearly depends on the integrity of individual authors.

At the same time as scientists and editors grapple with the issue of the ever-larger number of authors and the proper allocation of credit, the cost of doing science rises, and research budgets for scientific projects not related to national defense shrink. Thus science becomes more competitive. The need to obtain funding and attract other researchers and students and hence more funding, increases the pressure for individual scientists to seek publication that will advance their careers (Cohen 1995; Macrina 2000, p 50). This may mean competing to publish in more prestigious journals, to work with well-known scientists, or to be the primary author of important research. These two trends—toward increased collaboration and increased competition—are often in opposition to each other (CAIRE et al. 2002). Sometimes the human drive for personal recognition and the increased pressure for funding work against the "ethic of collaboration" that is at the heart of science. Some observers of the Pons and Fleischmann case described in Chapter 1 ascribe their "fall from grace" to just such a conflict (e.g., Crease and Samios 1989; Close 1992). Whether the conflict between collaboration and competition is a major contributing factor in the cold fusion debacle or not, scientists in all fields must continuously negotiate this very real tension between "the ethic of collaboration and the drive for glory" (Benditt 1995, p 1705).

## EXERCISE 9.3

Read the following two scenarios from Macrina (2000, p 66, 69), and prepare for class discussion your answers to the questions posed at the end of each scenario. For each case, also speculate about why events might have unfolded as they did. What are the practical and ethical ramifications of those possibilities for scientific publication?

A. Sara Nichols had a very productive postdoctoral training experience. With her mentor she coauthored four important papers on oncogene expression. She was the first author on all of these papers. Jacob Smith, her mentor, conceived of the ideas for the work, but Sara did all the experiments, interpreted the results, and wrote the papers. Sara is now an assistant professor struggling to get her first grant in order to continue her oncogene research. She reads a new review article on oncogene expression in which the author repeatedly cites her four papers as being very important. However, the author of the review continually refers to these papers as the contributions of "Smith and coworkers." Sara is offended and upset by this. There were no other coworkers who contributed to this work, and she believes that the papers should be referred to as the work of "Nichols and Smith." She is worried that the inappropriate references to her work will undermine her contributions and deprive her of credit that can promote her career advancement. She writes to you, the editor of the journal that published the review. How do you respond to her? What, if anything, will you do about this situation?

B. Marvin Brian, a faculty member at a major research university, is the advisor of an advanced doctoral student, Henry Ruth, and a beginning graduate student, Mark Butterworth. Henry serves as the lead investigator and prepares and presents reports to the funding agency. Mark works on the same project, sharing his data with Dr. Brian and Henry. After working on the project for about 2 years, Mark submits his master's thesis, which is reviewed and approved by Dr. Brian but not seen by Henry. A year or so later, when Henry is finishing the text of his doctoral dissertation, he discovers that Mark's thesis contains at least one complete table representing his work in exactly the same format that Henry has used to express his results. The master's thesis contains a general acknowledgment of Henry, among others, but no specific attribution is given for the verbatim table. All parties are aware that this research is supported by a contract with a defined work scope. Does this sponsorship justify duplicate publication in a master's thesis and a doctoral dissertation without explanation? If not, how should the matter be handled? Once duplicate publication occurs, what should be done and who is responsible for initiating remedial action?

# 9.3 Scientific Communication as Moral Responsibility

An ethical issue related to the allocation of credit is the sharing of information. If science is a social enterprise, the sharing of information becomes in effect a moral imperative for scientists. The CAIRE et al. report on "Integrity in Scientific Research" makes this clear: "Collegiality in scientific interactions, including communications and sharing resources requires that investigators report findings to the scientific community in a full, open, and timely manner" (2002, p 37). At the same time, we've just underscored the highly competitive nature of science. Thus, the drive for personal recognition that sometimes works against the ethic of collaboration also may work against the ethic of sharing. As Hoshiko notes, "Competition fosters self-protection and inhibits collegiality among investigators" (1991, p 11). In addition, conflicts between the need for scientists to cooperate and the need to "protect" one's findings can be complicated by the requirements of corporate secrecy or national security, competition between labs or universities, and even personal animosities. As shown in Chapter 1, some of these motives also may have come into play in the case of Pons and Fleischmann.

The conflict between the open sharing of scientific results obtained in research universities and the proprietary nature of results obtained in industry can be particularly apparent when the researchers are working on the same problem, as they often are, or when a university study is funded by industry, and the stakes—be they scientific, social, or monetary—are high. For instance, in 2000, *Science* reported a case in which Immune Response Corporation attempted to stop the publication of an immune system booster study by James Kahn at the University of Southern California; Kahn's study indicated the company's drug was not effective in inhibiting the progression of HIV. Immune Response Corporation demanded that Kahn and his fellow researchers include the company's own data, which they claimed indicated more positive effects of the drug for a small subset of people, and that the company review revisions prior to publication. "The researchers refused" (Morton 2000, p 1063).

But perhaps the largest and most obvious case of competition in recent memory between public and private sectors of scientific society has been the competition between the National Genome Research Institute, a federally funded project led by Dr. Francis Collins, and Celera Genomics, a public exchange corporation led by Dr. J. Craig Venter. The heated competition to be the first to map the human genome has erupted on several occasions into arguments about technique and technology, government versus corporate sponsorship, the shared public versus the competitive proprietary nature of gene data, and the ultimate "ownership" of the research—the right to profit from the human genome itself (Science 2000b). Despite their appearance together onstage for a tribute by President Bill Clinton on June 26, 2000, to celebrate a major milestone in the mapping of the human genome, the personal animosity between these two leading researchers has been obvious (Marshall 2000a). In 2000, *Science* waived its standard requirement that DNA data of a study be deposited in a public database in order to publish Celera's genome research, which the journal considered of primary importance to the scientific community; as a result, the National Genome Research Institute announced it would no longer publish their results in *Science* (Schultz 2000; Marshall 2000b).

**EXERCISE 9.4**

Read Chiba et al. (2002) in Chapter 10 (pages 280–286). Note the affiliation of each author. Also note the acknowledgment section at the end of the report, particularly the source of the funding for this project and the statement of competing interests.

Identify possible competing interests in this report. Do you think this study is impartial? Why or why not? Is the risk of bias in research in industry similar to or different from risks in other scientific settings? What safeguards are in place to protect the research from bias in industry? Are they similar to those in place to protect research in other scientific settings? In your opinion, are these safeguards adequate?

Many scientists believe that generosity and openness, as opposed to competition, are in the long run rewarded in science (Cohen 1995). Nevertheless, the social ethics of science are in a constant state of tension with the individualistic ethics of science. It is obvious that the time required by the peer review process works against competition. We saw in Chapter 1 how in the case of Pons and Fleischmann, competition and the rush to publicize undercut peer review (Huizenga 1992). "The need for skeptical review of scientific results is one reason why free and open communication is so important in science," says the National Academy of Sciences (1995, p 12). But this social mechanism, designed to maintain professional standards and foster the sharing of information, can work against the individual scientist's drive for priority and credit that is a major incentive in scientific research (NAS 1995; Macrina 2000).

One source of sought-after credit is publication in the most prestigious journals. While arguments in favor of electronic publication often point to the speed and openness of this medium (e.g., Harnad and Hey 1995; Smith 1997), one concern is that e-journals, which are less expensive to produce, will drive out more expensive print journals. The resulting decrease in the number of journals overall, coupled with the fear that electronic journals are less likely to be peer reviewed and thus less prestigious, raises concerns that this trend will reduce competition. While increasing the sharing of information, the net effect of more open and interactive publication in e-journals could be to undermine one of the major incentives in science: individual advancement (e.g., Butler and Wadman 1999; Hogan 1999).

The responsibility of scientists to share their work with each other, then, is another complicated issue. Scientists are entitled to a "period of privacy," especially in the early stages of research (NAS 1995, p 10). But during peer review or after the conference presentation or publication, scientists have a responsibility to share results openly, honestly, and accurately. Perhaps a reflection of the electronic openness of our age, arguments for electronic publication sometimes also cite as advantages the possibility of more interaction in the prepublication stage "in which most of the cognitive work is done" (Harnad 1990, p 342). However, pride, competitiveness, the desire for originality, the fear of being scooped, the need for professional advancement (and the research funds that go with it), the pressure for secrecy from corporations or government, and, in a few cases, the profit motive, all potentially work against the open, interactive nature of electronic communication, just as they work against the ethic of sharing. The complexity of these issues for scientists is best summed up in the CAIRE et al. report:

> Because science is a cumulative, interconnected, and competitive enterprise, with tensions among the various societies in which research is conducted, now more than ever researchers must balance collaboration and collegiality with competition and secrecy. Another result of these tensions is conflict-of-interest and intellectual property issues, which are increasingly important to administrators of research institutions. (CAIRE et al. 2002, p 25)

Ways to better handle some of these ethical conflicts—through improvements in science education, further articulation of policy by funding agencies and professional societies, instructions to authors, and more rigorous evaluation and monitoring—are continuously being developed (see Hoshiko 1991; NAS 1995; Scheetz 2000; CAIRE et al. 2002).

---

**EXERCISE 9.5**

In light of this discussion of competing ethics within science, review the case of Pons and Fleischmann in Figure 1.1 (pages 13–14). List the competing ethical factors in this case. Write a paragraph defending Pons and Fleischmann's actions; then write a paragraph criticizing them. Different groups may be assigned parts of this exercise in preparation for the kind of ethical debate scientists, science educators, and policy makers are increasingly engaging in (Benditt 1995). Does this ethical examination of conflicts of interest in science give you a better understanding of the motives or increased sympathy for the plight of Pons and Fleischmann? Why or why not?

---

## 9.4 Scientific Communication and Public Communication: An Ethical Conflict?

We discussed in Chapter 8 why scientists have an obligation to communicate with the public and why the public has both a need and a right to know what is going on in science. We have seen that sometimes, as perhaps in the Pons and Fleischmann case, the need for peer review and the need or desire to share information with the public come into conflict as well. This can become an ethical dilemma for the scientist. While there are strong incentives to keep research secret until publication, some of the same factors may provide incentive to release results to the public before peer review and publication—both positive factors such as the desire to respond to urgent social need and negative factors such as competition, pride, and the desire for credit and fame.

As we discussed in Chapter 1, in addition to violating the ethic of peer review embedded deep within the enterprise of science, there is also a stigma attached to reprinting material in scientific journals that has already been "prepublished" in the popular press. We also mentioned that the problem of prepublication applies to email, discussion lists, and message boards too, where information is exchanged more freely. Most editors will refuse papers that have been published elsewhere (as *Gastroenterology*'s guidelines indicate; see Figure 9.3). Some journals, like the *Journal of the American Chemical Society*, specifically prohibit papers that have been previously broadcast on the web.[6] In some cases, however, editors have been willing to reconsider the injunction against a press conference prior to scientific publication, or the embargo against publication of a study that has been released to the public (Roy 1993), especially if the research is urgent or important enough to warrant its early release (Maddox 1989; CBE 1995). The problem is that it is usually only after scientific peer review and publication that urgency or importance can be determined—too late for the unfortunate scientists who are wrong or the public that is misinformed and confused. As with most ethical issues, in specific cases the line in science between the obligation to monitor itself and the obligation to inform society is more often than not blurred.

## EXERCISE 9.6

The following case study is from the Report from the 1994 Annual Meeting of the Council of Biology Editors (CBE 1995, p 14), at which editors of scientific publications were asked to respond to different scenarios concerning the "prepublication release of information." This two-part scenario was addressed to Richard Horton, editor of the *Lancet*. What are the ethical issues involved? What information do you need in order to make a decision? How do you think Horton responded to each scenario, and why?

> Part 1. Dr. Gilbert believes he has found a new gene for Alzheimer's disease. He submits an abstract of his work to his professional society for presentation at its annual meeting. The abstract is not accepted, but he is invited to prepare a poster, which he does. The poster attracts a great deal of media attention at the meeting, and a story about it appears in the Sunday *New York Times*, complete with the figures and tables, as well as a summary of the study design and results. A similar story appears in *Newsweek*. A month later, the study report is submitted to the *Lancet* for publication. Horton believes the work is novel, interesting, and important. Is he disturbed by the prepublication release of information?

> Part 2. Dr. Sullivan believes she has found a new gene for bipolar disorder, and she presents her work orally at her professional society's annual meeting. The presentation is covered extensively in the popular media, although figures and tables are not available. Will this affect later publication in the *Lancet?* Suppose Sullivan elaborates on the study in a press conference after the presentation. Suppose she passes out copies of the paper she plans to submit to the *Lancet*. Suppose she publishes a slightly longer summary of the study in a book of proceedings of the meeting.

---

[6]Cited 2003 March 2, this policy is available at https://paragon.acs.org/paragon/application?pageid=content&parentid=authorchecklist&mid=ja_policy.html&headername=Policy%20_On%20Publication%20of%20materials%20Previously%20Posted%20on%20the%20Web. Cf. Wilkinson 2001.

It is not only the system of scientific checks and balances or the injunction against prior publication that prevents scientists from communicating with the public. Sometimes interaction with the public is impeded by the requirement of political or corporate secrecy. This sets up another ethical dilemma. The conflict between the need for secrecy in science and the public's right to know is not unlike the (unresolvable?) legal conflict over the need for national security versus the First Amendment right of free speech (e.g., former government officials are prevented from publishing what they know about classified government activity—even illegal activity—until the information has passed through government review). In fact, questions concerning secrecy versus the public's right to know have become further complicated and particularly acute following the 2001 terrorist attack on the U.S. mainland. Research, especially in sciences and technologies related to the development and/or detection of nuclear, chemical, or biological weapons, and those connected with surveillance, is perhaps more tightly interwoven with national security and thus secrecy than ever before. Scientists are working on projects involving "homeland security," which at the time of this writing has just received cabinet status and a massive infusion of funding. But at the same time scientists realize that good science requires open science, and open science is international; Neal Lane, Assistant to President George W. Bush for Science and Technology and Director of the Office of Science and Technology Policy, suggests that to balance these conflicting needs, the federal government must trust scientists themselves to handle and maintain security (Lane 2001).

Sometimes, scientific and other information is kept from the public for legitimate security reasons. But sometimes information is kept from the public for reasons other than legitimate ones. The problem is, how do we, the public, know? Politicians have been known to "spin" or distort the data of a scientific report, or even to change it, for economic or political purposes. In at least some of these cases, the press has uncovered government misconduct and has reported it. For example, Philip Shabecoff (1989) revealed that the White House's Office of Management and Budget (OMB) under President George H. W. Bush changed the text of a scientific report on global warming, which was to be delivered to the Senate subcommittee on Science, Technology, and Space by the director of NASA's Goddard Institute for Space Studies. The wording and conclusion of the report were substantially altered to soften projections and thus justify the administration's tempered approach to international policy on the greenhouse effect. What outraged scientists and congressional members was that the OMB, a political office, not only misrepresented a scientist's conclusions, but actually tampered with his words.[7]

The requirement of secrecy can raise serious ethical and political questions about the power to control scientific information. On what grounds should information be kept from the public? What kind of information should be kept from the public, and under what circumstances? Perhaps most important, who in a democracy will make these decisions? Scientific experts? The government? The

---

[7]As widely reported in the press, the section of the 2003 EPA report identifying global warming as a threat was completely deleted after the administration of George W. Bush removed and/or changed the language of the study by the National Academy of Sciences (see Sierra Club 2003).

Pentagon? The CIA? Executives of corporations? Average citizens? The press? Finally, how will these decisions be made? How much does the public need to know to make these decisions, and how can security simultaneously be maintained? Who does the scientist ultimately owe allegiance to? These are all vexing questions in our democratic society, and there are no clear answers.

Unfortunately, history also contains some extreme cases where those in power have used science for immoral ends. Katz (1992b, 1993) has examined what can happen when political, economic, or corporate expediency is allowed to subvert the interests of science: he argues that when "the ethic of expediency" becomes the only goal of science or technology—becomes the value in science that overrides all other human values, as it did in Nazi Germany (or the former Soviet Union)—scientific communication can become an ethical weapon that can be used by the unscrupulous *against* society. As the National Academy of Sciences points out, the result is invariably a deterioration of science itself (see NAS 1995, p 7). The painfully ambiguous situation of the atomic physicist Werner Heisenberg, who elected to remain in Germany during the Nazi years to protect science from Hitler, he argued (see Heisenberg 1971, esp. p 165–179), is poignantly illustrated in the play *Copenhagen* (Frayn 2000), an exploration of the enormous ethical and human dilemmas that this scientist faced in his lifetime.

## EXERCISE 9.7

Can you think of other scientists in history who have had to make momentous ethical decisions? Name them, identify the circumstances and issues they faced, and describe the possible consequences of their options and action. Outline the ethical guidelines they seemed to follow in making their decision(s). What can we learn about the relationship between science, ethics, and life from these extraordinary cases?

Fortunately, most of us probably never will have to face such extreme ethical situations. But in cases involving governmental or corporate misconduct, the scientist may be called upon to "blow the whistle." This presents another ethical dilemma in scientific communication. Scientists are often hesitant to act because of uncertainty about the appearance of wrongdoing, or from fear of reprisal. But according to the National Academy of Sciences, "someone who has witnessed misconduct [in science] has an unmistakable obligation to act" (1995, p 18). However, *how* to act is often a pressing question. When do you "tell," and whom? Scientists may wish to seek advice from friends, colleagues, or supervisors before they act (see NAS 1995). Aside from the employees at Enron and World.com, perhaps the most spectacular example of whistle-blowing in recent history has been the allegations in the media by former top scientists of tobacco companies that their employers, contrary to their sworn statements, studied the addictive process of nicotine and deliberately manipulated its levels in cigarettes.

No matter what your politics are, the world has become an ethically complex place. Science and technology are at the center of this world. As CAIRE et al.'s report on research integrity demonstrates, it is not only the scientific domain, but also the political world that scientists must now negotiate as a part of their normal environment (CAIRE et al. 2002).

**EXERCISE 9.8**

Find out whether there are specific policies, safeguards, and procedures for communicating ethical violations in your department, lab, university, or discipline. What kinds of ethical misconduct are covered by these policies? Note any specifics about the procedure, such as to whom you are to communicate misconduct, time periods for reporting, protection from reprisals, processes of adjudicating charges, and results of possible outcomes. In your opinion, are these policies, safeguards, and procedures adequate? Why or why not?

# 9.5 Scientific Style and Social Responsibility

Throughout this book we have been exploring how scientific knowledge is constructed in and through the social process of communication. Because scientific knowledge is a social construction, scientists have a social and moral responsibility for the knowledge they create. In the traditional view of science as a dispassionate description of truth, scientists did not have a responsibility for their discoveries: what they discovered was simply "there"; the job of the scientist was to objectively observe and report the "facts" (Bazerman 1988). In the traditional view of science, it is up to others in society to decide how to use science and how to deal with the consequences of the "facts" discovered by scientists.

However, if science is socially constructed, as we argued in Chapter 1, then facts are not simply there but are construed by scientists based on the paradigms and values within which scientists work. In *Science and Human Values* (1965), Jacob Bronowski, chemist and man of letters, argued that scientific "facts" are concepts that organize appearances into laws. Thus for Bronowski, science and ethics are not different because both are "systems of concepts" that are tested in experience (p 40). "When judgment is recognized as a scientific tool, it is easier to see how science can be influenced by values," says the National Academy of Sciences; "values cannot—and should not—be separated from science. The desire to do good work is a human value. So is the conviction that standards of honesty and objectivity need to be maintained" (1995, p 6).

In the contemporary view of science, scientists participate in the construction of "reality" through the process of communication (Brummett 1976). Whether it be an idea, a theory, an observation, a measurement, a method, a discovery, an invention, or an experiment, scientists help shape the beliefs and concepts of a culture and so bear responsibility for those beliefs and concepts. We have explored throughout this book how argumentation and persuasion are integral to the construction of scientific knowledge. Thus, scientists also have a moral imperative to create and use scientific arguments responsibly, by employing theories, methods, logic, and data acceptable in the field as well as to society at large (CAIRE et al. 2002).

In constructing responsible arguments, we need to carefully consider not only *what* we say but *how* we say it, for style has an ethical dimension as well. Take the traditional use of passive voice in science as a case in point. Passive voice has the effect of de-emphasizing the individual scientist or lab in favor of an *ethos* of neutrality and objectivity. Thus, the sentence "We collected the samples" becomes "The samples were collected" in passive construction. The effect of this style is that the agent, the human being conducting the experiment, is no longer apparent. This is not usually a problem. As we discussed in Chapter 3, for example, the AIP Style Manual (1990) recommends the use of passive voice in some methods descriptions where the agent is not important. Passive voice is used conventionally in many forums; using it appropriately is part of learning how to speak as a professional scientist. But the use of passive voice can become an ethical problem when it is used to hide responsibility, as sometimes happens when information is released to the public: "In the 1960s, the CIA tested LSD on civilians" can become "In the 1960s, LSD was tested on civilians."

In Chapter 8 we touched on the fact that miscommunication played decisive roles in the Challenger explosion, in the disarray that followed the Three Mile Island meltdown, and in failed attempts to site radioactive waste facilities around the country; in that chapter our focus was on scientists' ability to understand and appeal to audiences outside their discipline. Here our focus is on the role of style in these catastrophes; *how* something is said can become as much of an issue as what is said in scientific and technological accidents. On January 28, 1986, as a horrified nation watched the space shuttle Challenger explode, NASA's media spokesperson announced: "A major malfunction has occurred." Many found the *ethos* of objectivity, of nonresponsibility, of calm scientific detachment—created by abstraction and jargon—inappropriate if not offensive because it ran counter to the way people actually felt; the perception was reinforced when this style was continued by NASA officials in press conferences throughout the day (Havelock 1986), creating an impression of callousness that was difficult for NASA to overcome. NASA did not make this public relations mistake again when in February 2003 the Columbia disintegrated minutes before landing. Where public health, safety, and/or trust are at stake, and the accountability of science is put to the test, scientists must recognize and take pains to avoid even the appearance of evasiveness that can be created by style.

The use of scientific style by others in society, like the use of scientific findings, also can create an ethical problem for scientists. Public officials may not have adequate knowledge of the science; they also may not have complete control of their style, as Katz (2001) demonstrates in an analysis of a speech given by former Secretary of Agriculture, Dan Glickman. At a press conference about the benefits of biotechnology, Glickman's language was full of hidden negative words and metaphors. For example, he referred to recent developments in biotechnology as "the tip of the biotechnology iceberg" (Glickman 1999, paragraph 7). After the sinking of the Titanic, the iceberg metaphor is problematic, hinting that bigger problems lie beneath the surface; these kinds of stylistic choices create a negative tone that permeates much of the speech and undercuts Glickman's intended message, further complicating the relationship between biotech scientists and the public.

The scientific jargon sometimes used by government officials to support *political* positions also can create an *ethos* of scientific objectivity while obfuscating meaning, as in the following example: "Understanding the role natural site features play as part of an integrated system in protecting public health and safety is an important part of the comprehensive site assessment activities" (North Carolina Low-Level Radioactive Waste Management Authority 1996, p 1). What does this sentence really say? What is meant by "natural site feature"? How does it "play as part of an integrated system in protecting public health and safety"? What are "comprehensive site assessment activities"? These abstract words and noun compounds strung together as official terminology into a positive sounding sentence may seem scientific and impressive but actually are an exhibition of a false *ethos* of objectivity created by an inflated and vacuous style that stands entirely on the top rung of the ladder of abstraction discussed in Chapter 8.

The *ethos* of science is also used by advertisers to sell products ("9 out of 10 doctors recommend . . . ,"), a point that CAIRE et al. optimistically cites as one indication of the public's positive attitude toward science (CAIRE et al. 2002, p 16) but that Macrina points to as a problematic perception of science as infallible (2000, p 2–3). The use of the *ethos* of objectivity created by such devices as passive voice, abstraction, jargon, and "context-free" statistics helps foster a false public image of scientists as omniscient and infallible. The National Academy of Sciences warns scientists not to set science up as a form of knowledge superior to other forms of knowledge (1995, p 21). But the *ethos* of science created *by others* can raise some unforeseen social and political consequences for scientists as well.

As we explored in Chapter 3, scientists understand (and indicate with the use of qualifiers) that scientific knowledge is always open-ended, uncertain, and subject to change. "Scientists recognize that this is how science normally works, but in general, people outside of science do not have this same understanding," says Macrina (2000, p 3). Because the public expects scientists to always be right and scientific knowledge to be absolute, public audiences often become frustrated and confused when scientists qualify their claims or when scientists disagree. In such cases, the public often reverts back to their own beliefs and previous convictions (Nelkin 1975; 1979). As the National Academy of Sciences points out,

> Many people harbor misconceptions about the nature and aims of science. They believe it to be a cold, impersonal search for truth devoid of human values. Scientists know these misconceptions are mistaken, but the misconceptions can be damaging. They can influence the way scientists are treated by others, discourage young people from pursuing interests in science, and, at worst, distort the science-based decisions that must be made in a technological society. Scientists must work to counter these feelings. (NAS 1989, p 20)

**EXERCISE 9.9**

A. Nelkin (1979) discusses a case in which the California State Legislature, deliberating about the safety of nuclear power, abandoned their attempt to seek the advice of experts after hearing 120 of them argue about the issue, deciding instead to rely on the opinion of the voters. How can scientists who disagree help avoid this response from the public and its official representatives? How should scientists disagree with each other? How should they explain their differences to the public? Make a list of the ways scientists could help avoid this kind of response from the public. Be sure to consider issues of communication and style.

B. In his book *Scientific Integrity*, Francis Macrina states: "Disagreements, errors, and new interpretations of results are sometimes reported to the public by the media. It is easy for such reporting to be misinterpreted. The debate about emerging or evolving scientific knowledge can be seen as confusion or interpreted as accusation. This may even cause some to question the integrity of science" (2000, p 3). Find a particular example of such misinterpretation in your field caused by a report in the media. How should scientists respond to the confusion created by the media? Can they do so in a way that won't create more conflict, misinterpretation, or confusion when it is reported? Should scientists respond to the press? Should they deal with the media differently in the future? Make a list of possible approaches and/or responses. In groups or as a class, discuss the pros and cons of each approach.

---

The traditional image of science as objective and absolute can be misleading and can engender a false faith in science. Though scientists hold more temperate views concerning the certainty of the results of their work, they must realize how their work is perceived and interpreted by others in society, and must take responsibility for correcting misperceptions. As the National Academy of Sciences states, "concern and involvement with the broader uses of scientific knowledge are essential if scientists are to retain the public's trust" (1995, p 21).

A part of the solution entails understanding how the public's image of science is the result of communication processes. In Chapter 8 we touched on how scientific information changes as it moves from expert to nonexpert forums because of the need for scientific information to adapt to the knowledge, values, goals, and concerns of public audiences. Fahnestock (1986) also has demonstrated that in this process of accommodation based on appeals to practical application and wonder, the level of certainty asserted rises. Thus, while an official of the Nuclear Regulatory Commission called the accidental release of nuclear radiation from the Three Mile Island Nuclear Power Plant into the atmosphere a "serious contamination problem on site," an official from the Pennsylvania Emergency Management Agency put a more positive—and certain—spin on things: "people have nothing to worry about. The radiation level is what people would get if they played golf in the sunshine" (Farrell and Goodnight 1981, p 283). Whether they or journalists are fulfilling the social responsibility to inform the public of scientific developments, scientists must realize the image of science that is created in the process (NAS 1989, p 21).

## 9.6 The Ethics of Style as Socialization

There is another ethical dimension of style that you should be aware of, one that affects you personally as a new member of a scientific community. Because science is a social enterprise, governed both by explicit and tacit conventions, there is enormous pressure for new members of the field to abide by those conventions. As we mentioned in Chapter 2, in learning the conventions of your field, what you are engaged in is a process of socialization.

We all conform to some set of values; and as the National Academy of Sciences implies, the values of science have produced an unprecedented period of scientific progress (1995). Conformity facilitates communication; by adhering to conventions, writers and speakers in a community fulfill the expectations of readers and listeners in the community. In Chapter 3 we saw how conforming to the IMRAD form facilitates communication: by adhering to structural conventions, researchers place information where it is expected and thus make information more accessible and the structure of the overall argument more familiar and thus easier to follow. Conforming to conventions makes researchers in the scientific community more credible and better understood.

By the same token, however, conformity by its very nature constrains individuality, creativity, originality, and personal style. In adhering to these conventions, the engaged scientist does not necessarily say only what the community expects to hear, but does need to *acknowledge* what the community expects to hear when attempting to introduce new ideas. Conventions exist for a reason, but we need to maintain an awareness of the rhetorical role of convention in science and the fact that conventions continually evolve. The use of active voice and first person pronouns in some fields—the recognition of the personal voice and *ethos* of the researcher(s) in a formal scientific text—is a profound shift in the social conventions of those fields. So too is the increasing use of email for the conduct of scientific research, which is beginning to alter the role of peer review in science and, perhaps eventually, of journal publication itself (Brody 1996; Harnad 1997; Delmothe and Smith 1999).

We also should point out the possible consequences of violating the conventions of your field. As shown in the cases of Barry Marshall and of Pons and Fleischmann, failure to conform to the stylistic expectations of the field can result in difficulty being heard. In the long run, Marshall's case demonstrates that it is possible for individuals or nonconformists to be heard and ultimately to influence the community's "language"—but only if the person engages in free and open inquiry with the community, which Pons and Fleischmann apparently did not.

The ethical issue of style as socialization raises another question. In Chapter 8 and in this chapter we have talked about the need for scientists to communicate with the public. However, there is in some quarters in some fields professional pressure not to do so. Sometimes the efforts of scientists to communicate with the public damage their professional standing. Astronomer Carl Sagan was ridiculed and ostracized by some in his field after his television series and book, *Cosmos*.

Some astronomers felt that his demeanor pandered to the public, debased the field of astronomy, and tarnished their image. Other scientists, such as Linus Pauling and B. F. Skinner, also have come in for criticism by their peers for statements made in public (Wilkes 1990).

Thus, again, the individual scientist is faced with another ethical dilemma, another moral choice: How to balance the need and desire to address the public with the need and desire to establish and maintain a successful career as a professional scientist? Should scientists have to jeopardize their professional *ethos*, as Carl Sagan did, to generate interest and enthusiasm for science in the public? Is there a way to both maintain your professional *ethos* and communicate effectively with the public?

As we noted in Chapter 8, there are many examples of scientists from a variety of fields who have been able to write for a popular audience without damaging their professional *ethos*. Wilkes (1990) suggests that as long as scientists talk about their own field instead of presenting themselves as general authorities, they'll be on safe ground. However, the case of Carl Sagan, when contrasted with that of Jacob Bronowski, suggests that the answer may be more a matter of style than of subject matter. In his book and television program, Carl Sagan *was* talking about his own field! Jacob Bronowski, on the other hand, was a chemist who had a successful book and television series on Western civilization, *The Ascent of Man*, and remained well regarded by his professional colleagues. The difference in the reception of their "popularizations" by their respective fields apparently was not related to whether they talked about their own field. Rather, Sagan was criticized more for his "gee whiz" style—for his Mr. Wizard *ethos*—than for the scientific content of his presentations. Jacob Bronowski, on the other hand, was praised for his dignified style of presentation, although he spoke on subjects outside the field of chemistry.

While it's important to adapt your style to appeal to audience interests, moderation may be the key. It may be a matter of degree rather than of kind. But this is a complex issue. Both Albert Einstein and J. Robert Oppenheimer became outspoken advocates of world peace; but while Einstein was praised, Oppenheimer was destroyed by the politics of the cold war. Thus, context—the time and place and politics in which scientists live—as well as style undoubtedly plays a role in the professional reception of public pronouncements by peers. And as the case of Heisenberg illustrates, it is not only or even the professional reputation that may suffer: as a citizen of a German nation occupying Denmark, Heisenberg lost the personal friendship of his father figure and mentor, Neils Bohr, who was Danish and half-Jewish.

The taboo against communicating with the public is lessening, thanks in no small measure to the National Academy of Sciences, which has pointed out the importance of the public to science itself (1989; 1995). Other organizations, such as the National Research Council (administered by NAS) and the National Academy of Engineering, help foster a climate of communication and cooperation across social and professional boundaries by gathering scientists and leaders from all parts of society to address scientific issues and give advice to the government and the public.

Science is a social institution. As such, new scientists will want to learn the conventions their research communities have established. Those conventions also constitute a system of ethics in your field, and behaving according to those ethics constitutes a process of socialization and increasing conformity. Socialization and conformity do not necessarily preclude individuality, as the Marshall case illustrates; but the process of joining a community has both moral obligations and moral implications. The scientist who recognizes the ethical dimension of these conventions will be in a better position to understand not only what is expected of her or him as a professional but also how to act in conflicting or difficult situations. Equally important, scientists who are aware of the conventions governing their fields and the assumptions underlying them are better able to work for change in their community's professional practices—to modify conventions that have become outmoded and no longer reflect or serve the interests and values of scientists, science, or society.

# Activities and Assignments

1. For a class discussion or a short paper, go to the Department of Health and Human Services' website at http://ori.dhhs.gov/html/misconduct/casesummaries.asp and then click on the link to the Office on Research Integrity; this URL should take you directly to the Case Summaries under Handling Misconduct.[8] Click on and read one or more of the case summaries of scientists who, after investigations by their institutions and the Office on Research Integrity, have been found guilty of scientific misconduct. Then answer the following questions.

   a. According to the final findings, what acts of misconduct are the scientists guilty of? List the ethical criteria and conflicts involved. How did ORI classify these acts (i.e., falsification, fabrication, and/or plagiarism)? What are the communication dimensions of the misconduct? What were the immediate consequences of the misconduct for the research problem, the researcher's coauthors and colleagues, the publications and funding agencies, the public, the field? What were the administrative actions of ORI (i.e., the penalty for the misconduct)? What is the communication dimension of the penalty?

   b. Do you think the final finding was fair? Do you think the penalty was too strict or too lenient? What scientific conventions are guiding ORI's decisions? (You can write a paragraph or two on each side of the issue or debate it in small groups or with your class.) What are the short-term effects on the career of the scientist(s) found guilty of scientific misconduct? Do you think there will be any long-term effects on the career of the scientist(s)? If so, what might they be? What do you learn from the case(s) about the ethics of communication in science?

2. Choose one of the ethical dilemmas described in this book or a case you know from your own field. Read about it, and plan a presentation for your class. What issues were involved? How were they handled? Were they solved

---

[8]Alternatively, you may search the site by clicking on http://ori.dhhs.gov/, which will take you to the Office on Research Integrity homepage. Once there, click on Resources on the menu to the left, and then Findings of Scientific Misconduct under Breaking News.

by the institution, the discipline, or the government? What was the outcome? Was it satisfactory?

3. Watch a commercial on television that seems to involve the testimony of scientists or health practitioners. Is an *ethos* of objectivity being created by advertisers for the purposes of selling products? If so, how? Do you feel it is appealing? Accurate? Ethical? On what do you base your answer? Can commercials be accurate and still appeal to their viewers' needs, interests, and values? How would you change this commercial to balance these conflicting needs?

# PART 2

# Sample Research Cases

# CHAPTER 10

# Research on the "Ulcer Bug": From Theory to Clinical Application

At an international conference on *Campylobacter* infections in 1983, a young Australian internist, Dr. Barry Marshall, reported that he and Dr. Robin Warren had identified a bacterium responsible for stomach ulcers—a revolutionary claim, given the prevailing assumption in the medical community that ulcers were caused by stress and other psychological factors. Warren and Marshall announced their discovery in print that same year, in a pair of letters in the *Lancet*, the first items reprinted in this chapter. Their theory that ulcers and other gastric illnesses were related to bacterial infection in the stomach lining was greeted with both intense interest and skepticism by experts in the field. By the time Martin Blaser wrote a review of research on the topic for *Gastroenterology* in 1987, the next document included here, over one hundred studies had been published. The theory was much debated in a number of scientific forums, including the letters sections of the *Lancet* and *Annals of Internal Medicine*. Five of these research letters, published between 1988 and 1991, are included in this chapter. These selections are followed by a 1992 research report by David Graham and colleagues at the Houston Veterans Affairs Medical Center, one of many research teams to study Marshall and Warren's assertion of a relationship between *H. pylori* and ulcers. Their report appeared in the *Annals of Internal Medicine*.

Reports of the revolution in ulcer treatment began appearing in the popular press in the 1990s. We've included a second review piece, written by Martin Blaser in 1996, this time addressing the general readership of *Scientific American*. Commercial testing kits became available in this decade as well, offering techniques for easily detecting the presence of *H. pylori* from breath samples, a concept initially proposed in the late 1980s by research teams headed by Barry Marshall and by David Graham (Fallone et al. 2000). The clinical instructions for one such diagnostic kit from Alimenterics, Inc. are reprinted in this chapter. The final entry is a 2002 research report by a Canadian research team headed by Naoki

Chiba, who used one of these commercially available breath tests to detect the presence and eradication of *H. pylori* in an investigation of its role in common dyspepsia. This piece, published in the *British Medical Journal*, is representative of the field's current focus on clinical detection and treatment of *H. pylori*, twenty years after Marshall and Warren's startling announcement of its presence in the stomach.

**Note:** The varying terminology in this set of texts reflects the fact that, although originally identified as a member of the *Campylobacter* genus, the organism was reclassified in 1989 as a new genus, *Helicobacter*.

### UNIDENTIFIED CURVED BACILLI ON GASTRIC EPITHELIUM IN ACTIVE CHRONIC GASTRITIS

SIR,—Gastric microbiology has been sadly neglected. Half the patients coming to gastroscopy and biopsy show bacterial colonisation of their stomachs, a colonisation remarkable for the constancy of both the bacteria involved and the associated histological changes. During the past three years I have observed small curved and S-shaped bacilli in 135 gastric biopsy specimens. The bacteria were closely associated with the surface epithelium, both within and between the gastric pits. Distribution was continuous, patchy, or focal. They were difficult to see with haematoxylin and eosin stain, but stained well by the Warthin-Starry silver method (figure).

I have classified gastric biopsy findings according to the type of inflammation, regardless of other features, as "no inflammation", "chronic gastritis" (CG), or "active chronic gastritis" (ACG). CG shows more small round cells than normal while ACG is characterised by an increase in polymorphonuclear neutrophil leucocytes, besides the features of CG. It was unusual to find no inflammation. CG usually showed superficial oedema of the mucosa. The leucocytes in ACG were usually focal and superficial, in and near the surface epithelium. In many cases they only infiltrated the necks of occasional gastric glands. The superficial epithelium was often irregular, with reduced mucinogenesis and a cobblestone surface.

When there was no inflammation bacteria were rare. Bacteria were often found in CG, but were rarely numerous. The curved bacilli were almost always present in ACG, often in large numbers and often growing between the cells of the surface epithelium (figure). The constant morphology of these bacteria and their intimate relationship with the mucosal architecture contrasted with the heterogeneous bacteria often seen in the surface debris. There was normally a layer of mucous secretion on the surface of the mucosa. When this layer was intact, the debris was spread over it, while the curved bacilli were on the epithelium beneath, closely spread over the surface (figure).

The curved bacilli and the associated histological changes may be present in any part of the stomach, but they were seen most consistently in the gastric antrum. Inflammation, with no bacteria, occurred in mucosa near focal lesions such as carcinoma or peptic ulcer. In such cases, the leucocytes were spread through the full thickness of the nearby mucosa, in contrast to the superficial infiltration associated with the bacteria. Both the bacteria and the typical histological changes were commonly found in mucosa unaffected by the focal lesion.

The extraordinary features of these bacteria are that they are almost unknown to clinicians and pathologists alike, that they are closely associated with granulocyte infiltration, and that they are present in about half of our routine gastric biopsy specimens in numbers large enough to see on routine histology. The only other organism I have found actively growing in the stomach is *Candida*, sometimes seen in the floor of peptic ulcers. These bacteria were not mentioned in two major studies of gastrointestinal microbiology[1,2] possibly because of their unusual atmospheric requirements and slow growth in culture (described by Dr B. Marshall in the accompanying letter). They were mentioned in passing by Fung et al.[3]

How the bacteria survive is uncertain. There is a pH gradient from acid in the gastric lumen to near neutral in the mucosal vessels. The bacteria grow in close contact with the epithelium, presumably near the neutral end of this gradient, and are protected by the overlying mucus.

The identification and clinical significance of this bacterium remain uncertain. By light microscopy it resembles *Campylobacter jejuni* but cannot be classified by reference to *Bergey's Manual of*

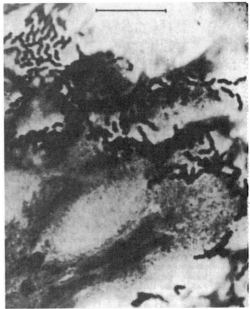

**Curved bacilli on gastric epithelium.**

Section is cut at acute angle to show bacteria on surface, forming network between epithelial cells. (Warthin-Starry silver stain; bar = 10 μm.)

*Determinative Bacteriology.* The stomach must not be viewed as a sterile organ with no permanent flora. Bacteria in numbers sufficient to see by light microscopy are closely associated with an active form of gastritis, a cause of considerable morbidity (dyspeptic disease). These organisms should be recognised and their significance investigated.

Department of Pathology,
Royal Perth Hospital,
Perth, Western Australia 6001                             J. ROBIN WARREN

SIR,—The above description of S-shaped spiral bacteria in the gastric antrum, by my colleague Dr J. R. Warren, raises the following questions: why have they not been seen before; are they pathogens or merely commensals in a damaged mucosa; and are they campylobacters?

In 1938 Doenges[1] found "spirochaetes" in 43% of 242 stomachs at necropsy but drew no conclusions because autolysis had rendered most of the specimens unsuitable for pathological diagnosis. Freedburg and Barron[2] studied 35 partial gastrectomy specimens and found "spirochaetes" in 37%, after a long search. They concluded that the bacteria colonised the tissue near benign or malignant ulcers as non-pathogenic opportunists. When Palmer[3] examined 1140 gastric suction biopsy specimens he did not use silver stains, so, not surprisingly, he found "no structure which could reasonably be considered to be of a spirochaetal nature". He concluded that the gastric "spirochaetes" were oral contaminants which multiplied only in post mortem specimens or close to ulcers. Since that time, the spiral bacteria have rarely been mentioned, except as curiosities,[4] and the subject was not reopened with the

1. Gray JDA, Shiner M. Influence of gastric pH on gastric and jejunal flora. *Gut* 1967; **8**: 574–81.

2. Drasar BS, Shiner M, McLeod GM. Studies on the intestinal flora I: The bacterial flora of the gastrointestinal tract in healthy and achlorhydric persons. *Gastroenterology* 1969; **56**: 71–79.

3. Fung WP, Papadimitriou JM, Matz LR. Endoscopic, histological and ultrastructural correlations in chronic gastritis *Am J Gastroenterol* 1979; **71**: 269–79

1. Doenges JL. Spirochaetes in the gastric glands of *Macacus rhesus* and humans without definite history of related disease. *Proc Soc Exp Med Biol* 1938; **38**: 536–38.

2. Freedburg AS, Barron LE. The presence of spirochaetes in human gastric mucosa. *Am J Dig Dis* 1940; **7**: 443–45.

3. Palmer ED. Investigation of the gastric spirochaetes of the human? *Gastroenterology* 1954; **27**: 218–20.

4. Ito S. Anatomic structure of the gastric mucosa. In: Heidel US, Cody CF, eds. Handbook of physiology, section 6: Alimentary canal, vol II· Secretion. Washington, DC: American Physiological Society, 1967: 705–41.

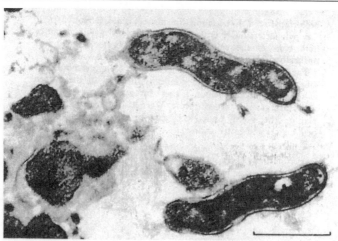

**Fig 1—Thin-section micrograph showing spiral bacteria on surface of a mucous cell in gastric biopsy specimen. (Bar = 1 μm.)**

advent of gastroscopic biopsy. Silver staining is not routine for mucosal biopsy specimens, and the bacteria have been overlooked.

In other mammals spiral gastric bacteria are well known and are thought to be commensals[5] (eg, Doenges[1] found them in all of forty-three monkeys). They usually have more than two spirals and inhabit the acid-secreting gastric fundus.[5] In cats they even occupy the canaliculi of the oxyntic cells, suggesting tolerance to acid.[6] The animal bacteria do not cause any inflammatory response, and no illness has ever been associated with them.

Investigation of gastric bacteria in man has been hampered by the false assumption that the bacteria were the same as those in animals and would therefore be acid-tolerant inhabitants of the fundus. Warren's bacteria are, however, shorter, with only one or two spirals and resemble campylobacters rather than spirochaetes. They live beneath the mucus of the gastric antrum well away from the

acid-secreting cells.

We have cultured the bacteria from antral biopsy specimens, using *Campylobacter* isolation techniques. They are microaerophilic and grow on moist chocolate agar at 37°C, showing up in 3–4 days as a faint transparent layer. They are about $0 \cdot 5 \, \mu m$ in diameter and $2 \cdot 5 \, \mu m$ in length, appearing as short spirals with one or two wavelengths (fig 1). The bacteria have smooth coats with up to five sheathed flagellae arising from one end (fig 2). In some cells, including dividing forms, flagellae may be seen at both ends and in negative stain preparations they have bulbous tips, presumably an artefact.[7]

These bacteria do not fit any known species either morphologically or biochemically. Similar sheathed flagellae have been described in vibrios[7] but micro-aerophilic vibrios have now

5 Lockard VG, Boler RK. Ultrastructure of a spiraled micro-organism in the gastric mucosa of dogs. *Am J Vet Res* 1970; **31:** 1453–62.

6 Vial JD, Orrego H. Electron microscope observations on the fine structure of parietal cells. *J Biophys Biochem Cytol* 1960; **7:** 367–72.

7. Glauert AM, Kerridge D, Horne RW. The fine structure and mode of attachment of the sheathed flagellum of *Vibrio metchnikovii. J Cell Biol* 1963; **18:** 327–36.

8. Shewan JM, Veron M. Genus I vibrio. In: Buchanan RE, Gibbons NE, eds. Bergey's manual of determinative microbiology, 8th ed. Baltimore: Williams & Wilkins, 1974: 341.

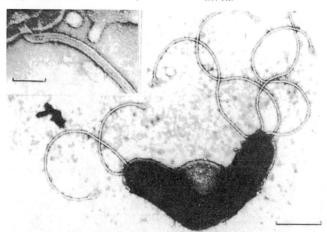

**Fig 2—Negative stain micrograph of dividing bacterium from broth culture.**

Multiple polar flagellae have terminal bulbs, (2% phosphotungstate, pH 6·8; bar = 1 μm.) Inset: detail showing sheathbed flagellum and basal disc associated with plasma membrane. (3% ammonium molybdate, pH 6·5; bar = 100 nm.)

been transferred to the family Spirillaceae genus *Campylobacter.*[8] Campylobacters however, have "a single polar flagellum at one or both ends of the cell" and the campylobacter flagellum is unsheathed.[9] Warren's bacteria may be of the genus *Spirillum.*

The pathogenicity of these bacteria remains unproven but their association with polymorphonuclear infiltration in the human antrum is highly suspicious. If these bacteria are truly associated with antral gastritis, as described by Warren, they may have a part to play in other poorly understood, gastritis associated diseases (ie, peptic ulcer and gastric cancer).

I thank Miss Helen Royce for microbiological assistance, Dr J. A. Armstrong for electronmicroscopy, and Dr Warren for permission to use fig 1.

Department of Gastroenterology,
Royal Perth Hospital,
Perth, Western Australia 6001        BARRY MARSHALL

# Gastric *Campylobacter*-like Organisms, Gastritis, and Peptic Ulcer Disease

MARTIN J. BLASER

Medical Service, Veterans Administration Medical Center, and Division of Infectious Diseases, Department of Medicine, University of Colorado School of Medicine, Denver, Colorado

*Although the presence of gastric bacteria has been long established, the recognition and isolation of Campylobacter pylori and similar organisms has opened a new era in the understanding of inflammatory gastroduodenal conditions. Visualization or isolation of gastric Campylobacter-like organisms (GCLOs) is significantly associated with histologic evidence of gastritis, especially of the antrum. Correlation with peptic ulceration also exists but probably is due to concurrent antral gastritis. Outbreaks of hypochlorhydria with concomitant gastritis have been attributed to GCLO infection, and a human volunteer became ill after ingesting C. pylori. Despite rapid microbiologic characterization of the organisms and the epidemiology, pathology, and serology of infection, the pathogenetic significance of GCLOs remains unknown. Whether GCLOs cause, colonize, or worsen gastritis must be considered an unanswered question at present. The efficacy of antimicrobial treatment of GCLO infection on the natural history of gastritis is not presently resolved. Nevertheless, GCLOs are at the least an important marker of inflammatory gastroduodenal disease, and attempts to ascertain their clinical significance are clearly warranted.*

Peptic ulcer disease and other inflammatory gastroduodenal conditions are among the most common maladies of humans throughout the world (1). In most cases, their etiologies cannot be discerned and, despite extensive investigation, the pathophysiology of these processes remains obscure. Although the presence of gastric bacteria has long been known, their significance has been uncertain. The development of fiberoptic endoscopy, permitting collection of fresh clinical specimens, has ushered in a new era for the management and investigation of gastroduodenal inflammatory conditions. Gastric bacteria now are being observed with regularity (2–4), and recently, Marshall and Warren (5,6) were able to isolate a spiral bacterium that had never been cultivated before. This organism, which they called *Campylobacter pyloridis*, has since been isolated by many other investigators (7–9). The field has moved quickly, and a review of its current status is appropriate.

## Historical Developments

After Bottcher's observation of bacteria in the human stomach in 1874, similar spiral organisms were identified in the stomachs of animals (10,11) and in patients with gastric carcinoma (12), but their visualization in patients with nonmalignant conditions was variable (13–17). By 1939, Doenges had found several types of spiral organisms in 43% of 242 stained human stomach autopsy specimens (18). The "encouraging results" of Gorham (cited in 16), who gave bismuth intramuscularly to treat chronic peptic ulcers, were considered to be due to its antibacterial action.

More than 30 yr later, using electron microscopy, Steer visualized curved bacteria on the surface of the gastric epithelium (2) in biopsy specimens obtained from patients with gastric ulceration but not from normal subjects (19). By light microscopy, Rollason and colleagues (4) observed spiral organisms in 42.6% of stained specimens of fresh gastric tissue from 310 consecutive endoscopic biopsies. Organisms were present on the surface of the gastric

Received September 12, 1986. Accepted February 2, 1987.

Address requests for reprints to: Martin J. Blaser, M.D., Infectious Disease Section (111L), Veterans Administration Medical Center, 1055 Clermont Street, Denver, Colorado 80220.

Dr. Blaser is a Clinical Investigator of the Veterans Administration.

The author thanks Dr. William R. Brown and Dr. Dennis Ahnen for review of the manuscript and Dr. Wen-lan L. Wang and Dr. Bruce Dunn for providing the clinical specimens.

0016-5085/87/$3.50

Abbreviation used in this paper: GCLO, gastric *Campylobacter*-like organism.

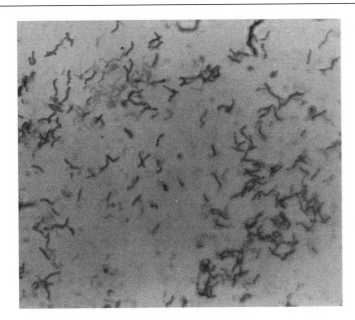

Figure 1. Gram stain of *C. pylori* in pure culture after 96-h incubation on chocolate agar (original magnification, ×1000).

mucosa, with similar frequency in the cardia, body, and antrum, were most readily seen in the necks and bases of the gastric glands, but did not invade the epithelial cells or lamina propria. Organisms were faintly gram-negative but were intensely stained using the Warthin–Starry silver impregnation method. By electron microscopy, they were a homogeneous population of curved rods, up to 6 μm in length, but they failed to grow under anaerobic culture.

In 1980, Dr. J. R. Warren in Perth, Australia observed similar curved and S-shaped bacilli, most often in the gastric antrum and associated with active gastritis (3); by light microscopy they resembled *Campylobacter jejuni*. As such, using techniques developed for culture of campylobacters, Marshall first isolated microaerophilic curved bacteria from gastric antral biopsy specimens. Subsequently, Marshall and Warren (6) studied biopsy specimens from 100 patients who had been referred for gastroscopy. Spiral bacteria were visualized by silver staining in 87% of 31 patients who had gastric or duodenal ulcers, most of whom also had gastritis. Bacteria were seen in 81% of 69 patients with acute or chronic gastritis, but in only 2 (6%) of 31 patients without gastritis. Of greatest significance is that 11 specimens from 96 patients again yielded a curved microaerophilic organism, which originally was called *C. pyloridis* (6,20) but is now known as *C. pylori*.

## Microbiologic Characteristics of *Campylobacter pylori* and Related Organisms

*Campylobacter pylori* are small nonsporulating gram-negative bacteria with flagellae at one end. They are curved rods, 3.5 μm long and 0.5–1 μm wide, with a spiral periodicity (Figure 1); Gram stain may show U-shaped or circular cells. These spiral organisms differ from spirochetes in having the rigid cell walls and flagella characteristic of most gram-negative bacteria. The guanine plus cytosine content of *C. pylori* (35.8–37.1 mol %) is within the range for *Campylobacter* species but less than for *Spirillum* or *Vibrio* (20). By electron microscopy, however, the structure of *C. pylori* is different from that of other campylobacters and more closely resembles *Spirillum* (21). Several other morphologic and biochemical characteristics (7,21–26) are markedly different from those of other *Campylobacter* species (Table 1) and other enterobacteria; most notably, *C. pylori* is a strong producer of the enzyme urease. *Campylobacter pylori* are fastidious and do not grow aerobically or anaerobically and when incubated below 30°C. Growth is poor in most liquid media; either a blood or hemin source appears essential (24,27–29). Best growth is on chocolate or blood agar plates and takes 2–5 days. Because the taxonomic status of *C. pylori* is not settled at present (30) and because of phenotypic variability, it is

*Table 1. Characteristics Distinguishing Gastric* Campylobacter-*like Organisms From Intestinal Campylobacters*

|  | C. pylori | GCLO-2 | Ferret GCLO | C. jejuni | C. laridis | C. fetus |
|---|---|---|---|---|---|---|
| Optimal growth temperature (°C) | 37 | 37 | ND | 42 | 42 | 37 |
| Hippurate hydrolysis | − | + | − | + | − | − |
| Susceptibility to cephalothin | S | S | R | R | R | R |
| Susceptibility to naladixic acid | R | R | S | S | R | R |
| Urease | + | − | + | − | − | − |
| C-19 cyclopropane on GLC | + | + | ND | + | − | − |
| Nitrate reductase | − | − | + | + | + | + |

C., *Campylobacter*; GLC, gas-liquid chromatography; GCLO, gastric *Campylobacter*-like organism; ND, not determined; R, resistant; S, sensitive.

preferable at this time to call this family of organisms gastric *Campylobacter*-like organisms (GCLOs) unless *C. pylori* is specifically isolated and identified.

Kasper and Dickgiesser (31) isolated *C. pylori* from 39% of 328 patients who had antral biopsies, but they isolated a different organism, which they called GCLO-2, from 6 (2%) other patients. Gastric *Campylobacter*-like organism-2 colonies are 1–5 mm in size, are similar in appearance to colonies of *C. jejuni*, and grow heavily within 48 h but have a number of biochemical differences (Table 1). Although deoxyribonucleic acid hybridization studies have not been performed, available data suggest that GCLO-2 represents a new species, closely related to but distinctive from *C. jejuni* (22,32). In addition to GCLO-2, cigar-shaped organisms have been seen on the surface of the mucus in biopsy specimens obtained from patients with peptic ulcers, but these organisms could not be cultivated; *C. pylori* were grown from all but one (33), which raises the possibility that these are aberrant forms of *C. pylori*. Strongly urease-positive GCLOs are apparently present in the stomachs of many (34), if not all (35), laboratory-raised ferrets, but these organisms also are distinct from *C. pylori* (Table 1). Other urease-positive thermophilic campylobacters have been isolated from river and sea water and shellfish, but not from the human specimens (36). Whether any of these organisms are related to *C. pylori* or are associated with gastritis remains to be determined.

The antigenic nature of *C. pylori* has not been completely defined. Most strains appear to share a major surface antigen with *C. jejuni* (33). Group antigens are present, but some antigenic heterogeneity clearly exists (37,38,39). By sodium dodecyl sulfate-polyacrylamide-gel electrophoresis, organisms appear homogeneous (26), with major bands in the regions of 14–21, 29–33, 50–55, 60–68, and 95–105 kilodaltons, and are dissimilar from other *Campylobacter* species (39), but relative proportions of proteins among strains vary (40). By Western blotting, essentially all these bands were antigenic to

naturally infected humans and immunized rabbits. A major surface-exposed 60-kilodalton antigen was detected by radioimmunoprecipitation and immunoblotting (40).

## Pathological Associations With Gastric *Campylobacter*-like Organism Infection

### Association of Gastric Campylobacter-*like Organisms With Gastritis*

Investigators on four continents have now identified GCLOs in gastric biopsy specimens and have shown an association between the presence of GCLOs and gastritis diagnosed by histology in adults (Table 2). Although methods employed in these studies to document the presence of GCLOs have varied, as have the definitions of gastritis used, it is notable that in all but one study the GCLO detection rate was significantly greater in patients with gastritis than in those without. The exception occurred in a small study in Australia in which nearly equal rates of GCLO detection were found in the two groups (44). Whether idiopathic antral gastritis in children is specifically associated with the presence of GCLOs is not yet settled (53–55). In attempts to answer the important question of whether GCLOs are present in healthy persons, three endoscopic studies of asymptomatic volunteers have been reported. Despite absence of symptoms or risk factors and the young age of the volunteers (mean ages 27–30 yr), 20.4% were found to have histologic gastritis; all of these subjects had GCLO present (Table 2). In contrast, no gastritis was found in 79.6% of the subjects and GCLOs were not detected in any of these cases. In total, the volunteer studies indicate that gastritis may be present in asymptomatic young adults, that this gastritis is associated with the presence of GCLOs, but that in the majority of subjects neither gastritis nor GCLOs are present.

*Table 2.* Association of Gastritis With Presence of Gastric Campylobacter-like Organisms in Adults

| | Method(s) for detecting GCLO | Patients with gastritis | | Patients without gastritis | |
|---|---|---|---|---|---|
| | | Total No. | With GCLO (%) | Total No. | With GCLO (%) |
| Symptomatic patients | | | | | |
| Eleven studies (6–9,24,33,41–45)[a] | Culture and histology | 523 | 76.9 | 191 | 9.4 |
| Five studies (46–50) | Histology alone | 169 | 72.2 | 35 | 5.7 |
| One study (51) | Culture alone | 86 | 70.9 | 133 | 11.3 |
| Total (17 studies) | Any of the above | 778 | 75.2 | 359 | 9.7 |
| Asymptomatic volunteers | | | | | |
| Three studies (7,45,52) | Culture and histology | 11 | 100 | 43 | 0 |

GLCO, gastric *Campylobacter*-like organism. [a] Parenthetical values represent reference numbers.

## Location of Gastric Campylobacter-like Organisms

Gastric *Campylobacter*-like organisms have been found on the luminal aspect of surface mucus-secreting cells and within the gastric pits, but in most studies not invading tissue (9,48) (Figure 2). Organisms always are beneath or within the mucus layer (56), but GCLOs are not seen in biopsy specimens in which mucus totally covers the specimen (33). Most organisms beneath the mucus layer, adjacent to the epithelial cell surface, appear to lie within 2 $\mu$m of an intercellular junction (56). Gastric *Campylobacter*-like organism colonization has been associated with epithelial cell flattening and de-creased intracellular mucin (42,43). By electron microscopy, organisms appear to be in close contact or partially fused with epithelial cell membranes; some bacteria appear to be covered by epithelial microvilli or engulfed within endocytotic vacuoles (48,57), similar to enteropathogenic strains of *Escherichia coli*. In an electron microscopic study of 1 patient, some neutrophils located between intact gastric epithelial cells had spiral bacteria within their phagocytic vacuoles (58). In no cases have GCLOs been seen invading beyond the epithelium. Clearly, GCLOs are predominantly luminal, but whether or not the reports of tissue involvement can be substantiated will have great bearing on our assessment of their pathogenetic role and clinical relevance.

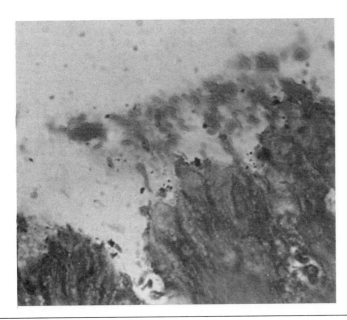

*Figure 2.* Gram stain (counterstained with carbol fuchsin) of gastric antral biopsy specimen from a patient with acute gastritis (magnification, ×1000). Numerous spiral bacteria can be seen adjacent to but not invading the mucosal tissue.

### Association With Gastric Versus Duodenal Mucosa

Gastric *Campylobacter*-like organisms are largely associated with abnormal gastric mucosa, but their distribution is irregular (27). Several investigators (9,47,59) have noted that GCLOs are present in the duodenum only in association with gastric metaplasia and not with normal duodenal mucosa. Conversely, GCLOs do not overlie intestinal metaplasia in the antrum but overlie the surrounding inflamed gastric mucosa (24,60). Patients with intestinal metaplasia have fewer GCLOs than patients with the same degree of gastritis but without intestinal metaplasia (24,43). These associations could be due to the presence of receptors for GCLOs on gastric cells and their absence from small intestinal cells, but no direct evidence for this hypothesis has been advanced.

### Associations by Type and Severity of Histology

Rathbone and colleagues (38) examined more than 150 antral and fundal biopsy samples and found that when *C. pylori* were present there always was abnormal mucosa somewhere in the stomach; *C. pylori* was present on normal fundal mucosa only when antral gastritis also was seen. In 103 Peruvian patients with gastritis, the cardia was as frequently and as densely colonized by GCLOs as was the antrum (43). In the study of healthy adult volunteers by Peterson et al. (50), GCLOs were commonly found adjacent to endoscopically and histologically normal fundal mucosa, but in the antrum, the presence of GCLOs was significantly associated with the histologic lesion of acute gastritis. These observations suggest that GCLOs are normal residents in healthy fundic mucosa and colonize the antrum when acute gastritis develops. Alternatively, if GCLOs are indeed pathogens, when introduced into the stomach they may only be capable of inducing acute gastritis in the antrum.

In autoimmune pernicious anemia, the principal lesion is atrophic gastritis of the fundus (type A) with loss of normal epithelial cells, whereas antral gastritis (type B) is typically considered "peptic" in origin (61). Gastric *Campylobacter*-like organisms were observed only in small numbers and only in the fundus in 3 of 14 patients with pernicious anemia, but in large numbers and in the antrum in 13 of 14 age- and sex-matched controls with peptic ulceration (62). The rarity of GCLOs in pernicious anemia argues against the possibility that GCLOs secondarily colonize inflamed gastric tissue; however, 11 of 14 patients had intestinal metaplasia, which might also explain the low detection rate (24,60). Gastric

*Campylobacter*-like organisms are present significantly more often in patients with chronic superficial gastritis than in patients with gastritis due to alkaline bile reflux from the duodenum (63), especially when it occurs after major resections for duodenal ulcers (64).

Attempts to correlate the severity or chronicity of gastritis with the number of GCLOs present have yielded variable results. Among 51 patients who underwent endoscopy, GCLOs were isolated from 68% of biopsy specimens classified as actively inflamed and in 62% classified as "quiescent" (33); similar results were reported by Jones et al. (9). In contrast, Marshall and Warren (6) found GCLOs in 95% of patients with active gastritis but only 58% of patients with chronic or quiescent gastritis, and McNulty and Watson (8) found a strong correlation between the number of bacteria present and the severity of gastritis. In other studies, the presence of GCLOs was best associated with severe (65) or active gastritis [defined as having neutrophils present in the specimen (66)] and less well with chronic [no neutrophils present (66)] or moderately severe (65) gastritis. Acute, purulent, antral gastritis associated with the presence of GCLOs also has been reported (67).

### Relation of Gastric *Campylobacter-like* Organisms to Peptic Ulcer Disease

Antral gastritis is nearly always present in patients with benign gastric ulcers (61,68), and usually persists after ulcer healing (69), but the pathogenic relationship is controversial (61,68,69). Duodenal ulceration also is highly associated with antral gastritis and with gastric metaplasia in the duodenal mucosa (68). By scanning electron microscopy of biopsy specimens from patients with duodenal ulcers, spiral (and comma) shaped bacteria are present in the antrum and in areas of gastric metaplasia in the duodenum (70). Gastric *Campylobacter*-like organisms have now been identified in the gastric antra of a substantial proportion of patients with either gastric or duodenal ulcers (Table 3), and in the latter condition the frequency of GCLO infections is significantly greater than that in patients with gastritis alone; however, this association of GCLOs with duodenal ulceration may be due to coexisting severe antral gastritis (65). In a study of 108 unselected endoscopy patients, isolation of GCLOs was associated with a past history of ulcer disease or active ulceration, and with histologic appearance of gastritis (26). In one study (33), 20 of 21 persons with duodenal ulcers had an abnormal antral biopsy specimen, and GCLOs were found in 17; GCLOs were absent from the only normal antral specimen. At the

*Table 3. Association of Gastric* Campylobacter-like *Organism in the Antrum and Presence of Peptic Ulceration[a]*

| Condition | No. of patients | Percentage with GCLO[b] |
|---|---|---|
| Duodenal ulcer | 119 | 77.0 ± 10[c] |
| Gastric ulcer | 111 | 65.5 ± 5 |
| Gastritis without ulceration | 238 | 60.5 ± 9[c] |

GCLO, gastric *Campylobacter*-like organism. [a] From References 6–8, 24, 41, 42, 49, and 71. [b] Values expressed as mean ± SEM. [c] p = 0.003, two-tailed paired *t*-test.

least, the presence of GCLOs may be a marker of those persons who are at high risk for duodenal ulcer.

## Epidemic Hypochlorhydria

Sixteen healthy persons and 1 person with the Zollinger–Ellison syndrome undergoing gastric secretion studies in Dallas, Texas became rapidly and profoundly hypochlorhydric (72). Nine subjects had had a mild illness with abdominal pain, nausea, and vomiting a few days before hypochlorhydria was detected. Severe fundal and, to a lesser extent, antral gastritis was present, but parietal cells appeared normal and parietal cell antibodies were not found. No duodenitis was found. Serum gastrin levels were normal or high but pepsinogen levels were elevated. Acid secretion returned to baseline levels after a mean of 126 days in parallel with an improved histologic appearance of the mucosa. Serologic, cultural, and microscopic studies were unable to define an etiology for this outbreak, which was believed to be due to transmission of an infectious agent from a contaminated pH electrode. Apparently, Dr. Marshall has since observed GCLOs on biopsy specimens from these subjects (73).

Similarly, in a British study of gastric secretion, 4 of 6 previously healthy subjects developed hypochlorhydria after a transient illness with nausea, vomiting, and abdominal pain (74). Biopsy specimens showed active gastritis with polymorphonuclear leukocyte infiltration, which was most severe in the antrum. Decreased basal and peak acid output was observed during an 8-mo follow-up, with evidence of progressive recovery of function. Follow-up biopsy at 8 mo showed clearing of inflammation in 1 patient, but chronic gastritis with plasma cells and lymphocytes in the other 3. No conventional microbial agents were found, but, again, GCLOs were observed on later staining of tissues (cited by Dr. Marshall). Although these two outbreaks strongly point to a transmissable agent as the cause of gastritis leading to hypochlorhydria, and are compatible with

GCLOs being the transmitted agents, they do not rule out the possibility that another agent caused this illness and GCLOs were secondary colonizers.

## Serology of Gastric *Campylobacter*-like Organism Infection

Serologic studies of *C. pylori* infection have been performed using a variety of antigens and antibody-detection techniques (9,38,65,75–85; Perez GP and Blaser MJ, unpublished observations). Although the results of all these studies are not in full agreement, the major findings are as follows: (a) patients with gastritis or peptic ulcer disease or dyspeptic symptoms have elevated *C. pylori*-specific serum antibody levels compared with healthy controls; (b) specific serum immunoglobulin A and immunoglobulin G levels appear to correlate best with infection; (c) specific antibodies may be found in gastric secretions; (d) gastritis severity and level of antibody response do not correlate; (e) both in patients with gastritis and in control populations, the percentage of persons with antibodies present rises with age; and (f) once present, specific antibodies persist for years.

## Epidemiology of Gastric *Campylobacter*-Like Organism Infections

The epidemiology of GCLO infections is gradually becoming clearer. Studies on four continents suggest that GCLOs are ubiquitous (6–8,24,42,43), usually in association with gastritis. Affected populations have ranged from well-to-do persons in developed countries to indigent persons in developing countries (43). During the outbreak of hypochlorhydria in Dallas, among 37 subjects exposed to the contaminated pH electrode, those who developed hypochlorhydria were significantly younger than those who remained normal (72). These data suggest an age-related immunity to the etiologic agent. The prevalence of GCLO infection, as documented by histologic (66) and serologic studies (77,80–82), rises with age, as does gastritis (86). The source and transmission of GCLOs are unknown. Spiral organisms are common in the mouth and the lower intestinal tract in all mammals, and thus are potential reservoirs for GCLOs. However, neither culture nor histologic studies have yet identified these organisms outside the stomach and duodenum (24). Attempts to isolate GCLOs from bile, saliva, gingival mucosa, colonic contents, feces, duodenal or jejunal fluid, and urethral and vaginal swabs have been negative (27,42,87). Gastric *Campylobacter*-like organisms were not isolated from the nasopharynges or saliva of patients with known *C. pyloridis* infection or gastritis (42,57).

Among 9 GCLO-positive patients, both gastritis and organisms still were present after a mean follow-up of 17 wk; however, 22 GCLO-negative persons remained negative after a mean follow-up of 11 wk (88). Restriction endonuclease analysis of *C. pylori* deoxyribonucleic acid revealed that isolates from 16 patients showed different patterns, indicating that the sources of infection are heterogeneous (89). Multiple isolates from the same patient, however, had identical patterns regardless of colony type, interval between collecting specimens (up to 2 yr), and antimicrobial treatment. After antimicrobial treatment and apparent elimination of GCLOs, recurrence of organisms thus represented recrudescence of the original infection rather than new infection. These and the serologic data suggest that exposure to an infective dose of GCLOs does not occur frequently, and that when organisms are present, they persist for at least several months and probably for years.

### Pathophysiology of Gastric *Campylobacter*-Like Organism Infection

The pathophysiology of GCLO infection is not well understood, but preliminary studies have shed light on the organism's niche and relationship to tissue damage. That large numbers of organisms are present adjacent to inflamed tissues suggests that the bacteria are not just passively traversing the stomach from the oropharynx (90); large numbers suggest in situ multiplication. In comparison to *Escherichia coli*, *C. pylori* are able to remain motile in a highly viscous environment, suggesting that these organisms are adapted to the mucus layer of the gastrointestinal tract (56). Gastric *Campylobacter*-like organisms, which are exquisitely sensitive to gastric acidity (24,42), live within mucus and between the mucus layer and the gastric mucosal surface in an environment that has a nearly neutral pH. Colonization of the mucosa appears to be independent of gastric pH (43), possibly because the mucus layer protects GCLOs from gastric acid. Because generation of ammonia may buffer gastric acid, the high urease activity of these organisms also may be adaptive. The presence of these organisms under gastric mucus and their marked sensitivity to acid suggest that these are not directly implanted onto the mucosa after ingestion of contaminated food. An alternative hypothesis is that these organisms colonize esophageal or small intestinal epithelium and migrate to the stomach underneath the protective mucus layer.

In contrast to *C. jejuni*, *C. pylori* does not invade HeLa cells in tissue culture (70). By immunofluorescent staining of clinical specimens, *C. pylori* always is confined to the luminal surface of the gastric mucus-secreting cells (70), but many of the adjacent epithelial cells have lost their microvilli (91). Gastric inflammation without bacteria present occurs in mucosa near focal lesions, such as carcinoma or peptic ulcer. In those cases, leukocytes are seen in the full thickness of the mucosa, whereas only superficial infiltration is seen in association with the bacteria (3). The number of GCLOs and of intraepithelial polymorphonuclear cells present in a biopsy specimen appear highly associated (70), and neutrophils present in the gastric lumen may ingest GCLOs (91). Colonization of affected areas is patchy, with heavily colonized areas adjacent to those with no colonization. Depletion of mucin in the epithelial glands is found in colonized areas but not in noncolonized areas (43). Whether this association is causal is not known, but mucus depletion could expose the gastric epithelium to gastric acid, pepsin, and bile salts.

An intraperitoneal injection of *C. pylori* into 2 rats caused no ill effects (92). No other animal studies have been reported, but a human volunteer ingested 100 million colony forming units of a *C. pylori* strain originally isolated from a patient with nonulcer dyspepsia (92). After 7 days he noted epigastric fullness, vomited once, and had "putrid breath and morning hunger" until taking an antimicrobial agent (tinidazole) to which the organism was susceptible; no further symptoms were noted. Gastric biopsy before the ingestion appeared entirely normal but follow-up biopsy on day 10 showed large numbers of GCLOs, depletion of mucus in epithelial cells, and apparent infiltration of the lamina propria with neutrophils. Repeat biopsy 4 days later, before treatment, showed normalization of findings. No serologic response to infection was detected.

### Diagnosis of Gastric *Campylobacter*-like Organism Infection

Diagnosis of GCLO infection can now be made by isolation of the organisms, visualization of organisms in gastric or duodenal specimens, serologic testing, or biochemical assays based on their metabolic activities.

Isolation of GCLOs primarily depends on culture of mucosal tissue, owing to their location under the gastric mucus, rather than from gastric juice where they are present only in very low numbers (7). Yields from culture are highest when specimens are transported in glucose (57,93), are ground (27), or are incubated within 2 h of endoscopy (44). Optimal incubation is at 37°C in an atmosphere containing 5% oxygen, 7% carbon dioxide, 8% hydrogen, and 80% nitrogen with high humidity (27). On subculture, candle jars may be used (33). When gastric antral biopsy specimens were incubated on *Campylobacter*-selective and -nonselective media, GCLOs were isolated (24) in concentrations ranging from about 10,000 to 10 million per gram. Although GCLOs were present in pure culture in 65% of specimens yielding any growth, the other organisms (streptococci, *Enterobacteriaceae*, diphtheroids, and occasional anaerobic bacteria), present in 35% of specimens, overgrew the more fastidious and slow-growing GCLOs in the nonselective medium. Thus, use of a selective medium is justified for attempts at primary isolation (27) even though some *C. pylori* strains may be susceptible to the antimicrobial agents present. One such medium contains brain-heart infusion agar, 7% horse blood, 6 mg/L vancomycin, 20 mg/L naladixic acid, and 2 mg/L amphotericin B (27). Colonial morphology and urease production can be used to make a rapid presumptive identification of *C. pylori* (56). Microbiologic characteristics of the GCLOs for use in identifying the organisms to a species level include vibrio-like morphology on Gram stain, assessment of oxidase, catalase, and urease tests, and antimicrobial susceptibility (Table 1).

Histologic identification of GCLOs in gastric mucosal specimens is highly correlated with culture-positivity (7–9,28,29,33,42,66). Best visualization of GCLOs is in the prepyloric region of the stomach (70), and can be seen on the surface of the gastric epithelium by means of H&E, Giemsa, acridine-orange, and Gram or silver stains. Gastric *Campylobacter*-like organisms stain intensely with silver, such as in the Warthin–Starry technique, but these methods are difficult to standardize. In contrast, the H&E and Giemsa stains are easy to perform but examination of the specimens are slow and tedious. The acridine-orange stain, which is easy to perform and can be done on formalin-fixed paraffin-embedded specimens (94), is sensitive and specific (66). Gram staining of cytology specimens obtained by brushing the gastric mucosa also is simple and may be more effective at showing GCLOs than silver staining of biopsy specimens (43). Examination of minced tissue specimens by phase contrast microscopy is another rapid, simple, sensitive, and inexpensive test (29). The development of a monoclonal antibody specific for *C. pylori* should aid in the use of rapid immunofluorescent detection of these organisms in tissue (95).

*Campylobacter pylori* are rapidly and strongly urease-positive (7,24,96,97), and there is an excellent correlation between isolation of *C. pylori* and the presence of a positive urease reaction when gastric biopsy specimens are placed in Christenson's medium (96). Whereas *C. pylori* are positive, intestinal campylobacters are negative. A commercially available test in Australia (CLO-test; Delta West Ltd., Canning Vale, Western Australia) detects preformed urease in gastric or duodenal biopsy specimens in 15 min to 3 h (98). Alternatively, a biopsy specimen may be crushed and immediately placed in a urea broth test: most are positive within 1 h (99). The high urease activity is the basis for detection in a noninvasive assay; after a liquid test meal, given to delay gastric emptying, [$^{13}$C]urea was administered orally to healthy control subjects and persons with GCLO identified by culture (100). In patients with GCLO infection, urea-derived $^{13}CO_2$ appeared in the breath within 20 min and accumulated for more than 100 min, whereas normal subjects showed no release of labeled $CO_2$. Similarly, *C. pylori*-infected patients have low urea and high ammonia levels in gastric juice compared with uninfected controls (101).

## Treatment of Gastric *Campylobacter*-like Organism Infections

Because of the limited effectiveness of acid reduction in the long-term treatment of peptic conditions (102), attention has turned to antimicrobial treatment of GCLOs as an alternative approach. If indeed GCLOs are causative, antimicrobial treatment should be effective; moreover, its efficacy would provide further evidence that these organisms have an etiologic role. *Campylobacter pylori* are susceptible to erythromycin, tetracycline, penicillin, ampicillin, cefoxitin, ciprofloxacin, gentamicin, cephalothin, clindamycin, and kanamycin, but resistant to naladixic acid, sulfonamides, and trimethoprim; some strains are susceptible to metronidazole (6,9,103,104). Spiramycin, cimetidine, ranitidine, and sucralfate have no apparent effect on GCLOs at clinically achievable levels (88,105), but the *C. pylori* isolation rate from patients with peptic ulcer and dyspepsia was significantly lower in those treated with cimetidine than those not treated (105).

Two bismuth salts, tripotassium dicitratobismuthate and bismuth sodium tartrate (Pepto-Bismol, Procter and Gamble, Cincinnati, Ohio), inhibited *C. pylori* at concentrations (between 4 and 32 $\mu$g/ml) that may be reached in the gastric lumen (104). Although bismuth compounds have been used for the relief of gastric disorders for >200 yr, the basis of their activity is poorly understood. Bismuth salts have a wide variety of actions that include an antacid effect, coating of the gastric mucosa, decreasing gastric and intestinal motility, increasing mucus secretion, absorbing fluids, and inhibiting the growth of a variety of microorganisms (106). Bismuth preparations are effective in the treatment of traveler's diarrhea (107), and upper gastrointestinal disturbances including nausea, heartburn, or pain. A variety of blinded and placebo-controlled trials have shown the efficacy of bismuth salts for treatment of everything from acute indigestion to duodenal or gastric ulcers (108–112). All of these studies were performed before the identification of GCLOs and thus did not include microbiologic studies.

Lambert and colleagues (113) treated 44 patients who had *C. pylori* present in the gastric antrum with tripotassium dicitratobismuthate for either 4 or 8 wk, or with cimetidine or placebo. Antral biopsy specimens obtained before the start of and 1 wk after the conclusion of the trial were assessed blindly for the presence of inflammatory cells, atrophy, metaplasia, and dysplasia. The pretrial gastritis scores for all four groups were similar, and the presence of GCLOs was strongly correlated with inflammation. Bacteria were eliminated from 69% and 73% of patients treated with tripotassium dicitratobismuthate for 4 and 8 wk, respectively, but from no patients in the other two groups. The gastritis score improved significantly in patients from whom the bacteria were eliminated, but no histologic improvement was seen in the other patients. Although supporting the role of GCLOs in the pathogenesis of chronic gastritis, the limited data presented do not indicate whether the trial was randomized or blinded. Furthermore, the effect of tripotassium dicitratobismuthate on gastritis in patients in whom GCLOs could not be detected was not examined. This is an important control group in helping to determine whether elimination of *C. pylori* causes or follows healing of inflammation.

In another preliminary study, treatment with either bismuth subcitrate or amoxicillin was associated with the elimination of GCLOs and histologic improvement of gastritis in the majority of subjects (88). However, relapse was common in the weeks after therapy. In another study, Pepto-Bismol treatment was associated with both clearance of organisms and histologic improvement in gastritis, whereas neither occurred after erythromycin or placebo treatment (114). In a third study, electron microscopy performed 40–100 min after tripotassium dicitratobismuthate treatment showed that GCLOs had detached from the gastric epithelium and that many of the organisms had lysed (115). In a fourth study (116), neither cimetidine nor erythromycin treatment changed the number of GCLOs in the gastric mucosa, whereas tripotassium dicitratobismuthate was effective. Gastric *Campylobacter*-like organisms remained present 6 mo later, however, and serum antibody levels were unaffected by treatment. The use of amoxycillin and bismuth for combination therapy also has been explored (115).

Furazolidone is a furan derivative with broad-spectrum antibacterial activity including activity toward many *Campylobacter* species (117) and *C. pylori* specifically (118). Direct inhibition of gastric secretion also has been reported. Zheng and colleagues (119) treated 70 patients who had endoscopically confirmed peptic ulcers with either furazolidone or placebo for 2 wk. Ulcer healing and reduced upper gastrointestinal pain were significantly more common in the furazolidone-treated group. From an earlier uncontrolled series, 74% responded favorably to furazolidone, and relapse rates over the next 4 yr were considerably lower than in an untreated group. Similarly, metronidazole, to which many *C. pylori* strains are susceptible, also has been reported as having antiulcer activity (120,121).

## Conclusions

The best information on the natural history of primary GCLO infection comes from the voluntary ingestion of *C. pylori* by a single subject. The results suggested that he suffered an acute but self-limited gastritis, and that *C. pylori* was causative. Indirect support for this concept comes from the two epidemics of hypochlorhydria in volunteers who apparently ingested contaminated gastric juices. It is clear that these subjects suffered from an acute gastritis, and apparently GCLOs were visualized on smears of injured tissue. Their gastritis, however, was more fundal than antral, and although *C. pylori* were present, it is not certain that they were causal.

There is unquestionably a striking association between the presence of GCLOs and chronic antral gastritis. In favor of a pathogenic role for the GCLO is (a) the serologic response of infected hosts to *C. pylori*, (b) that ingestion of GCLOs by polymorphonuclear leukocytes has been observed, and (c) that *C. pylori* is apparently able to cause an acute gastritis. None of these observations, however, proves causality of chronic gastritis, and finding similar organisms in healthy ferrets (34,35) suggests that they may not be pathogenic. That the human volunteer was apparently able to clear his infection, and that the presence of these organisms appears age-related argues that these may be secondary colonizers in persons with reduced gastric acidity or other deficits. Overgrowth of conventional bacterial organisms in the stomachs of patients with reduced acidity is a well-described phenomenon (122,123). Although these acid-sensitive organisms appear to reside in a niche largely protected from low pH, it may be that hypochlorhydria of any etiology removes a major inhibitory factor on their growth. Hypochlorhydria probably is not the only factor that permits multiplication of these organisms in vivo; the presence of tissue damaged by other causes may be another. That cimetidine-treated patients with peptic disease had a lower isolation rate of *C. pylori* than untreated patients (105) also suggests that the organisms may be colonizers of damaged mucosa. Clearly, GCLOs have a tropism for gastric mucosa, as confirmed in a complementary fashion by the observations concerning gastric metaplasia into the duodenum and intestinal metaplasia into the stomach. At present, however, there is insufficient evidence to conclude that the GCLOs cause chronic gastritis or peptic ulcer disease. Nevertheless, GCLOs are certainly a marker for these processes, and ascertainment of their presence is at least potentially valuable in that regard.

For the question of causality to be fairly tested, better definition and characterization of the term "gastritis" is needed. It is not clear from the literature how many different histologic types have been recognized, what their natural histories are, and with which types GCLOs are associated. The presence of GCLOs in patients who have gastritis due to known causes, such as ingestion of nonsteroidal antiinflammatory agents, would suggest that colonization is secondary to injury; their absence over time would suggest that injury does not necessarily predispose to colonization. Results of such studies have not yet been reported. Further volunteer studies will probably best help determine the role of GCLOs in acute gastritis and provide some information on the natural history of these infections. Before such studies are done, it will be necessary to ensure that the organisms can be eradicated from their gastric niche. Finally, double-blinded placebo-controlled studies using treatment modalities that only have antibacterial effects, that are effective against most GCLOs, and that are not inactivated in the gastric environment can help solve the problem of causality for chronic gastritis. Bismuth compounds, because of their varied effects, are probably not satisfactory for such trials.

The rediscovery of spiral gastric organisms and the cultivation of *Campylobacter pylori* has opened a new era in gastric microbiology that has great clinical relevance. The next few years may provide answers to the perplexing problem of chronic idiopathic gastritis and possibly provide insights into the pathogenesis of peptic ulcer disease. Studies to clarify the role of GCLOs in the pathogenesis of these conditions should be a high priority. At the present, however, clinicians might wait for more definitive clinical studies before attempting to obtain cultures from affected patients or initiating specific antimicrobial treatment.

## References

1. Langman MJS. The epidemiology of chronic digestive diseases. London: Edward Arnold, 1979.
2. Steer HW. Ultrastructure of cell migration through the gastric epithelium and its relationship to bacteria. J Clin Pathol 1975;28:639–46.
3. Warren JR, Marshall B. Unidentified cured bacilli on gastric epithelium in active chronic gastritis. Lancet 1983;i:1273.
4. Rollason TP, Stone J, Rhode JM. Spiral organisms in endoscopic biopsies of the human stomach. J Clin Pathol 1984; 37:23–6.
5. Marshall BJ. Unidentified curved bacilli on gastric epithelium in active chronic gastritis. Lancet 1983;ii:1273–5.
6. Marshall BJ, Warren JR. Unidentified curved bacilli in the stomach of patients with gastritis and peptic ulceration. Lancet 1984;i:1311–4.
7. Langenberg M-L, Tytgat GNJ, Schipper MEI, Rietra PJGM, Zanen HC. Campylobacter-like organisms in the stomach of patients and healthy individuals. Lancet 1984;i:1348.
8. McNulty CMA, Watson DM. Spiral bacteria of the gastric antrum. Lancet 1984;i:1068–9.
9. Jones DM, Lessells AM, Eldridge J. Campylobacter-like organisms on the gastric mucosa: culture, histological, and serological studies. J Clin Pathol 1984;37:1002–6.
10. Bizzozero G. Ueber die schlauchformigen drusen des magendarmkanals und die bezienhungen ihres epithels zu dem oberflachenepithel der schleimhaut. Arch f mikr Anast 1893;42:82.
11. Salomon H. Ueber das spirillam des Saugertiermagens und sein Verhalten zu den Belegzellen. Zentralblatt fur Bacteriologie 1896;19:433.
12. Krienitz W. Ueber Jas Auftreten von mageninhalt bei carcinoma ventriculi. Dtsch Med Wochenschr 1906;22:872.
13. Rosenow EC, Sandford AH. The bacteriology of ulcer of the stomach and duodenum in man. J Infect Dis 1915;17:210–6.
14. Celler HL, Thalheimer W. Bacteriological and experimental studies on gastric ulcer. J Exp Med 1916;23:791–800.
15. Appelmans R, Vassiliadis P. Etude sur la flore microbienne des ulcers gastro-duodeneaux et des cancers gastriques. Rev Belge Sci Med 1932;4:198–203.
16. Freedburg AS, Barron LE. The presence of spirochetes in the human gastric mucosa. Am J Dig Dis 1940;7:443–5.
17. Seeley GP, Colp R. The bacteriology of peptic ulcers and gastric malignancies: possible bearing on complications following gastric surgery. Surgery 1941;10:369–80.
18. Doenges JL. Spirochaetes in the gastric glands of *Macacus rhesus* and of man without related disease. Arch Pathol 1939;27:469.
19. Steer HW, Colin-Jones DG. Mucosal changes in gastric ulceration and their response to carbenoxolone sodium. Gut 1975;16:590–7.
20. Marshall BJ, Royce H, Annear DI, et al. Original isolation of *Campylobacter pyloridis* from human gastric mucosa. Microbios Lett 1984;25:83–8.
21. Jones DM, Curry A, Fox AJ. An ultrastructural study of the gastric Campylobacter-like organism "*Campylobacter pyloridis*". J Gen Microbiol 1985;131:2335–41.
22. Goodwin S, Blincow E, Armstrong J, McCulloch R, Collins D. *Campylobacter pyloridis* is unique: GCLO-2 is an ordinary campylobacter. Lancet 1985;ii:39.
23. Goodwin CS, McCulloch RK, Armstrong JA, Wee SH. Unusual cellular fatty acids and distinctive ultrastructure in a new spiral bacterium (*Campylobacter pyloridis*) from the human gastric mucosa. J Med Microbiol 1985;19:257–67.
24. Buck GE, Gourley WK, Lee WK, Subramanyan K, Latimer

JM, DiNuzzo AR. Relation of *Campylobacter pyloridis* to gastritis and peptic ulcer. J Infect Dis 1986;153:664–9.

25. Hudson MJ, Wait R. Cellular fatty acids of campylobacter species with particular reference to *Campylobacter pyloridis*. Pearson AD, Skirrow MB, Lior H, Rowe B, eds. *Campylobacter III*. Proceedings of the third international workshop on campylobacter infections. London: PHLS, 1985:198–9.

26. Pearson AD, Bamforth J, Booth L, et al. Polyacrylamide gel electrophoresis of spiral bacteria from the gastric antrum. Lancet 1984;i:1349–50.

27. Goodwin CS, Blincow E, Warren JR, Waters TE, Sanderson CR, Easton L. Evaluation of cultural techniques for isolating *Campylobacter pyloridis* from endoscopic biopsies of gastric mucosa. J Clin Pathol 1985;138:1127–31.

28. Kasper G, Dickgiesser N. Isolation of campylobacter-like bacteria from gastric epithelium. Infection 1984;12:179–80.

29. Pinkard KJ, Harrison B, Capstick JA, Medley G, Lambert JR. Detection of *Campylobacter pyloridis* in gastric mucosa by phase contrast microscopy. J Clin Pathol 1986;39:112–3.

30. Curry A, Jones DM, Eldridge J, Fox AJ. Ultrastructure of *Campylobacter pyloridis*—not a campylobacter? Pearson AD, Skirrow MB, Lior H, Rowe B, eds. Campylobacter III. Proceedings of the Third International Workshop on Campylobacter infections. London: PHLS, 1985:195.

31. Kasper G, Dickgiesser N. Isolation from gastric epithelium of *Campylobacter*-like bacteria that are distinct from "*Campylobacter pyloridis*". Lancet 1985;i:111–2.

32. Kasper G, Owen RJ. Characteristics of a new group of campylobacter-like organisms (CLOs) from gastric epithelium. Pearson AD, Skirrow MB, Lior H, Rowe B, eds. *Campylobacter III*. Proceedings of the Third International Workshop on *Campylobacter* infections. London: PHLS, 1985:203.

33. Price AB, Levi J, Dolby, et al. *Campylobacter pyloridis* in peptic ulcer disease: microbiology, pathology, and scanning electron microscopy. Gut 1985;26:1183–8.

34. Fox JG, Edrise BM, Cabot EB, Beaucage C, Murphy JC, Prostak KS. *Campylobacter*-like organisms isolated from gastric mucosa of ferrets. Am J Vet Res 1986;47:236–9.

35. Rathbone BJ, West AP, Wyatt JI, Johnson AW, Tompkins DS, Heatley RV. *Campylobacter pyloridis*, urease and gastric ulcers. Lancet 1986;ii:400–1.

36. Bolton FJ, Holt AV, Hutchinson DN. Urease-positive thermophilic campylobacters. Lancet 1985;i:1217.

37. Lior H, Pearson AD, Woodward DL, Hawtin P. Biochemical and serological characteristics of *Campylobacter pyloridis*. Pearson AD, Skirrow MB, Lior H, Rowe B, eds. Campylobacter III. Proceedings of the Third International Workshop on Campylobacter infections. London: PHLS, 1985:196–7.

38. Rathbone BJ, Wyatt JI, Worsley BW, Trejdosiewicz LK, Heatley RV, Losowsky MS. Immune response to *Campylobacter pyloridis*. Lancet 1985;i:1217.

39. Perez-Perez GI, Blaser MJ. Conservation and diversity of *Campylobacter pyloridis* major antigens. Infect Immun 1987;55:1256–63.

40. Newell DG. The outer membrane proteins and surface antigens of *Campylobacter pyloridis*. Pearson AD, Skirrow MB, Lior H, Rowe B, eds. Campylobacter III. Proceedings of the Third International Workshop on Campylobacter infections. London: PHLS, 1985:199–200.

41. Burnett RA, Forrest JAH, Girdwood RWA, Fricker CR. *Campylobacter*-like organisms in the stomach of patients and healthy individuals. Lancet 1984;i:1349.

42. Marshall BJ, McGechie DB, Rogers PA, Glancy RJ. Pyloric campylobacter infection and gastroduodenal disease. Med J Aust 1985;142:439–44.

43. Gilman RJ, Leon-Barua R, Koch J, et al. Rapid identification of pyloridisc campylobacter in Peruvians with gastritis. Dig Dis Sci 1986;31:1089–94.

44. Arnot RS. Gastritis and campylobacter infection. Med J Aust 1985;142:100–1.

45. Pettross CW, Cohen H, Appleman MD, Valenzuela JE, Chandrasoma P. *Campylobacter pyloridis*: relationship to peptic disease, gastric inflammation and other conditions (abstr). Gastroenterology 1986;90:1585.

46. Lopez-Brea M, Jimenez ML, Blanco M, Pajares JM. Isolation of *Campylobacter pyloridis* from patients with and without gastroduodenal pathology. Pearson AD, Skirrow MB, Lior H, Rowe B, eds. Campylobacter III. Proceedings of the Third International Workshop on Campylobacter infections. London: PHLS, 1985:193–194.

47. Thomas JM, Poynter D, Gooding C, et al. Gastric spiral bacteria. Lancet 1984;ii:100.

48. Tricottet V, Bruneval P, Vire O, Camilleri JP. *Campylobacter*-like organisms and surface epithelium abnormalities in active, chronic gastritis in humans: an ultrastructural study. Ultrastruct Pathol 1986;10:113–22.

49. Kalenic S, Faliseva V, Scukanec-Spoljar M, Vodopija I. *Campylobacter pyloridis* in the gastric mucosa of patients with gastritis and peptic ulcer. Pearson AD, Skirrow MB, Lior H, Rowe B, eds. Campylobacter III. Proceedings of the Third International Workshop on Campylobacter infections. London: PHLS, 1985:193.

50. Peterson WL, Lee EL, Feldman M. Gastric *Campylobacter*-like organisms in healthy humans: correlation with endoscopic appearance and mucosal histology (abstr). Gastroenterology 1986;90:1585.

51. Pearson AD, Ireland A, Holdstock G, et al. Clinical and pathological correlates of *Campylobacter pyloridis* isolated from gastric biopsy specimens. Pearson AD, Skirrow MB, Lior H, Rowe B, eds. Campylobacter III. Proceedings of the Third International Workshop on Campylobacter infections. London: PHLS, 1985:181–2.

52. Barthel JS, Westblum TU, Havey AD, Gonzalez FJ, Everett ED. Pyloric *Campylobacter*-like organisms (PCLOs) in asymptomatic volunteers (abstr). Gastroenterology 1986;90:1338.

53. Drumm B, O'Brien A, Cutz E, Sherman P. *Campylobacter pyloridis* are associated with primary antral gastritis in the pediatric population (abstr). Gastroenterology 1986;90:1399.

54. Cadranel S, Goosens H, DeBoeck M, Malengreau A, Rodesch P, Butzler JP. *Campylobacter pyloridis* in children. Lancet 1986;i:735–6.

55. Hill R, Pearman J, Worthy P, Caruso V, Goodwin S, Blincow E. *Campylobacter pyloridis* and gastritis in children. Lancet 1986;i:387.

56. Hazell SL, Lee A, Brady L, Hennessey W. *Campylobacter pyloridis* and gastritis: association with intracellular spaces and adaptation to an environment of mucus as important factors in colonization of the gastric epithelium. J Infect Dis 1986;153:658–63.

57. Goodwin CS, Armstrong JA, Marshall BJ. *Campylobacter pyloridis*, gastritis, and peptic ulceration. J Clin Pathol 1986;39:353–65.

58. Shoushia S, Bull TR, Parkins RA. Gastric spiral bacteria. Lancet 1984;ii:101.

59. Phillips AD, Hine KR, Holmes GKT, Woodings DF. Gastric spiral bacteria. Lancet 1984;ii:101.

60. Thomas JM. *Campylobacter*-like organisms in gastritis. Lancet 1984;ii:1217.

61. Strickland RG, Mackay IR. A reappraisal of the nature and significance of chronic atrophic gastritis. Am J Dig Dis 1973;18:426–40.

62. O'Connor HJ, Axon ATR, Dixon MF. Campylobacter-like

organisms unusual in type A (pernicious anaemia) gastritis. Lancet 1984;ii:1091.

63. O'Connor JH, Wyatt JI, Dixon MF, Axon ATR. *Campylobacter*-like organisms and reflux gastritis. J Clin Pathol 1986; 39:531–4.

64. O'Connor HJ, Dixon MF, Wyatt JI, et al. Effect of duodenal ulcer surgery and enterogastric reflux on *Campylobacter pyloridis*. Lancet 1986;ii:1178–81.

65. von Wulffen H, Heesemann J, Butzow GH, Loning T, Laufs R. Detection of *Campylobacter pyloridis* in patients with antrum gastritis and peptic ulcers by culture, complement fixation test, and immunoblot. J Clin Microbiol 1986;24: 716–20.

66. Rawles JW, Paull G, Yardley JH, et al. Gastric *Campylobacter*-like organisms (CLO) in a U.S. hospital population (abstr). Gastroenterology 1986;90:1599.

67. Salmeron M, Desplaces N, Lavergne A, Houdart R. *Campylobacter*-like organisms and acute purulent gastritis. Lancet 1986;ii:975.

68. Greenlaw R, Sheahan DG, DeLuca V, Miller D, Myerson D, Myerson P. Gastroduodenitis. A broader concept of peptic ulcer disease. Dig Dis Sci 1980;25:660–72.

69. Gear MWL, Truelove SC, Whitehead R. Gastric ulcer and gastritis. Gut 1971;12:639–42.

70. Steer H, Newell DG. Mucosa-related bacteria in benign peptic ulceration. Pearson AD, Skirrow MB, Lior H, Rowe B, eds. Campylobacter III. Proceedings of the Third International Workshop on Campylobacter infections. London: PHLS, 1985:173–4.

71. Bohnen J, Krajden S, Kempston J, Andeson J, Karmali M. *Campylobacter pyloridis* in Toronto. Pearson AD, Skirrow MB, Lior H, Rowe B, eds. Campylobacter III. Proceedings of the Third International Workshop on Campylobacter infections. London: PHLS, 1985:175–7.

72. Ramsey EJ, Carey KV, Peterson WL, et al. Epidemic gastritis with hypochlorhydria. Gastroenterology 1979;76:1449–57.

73. Marshall BJ. *Campylobacter pyloridis* and gastritis. J Infect Dis 1986;153:650–7.

74. Gledhill T, Leicester RJ, Addis B, et al. Epidemic hypochlorhydria. Br Med J 1985;290:1383–6.

75. Morris A, Nicholson G, Lloyd G, Haines D, Rogers A, Taylor D. Seroepidemiology of *Campylobacter pyloridis*. NZ Med J 1986;99:657–9.

76. Rathbone BJ, Trejdosiewicz LK, Heatley RV, Losowsky MS, Wyatt JI, Worsley BW. Epidemic hypochlorhydria. Br Med J 1985;291:52–3.

77. Kaldor J, Tee W, McCarthy P, Watson J, Dwyer B. Immune response to *Campylobacter pyloridis* in patients with peptic ulceration. Lancet 1985;i:921.

78. Eldridge J, Lessells AM, Jones DM. Antibody to spiral organisms on gastric mucosa. Lancet 1984;i:1237.

79. McNulty CAM, Crump B, Gearty J, et al. The distribution of and serological response to *Campylobacter pyloridis* in the stomach and duodenum. Pearson AD, Skirrow MB, Lior H, Rowe B, eds. Campylobacter III. Proceedings of the Third International Workshop on Campylobacter infections. London: PHLS, 1985:174–5.

80. Marshall BJ, McGechie DB, Francis GJ, Utley PJ. Pyloric campylobacter serology. Lancet 1984;ii:281.

81. Hawtin P, Pearson AD, McBride H, Gibson J, Booth L. Specific IgG and IgA responses to *Campylobacter pyloridis* in man. Pearson AD, Skirrow MB, Lior H, Rowe B, eds. Campylobacter III. Proceedings of the Third International Workshop on Campylobacter infections. London: PHLS, 1985:186–7.

82. Hutchinson DN, Bolton FJ, Hinchliffe PM, Holt AV. Distribution in various clinical groups of antibody to *Campylobacter pyloridis* detected by ELISA, complement fixation and microagglutination tests. Pearson AD, Skirrow MB, Lior H, Rowe B, eds. Campylobacter III. Proceedings of the Third International Workshop on Campylobacter infections. London: PHLS, 1985:185.

83. Marshall BJ, Whisson M, Francis G, McGechie D. Correlation between symptoms of dyspepsia and *Campylobacter pyloridis* serology in Western Australian blood donors. Pearson AD, Skirrow MB, Lior H, Rowe B, eds. Campylobacter III. Proceedings of the Third International Workshop on Campylobacter infections. London: PHLS, 1985:188–9.

84. Morris A, Lloyd G, Nocholson G. *Campylobacter pyloridis* serology among gasteoendoscopy clinic staff. NZ Med J 1986;99:820–1.

85. Ishii E, Inoue H, Tsuyuguchi T, et al. *Campylobacter*-like organisms in cases of stomach diseases in Japan. Pearson AD, Skirrow MB, Lior H, Rowe B, eds. Campylobacter III. Proceedings of the Third International Workshop on Campylobacter infections. London: PHLS, 1985:179–180.

86. Siurala M, Isokoski M, Varis K, Kekki M. Prevalence of gastritis in a rural population. Bioptic study of subjects selected at random. Scand J Gastroenterology 1968;3: 211–23.

87. Eldridge J, Jones DM, Sethi P. The occurrence of antibody to the *Campylobacter pyloridis* in various groups of individuals. Pearson AD, Skirrow MB, Lior H, Rowe B, eds. Campylobacter III. Proceedings of the Third International Workshop on Campylobacter infections. London: PHLS, 1985: 183–4.

88. Langenberg M-L, Rauws EAJ, Schipper MEI, et al. The pathogenic role of *Campylobacter pyloridis* studied by attempts to eliminate these organisms. Pearson AD, Skirrow MB, Lior H, Rowe B, eds. Campylobacter III. Proceedings of the Third International Workshop on Campylobacter infections. London: PHLS, 1985:162–3.

89. Langenberg W, Rauws EA, Widjojokusumo A, Tytgat GNJ, Zanen HC. Identification of *Campylobacter pyloridis* isolates by restriction endonuclease DNA analysis. J Clin Microbiol 1986;24:414–7.

90. Fricker CR. Adherence of bacteria associated with active chronic gastritis to plastics used in the manufacture of fibreoptic endoscopes. Lancet 1984;i:800.

91. Lee WK, Gourley WK, Buck GE, Subraanyam K. A study of *Campylobacter pyloridis* in patients with gastritis and gastric ulcer. Pearson AD, Skirrow MB, Lior H, Rowe B, eds. Campylobacter III. Proceedings of the Third International Workshop on Campylobacter infections. London: PHLS, 1985:178–9.

92. Marshall BJ, Armstrong JA, McGechie DB, Glancy RJ. Attempt to fulfil Koch's postulates for pyloridisc campylobacter. Med J Aust 1985;142:436–9.

93. Morris A, Gillies M, Patton K. Detection of *Campylobacter pyloridis* infection. NZ Med J 1986;99:336.

94. Walters LL, Budin RE, Paull G. Acridine-orange to identify *Campylobacter pyloridis* in formalin fixed paraffin-embedded gastric biopsies. Lancet 1986;i:42.

95. Engstrand L, Pahlson C, Gustavsson S, Schwan A. Monoclonal antibodies for rapid identification of *Campylobacter pyloridis*. Lancet 1986;ii:1403–4.

96. McNulty CAM, Wise R. Rapid diagnosis of *Campylobacter*-associated gastritis. Lancet 1985;i:1443–4.

97. Owen RJ, Martin SR, Borman P. Rapid urea hydrolysis by gastric campylobacters. Lancet 1985;i:111.

98. Morris A, McIntyre D, Rose T, Nicholson G. Rapid diagnosis of *Campylobacter pyloridis* infection. Lancet 1986;i:149.

99. McNulty CAM, Wise R. Rapid diagnosis of *Campylobacter pyloridis* gastritis. Lancet 1986;i:387.

100. Graham DY, Klein PD, Evans DG, et al. Rapid noninvasive diagnosis of gastric *Campylobacter* by a $^{13}$C urea breath test (abstr). Gastroenterology 1986;90:1435.

101. Marshall BJ, Langton SR. Urea hydrolysis in patients with *Campylobacter pyloridis* infection. Lancet 1986;i:965–6.

102. Nyren O, Adami H-O, Bates S, et al. Absence of therapeutic benefit from antacids or cimetidine in non-ulcer dyspepsia. N Engl J Med 1986;314:339–43.

103. Kasper G, Dickgiesser N. Antibiotic sensitivity of "*Campylobacter pyloridis*". Eur J Clin Microbiol 1984;3:444.

104. McNulty CAM, Dent J, Wise R. Susceptibility of clinical isolates of *Campylobacter pyloridis* to 11 antimicrobial agents. Pearson AD, Skirrow MB, Lior H, Rowe B, eds. Campylobacter III. Proceedings of the Third International Workshop on Campylobacter infections. London: PHLS, 1985:169–70.

105. Goodwin CS, Blake P, Blincow E. The minimum inhibitory and bactericidal concentrations of antibiotics and anti-ulcer agents against *Campylobacter pyloridis*. J Antimicrob Chemother 1986;17:309–14.

106. Wilson TR. The pharmacology of tri-potassium di-citrato bismuthate (TDB). Postgrad Med J 1975;51 (Suppl 5):18–21.

107. Ericsson CD, DuPont HL, Johnson PC. Nonantibiotic therapy for travelers' diarrhea. Rev Infect Dis 1986;8:S202–6.

108. Hailey FJ, Newsom JH. Evaluation of bismuth subsalicylate in relieving symptoms of indigestion. Arch Intern Med 1984;144:269–72.

109. Weiss G, Serfontein WJ. The efficacy of a bismuth-protein-complex compound in the treatment of gastric and duodenal ulcers. S Afr Med J 1971;45:467–70.

110. Moshal MG. A double-blind gastroscopic study of a bismuth-peptide complex in gastric ulceration. S Afr Med J 1974;48:1610–1.

111. Salmon PR, Brown P, Williams R, Read AE. Evaluation of colloidal bismuth (De-Nol) in the treatment of duodenal ulcer employing endoscopic selection and follow-up. Gut 1974;15:189–93.

112. Martin DF, Hollanders D, May SJ, Ravenscroft MM, Tweedle DEF, Miller JP. Differences in relapse rates of duodenal ulcer after healing with cimetidine or tripotassium dicitrato bismuthate. Lancet 1981;i:7–10.

113. Lambert JR, Dunn KL, Turner H, Korman MG. Effect on histological gastritis following eradication of *Campylobacter pyloridis* (abstr). Gastroenterology 1986;90:1509.

114. McNulty CAM, Gearty JC, Crump B, et al. *Campylobacter pyloridis* and associated gastritis: investigator blind, placebo controlled trial of bismuth salicylate and erythromycin ethylsuccinate. Br Med J 1986;293:645–9.

115. Marshall BJ, Armstrong JA, McGechie DB, Francis GJ. The antibacterial action of bismuth: early results of antibacterial regimens in the treatment of duodenal ulcer. Pearson AD, Skirrow MB, Lior H, Rowe B, eds. Campylobacter III. Proceedings of the Third International Workshop on Campylobacter infections. London: PHLS, 1985:165–6.

116. Jones DM, Eldridge J, Whorwell PJ, Miller JP. The effects of various anti-ulcer regimens and antibiotics on the presence of *Campylobacter pyloridis* and its antibody. Pearson AD, Skirrow MB, Lior H, Rowe B, eds. Campylobacter III. Proceedings of the Third International Workshop on Campylobacter infections. London: PHLS, 1985:161.

117. Wang W-LL, Reller LB, Blaser MJ. Comparison of antimicrobial susceptibility patterns of *Campylobacter jejuni* and *Campylobacter coli*. Antimicrob Agents Chemother 1984; 26:351–3.

118. Howden A, Boswell P, Tovey F. In-vitro sensitivity of *Campylobacter pyloridis* to furazolidone. Lancet 1986; ii:1035.

119. Zheng Z-T, Wang Z-Y, Chu YX, et al. Double-blind short term trial of furazolidone in peptic ulcer. Lancet 1985; i:1048–9.

120. Quintero Diaz M, Sotto Escobar A. Metronidazole versus cimetidine in the treatment of gastroduodenal ulcer. Lancet 1986;i:907.

121. Shirokova KI, Filomonow RM, Poliakova LV. Metronidazole in the treatment of peptic ulcer. Klin Med (Mosk) 1981; 59:48–50.

122. Drasar BS, Shiner M, McLoed GM. The bacterial flora of the gastrointestinal tract in healthy and achlorhydric persons. Gastroenterology 1969;56:71–9.

123. Reed PL, Haines K, Smith PLR, House FR, Walters CL. Gastric juice nitrosamines in health and gastroduodenal disease. Lancet 1981;ii:550–2.

## *Campylobacter pylori* Infection

*To the Editor:* We have difficulty with the conclusions reached by Perez-Perez and colleagues (1) on the use of *Campylobacter pylori* antibodies to determine the prevalence of *C. pylori* infection in healthy persons. The results of their IgG and IgA enzyme-linked immunosorbent assay (ELISA) tests have been used as another way to diagnose infection with *C. pylori*,but the current accepted "gold standard" for detecting this organism is the histologic examination or culture of endoscopically obtained gastric biopsy specimens. The $^{13}$C-urea and $^{14}$C-urea breath tests are other tests that can be used as a gold standard for detecting *C. pylori*; the advantage of these tests is that they probably do not depend on the sampling error that endoscopic biopsies may have (2-4).

Methodologic standards for determining the accuracy of diagnostic tests are well established (5, 6). The criterion that the results of the new test be compared with those of the old gold standard test was not fulfilled in all patients. Of the healthy controls, 166 never had endoscopy; in 29 patients who did have endoscopy, the antrum appeared macroscopically normal, and biopsy specimens were not obtained. *Campylobacter pylori* infection can be present in mucosa that appears endoscopically normal.

In Figure 2, the authors show the prevalence of *C. pylori* –specific serum antibodies based on the results of the ELISA tests. However, the data in this figure for the 166 healthy persons and the 29 persons who did not have endoscopy are extrapolations from the results in the patients shown to be positive for *C. pylori* who did have biopsy samples obtained. Whether most of these persons were indeed shown to be negative for *C. pylori* was never proved by any other test.

Although the ELISA tests the authors used may be an excellent noninvasive method for detecting *C. pylori* infection, we believe that the conclusions drawn by Perez-Perez and colleagues cannot be derived from their data.

*S. J. O. Veldhuyzen van Zanten, MD, MPH*
*J. Goldie*
*R.H. Riddell, MD*
*Richard H. Hunt, MD*
McMaster University Health Sciences Centre
Hamilton, Ontario L8N 3Z5

**References**
1. Perez-Perez GI, Dworkin BM, Chodos JE, Blaser MJ. *Campylobacter pylori* antibodies in humans. *Ann Intern Med.* 1988;109:11-7.
2. Graham DY, Klein PD, Evans DJ, et al. *Campylobacter pylori* detected noninvasively by the $^{14}$C-urea breath test. *Lancet.* 1985;1:1174-7.
3. Bell GD, Weil J, Harrison G, et al. $^{14}$C-urea breath analysis: a noninvasive test for *Campylobacter pylori* in the stomach. *Lancet.* 1987;1:1367-8.
4. Rauws EAJ, Tytgat GNJ, Langenberg W, van Royen E. Experience with $^{14}$C-urea breath test in detecting *Campylobacter pylori.* In: Menge H, Gregor M, Tytgat GNJ, Marshall BJ. *Campylobacter pylori: Proceedings of the First International Symposium on Campylobacter pylori.* Berlin: Springer-Verlag; 1988:151-3.
5. Department of Clinical Epidemiology and Biostatistics, McMaster University Health Sciences Centre. How to read clinical journals: II. To learn about a diagnostic test. *Can Med Assoc J.* 1981;124:703-10.
6. Griner PF, Mayewski R, Mushlin AI, et al. Selection and interpretation of diagnostic tests and procedures: principles and applications. *Ann Intern Med.* 1981;94:557-92.

## DUODENAL ULCER RELAPSE AFTER
## ERADICATION OF CAMPYLOBACTER PYLORI

SIR,—Dr Marshall and colleagues (Dec 24/31, p 1438) conclude that eradication of *Campylobacter pylori* led to a higher healing rate of duodenal ulcer and a longer remission. Both conclusions are unacceptable.

The healing rate of duodenal ulcer after 8 weeks of treatment with cimetidine 400 mg twice a day alone was 13/22 (59%) and, after colloidal bismuth subcitrate one tablet four times a day, 13/20 (65%). Such low 8-week healing rates have not been reported before—the expected figures should be about 90%. Marshall and colleagues did endoscopy at week 10 instead of week 8, and argue that the low healing rates were because of relapse during the 2 week interval. Their argument is not substantiated and is difficult to understand. It suggests that the 2-week relapse occurs equally quickly for bismuth as for cimetidine, and yet the 12-month relapse rates are different, as they have shown. The series of patients they studied appear unique.

The treatment with colloidal bismuth or cimetidine was not blinded to the patients or investigators. Hence, the title of the paper is misleading. Ulcer relapse was defined not only as the demonstration of an ulcer at scheduled endoscopy at weeks 14, 26, and 54 after treatment, but also as any recurrence of ulcer symptoms. Because bismuth but not cimetidine is associated with clearance of campylobacter, the design of this study does not safeguard against investigator or patient bias. The use of tinidazole and its placebo did not help because this drug would not of course differentiate cimetidine and bismuth, and because tinidazole was ineffective in the cimetidine group but effective in the bismuth group.

Eradication is a biased term. The distribution of campylobacter is patchy[1] and yet the two biopsy specimens taken for this purpose were regarded as adequate to justify the term "eradication".

Department of Medicine,
University of Hong Kong,
Queen Mary Hospital,
Hong Kong                                           S. K. LAM

1. Goodwin CS, Blincow E, Warren JR, et al. Evaluation of cultural techniques for isolating *Campylobacter pyloridis* from endoscopic biopsies of gastric mucosa. *J Clin Pathol* 1985; 138: 1127–31.

## DUODENAL ULCER RELAPSE AFTER ERADICATION OF CAMPYLOBACTER PYLORI

SIR,—Professor Lam (Feb 18 p 384) criticises our study. We wish to reply.

Eradication of *Campylobacter pylori* in our patients was associated with a much longer remission of healed duodenal ulcers and the apparent rate of healing was also higher. The apparent ulcer-healing rate at follow-up endoscopy was low in our patient group in which *C pylori* was not eradicated (CP + ve patients), irrespective of the therapy. Our Table II shows healing in 44/72 (61%) CP + ve patients. Since 27/44 (61%) of these healed ulcers relapsed before 3 months, we cannot see why it is difficult to understand our suggestion that others probably relapsed before the first follow-up 2 weeks after therapy ended. Even if the argument is not substantiated, it is not unreasonable. The failure rates at 12 months were the same for both cimetidine (46/50, 92%) and colloidal bismuth subcitrate (CBS) (19/22, 86%). We agree that our patients are unique, as Lam says, since all patients are unique.

Every effort was taken to keep the trial blind. Follow-up was timed 2 weeks after therapy ceased (so that mouth-staining would not be seen by the endoscopist) and communication was avoided between the clinician and the patient before gastroscopy. More importantly the pathologist and microbiologist were independent; they had no knowledge of the therapy and neither the clinician nor the patient knew of the presence or absence of bacteria. Thus we do not agree that the title is misleading.

Lam mentions that our definition of ulcer relapse includes "any recurrence of ulcer symptoms". This is almost correct—our patients were treated as in normal practice. If they complained of symptoms requiring therapy, they were considered to have relapsed and usually (32/43, 74%) an ulcer was found on gastroscopy. We thought it unreasonable to call such patients "successfully treated" if they had a recurrence of symptoms but did not have an ulcer crater at endoscopy.

We agree that "bismuth not cimetidine is associated with clearance of campylobacter", but if Lam really believes we did not "safeguard against investigator or patient bias", we suggest he considers the following. There is no way to demonstrate CBS therapy on histology or microbiology, unless Lam thinks that the absence of gastritis and campylobacter, only seen with CBS therapy, causes observer bias. All bias can thus be eliminated by considering only the group of patients treated with CBS. This shows 22 CP + ve patients, 10 (45%) healed at follow-up and 3 (14%) still healed at 12 months; compare this with 24 CP − ve patients, 22 (92%) healed and 17 (71%) still healed at 12 months.

Tinidazole was not intended to "differentiate cimetidine and bismuth", nor for that matter was our study. If Lam is interested in tinidazole, he should examine our results for CBS with or without tinidazole, which show an improvement with tinidazole. We were investigating the effect of the eradication of *C pylori*, not the effect of cimetidine, CBS, or tinidazole per se—our results should be viewed from this aspect.

The distribution of campylobacter is patchy, as stated by Lam, but only at a microscopic level and, with improved methods, we rarely have contradictory results. In this series no CP + ve patient gave negative histological findings and culture without treatment and only 1 patient gave two successive negative biopsy specimens followed by a *C pylori* positive culture (probably a true re-infection). The series includes about 450 biopsies on 100 patients. Under these circumstances we consider "eradication" a correct and justified term.

Our work was designed to show the effect of the eradication of *C pylori* on duodenal ulcer relapse. It was never intended as a drug trial. Cimetidine is an ulcer-healing agent but will not eradicate *C pylori*. Our results suggest that eradication of *C pylori* is associated with a dramatic reduction in the relapse of healed ulcers.

Internal Medicine,
Health Sciences Center,
University of Virginia,
Charlottesville, Virginia 22908, USA

Royal Perth Hospital,
Perth, Western Australia

BARRY J. MARSHALL

J. ROBIN WARREN
C. STEWART GOODWIN

SIR,—We agree that serological tests with sonicates or crude extracts of *Campylobacter pylori* suffer from cross-reaction with other microorganisms, notably *C jejuni*. Our assay, however, did not show such cross-reaction when tested by indirect methods. We see no reason why direct tests should have been done.

Dr Graham and his colleagues do not seem to understand how our cut-off values were established. We compared two well-defined groups of patients with non-ulcer dyspepsia with and without campylobacter-associated gastritis, and found clear cut-offs indicating 100% positive and negative predictive values, respectively, leaving circumvention of the problem of cross-reactivity totally aside.

We are not aware of having stumbled into the trap of assuming that optical density is proportional to antibody titre. Extinction values do reflect the amount of antibody in serum. Measurements on test sera were divided by control values to calculate the P/N ratio on a linear scale. It might have been better if we had mentioned this P/N ratio explicitly in the legend of fig 1 in our paper.

Our discussion was indeed based on the premise that there is a safe and effective therapy for *C pylori* and that patients with *C pylori* infection should be treated. While we agree with Graham's criticism on the second part of this premise, we maintain that *C pylori* infection can be safely and effectively treated with colloidal bismuth subcitrate.

Despite his criticisms, Graham's line of thinking—as indicated by his quotations from the 1989 *Gastroenterology* paper with which his letter ends—almost entirely coincides with the views expressed in our paper. We tried to estimate, on the basis of morbidity figures in our hospital, the decrease in endoscopies that might result from adoption of the proposed strategy and we expressed doubts about the need to treat *C pylori* infection in the last sentence of our paper. So what conceptual errors did we make—or have Dr Graham's views been subject to dramatic change recently?

R. J. L. F. LOFFELD
E. STOBBERINGH
J. W. ARENDS

University Hospital Maastricht,
6210 BX Maastricht, Netherlands

## Anti-*Helicobacter pylori* therapy: clearance, elimination, or eradication?

SIR,—The proceedings of the third workshop of the European *Helicobacter pylori* study group meeting held in Toledo, Spain, have been published as abstracts of the 16 oral communications and the 298 posters.[1] Not surprisingly no less than 61 abstracts relate to reporting of results of treatment of *H pylori* positive patients with various forms of antihelicobacter therapy. This includes various bismuth compounds, antibiotics, histamine antagonists, and omeprazole used either alone or in combination for periods ranging from a few days to several weeks.

Comparison of the results from various groups of workers is made even more difficult, or impossible, by the fact that the groups have either assessed their "success" rates while the patient is still on antihelicobacter treatment or is anything from 1 day to many months post therapy.

There is now general agreement[2,3] that "eradication" should only be applied if tests for *H pylori* are negative when repeated at least four weeks after treatment has been discontinued. In our experience with the [14]C-urea breath test,[4,5] and that of others with the [13]C-urea breath test,[6] within as little as 24–48 h of completing a course of, for instance, colloidal bismuth subcitrate, the organism that had been temporarily suppressed to undetectable levels rapidly multiplies to levels that again allow its detection. These rapid returns to positivity have been shown by restriction endonuclease DNA analysis to result from recrudescence and not re-infection.[7]

Several workers who report their results for success while on active treatment (or within 24 h of discontinuing it) use "clearance" or "elimination" of *H pylori* to describe their findings. I suggest both terms are inaccurate and misleading since what is in truth being measured is temporary suppression of the bacteria. Someone reading an article in which *H pylori* is described as cleared or eliminated could be forgiven for thinking both terms are synonomous with eradicated, which they certainly are not. If figures for clearance are to be reported then the figures for eradication should also be provided. How have the authors of the 61 abstracts from Toledo described their results?

11 abstracts provide only clearance data whereas 36 correctly report eradication figures. 4 abstracts show both clearance and eradication results, but in 1 these are described in terms of elimination of *H pylori* while on treatment and again elimination of the organism when re-assessed one month later. In 4 abstracts authors talk about eradication when they mean clearance and in 3 clearance is used when what they are describing is eradication. Finally 3 abstracts provide success rates at between 3 and 4 weeks after treatment and are probably describing eradication, but cannot be classified.

The importance of having a firm definition of clearance as well as of eradication is well illustrated in two papers from the same group[6,8]. in which serial [13]C-urea breath tests were used to look at the effect of colloidal bismuth subcitrate on *H pylori* status. In the first[8] the clearance rate was shown as only 5/28 (18%) but the breath test was delayed until 3 days after treatment was stopped. In contrast, in a later publication this group reported their results[6] following 1, 2, or 4 weeks of bismuth treatment in which clearance of *H pylori* was described as being 10/15 (67%), 8/12 (75%), and 17/20 (85%), respectively, but here the post-treatment test was done within 24 h of stopping antihelicobacter treatment. Within a week of stopping treatment in all but 1 patient the urea breath test was again positive. If clearance had been assessed at this stage a figure of 1/47 (2·1%) would have been recorded.[6]

I therefore propose that we all concentrate on eradication results rather than clearance figures, which are of little clinical relevance. Results obtained more than 24 h and less than 4 weeks after anti-helicobacter treatment is stopped are by definition neither clearance nor eradication and should not be published.

Ipswich Hospital,
Ipswich IP4 5PD, UK
G. D. BELL

1. Gastroduodenal pathology and *Helicobacter pylori*. Third workshop of European *Helicobacter pylori* study group. *Rev Esp Enf Digest* 1990; **78** (suppl 1): 3–140.
2. Weil J, Bell GD, Jones PH, Gant P, Trowell JE, Harrison G. "Eradication" of *Campylobacter pylori*: are we being misled? *Lancet* 1988; ii: 1245.
3. Tytgat GNJ, Axon ATR, Dixon MF, Graham DY, Lee A, Marshall BJ. *Helicobacter pylori*: causal agent in peptic ulcer disease? World Congress of Gastroenterology Working Party report, 1990; 36–45.
4. Bell GD, Weil J, Harrison G, et al. [14]C-urea breath test analysis, a non-invasive test for *Campylobacter pylori* in the stomach. *Lancet* 1987; i: 1367–68.
5. Weil J, Bell GD, Harrison G, Trowell JE, Gant P, Jones PH. *Campylobacter pylori* survives high dose bismuth subcitrate (De-Nol) therapy. *Gut* 1988; **29**: A1437.
6. Logan RPH, Gummett PA, Polson RJ, Baron JH, Misiewicz JJ. What length of treatment with tripotassium dicitrato-bismuthate (TDB) for *Helicobacter pylori*? *Gut* 1990; **31**: A1178.
7. Langenberg W, Rauws EA, Widjojokusumo A, et al. Identification of *Campylobacter*

# Effect of Treatment of *Helicobacter pylori* Infection on the Long-term Recurrence of Gastric or Duodenal Ulcer

## A Randomized, Controlled Study

David Y. Graham, MD; Ginger M. Lew, PA-C; Peter D. Klein, PhD; Dolores G. Evans, PhD; Doyle J. Evans, Jr., PhD; Zahid A. Saeed, MD; and Hoda M. Malaty, MD

■ *Objective:* To determine the effect of treating *Helicobacter pylori* infection on the recurrence of gastric and duodenal ulcer disease.

■ *Design:* Follow-up of up to 2 years in patients with healed ulcers who had participated in randomized, controlled trials.

■ *Setting:* A Veterans Affairs hospital.

■ *Participants:* A total of 109 patients infected with *H. pylori* who had a recently healed duodenal (83 patients) or gastric ulcer (26 patients) as confirmed by endoscopy.

■ *Intervention:* Patients received ranitidine, 300 mg, or ranitidine plus triple therapy. Triple therapy consisted of tetracycline, 2 g; metronidazole, 750 mg; and bismuth subsalicylate, 5 or 8 tablets (151 mg bismuth per tablet) and was administered for the first 2 weeks of treatment; ranitidine therapy was continued until the ulcer had healed or 16 weeks had elapsed. After ulcer healing, no maintenance antiulcer therapy was given.

■ *Measurements:* Endoscopy to assess ulcer recurrence was done at 3-month intervals or when a patient developed symptoms, for a maximum of 2 years.

■ *Results:* The probability of recurrence for patients who received triple therapy plus ranitidine was significantly lower than that for patients who received ranitidine alone: for patients with duodenal ulcer, 12% (95% CI, 1% to 24%) compared with 95% (CI, 84% to 100%); for patients with gastric ulcer, 13% (CI, 4% to 31%) compared with 74% (44% to 100%). Fifty percent of patients who received ranitidine alone for healing of duodenal or gastric ulcer had a relapse within 12 weeks of healing. Ulcer recurrence in the triple therapy group was related to the failure to eradicate *H. pylori* and to the use of nonsteroidal anti-inflammatory drugs.

■ *Conclusions:* Eradication of *H. pylori* infection markedly changes the natural history of peptic ulcer in patients with duodenal or gastric ulcer. Most peptic ulcers associated with *H. pylori* infection are curable.

*Annals of Internal Medicine.* 1992;116:705-708.

From Baylor College of Medicine, the Veterans Affairs Medical Center, and the U.S. Department of Agriculture/Agricultural Research Center Children's Nutrition Research Center. For current author addresses, see end of text.

Peptic ulcer disease is a chronic disease characterized by frequent recurrences. The continuation of antiulcer therapy after ulcer healing results in a reduced rate of ulcer recurrence but does not affect the natural history of the disease, because the expected pattern of rapid recurrence resumes when maintenance therapy is discontinued (1). Recent studies have suggested that the eradication of *Helicobacter pylori* infection affects the natural history of duodenal ulcer disease such that the rate of recurrence decreases markedly (2-6). However, the interpretation of these results has been complicated by the fact that several of the larger studies did not use control groups or any form of blinding (3, 5, 6). In addition, studies of the effect of *H. pylori* eradication in patients with gastric ulcer have not been done. We report the results of a randomized, controlled trial in which we evaluated the effect of therapy designed to eradicate *H. pylori* on the pattern of ulcer recurrence in patients with duodenal or gastric ulcer.

## Methods

Our study took place between September 1988 and October 1990 at a single Veterans Affairs hospital. All patients whose *H. pylori*-associated active duodenal or gastric ulcer had healed during randomized trials comparing ranitidine and ranitidine plus "triple therapy" were invited to participate in this follow-up study. During the initial studies, patients were randomly assigned to either ranitidine alone (300 mg once daily in the evening) or to ranitidine plus triple therapy. Triple therapy consisted of bismuth subsalicylate and two antimicrobial agents: tetracycline hydrochloride, 500 mg four times a day and metronidazole, 250 mg thrice daily. Bismuth subsalicylate tablets containing 151 mg bismuth per tablet (Pepto-Bismol, Proctor & Gamble, Cincinnati, Ohio) were administered for the first 2 weeks of therapy, and patients received 5 or 8 tablets.

Two groups of patients were entered into our follow-up study. Patients with healed duodenal ulcers came from a randomized study of 146 patients, 105 of whom have been previously described (7); of these 146 patients, 112 experienced ulcer healing, 24 were lost to follow-up during the 16 weeks of therapy, and 10 had no ulcer healing after 16 weeks of treatment. Of the 112 patients in whom documented healing occurred, 83 (74%) agreed to enter the follow-up study. Patients with healed gastric ulcers came from a randomized trial of 41 patients; of these 41 patients, 31 had ulcer healing, 9 were lost to follow-up during the 16 weeks of therapy, and 1 patient had no ulcer healing after 16 weeks of treatment. Of the 31 patients in whom documented healing occurred, 26 (84%) agreed to enter the follow-up study. In sum, 109 patients with healed

**Table 1. Demographic and Clinical Characteristics of Patients**

| Variable | Patients with Duodenal Ulcer | | Patients with Gastric Ulcer | |
| --- | --- | --- | --- | --- |
| | Ranitidine Alone | Triple Therapy plus Ranitidine | Ranitidine Alone | Triple Therapy plus Ranitidine |
| Patients, *n* | 36 | 47 | 11 | 15 |
| Median age (range), *y* | 61 (31-85) | 58 (29-79) | 66 (43-76) | 60 (27-67) |
| Male gender, % | 97 | 100 | 100 | 100 |
| Race, *n(%)* | | | | |
|   White | 25 (69) | 30 (64) | 7 (64) | 8 (53) |
|   Black | 11 (30) | 17 (36) | 3 (27) | 7 (47) |
|   Other | 0 | 0 | 1 (9) | 0 |
| Recent NSAID* use, *n(%)* | 6 (17) | 11 (23) | 2 (18) | 5 (33) |
| Daily aspirin (1 tablet), *n(%)* | 3 (8) | 4 (8.5) | 1 (9) | 1 (6) |
| Smoker†, *n(%)* | 17 (47) | 34 (72) | 6 (54) | 11 (73) |
| Alcohol use | | | | |
|   1 or more drinks/wk, *n(%)* | 7 (19) | 15 (32) | 2 (18) | 4 (27) |
| *H. pylori* infection, %‡ | 100 | 100 | 100 | 100 |

\* NSAID = nonsteroidal anti-inflammatory drugs.
† *P* = 0.03 for the difference between the ranitidine and triple therapy groups among patients with duodenal ulcer.
‡ Infection at entry into the ulcer healing study was confirmed by at least two of the following: urea breath test, histologic evaluation, culture, and serologic testing.

peptic ulcers (83 with duodenal ulcers and 26 with gastric ulcers) were included in our long-term follow-up study.

Patients were followed for up to 2 years. During this time, patients received no antiulcer medications (including antacids). Twenty-four patients regularly receiving nonsteroidal anti-inflammatory drugs were allowed to continue them if they wished. Patient follow-up visits were scheduled for 1 month

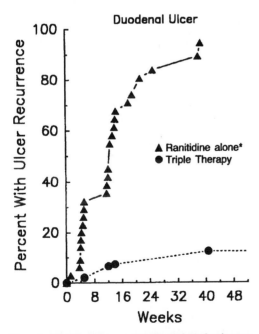

**Figure 1. Lifetable recurrence of duodenal ulcers for the year after successful healing with ranitidine alone or triple therapy plus ranitidine.** No maintenance therapy was given; the recurrence rate of ulcers in patients healed with ranitidine alone was significantly greater (*P* < 0.01) than in those who received triple therapy plus ranitidine. * The only patient in the ranitidine alone group who had not developed recurrent ulcer by October 1990 was then followed for a total of 16 months without ulcer recurrence (*see* text).

after therapy, 3 months after therapy, and every 3 months for up to 2 years. Patients were also instructed to return for endoscopy if symptoms recurred. The endoscopist was blinded to the treatment status of the patients.

Patients originally assigned to receive ranitidine alone who experienced ulcer recurrence were crossed over to receive triple therapy plus ranitidine; after ulcer healing occurred with this latter therapy, patients were offered follow-up using the protocol described above. These patients were termed "crossover follow-up" patients.

Ulcers were identified by endoscopy using Fujinon videoendoscopes (Fujinon, Inc., Wayne, New Jersey). A video still (ProMavica, Sony Corporation of America, Sony Park Ridge, New Jersey) of each ulcer was made so that the site and characteristics of the ulcer could be reviewed before subsequent endoscopic procedures. One video disk was assigned to each patient. An ulcer was defined as a circumscribed break in the duodenal mucosa that measured at least 5 mm in diameter, had apparent depth, and was covered by an exudate.

All patients were assessed for *H. pylori* infection by the $^{13}$C-urea breath test (8, 9); by a sensitive and specific enzyme-linked immunosorbent assay (ELISA) for IgG antibody against the high-molecular-weight, cell-associated proteins of *H. pylori* (10); by culture; and by histologic evaluation of antral mucosal biopsy specimens. Eradication was defined by no evidence of *H. pylori* infection (by urea breath test, culture, or histologic evaluation) 1 or more months after discontinuing triple therapy. Patients were tested every 3 months and when symptomatic.

The protocol was approved by the Institutional Review Board at the Veteran Affairs Medical Center and Baylor College of Medicine. Written informed consent was obtained before patient entry.

Statistical Analysis

Ulcer recurrence was calculated by the lifetable method (Lifetest procedure, SAS/STAT software release 6.04, SAS Institute, Inc., Cary, North Carolina). Categorical data were evaluated by chi-square test with the Yates correction or by the Fisher exact test. All *P* values ≤ 0.05 (two-tailed) were considered to be significant. Ninety-five percent confidence intervals are given when appropriate.

Results

We followed 83 patients with duodenal ulcer and 26 patients with gastric ulcer (median age, 62 years). The sample was 98% men. The two groups of patients (assigned to ranitidine alone or ranitidine plus triple therapy) had similar demographic and clinical characteristics (Table 1). The percentage of smokers and alcohol users was higher in the group receiving triple therapy plus ranitidine than in the group receiving ranitidine alone. Only one significant difference was found between the treatment groups: Among patients with duodenal ulcer, the group receiving triple therapy had more smokers than the group receiving ranitidine alone ($P$ = 0.03). All patients had active *H. pylori* infection before the start of ulcer therapy. All 47 patients treated with ranitidine alone were still infected at the end of therapy. In contrast, *H. pylori* was eradicated in 55 of 62 patients (89%) receiving triple therapy.

The lifetable probability of ulcer recurrence 1 year after ulcer healing (Figures 1 and 2) was significantly lower for patients who received triple therapy plus ranitidine (12% [CI, 1% to 24%] for patients with duodenal ulcer and 13% [CI, 4% to 31%] for patients with gastric ulcer) compared with those who received ranitidine alone (95% [CI, 84% to 100%] for patients with duodenal ulcer and 74% [CI, 44% to 100%] for patients with gastric ulcer) ($P$ = 0.001). The median duration of follow-up for patients who had received triple therapy plus ranitidine was 38 weeks (range, 4 to 108 weeks) for patients with duodenal ulcer and 52 weeks (range, 12 to 95 weeks) for patients with gastric ulcer.

Fifty percent of patients with either duodenal or gastric ulcer who experienced healing with ranitidine alone had a recurrence within 12 weeks (*see* Figures 1 and 2). Seventy-five percent of recurrences were symptomatic. At the end of the study period, only three patients in the ranitidine alone group (one with duodenal ulcer and two with gastric ulcer) had not had ulcer recurrence. Follow-up on these three patients after the study was completed showed the following: One patient with duodenal ulcer was lost to follow-up after 16 months, and two patients with gastric ulcer were last seen at the 15-month follow-up visit (one was lost to follow-up and the other died of an unrelated illness).

Infection with *H. pylori* was a strong predictor of ulcer recurrence. All 47 patients whose ulcers healed while receiving ranitidine alone still had *H. pylori* infection at the end of therapy, and, by lifetable analysis, 95% of them developed recurrent ulcers by the end of 1 year. None of the patients in whom *H. pylori* was eradicated became reinfected during the study period. *Helicobacter pylori* infection was not eradicated in seven patients who received triple therapy. Of these seven patients, four experienced ulcer recurrence and three were lost to follow-up (two patients after 6 months and one patient after 1 year).

Three patients with duodenal ulcer in whom *H. pylori* infection was eradicated after triple therapy still developed recurrent duodenal ulcers (two patients after 3 months and one patient after 9 months). All three patients were using nonsteroidal anti-inflammatory drugs (ibuprofen, piroxicam, or salsalate). In addition, two patients with recurrent gastric ulcer who were also receiving such drugs had persistent *H. pylori* infection.

Ten patients with duodenal ulcer who experienced ulcer recurrence after healing with ranitidine alone were crossed over to receive ranitidine plus triple therapy after the completion of the randomized trial. Triple ther-

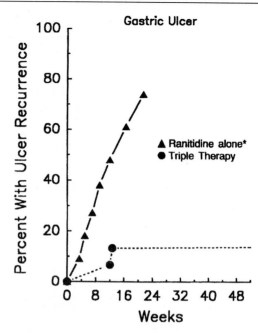

**Figure 2. Lifetable recurrence of gastric ulcers for the year after successful healing with ranitidine alone or triple therapy plus ranitidine.** No maintenance therapy was given; the recurrence rate of ulcers in patients healed with ranitidine alone was significantly greater ($P < 0.01$) than in those who received triple therapy plus ranitidine. * The two patients in the ranitidine alone group who had not developed recurrent ulcers by October 1990 were then followed for a total of 15 months without ulcer recurrence (*see* text).

apy resulted in the eradication of *H. pylori* infection in these patients. After ulcer healing, the patients were followed for a median of 44 weeks (range, 23 to 116 weeks), and none experienced ulcer recurrence. Four other patients with duodenal ulcer refractory to ranitidine alone were crossed over to receive ranitidine plus triple therapy. *Helicobacter pylori* infection was eradicated in all four patients. After ulcer healing, the patients were followed for a median of 40 weeks, and none experienced ulcer recurrence.

### Discussion

Recent studies have shown that the eradication of *H. pylori* infection is associated with healing of gastritis (11) and a marked reduction in the rate of recurrence of duodenal ulcers (2-6). The protocols of these studies have varied, but the results have been the same; the eradication of *H. pylori* infection changes the natural history of duodenal ulcer disease, and factors that contribute to rapid ulcer recurrence, such as smoking, seem to no longer to pose a risk (6). Our study confirms previous findings in patients with duodenal ulcers and extends the findings to patients with gastric ulcers. Patients in whom *H. pylori* infection was eradicated remained asymptomatic and ulcer free.

Smoking, alcohol use, and male gender have all been

described as risk factors for ulcer recurrence (1, 12). In our study, smoking and alcohol use were more frequent in the group that received triple therapy plus ranitidine, a factor that could have biased our results. However, our study confirms the observation that smoking is not a risk factor for ulcer recurrence after the eradication of *H. pylori* infection (6). In our patients, the only factors associated with ulcer recurrence were *H. pylori* infection and the continued use of nonsteroidal anti-inflammatory drugs.

Our study was single-blind, with the endoscopist blinded to initial therapy. Although some may argue that the lack of double-blinding introduced an important bias into our study, no objective data support such a contention, and we believe such a scenario extremely unlikely, especially considering the equipment now available for studying the gastroduodenal mucosa.

Three previous reports have used the word "cure" in the title or discussion (4-6). These studies, taken together, provide compelling evidence for the hypothesis that peptic ulcer, either duodenal or gastric, is the end result of a bacterial infection. We believe that, eventually, anti-*H. pylori* agents will be part of the therapy for *H. pylori*-associated ulcer disease. The universal introduction of such therapy may be delayed because several safe and effective therapies are currently available for healing peptic ulcers and because ulcer relapse can be greatly reduced by maintenance therapy with histamine-2-receptor antagonists (1, 12). In addition, there are still concerns that the benefits of therapy (that is, reduced recurrence) may not yet outweigh such side effects as antibiotic-associated diarrhea or the development of widespread antibiotic resistance in *H. pylori* and other bacteria (13). We have observed that triple therapy is often not effective in patients who have previously received metronidazole (unpublished data), and compliance with the complicated treatment protocols remains a major problem (14). Simpler protocols and improved therapies are needed. The eradication of the infection may also not yield a true cure because the patient vulnerable to additional *H. pylori* encounters may acquire a new infection and experience recurrence of the original disease. We recommend that patients with resistant ulcers (defined as failure to heal in 12 weeks), those with ulcer-associated complications, and those with symptoms severe enough to be candidates for surgery receive triple therapy for *H. pylori* infection.

*Grant Support:* In part by the Department of Veterans Affairs, by grant DK 39919 from the National Institute of Diabetes and Digestive and Kidney Diseases, by the U.S. Department of Agriculture/Agricultural Research Service Children's Nutrition Research Center, and by Hilda Schwartz.

*Requests for Reprints:* David Y. Graham, MD, Veterans Affairs Medical Center (111D), 2002 Holcombe Boulevard, Houston, TX 77030.

*Current Author Addresses:* Drs. Graham, Evans, Evans, Jr., Saeed, Malaty, and Ms. Lew: Veterans Affairs Medical Center (111D), 2002 Holcombe Boulevard, Houston, TX 77030.
Dr. Klein: Children's Nutrition Research Center, 1100 Bates Street, Houston, TX 77030.

**References**

1. **Sontag SJ.** Current status of maintenance therapy in peptic ulcer disease. Am J Gastroenterol. 1988;83:607-17.
2. **Coghlan JG, Gilligan D, Humphries H, McKenna D, Dooley C, Sweeney E, et al.** Campylobacter pylori and recurrence of duodenal ulcers—a 12-month follow-up study. Lancet. 1987;2:1109-11.
3. **Lambert JR, Borromeo M, Korman MG, Hansky J, Eaves ER.** Effect of colloidal bismuth (De-Nol) on healing and relapse of duodenal ulcers-role of Campylobacter pyloridis [Abstract]. Gastroenterology. 1987;92:1489.
4. **Marshall BJ, Goodwin CS, Warren JR, Murray R, Blincow ED, Blackbourn SJ, et al.** Prospective double-blind trial of duodenal ulcer relapse after eradication of Campylobacter pylori. Lancet. 1988;2:1437-42.
5. **Rauws EA, Tytgat GN.** Cure of duodenal ulcer associated with eradication of Helicobacter pylori. Lancet. 1990;335:1233-5.
6. **George LL, Borody TJ, Andrews P, Devine M, Moore-Jones D, Walton M, et al.** Cure of duodenal ulcer after eradication of Helicobacter pylori. Med J Aust. 1990;153:145-9.
7. **Graham DY, Lew GM, Evans DG, Evans DJ Jr, Klein PD.** Effect of triple therapy (antibiotics plus bismuth) on duodenal ulcer healing with ranitidine. A randomized controlled trial. Ann Intern Med. 1991;115:266-9.
8. **Graham DY, Klein PD, Evans DJ Jr., Evans DG, Alpert LC, Opekun AR, et al.** Campylobacter pylori detected noninvasively by the 13C-urea breath test. Lancet. 1987;1:1174-7.
9. **Klein PD, Graham DY.** Campylobacter pylori detection by the $^{13}$C-urea breath test. In: Campylobacter pylori and Gastroduodenal Disease. Rathbone BJ, Heatley V, eds. Blackwell Scientific Publications, Oxford, 1989, pp. 94-106.
10. **Evans DJ Jr, Evans DG, Graham DY, Klein PD.** A sensitive and specific serologic test for detection of Campylobacter pylori infection. Gastroenterology. 1989;96:1004-8.
11. **Rauws EA, Langenberg W, Houthoff HJ, Zanen HC, Tytgat GN.** Campylobacter pyloridis-associated chronic active antral gastritis: a prospective study of its prevalence and the effects of antibacterial and antiulcer treatment. Gastroenterology. 1988;94:33-40.
12. **Van Deventer GM, Elashoff JD, Reedy TJ, Schneidman D, Walsh JH.** A randomized study of maintenance therapy with ranitidine to prevent the recurrence of duodenal ulcer. N Engl J Med. 1989;320:1113-9.
13. **Graham DY, Börsch GM.** The who's and when's of therapy for Helicobacter pylori [Editorial]. Am J Gastroenterol. 1990;85:1552-5.
14. **Graham DY, Lew GM, Malaty HM, Evans DG, Evans DJ Jr, Klein PD, et al.** Factors influencing the eradication of Helicobacter pylori with triple therapy. Gastroenterology. 1992;102:493-6.

# The Bacteria behind Ulcers

*One half to one third of the world's population harbors* Helicobacter pylori, *"slow" bacteria that infect the stomach and can cause ulcers and cancer there*

by Martin J. Blaser

In 1979 J. Robin Warren, a pathologist at the Royal Perth Hospital in Australia, made a puzzling observation. As he examined tissue specimens from patients who had undergone stomach biopsies, he noticed that several samples had large numbers of curved and spiral-shaped bacteria. Ordinarily, stomach acid would destroy such organisms before they could settle in the stomach. But those Warren saw lay underneath the organ's thick mucus layer—a lining that coats the stomach's tissues and protects them from acid. Warren also noted that the bacteria were present only in tissue samples that were inflamed. Wondering whether the microbes might somehow be related to the irritation, he looked to the literature for clues and learned that German pathologists had witnessed similar organisms a century earlier. Because they could not grow the bacteria in culture, though, their findings had been ignored and then forgotten.

Warren, aided by an enthusiastic young trainee named Barry J. Marshall, also had difficulty growing the unknown bacteria in culture. He began his efforts in 1981. By April 1982 the two men had

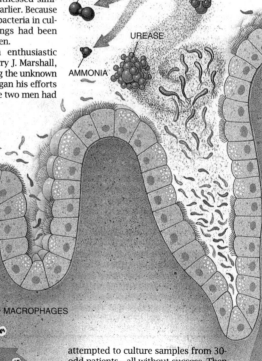

ULCER-CAUSING BACTERIA (*Helicobacter pylori*) live in the mucus layer (*pale yellow*) lining the stomach. There they are partially protected from the stomach's acid (*pink*). The organisms secrete proteins that interact with the stomach's epithelial cells and attract macrophages and neutrophils, cells that cause inflammation (*left*). The bacteria further produce urease, an enzyme that helps to break down urea into ammonia and carbon dioxide; ammonia can neutralize stomach acid (*center*). *H. pylori* also secrete toxins that contribute to the formation of stomach ulcers (*right*). The microbes typically collect in the regions shown in the diagram at the far right.

attempted to culture samples from 30-odd patients—all without success. Then the Easter holidays arrived. The hospital laboratory staff accidentally held some of the culture plates for five days instead of the usual two. On the fifth day, colonies emerged. The workers christened them *Campylobacter pyloridis* because they resembled pathogenic bacteria of the *Campylobacter* genus found in the intestinal tract. Early in 1983 Warren and Marshall published their first report, and within months scientists around the world had isolated the bacteria. They found that it did not, in fact, fit into the *Campylobacter* genus, and so a new genus, *Helicobacter*, was created. These researchers also confirmed Warren's initial finding, namely that *Helicobacter pylori* infection is strongly associated with persistent stomach inflammation, termed chronic superficial gastritis.

The link led to a new question: Did

acquire chronic superficial gastritis. Left untreated, both the infection and the inflammation last for decades, even a lifetime. Moreover, this condition can lead to ulcers in the stomach and in the duodenum, the stretch of small intestine leading away from the stomach. *H. pylori* may be responsible for several forms of stomach cancer as well.

More than 40 years ago doctors rec-

essary for ulcers to form, it is not sufficient to explain their occurrence—most patients with ulcers have normal amounts of stomach acid, and some people who have high acid levels never acquire ulcers.

Nevertheless, the stress-acid theory of ulcers gained further credibility in the 1970s, when safe and effective agents to reduce gastric acid were introduced.

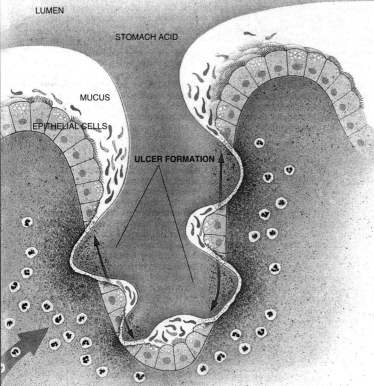

LUMEN

STOMACH ACID

MUCUS

EPITHELIAL CELLS

ULCER FORMATION

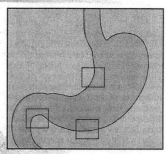

Many patients felt free of pain for the first time while taking these medications, called histamine 2-receptor blockers ($H_2$-receptor blockers). The drugs often healed ulcers outright. But when patients stopped taking them, their ulcers typically returned. Thus, patients were consigned to take $H_2$-receptor blockers for years. Given the prevalence of ulcer disease—5 to 10 percent of the world's population are affected at some point during their lifetime—it is not surprising that $H_2$-receptor blockers became the most lucrative pharmaceutical agents in the world. Major drug companies felt little incentive to explore or promote alternative models of peptic ulcer disease.

### The Bacteria and Ulcer Disease

the inflamed tissue somehow invite *H. pylori* to colonize there, or did the organisms actually cause the inflammation? Research proved the second hypothesis correct. In one of the studies, two male volunteers—Marshall included—actually ingested the organisms [*see box on next page*]. Both had healthy stomachs to start and subsequently developed gastritis. Similarly, when animals ingested *H. pylori*, gastritis ensued. In other investigations, antibiotics suppressed the infection and alleviated the irritation. If the organisms were eradicated, the inflammation went away, but if the infection recurred, so did the gastritis. We now know that virtually all people infected by *H. pylori*

ognized that most people with peptic ulcer disease also had chronic superficial gastritis. For a variety of reasons, though, when the link between *H. pylori* infection and gastritis was established, the medical profession did not guess that the bacteria might prompt peptic ulcer disease as well. Generations of medical students had learned instead that stress made the stomach produce more acid, which in turn brought on ulcers. The theory stemmed from work carried out by the German scientist K. Schwartz. In 1910, after noting that duodenal ulcers arose only in those individuals who had acid in their stomachs, he coined the phrase "No acid, no ulcer." Although gastric acidity is nec-

In fact, ulcers can result from medications called nonsteroidal anti-inflammatory agents, which include aspirin and are often used to treat chronic arthritis. But all the evidence now indicates that *H. pylori* cause almost all cases of ulcer disease that are not medication related. Indeed, nearly all patients having such ulcers are infected by *H. pylori*, versus some 30 percent of age-matched control subjects in the U.S., for example. Nearly all individuals with ulcers in the duodenum have *H. pylori* present there. Studies show that *H. pylori* infection and chronic gastritis increase from three to 12 times the risk of a peptic ulcer developing within 10 to 20 years of infection with the bacte-

## Don't Try This at Home

**B**arry J. Marshall (*left*) of the Royal Perth Hospital in Australia made headlines after he announced in 1985 that he had ingested *Helicobacter pylori*. Marshall hoped to demonstrate that the bacteria could cause peptic ulcer disease. Marshall did in fact develop a severe case of gastritis, but the painful inflammation vanished without treatment.

Two years later Arthur J. Morris and Gordon I. Nicholson of the University of Auckland in New Zealand reported the case of another volunteer who wasn't so lucky. This man, a healthy 29-year-old, showed signs of infection for only 10 days, but his condition lasted much longer. On the 67th day of infection, the volunteer started treatment with bismuth subsalicylate (Pepto-Bismol). Five weeks later a biopsy indicated that the medication had worked. But a second biopsy taken nine months after the first showed that both the infection and the gastritis had recurred. Only when the subject received two different antibiotics as well as bismuth subcitrate was his infection finally cured three years later.                              —*M.J.B.*

ria. Most important, antimicrobial medications can cure *H. pylori* infection and gastritis, thus markedly lowering the chances that a patient's ulcers will return. But few people can overcome *H. pylori* infection without specific antibiotic treatment.

When someone is exposed to *H. pylori*, his or her immune system reacts by making antibodies, molecules that can bind to and incapacitate some of the invaders. These antibodies cannot eliminate the microbes, but a blood test readily reveals the presence of antibodies, and so it is simple to detect infection. Surveys consistently show that one third to one half of the world's population carry *H. pylori*. In the U.S. and western Europe, children rarely become infected, but more than half of all 60-year-olds have the bacteria. In contrast,

60 to 70 percent of the children in developing countries show positive test results by age 10, and the infection rate remains high for adults. *H. pylori* infection is also common among institutionalized children.

Although it is as yet unclear how the organisms pass from one person to another, poor sanitation and crowding clearly facilitate the process. As living conditions have improved in many parts of the world during the past century, the rate of *H. pylori* infection has decreased, and the average age at which the infection is acquired has risen. Gastric cancer has also become progressively less common during the past 80 years. At the start of the 20th century, it was the leading cause of death from cancer in the U.S. and many other developed countries. Now it is far down on

the list. The causes for its decline are not well understood, but we have reason to believe that the drop in *H. pylori* infection rates deserves some credit.

### A Connection to Cancer

**I**n the 1970s Pelayo Correa, now at Louisiana State University Medical Center, proposed that gastric cancer resulted from a series of changes in the stomach taking place over a long period. In Correa's model, a normal stomach would initially succumb to chronic superficial gastritis for unknown reasons. We now know that *H. pylori* are to blame. In the second step—lasting for perhaps several decades—this gastritis would cause more serious harm in the form of a lesion, called atrophic gastritis. This lesion might then lead to fur-

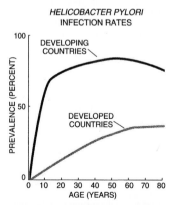

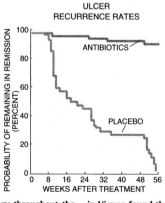

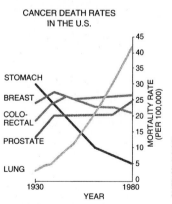

**RATES OF INFECTION** with *H. pylori* vary throughout the world. In developed countries, the infection is rare among children, but its prevalence rises with age. In developing countries, far more people are infected in all age groups (*left*). Supporting the fact that such infections cause ulcer disease, Enno Hentschel and his colleagues at Hanusch Hospital in Vienna found that antimicrobial therapy dramatically decreased the chance that a duodenal ulcer would recur (*center*). As infection rates have declined during the past century in the U.S., so, too, have the number of deaths from stomach cancer (*right*)—suggesting that *H. pylori* infection can, under some circumstances, cause that disease as well.

ther changes, among them intestinal metaplasia and dysplasia, conditions that typically precede cancer. The big mystery since finding *H. pylori* has been: Could the bacteria account for the second transition—from superficial gastritis to atrophic gastritis and possibly cancer—in Correa's model?

The first real evidence linking *H. pylori* and gastric cancer came in 1991 from three separate studies. All three had similar designs and reached the same conclusions, but I will outline the one in which I participated, working with Abraham Nomura of Kuakini Medical Center in Honolulu. I must first give some background. In 1942, a year after the bombing of Pearl Harbor, the selective service system registered young Japanese-American men in Hawaii for military service. In the mid-1960s medical investigators in Hawaii examined a large group of these men—those born between 1900 and 1919—to gain information on the epidemiology of heart disease, cancer and other ailments. By the late 1960s they had assembled a cohort of about 8,000 men, administered questionnaires and obtained and frozen blood samples. They then tracked and monitored these men for particular diseases they might develop.

For many reasons, by the time we began our study we had sufficient information on only 5,924 men from this original group. Among them, however, 137 men, or more than 2 percent, had acquired gastric cancer between 1968 and 1989. We then focused on 109 of these patients, each of whom was matched with a healthy member of the cohort. Next, we examined the blood samples frozen in the 1960s for antibodies to *H. pylori.* One strength of this study was that the samples had been taken from these men, on average, 13 years before they were diagnosed with cancer. With the results in hand, we asked the critical question: Was evidence of preexisting *H. pylori* infection associated with gastric cancer? The answer was a strong yes. Those men who had a prior infection had been six times more likely to acquire cancer during the 21-year follow-up period than had men showing no signs of infection. If we confined our analysis to cancers affecting the lower part of the stomach—an area where *H. pylori* often collect—the risk became 12 times as great.

The other two studies, led by Julie Parsonnet of Stanford University and by David Forman of the Imperial Cancer Research Fund in London, produced like findings but revealed slightly lower risks. Over the past five years, further epidemiological and pathological investigations have confirmed the associa-tion of *H. pylori* infection and gastric cancer. In June 1994 the International Agency for Research in Cancer, an arm of the World Health Organization, declared that *H. pylori* is a class-1 carcinogen—the most dangerous rank given to cancer-causing agents. An uncommon cancer of the stomach, called gastric lymphoma, also appears to be largely caused by *H. pylori.* Recent evidence suggests that antimicrobial treatment to cure *H. pylori* infection may bring about regression in a subset of tumors of this kind, which is an exciting development in both clinical medicine and cancer biology.

### How Persistence Takes Place

Certainly most bacteria cannot survive in an acidic environment, but *H. pylori* are not the only exception. Since that bacteria's discovery, scientists have isolated 11 other organisms from

dioxide. Fostering the production of ammonia may be one way helicobacters neutralize the acid in their local environment, further securing their survival.

An interesting puzzle involves what *H. pylori* eat. There are two obvious guesses: the mucus in which it lives and the food its human host ingests. But Denise Kirschner of Texas A&M University and I constructed a mathematical model showing that *H. pylori* would not be able to persist for years relying on those nutrient sources. In our model, the mathematics of persistence in the stomach requires some regulated interaction between the host cells and the bacteria. Inflammation provides one such interaction, and so I have proposed that *H. pylori* might trigger inflammation for the purpose of acquiring nutrients. An apparent paradox in *H. pylori* biology is that although the organisms do not invade the gastric tissue, they can cause irritation there. Rather, as we

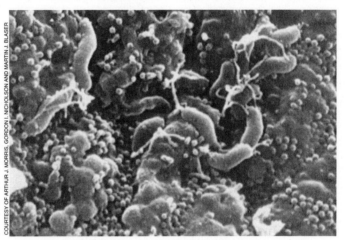

CURVED ORGANISMS, magnified 8,000 times, are *H. pylori* in the stomach of the second human volunteer to ingest the bacteria. The man had consumed the microbes 463 days earlier and developed chronic superficial gastritis as a result.

COURTESY OF ARTHUR J. MORRIS, GORDON I. NICHOLSON AND MARTIN J. BLASER

the stomachs of other primates, dogs, cats, rodents, ferrets and even cheetahs. These bacteria, for now considered to be members of the *Helicobacter* family, seem to have a common ancestor. All are spiral-shaped and highly motile (they swim well)—properties enabling them to resist the muscle contractions that regularly empty the stomach. They grow best at oxygen levels of 5 percent, matching the level found in the stomach's mucus layer (ambient air is 21 percent oxygen). In addition, these microbes all manufacture large amounts of an enzyme called urease, which cleaves urea into ammonia and carbon

and others have found, the microbes release chemicals that the stomach tissue absorbs. These compounds attract phagocytic cells, such as leukocytes and macrophages, that induce gastritis.

The host is not entirely passive while *H. pylori* bombard it with noxious substances. Humans mount an immune response, primarily by making antibodies to the microbe. This response apparently does not function well, though, because the infection and the antibodies almost inevitably coexist for decades. In essence, faced with a pathogen that cannot be easily destroyed, humans had two evolutionary options: we could have evolved to fight *H. pylori* infection to its death, possibly involving the abrogation of normal gastric function, or we could have become tolerant and tried to ignore the organisms. I believe the choice was made long ago in favor of tolerance. The response to other persistent pathogens—such as the microbes responsible for malaria and leprosy—may follow the same paradigm, in which it is adaptive for the host to dampen its immune reaction.

Fortunately, it is not in *H. pylori's* best interest to take advantage of this passivity, growing to overwhelming numbers and ultimately killing its host. First, doing so would limit the infection's opportunity to spread. Second, even in a steady state, *H. pylori* reaches vast numbers (from $10^7$ to $10^{10}$ cells) in the stomach. And third, further growth might exhaust the mechanisms keeping the immune system in check, leading to severe inflammation, atrophic gastritis and, eventually, a loss of gastric acidity. When low acidity occurs, bacteria from the intestines, such as *Escherichia coli*, are free to move upstream and colonize the stomach. Although *H. pylori* can easily live longer than *E. coli* in an acid environment, *E. coli* crowds *H. pylori* out of more neutral surroundings. So to avoid any competition with intestinal bacteria, *H. pylori* must not cause too much inflammation, thereby upsetting the acid levels in the stomach.

Are *H. pylori* symbionts that have only recently evolved into disease-causing organisms? Or are they pathogens on the long and as yet incomplete road toward symbiosis? We do not yet know, but we can learn from the biology of *Mycobacterium tuberculosis,* the agent responsible for tuberculosis. It, too, infects about one third of the world's population. But as in *H. pylori* infection, only 10 percent of all infected people become sick at some point in their life; the other 90 percent experience no symptoms whatsoever. The possible explanations fall into several main categories. Differences among microbial strains or among hosts could explain why some infected people acquire certain diseases and others do not. Environmental cofactors, such as whether someone eats well or smokes, could influence the course of infection. And the age at which someone acquires an infection might alter the risks. Each of these categories affects the outcome of *H. pylori* infection, but I will describe in the next section the microbial differences.

### Not All Bacteria Are Created Equal

Given its abundance throughout the world, it is not surprising that *H. pylori* are highly diverse at the genetic level. The sundry strains share many structural, biochemical and physiological characteristics, but they are not all equally virulent. Differences among them are associated with variations in two genes. One encodes a large protein that 60 percent of all strains produce. Our group at Vanderbilt University, comprising Murali Tummuru, Timothy L. Cover and myself, and a group at the company Biocine in Italy, led by Antonello Covacci and Rino Rappuoli, identified and cloned the gene nearly simultaneously in 1993 and by agreement called it *cagA*. Among patients suffering from chronic superficial gastritis alone, about 50 to 60 percent are infected by *H. pylori* strains having the *cagA* gene. In contrast, nearly all individuals with duodenal ulcers bear cagA strains. Recently we reexamined the results of the Hawaiian study and found that infection by a cagA strain was associated with a doubled risk of gastric cancer. Research done by Jean E. Crabtree of Leeds University in England and by the Vanderbilt group has shown that persons infected by cagA strains experience more severe inflammation and tissue injury than do those infected by strains lacking the *cagA* gene.

The other *H. pylori* gene that seems to influence disease encodes for a toxin. In 1988 Robert D. Leunk, working for Procter & Gamble—the makers of bismuth subsalicylate (Pepto-Bismol)—reported that a broth containing *H. pylori* could induce the formation of vacuoles, or small holes, in tissue cultures. In my group, Cover had clearly shown that a toxin caused this damage and that it was being made not only by *H. pylori* grown in the laboratory but also by those residing in human hosts. In 1991 we purified the toxin and confirmed Leunk's finding that only 50 to 60 percent of *H. pylori* strains produced it. Our paper was published in May 1992 and included a brief sequence of some of the amino acids that encode for the mature toxin. Based on that scanty information, within the next year four groups—two in the U.S., including our own, one in Italy and one in Germany—were able to clone the gene, which we all agreed to name vacA. The race to publish was on. Each of our four papers appeared in separate journals within a three-month period.

Lest this sounds like duplicated labor, I should point out that each team had in fact solved a different aspect of the problem. We learned, for example, that virtually all *H. pylori* strains possess *vacA*, whether or not they produce the toxin when grown in culture. We also discovered that there is an extraordinary amount of strain-to-strain variability in *vacA* itself. In addition, broth from toxin-producing strains inoculated directly into the stomach of mice brought

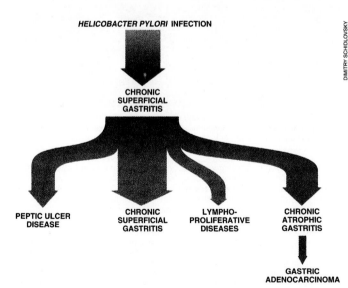

**HELICOBACTER PYLORI INFECTION**

**CHRONIC SUPERFICIAL GASTRITIS**

**PEPTIC ULCER DISEASE**

**CHRONIC SUPERFICIAL GASTRITIS**

**LYMPHO-PROLIFERATIVE DISEASES**

**CHRONIC ATROPHIC GASTRITIS**

**GASTRIC ADENOCARCINOMA**

DIMITRY SCHIDLOVSKY

*H. PYLORI* INFECTION PROGRESSES to chronic superficial gastritis within months. Left untreated, the condition persists for life in most people. A small fraction, however, may develop peptic ulcer disease, lymphoproliferative diseases or severe chronic atrophic gastritis, leading to adenocarcinoma of the stomach.

## Which Treatment Strategy Should You Choose?

| | OLD MODEL | NEW MODEL |
|---|---|---|
| **CAUSE** | **Excess stomach acid** eats through tissues and causes inflammation | *Helicobacter pylori* bacteria secrete toxins and cause inflammation in the stomach, bringing about damage |
| **TREATMENTS** | **Bland diet**, including dairy products every hour, small meals, no citrus or spicy foods and no alcohol or caffeine<br>**H₂-receptor blockers** lessen blood levels of histamine, which increases the production of stomach acid<br>**Surgery** to remove ulcers that do not respond to medication or that bleed uncontrollably. In the 1970s, it was the most common operation a surgical resident learned. Now it is increasingly rare | **Antibiotic regimen.** In February 1994 an NIH panel endorsed a two-week course of antibiotics for treating ulcer disease: amoxicillin or tetracycline, metronidazole (Flagyl) and bismuth subsalicylate (Pepto-Bismol). In December 1995 an FDA advisory committee recommended approval of two new four-week treatments, involving clarithromycin (Biaxin) with either omeprazole (Prilosec) or ranitidine bismuth citrate (Tritec). One-week therapies are also highly effective |
| **SUCCESS** | Patients who stop taking H₂-receptor blockers face a **50 percent chance** that their ulcers will recur within six months and a 95 percent chance that they will reappear within two years | **No recurrence** after the underlying bacterial infection is eliminated |
| **COST** | H₂-receptor blockers cost from $60 to $100 per month, adding up to **thousands of dollars** over decades of care. Surgery can cost as much as $18,000 | **Less than $200** for a standard one-week therapy |

ROBERTO OSTI (*drawings*); LISA BURNETT

about substantial injury. Strains that produce the toxin are some 30 to 40 percent overrepresented in ulcer patients compared with those having gastritis alone. And toxigenic strains usually but not always contain *cagA*, which is located far away from *vacA* on the chromosome.

### Slow-Acting Bacteria and Disease

Over the past 15 years, researchers and physicians have learned a good deal about *H. pylori*. This knowledge has revolutionized our understanding of gastritis, formerly thought to represent the aging stomach, and of peptic ulcer disease and gastric cancer. It has made possible new treatments and screening methods. In addition, a new field of study has emerged—the microbiology and immunology of the human stomach—that will undoubtedly reveal more about persistent infections within mucosal surfaces.

But let us extrapolate from these findings. Consider that slow-acting bacteria, *H. pylori*, cause a chronic inflammatory process, peptic ulcer disease, that was heretofore considered metabolic. And also keep in mind that this infection greatly enhances the risk of neo-

plasms developing, such as adenocarcinomas and lymphomas. It seems reasonable, then, to suggest that persistent microbes may be involved in the etiology of other chronic inflammatory diseases of unknown origin, such as ulcerative colitis, Crohn's disease, sarcoidosis, Wegener's granulomatosis, systemic lupus erythematosus and psoriasis, as well as various neoplasms, including carcinomas of the colon, pancreas and prostate. I believe *H. pylori* are very likely the first in a class of slow-acting bacteria that may well account for a number of perplexing diseases that we are facing today.

### The Author

MARTIN J. BLASER has been the Addison B. Scoville Professor of Medicine and director of the Division of Infectious Diseases at Vanderbilt University and at the Nashville Veterans Affairs Medical Center since 1989. He has worked at the Rockefeller University, the University of Colorado, the Denver Veterans Administration Medical Center, the Centers for Disease Control and Prevention, and St. Paul's Hospital in Addis Ababa, Ethiopia. He received a B.A. with honors in economics from the University of Pennsylvania in 1969 and an M.D. from New York University in 1973. He holds several patents and is a member of numerous professional societies and editorial boards. He has written more than 300 articles and edited several books.

### Further Reading

UNIDENTIFIED CURVED BACILLI IN THE STOMACH OF PATIENTS WITH GASTRITIS AND PEPTIC ULCERATION. Barry J. Marshall and J. Robin Warren in *Lancet*, No. 8390, pages 1311-1315; June 16, 1984.

HELICOBACTER PYLORI INFECTION AND GASTRIC CARCINOMA AMONG JAPANESE AMERICANS IN HAWAII. A. Nomura, G. N. Stemmermann, P.-H. Chyou, I. Kato, G. I. Perez-Perez and M. J. Blaser in *New England Journal of Medicine*, Vol. 325, No. 16, pages 1132-1136; October 17, 1991.

HUMAN GASTRIC CARCINOGENESIS: A MULTISTEP AND MULTIFACTORIAL PROCESS. Pelayo Correa in *Cancer Research*, Vol. 52, No. 24, pages 6735-6740; December 15, 1992.

EFFECT OF RANITIDINE AND AMOXICILLIN PLUS METRONIDAZOLE ON THE ERADICATION OF HELICOBACTER PYLORI AND THE RECURRENCE OF DUODENAL ULCER. Enno Hentschel et al. in *New England Journal of Medicine*, Vol. 328, No. 5, pages 308-312; February 4, 1993.

REGRESSION OF PRIMARY LOW-GRADE B-CELL GASTRIC LYMPHOMA OF MUCOSA-ASSOCIATED LYMPHOID TISSUE TYPE AFTER ERADICATION OF HELICOBACTER PYLORI. A. C. Wotherspoon et al. in *Lancet*, Vol. 342, No. 8871, pages 575-577; September 4, 1993.

PARASITISM BY THE "SLOW" BACTERIUM HELICOBACTER PYLORI LEADS TO ALTERED GASTRIC HOMEOSTASIS AND NEOPLASIA. Martin J. Blaser and Julie Parsonnet in *Journal of Clinical Investigation*, Vol. 94, No. 1, pages 4-8; July 1994.

# ALIMENTERICS

# PYLORI-CHEK
# BREATH TEST KIT

08-0190 Rev 0C

 **Alimenterics Inc.**
**301 American Road**
**Morris Plains, NJ 07950 U.S.A.**

Alimenterics Inc., 1999

## ALIMENTERICS PYLORI-CHEK BREATH TEST KIT

### I. INTENDED USE

The Pylori-Chek Test System is intended for use with the LARA Laser Assisted Ratio Analyzer for the qualitative detection of urease associated with Helicobacter pylori infection in the human stomach and as an aid in the diagnosis of H. pylori infection in symptomatic adult patients. The Pylori-Chek Test system consists of a Pylori-Chek test kit for the collection of breath samples and a LARA Laser Assisted Ratio Analyzer for the measurement of the ratio of $^{13}CO_2$ to $^{12}CO_2$ in the breath samples.

For use by healthcare professionals. To be administered under a physician's supervision.

### II. SUMMARY AND EXPLANATION OF THE TEST

Peptic ulcer disease is a chronic inflammatory condition of the stomach and duodenum. Despite the fact that the disease has relatively low mortality, it results in substantial suffering of those affected.

A strong association between *H. pylori*, and chronic superficial gastric and gastrointestinal disease has been well established. It is associated with type B gastritis, [2,3] duodenal ulcer, [4,5] gastric ulcer, [6,7] gastric cancer [8] and non-Hodgkin's lymphoma.[8]

H. pylori was first cultured from human gastric mucosa in 1982.[1] The organisms, spiral gram negative bacteria, are found in the human stomach between the gastric epithelium and the mucosa. Isolates implicated in the above mentioned disease states are distinguished by the production of copious amounts of endogenous urea amidohydrolase (urease). [9,10] The enzyme catalyzes the breakdown of urea to carbon dioxide and ammonia, which are absorbed into the bloodstream.

Several methods are employed to determine the presence of *H.pylori* in the gastrointestinal tract. Histologic staining of biopsy tissue using various stains has been shown to give adequate results with specificity of over 90%.[11] Mucosal biopsy samples can be cultured using non-selective enriched media. However, due to the exacting needs of the organism, culture is the least sensitive (70-80%)[11] of all available techniques. Direct detection of urease activity of biopsy specimens is achieved by placing tissue in Christensen's urea agar and observing a color change. Biopsy and its associated analytic techniques are invasive and often not well tolerated by patients. Non-invasive tests consist of serum assays for IgG antibodies against *H. pylori*.[11]

However, these are not a reliable indicator of current infection. Urea breath tests using the radioactive isotope $^{14}C$ or the non-radioactive isotope $^{13}C$ labeled urea can detect current infection with *H. pylori*, and have been shown to be highly sensitive.[12,13]

## III.   PRINCIPLE OF TEST

The Alimenterics Pylori-Chek Breath Test is based on the ability of the organism, *H. pylori*, to produce large amounts of the enzyme urease, which hydrolyses urea to $NH_4^+$ and $HCO_3$, the latter being exhaled as $CO_2$. Using non-radioactive $^{13}C$-labeled urea, an increase in the ratio of $^{13}CO_2$ to $^{12}CO_2$ over time is an indication of the presence of *H. pylori*. This change can be detected by isotope ratio analyzers that measure pre-ingestion and post-ingestion samples of the patient's breath.

Each Pylori-Chek test kit comes with a jar containing 100 mg of white crystalline powder which is reconstituted using water just prior to taking the test. The powder is urea (see structure and molecular weight given below), labeled with the non-radioactive isotope $^{13}$carbon ($^{13}C$-urea).

Name and structure:

Urea (non-radioactive $^{13}$Carbon-labeled ($^{13}C$), 99%) $^{13}CH_4N_2O$

$$\underset{\displaystyle H_2N-\overset{\displaystyle 13}{C}-NH_2}{\overset{\displaystyle O}{\overset{\displaystyle \|}{\phantom{x}}}}$$

Molecular Weight 61.05

The non-radioactive $^{13}C$-labeled urea preparation contains no additives or bulking compounds.

The test procedure involves collection of a sample of the patient's breath to determine the baseline ratio of $^{13}CO_2$ and $^{12}CO_2$. Following ingestion of a test meal and an aqueous solution containing 100 mg non-radioactive $^{13}C$-labeled urea, two additional breath samples are collected, one at thirty and one at sixty minutes. The samples are then introduced into the LARA System which measures the ratio of $^{13}CO_2$ to $^{12}CO_2$. The results of the three readings are analyzed concurrently by the LARA System. Test values are generated and a report is printed for each patient as indicated in the test report section.

## IV. WARNINGS AND PRECAUTIONS

1. For <u>in vitro</u> diagnostic use only. The Pylori-Chek urea solution is ingested as part of the diagnostic procedure.

2. A negative Pylori-Chek breath test result alone does not rule out the possibility of *H.pylori* infection. Typically, with this procedure, false negative results occur at a rate less than 1.5%. Always evaluate the Pylori-Chek Breath Test results along with clinical signs and patient history when diagnosing *H.pylori* related disease. If clinical signs and patient history are suggestive of *H.pylori* infection and the Pylori-Chek Breath Test result is negative, retest with a new sample or an alternative method.

3. The validity of the Pylori-Chek test depends on using the LARA to analyze breath samples captured within collectors supplied with the test kit. Since using other methods may change the precision and accuracy of the results, the manufacturer does not recommend using any alternative analytical methods or other urea breath test kits in conjunction with the Pylori-Chek Test Kit. The LARA software is only valid for analysis of Pylori-Chek tests.

4. Failure of the patient to fast as directed may affect the test results.

5. Antimicrobials, omeprazole, and bismuth preparations suppress *H.pylori*. Ingestion of these substances within four (4) weeks prior to performing the Alimenterics Pylori-Chek Breath Test may lead to false negative results.

6. False positive results may occur in patients with achlorhydria or who have gastric spiral organisms such as *Helicobacter hominis*.

7. Lack of sufficient $CO_2$ (less than 2.0%) or too much $CO_2$ (more than 6.0%) in any sample may lead to a non-evaluable test. This can be avoided by insuring that the breath collector is closed tightly after obtaining the samples.

## V. SHELF LIFE, STORAGE AND HANDLING OF KIT

The shelf life for the urea powder is twenty (20) months after date of manufacture. Use the urea solution within five minutes after reconstitution.

Store the Pylori-Chek Breath Test Kit between 20°C and 25°C (68$^0$F – 77$^0$F). The non-radioactive $^{13}$C-urea and the sterile water have expiration dates. Do not use these materials beyond the expiration date stated on their labels.

**VI.**     **PATIENT PREPARATION**

It is important that the patient fast 6 (six) hours before performing the Alimenterics Pylori-Chek Breath Test.  In addition, in the four weeks prior to performing the test, the patient must avoid the use of antimicrobials, omeprazole, and bismuth preparations, which are known to suppress *H.pylori*.

**VII.**     **LIST OF REQUIRED MATERIALS PROVIDED**

The Alimenterics Pylori-Chek Breath Test Kit, with sufficient material for one determination, consists of:

1.     One vial containing 100 mg non-radioactive $^{13}$C-labeled urea (powder)
2.     One bottle containing 50 mL purified water for reconstituting the urea.
3.     Three breath collectors, each with an affixed bar code label
4.     Three patient labels (labeled 1, 2, and 3)
5.     One reusable pouch (containing the three breath collectors and 3 patient labels)
6.     Package insert

*Note: A LARA Analyzer is required for analysis of breath samples.*

**VIII.**     **LIST OF REQUIRED MATERIALS NOT PROVIDED**

1.     One 8 oz. test meal (Ensure vanilla-flavored liquid)
2.     A timer capable of timing an interval up to thirty (30) minutes

**IX.**     **STEP-BY-STEP PROCEDURE**

*Note: Time intervals indicated in this procedure are critical.*

1.     Verify that the patient has been prepared as specified in Section VII.

2.     Open the Alimenterics Pylori-Chek Breath Test Kit, which should contain all materials listed in Section VIII.

3.     Remove one breath collector from reusable pouch.

4.     Ask the patient to provide a baseline measurement.  Note that there is a narrow and a wide end of the breath collector.  Instruct the patient to:

- Take a deep breath
- Pause momentarily
- Place the lips firmly around the narrow end of the breath collector
- Exhale into the breath collector
- Close the breath collector toward the end of exhalation by twisting the far end of the breath collector in a counter-clockwise direction.

Make sure the collector is closed tightly after it is removed from the mouth.

5.  Record patient name, date and time on the No. 1 patient label and affix patient label to the breath collector that now contains the baseline specimen. **Do not cover the bar code on the breath collector.**

6.  Instruct the patient to consume an entire 8 oz. test meal (Ensure-vanilla flavored liquid).

7.  Prepare the Pylori-Chek solution by:

- Carefully opening the 50 ml bottle of purified water and adding it to the jar containing the non-radioactive $^{13}C$-urea powder.

- Recapping the jar and mixing by inversion until all powder is dissolved.

*Note: Reconstituted urea must be used within 5 minutes. Urea slowly decomposes in solution.*

8.  Instruct the patient to drink all of the Pylori-Chek Urea solution. After the patient ingests the solution, set the timer for 30 minutes. Instruct the patient to sit calmly for the 30-minute interval. The patient should not eat, drink, or smoke during this time period.

9.  Remove second breath collector from the reusable pouch.

10. Thirty minutes after the patient has ingested the urea solution, ask the patient to provide a second breath sample in breath collector number 2 in the same manner as outlined in paragraph 4 above. After the patient has provided the second breath sample, re-set the timer for 30 minutes.

Instruct the patient to sit calmly for the 30-minute interval. The patient should not eat, drink, or smoke during this time period.

11.     Record patient name, date and time on the No. 2 patient label and affix patient label to the breath collector that now contains the 30-minute specimen. Do not cover the bar code on the breath collector.

12.     Remove third breath collector from the reusable pouch.

13.     Thirty minutes after the second sample, ask the patient to provide a third breath sample in breath collector number 3 in a similar manner as outlined in paragraph 4 above.

14.     Record patient name, date and time on the No. 3 patient label and affix patient label to the breath collector that now contains the 60-minute specimen. Do not cover the bar code on the breath collector.

15.     Place the three breath samples in the reusable pouch and seal and deliver to laboratory for analysis.

16.     Directions for analyzing the samples using the LARA Analyzer are provided in the LARA Analyzer Operator's Manual.

**X.     QUALITY CONTROL**

As detailed in the LARA Analyzer Operator's Manual, the accuracy of system results is assured by several quality controls that are designed to eliminate or detect measurement errors. Internal calibrations are performed continuously when the system is not in use. Positive and negative control samples are run periodically to verify proper calibration of the instrument.

Both sets of controls are labeled with expected values which vary slightly from one lot to the next. Typically, positive controls have $\delta$ values above 12 and negative controls have $\delta$ values near zero. See the next section (Interpretation of Results) for a definition of delta value. Each control kit is labeled with its expected $\delta$ value. A control measurement fails when the value measured by the LARA falls outside $\pm 3.0$ $\delta$ units from the control's expected value. When a control measurement fails, the system will alert the operator and issue a warning if the operator attempts to run patient samples. The operator should recalibrate the LARA system and then run another control test. If the second control measurement fails, the operator should mark the instrument "Out of Service" and call Alimenterics for repair.

Each breath specimen must contain at least 2% $CO_2$ and no more than 6% $CO_2$. Breath Specimen levels outside this range are rejected with the output message "$CO_2$ range error."

## XI.    INTERPRETATION OF RESULTS

### 1.    Calculation of Test Results

The LARA Analyzer takes one reading from each breath collector to determine a $\delta$ value for each sample. The $\delta$ value is the ratio of $^{13}CO_2$ to $^{12}CO_2$ within gas extracted from the breath collector, expressed in parts per thousand. The baseline reading is then compared to those at 30 and 60 minutes ($\delta_{30}$ and $\delta_{60}$).

There are three possible outcomes from a test:

1) A test is considered positive for the presence of urease associated with H. pylori infection when either $\delta_{30}$ or $\delta_{60}$ are greater than 6.7;
2) A test is considered negative for the presence of urease associated with H. pylori infection when both $\delta_{30}$ and $\delta_{60}$ are less than or equal to 5.5;
3) A test is considered indeterminate for the presence of urease associated with H. pylori infection when both $\delta_{30}$ and $\delta_{60}$ are between 5.5 and 6.7 or when one value is between 5.5 and 6.7 and the other is less than or equal to 5.5. If possible, the patient should be asked to return and repeat the test.

These possibilities are summarized in the table shown below:

| $\delta_{30}$ | $\delta_{60}$ | determination |
|---|---|---|
| > 6.7 | any value | positive |
| any value | > 6.7 | positive |
| < = 5.5 | < = 5.5 | negative |
| between 5.5 and 6.7 | between 5.5 and 6.7 | indeterminate |
| between 5.5 and 6.7 | < = 5.5 | indeterminate |
| < = 5.5 | between 5.5 and 6.7 | indeterminate |

Sample Calculation 1:

    Test results:        $\delta_{base} = 1.0\ \delta$
                               $\delta_{30} = 4.0\ \delta$
                                     $\delta_{60} = 9.0\ \delta$

    Calculation:

                    $\delta_{60} - \delta_{base} = 8.0\ \delta$

    Result:        Positive (one or both values above 6.7 $\delta$).
                      Note: When one value is greater than 6.7 $\delta$, the test is automatically positive. The other value may be in or below the indeterminate zone or may not be defined in the case of a $CO_2$ range error.

Sample Calculation 2:

    Test results:        $\delta_{base} = 1.0\ \delta$
                               $\delta_{30} = 4.0\ \delta$
                               $\delta_{60} = 3.0\ \delta$

    Calculation:

                    $\delta_{30} - \delta_{base} = 3.0\ \delta$
                    $\delta_{60} - \delta_{base} = 2.0\ \delta$

    Result:       Negative (both values below or equal to 5.5 $\delta$)

Sample Calculation 3:

    Test results:        $\delta_{base} = 0.0$
                               $\delta_{30} = 5.7\ \delta$
                               $\delta_{60} = 3.0\ \delta$

    Calculation:

                    $\delta_{30} - \delta_{base} = 5.7\ \delta$
                    $\delta_{60} - \delta_{base} = 3.0\ \delta$

    Result:                  Indeterminate (one value in grey zone, 5.5 to 6.7 $\delta$, the other below 5.5 $\delta$)

Sample Calculation 4:

Test Results:          $\delta_{base} = 0.0$
                       $\delta_{30} = 5.7\ \delta$
                       $\delta_{60} = 6.3\ \delta$

Calculation:
                       $\delta_{30} - \delta_{base} = 5.7\ \delta$
                       $\delta_{60} - \delta_{base} = 6.3\ \delta$

Result:                Indeterminate (both values in grey zone, 5.5 to 6.7).

## 2. Test Report

The LARA Analyzer prints a report that records the type of test performed, the sample identification, the date and time, the $\delta$ values for the 30 and 60 minute samples and the test result; positive, negative or indeterminate. A negative test result alone does not rule out the possibility of *H.pylori* infection. For a negative result, if clinical signs and patient history are suggestive of *H.pylori* infection, and for an indeterminate result, the patient should be asked to return and repeat the test.

An unevaluable test result may be obtained when the percent $CO_2$ in a sample falls below 2% or above 6%. Such a result would be indicated as UAP (unable to process) on the report. These patients should be retested if possible, by obtaining fresh samples.

## 3. Determination of a Cut-Off Point

The cut-off point is the Alimenterics Pylori-Chek Breath Test Result ($\delta$ value) above which patients are considered to be infected with *H.pylori*. For the Alimenterics Pylori-Chek Breath Test, the delta cut-off point was determined to be 6.1 $\delta$ in a clinical study of 395 evaluable patients (including 235 infected patients and 160 uninfected patients). The reference standards were bacterial culture, histopathology and CLO test. The cut-off point was determined by calculating the breath test result ($\delta$ value) which best distinguished those patients that were determined to be negative and positive using the reference standard.

**XII.     EXPECTED VALUES**

The range of Alimenterics Pylori-Chek Breath Test results for the uninfected group was less than 5.5 δ.  Figure 1 is a histogram for the distribution of results from the uninfected and infected patients.

## XIII. PERFORMANCE CHARACTERISTICS

### 1.     Precision of the LARA Analyzer

To estimate precision, a LARA continuously measured control samples over an 8 hour period.  This experiment simulates worst-case conditions of continuous operation without recalibration.  Sixty (60) negative and 60 positive controls were measured alternately, at levels of 0.0 δ and 12.7 δ, respectively.

The mean, standard deviation and coefficient of variation for all measurements during the 8 hour period are shown below:

| CONTROL TYPE | MEAN (δ) | STANDARD DEVIATION | COEFFICIENT OF VARIATION |
|---|---|---|---|
| Negative | -0.24 | 0.62 | undefined |
| Positive | 12.31 | 0.74 | 6.0% |

### 2.     Clinical Trials

#### A.     Study Protocol

The data presented in this section were collected from a clinical trial conducted at seven sites in both Europe and the United States.  Patients who were referred for upper gastrointestinal endoscopy were eligible to enter the study, regardless of whether the patients had a history of ulcer.  Of the 432 patients who were enrolled in the study, 398 successfully completed the Pylori-Chek breath test.  Of these, 31 were diagnosed with duodenal ulcers, 19 were diagnosed with gastric ulcers, 393 were diagnosed with gastritis and 42 were diagnosed as normal.  Including the breath test, patients who entered the study were tested for *H.pylori* infection using the following four diagnostic methods:

<u>Histopathology</u>.  Two biopsy specimens obtained from endoscopy were evaluated by an experienced pathologist using hematoxylin-eosin stain and the Warthin-Starry methods.

Bacterial Culture.  Biopsy tissue was cultured using selective and nonselective media at 37 °C.  Samples were examined every 3 days for 12 days.  *H.pylori* were identified on the basis of gram morphology and production of cytochrome oxidase, catalase and urease.

CLOtest®.  A biopsy specimen obtained from endoscopy was tested for urease activity in accordance with the CLOtest® instructions.

Alimenterics Pylori-Chek Breath Test.  The breath test was performed in accordance with the instructions described in this package insert.

B.      Study Results

This section compares the results of the Alimenterics Pylori-Chek Breath Test to the results obtained with other reference standards: histology, bacterial culture and CLOtest® for urease activity.  Tables 1 and 2 compare the Alimenterics Pylori-Chek Breath Test results to histology and CLOtest®, respectively.  Table 3 compares the Alimenterics LARA™ Breath Test results to the result determined by a combination of reference methods (i.e., a patient was considered positive if either the culture was positive or both the CLOtest® and histology were positive).

**Table 1**
**Comparison of Pylori-Chek Breath Test**
**to Histology**

| Histology Results | Pylori-Chek Breath Test Results | | | |
|---|---|---|---|---|
| | Positive | Negative | Indeterminate | Total |
| Positive | 215 | 12 | 4 | 231 |
| Negative | 6 | 143 | 8 | 157 |
| Total | 221 | 155 | 12 | 388   (Note 1) |

Sensitivity: 94.7%    CI(97.2-90.9%)
Specificity: 96.0%    CI(98.5-91.4%)

Note 1 - Different numbers of patients successfully completed histopathology, bacterial culture and CLOtest®.  Depending on the definition of the reference, this excluded some of the 398 patients who successfully completed the breath test.

**Table 2**
**Comparison of Pylori-Chek Breath Test**
**to CLOtest®**

| CLOtest® Results | Pylori-Chek Breath Test Results | | | |
|---|---|---|---|---|
| | Positive | Negative | Indeterminate | Total |
| Positive | 219 | 10 | 3 | 232 |
| Negative | 6 | 146 | 9 | 161 |
| Total | 225 | 156 | 12 | 393  (Note 1) |

Sensitivity: 95.6%    CI(97.9-92.1%)
Specificity: 96.1%    CI(98.5-91.6%)

Note 1 - Different numbers of patients successfully completed histopathology, bacterial culture and CLOtest®.  Depending on the definition of the reference, this excluded some of the 398 patients who successfully completed the breath test.

**Table 3**
**Comparison of Pylori-Chek Breath Test**
**to Combined Reference Methods**

| Culture Positive or Histology and CLO Test® Positive | Pylori-Chek Breath Test Results | | | |
|---|---|---|---|---|
| | Positive | Negative | Indeterminate | Total |
| Positive | 221 | 14 | 4 | 239 |
| Negative | 5 | 144 | 8 | 157 |
| Total | 226 | 158 | 12 | 396  (Note 1) |

Sensitivity: 94.0%    CI(96.7-90.2%)
Specificity: 96.6%    CI(98.9-92.3%)

Note 1 - Different numbers of patients successfully completed histopathology, bacterial culture and CLOtest®. Depending on the definition of the reference, this excluded some of the 398 patients who successfully completed the breath test.

## XIV. UNABLE TO PROCESS

An "Unable to Process" result may occur when insufficient $CO_2$ (less than 2%) is found in the patient breath collector.

This condition is usually caused by either improper usage or closure of the breath collector by the patient.

In a study of 430 patients, the results for 30 patients or 7% were unavailable because of "Unable to Process" samples. Nearly half of those samples were from baseline breath collectors where patients were most unfamiliar with their use.

In a few instances, too much $CO_2$ (greater than 6%) was provided by the patient, again yielding an "Unable to Process" result. This can be avoided by insuring that the patient does not smoke before or during the test.

If an "unable to process" result occurs, the physician may ask the patient to return for a second test.

## XV. LIMITATIONS OF THE TEST

1. A correlation between the number of *H. pylori* organisms in the stomach and the $\delta$ values has not been established.

2. The performance characteristics of the test have not been established for monitoring the efficacy of antimicrobial therapies for the treatment of *H. pylori* infection.

3. The performance characteristics of this test for persons under the age of 18 and over the age of 75 have not been established.

4. The breath specimen integrity due to storage of breath samples in breath collectors under ambient conditions has not been determined beyond 30 days.

5. Pylori-Chek Breath Test Kit should be used only to evaluate patients

with clinical signs and symptoms suggestive of gastrointestinal disease and is not intended for use with asymptomatic patients.

## XVI.   BIBLIOGRAPHY

1.    Marshall BJ, Royce H, Annear DI, Goodwin CS, Pearman JW, Armstrong JA.  Original isolation of *Campylobacter pyloridis* from human gastric mucosa.        Microbios. Lett. 25: 83088, 1983.

2.    Andersen LP, Holck S, Poulsen CO, Elsborg L, Justesen T. *Campylobacter pyloridis* in peptic ulcer disease.  Scand J. Gastroenterol. 22 : 219-224, 1987.

3.    Blaser MJ. Gastric *Campylobacter*-like organisms, gastritis and peptic ulcer disease. Gastroenterol. 93 : 971-983, 987.

4.    Langenberg ML, Tytgat GNJ, Schipper MEI, Rietra PJGM, Zanen HC. Campylobacter-like organisms in the stomach of patients and healthy individuals.  Lancet I: 1348, 1984.

5.    Marshall BJ, Guerrant RL, Plankey MW, *et al.* Comparison of $^{14}$C-urea breath test, microbiology and histology for the diagnosis of *Campylobacter pylori* (Abstract). Gastroenterol; 94: A284, 1988.

6.    Graham DY, Klein PD, Opekun AR, Boutton TW.  Effect of age on the frequency of active *Campylobacter pylor*i diagnosed by $^{13}$C urea breath test in normal subjects and patients with peptic ulcer disease.  J. Infect. Dis. 157: 777-780, 1988.

7.    Marshall BJ, Warren JR.  Unidentified curved bacilli in the stomach of patients with gastritis and peptic ulceration.  Lancet 1: 1311-15, 1984.

8.    Parsonnet J, Hansen S. RodriguexL, Gelb AB, Warnke RA, Jellum E. Orentreich N, Vogelman JH, Freidman GD.  *Helicobacter pylori* infection and gastric lymphoma.  The New England Journal of Medicine 330:  1267-1271, May 1994.

9.    Hazell SL, Borody TJ, Gal A, Lee A. *Campylobacter pyloridis* gastritis: detection of urease as a marker of bacterial colonization.  Am. J. Gastroenterol. 82: 292-296, 1987.

10.  Marshall BJ, Warren JR, Grancis GS, *et al.* Rapid urease test in the management of *Campylobacter pyloridis*-associated gastritis. Am. J. Gastroenterol. 82: 200-210, 1987.

11.  Helicobacter pylori in peptic ulcer disease. Consensus Development Conference Statement, National Institutes of Health. Feb. 7-9, 1994.

12.  Rauws EAJ,. Van Royen EA, Tytgat GNJ, et al. $^{14}$C-urea breath test in C. *pylori* gastritis. Gut 30: 7908-803, 1989.

13.  Eggers RH, Kulp A, Ludtke FE, Bauer FEE. Characterization of the LARA™ breath test for the diagnosis of *Campylobacter pylori* infections. Stable Isotopes in Pediatric Nutritional and Metabolic Research: 295-301, 1990.

Alimenterics Inc.
301 American Road
Morris Plains, N.J.  07950

Revision No:_____                              Rev. Approval Date: _____

# Primary care

## Treating *Helicobacter pylori* infection in primary care patients with uninvestigated dyspepsia: the Canadian adult dyspepsia empiric treatment–*Helicobacter pylori* positive (CADET-*Hp*) randomised controlled trial

Naoki Chiba, Sander J O Veldhuyzen van Zanten, Paul Sinclair, Ralph A Ferguson, Sergio Escobedo, Eileen Grace

## Abstract

**Objective** To determine whether a "test for *Helicobacter pylori* and treat" strategy improves symptoms in patients with uninvestigated dyspepsia in primary care.

**Design** Randomised placebo controlled trial.

**Setting** 36 family practices in Canada.

**Participants** 294 patients positive for *H pylori* ([13]C-urea breath test) with symptoms of dyspepsia of at least moderate severity in the preceding month.

**Intervention** Participants were randomised to twice daily treatment for 7 days with omeprazole 20 mg, metronidazole 500 mg, and clarithromycin 250 mg or omeprazole 20 mg, placebo metronidazole, and placebo clarithromycin. Patients were then managed by their family physicians according to their usual care.

**Main outcome measures** Treatment success defined as no symptoms or minimal symptoms of dyspepsia at the end of one year. Societal healthcare costs collected prospectively for a secondary evaluation of actual mean costs.

**Results** In the intention to treat population (n=294), eradication treatment was significantly more effective than placebo in achieving treatment success (50% *v* 36%; P=0.02; absolute risk reduction=14%; number needed to treat=7, 95% confidence interval 4 to 63). Eradication treatment cured *H pylori* infection in 80% of evaluable patients. Treatment success at one year was greater in patients negative for *H pylori* than in those positive for *H pylori* (54% *v* 39%; P=0.02). Eradication treatment reduced mean annual cost by $C53 (−86 to 180) per patient.

**Conclusions** A "test for *H pylori* with [13]C-urea breath test and eradicate" strategy shows significant symptomatic benefit at 12 months in the management of primary care patients with uninvestigated dyspepsia.

## Introduction

Dyspepsia is a common condition that affects up to 40% of the general population and has adverse effects on quality of life.[1] In Canada, 7% of visits to family practitioners are for dyspepsia.[2] Most patients presenting with upper gastrointestinal symptoms in primary care are uninvestigated, and the cause of the symptoms is usually unknown. The differential diagnoses include functional dyspepsia, peptic ulcer disease, gastro-oesophageal reflux disease, and (rarely) gastric cancer. Family practitioners are comfortable treating patients without an initial diagnosis, prescribing up to 2.5 courses of empirical drug treatment before referring the patient for investigations.[2] In most (up to 60%) of these patients, results of investigations are normal and the diagnosis is functional dyspepsia.[3] Whether treatment to eradicate *Helicobacter pylori* in functional (that is, investigated) dyspepsia is beneficial has been controversial; positive and negative trials have been reported.[4][5]

A suggested strategy for managing uninvestigated dyspepsia is to screen patients aged under 50 without alarm symptoms with a non-invasive test for *H pylori* and to treat patients with positive results with drugs to eradicate *H pylori*.[6] As this recommendation is not based on evidence from randomised controlled trials, we undertook a study to determine whether a non-invasive *H pylori* "test and treat" strategy in primary care for adult patients of any age with uninvestigated dyspepsia would result in improvement or cure of dyspepsia over one year.

## Methods

This was a double blind placebo controlled parallel group multicentre randomised trial, performed in 36 family practitioner centres across Canada between September 1997 and April 1999. Local ethics committees approved the study protocol, and each participant gave written informed consent.

### Selection of patients

Patients were eligible if they were aged 18 years or over with uninvestigated symptoms of dyspepsia for at least the previous three months. We defined dyspepsia as a symptom complex of epigastric pain or discomfort thought to originate in the upper gastrointestinal tract

Division of Gastroenterology, McMaster University, Hamilton, ON, Canada L8N 3Z5
Naoki Chiba
*associate clinical professor of medicine*

Division of Gastroenterology, Dalhousie University, Halifax, NS, Canada B3H 2Y9
Sander J O Veldhuyzen van Zanten
*professor of medicine*

AstraZeneca Canada Inc, 1004 Middlegate Road, Mississauga, ON, Canada L4Y 1M4
Paul Sinclair
*research scientist*
Ralph A Ferguson
*research scientist*
Sergio Escobedo
*statistician*
Eileen Grace
*health economist*

Correspondence to:
N Chiba, Surrey GI Clinic/Research, 105-21 Surrey Street West, Guelph, ON, Canada N1H 3R3
chiban@on.aibn.com

bmj.com 2002;324:1012

and including any of the following additional symptoms: heartburn, acid regurgitation, excessive burping or belching, increased abdominal bloating, nausea, feeling of abnormal or slow digestion, or early satiety.[7 8] Patients with only heartburn, regurgitation, or both were considered to have a diagnosis of gastro-oesophageal reflux disease and were excluded. We also excluded patients investigated by upper gastrointestinal endoscopy, barium study, or both less than six months before randomisation or on more than two separate occasions within the preceding 10 years and patients given eradication therapy for *H pylori* less than six months before randomisation.

We excluded patients who had previous gastric surgery, previously documented ulcer disease or endoscopic oesophagitis, irritable bowel syndrome, or clinically significant laboratory abnormalities. We did not permit a course of treatment within 30 days before randomisation or during the treatment period with a non-steroidal anti-inflammatory drug, aspirin (>325 mg/day), antibiotic, $H_2$ receptor antagonist, proton pump inhibitor, misoprostol, sucralfate, prokinetic agent, or bismuth compound. Women of childbearing potential had to have a negative pregnancy test at baseline and maintain effective contraception.

We performed the Helisal rapid blood test (Cortecs Diagnostics, Deeside, UK) at the pre-entry visit as an initial screening test to exclude patients negative for *H pylori*.[9] Patients had to have both a positive Helisal test result and a positive $^{13}C$-urea breath test result before randomisation.[10]

## Randomisation and interventions

A computer randomisation was generated in blocks of four consecutive patients and given to each centre in sealed, sequentially numbered envelopes. Active and placebo medications were identical in appearance and were packaged into blister packages placed in a sealed box by non-study personnel. The randomisation code was broken only at the end of the study after the database was locked.

We allocated patients randomly to either omeprazole 20 mg, metronidazole 500 mg, and clarithromycin 250 mg ("eradication arm") or omeprazole 20 mg, placebo metronidazole, and placebo clarithromycin ("placebo arm") twice daily for seven days. The follow up period was 12 months, with assessments at monthly intervals. During these clinic and telephone visits, the study coordinator interviewed the patients. We did not include these scheduled visits in the economic analysis. We repeated the $^{13}C$-urea breath test at three months and 12 months after the end of treatment to determine *H pylori* status. Investigators remained blinded to results of breath tests throughout the study.

During follow up, patients were managed by their family practitioners according to their usual clinical practice. Recurrent dyspepsia during follow up did not result in discontinuation from the study. Endoscopy or barium radiography was not performed at the beginning of the study but could be done during follow up at the family practitioners' discretion. Family practitioners could prescribe *H pylori* eradication treatment and other treatments such as $H_2$ antagonists or proton pump inhibitors as clinically indicated. Information about drugs consumed, tests performed, and all adverse events was recorded.

### Adherence to drugs

Patients were considered adherent by pill count if 12 of the 14 doses were taken during the treatment phase. No patient was withdrawn as a result of poor adherence.

### Outcome measures

*Global overall symptoms of dyspepsia*
We assessed the global overall severity of dyspepsia symptoms over the preceding four weeks by using the following seven point Likert-type scale (GOS scale): (1) no problem; (2) minimal problem—can be easily ignored without effort; (3) mild problem—can be ignored with effort; (4) moderate problem—cannot be ignored but does not influence daily activities; (5) moderately severe problem—cannot be ignored and occasionally limits daily activities; (6) severe problem—cannot be ignored and often limits concentration on daily activities; (7) very severe problem—cannot be ignored, markedly limits daily activities, and often requires rest. This seven point scale was amended from previously validated five point and seven point scales.[11 12]

All enrolled patients had epigastric pain or discomfort and a symptom score of at least moderate severity ($\geq 4/7$) over the previous month. For the primary outcome measure, we defined treatment success as a score of either 1 (none) or 2 (minimal) on the symptom scale at the final visit.[13] As secondary outcome measures, we determined the proportion of patients becoming completely asymptomatic and treatment success according to *H pylori* status.

*Other symptoms and subgroups of dyspepsia*
At each visit, patients were asked to rate the severity of specific symptoms of dyspepsia over the previous month with the same seven point scale as for global overall symptoms. We carried out retrospective analysis of treatment success for patients with reflux predominant symptoms compared with those for whom the reflux symptoms were not predominant (non-reflux predominant).

*Quality of life questionnaire*
We assessed quality of life by using the validated, self administered quality of life in reflux and dyspepsia (QOLRAD) instrument.[14] This disease specific instrument uses a seven point Likert-type scale in which higher scores indicate better quality of life. Results are reported as average change in each of five dimensions.

*Gastrointestinal symptom rating scale questionnaire*
The gastrointestinal symptom rating scale (GSRS) questionnaire is a well validated and self administered instrument. It includes 15 questions on different gastrointestinal symptoms, with a seven point Likert-type scale in five dimensions.[15] The severity of symptoms reported increases with decreasing score.

*Dyspepsia related health utilisation costs*
Our objective was to compare the mean annual cost of *H pylori* eradication treatment with that of placebo. Study personnel measured dyspepsia related use of health resources prospectively at monthly intervals by telephone and clinic interviews with a health resource

utilisation questionnaire. Direct costs included visits to the physician (specialist, family physician) and other healthcare professionals, drugs (prescription, over the counter), and investigations (for example, laboratory tests, radiography, endoscopy). Indirect costs of decreased productivity as a consequence of days lost through dyspepsia took into consideration whether the patient was employed, unemployed, or a senior citizen (aged over 65) and were calculated from Canadian labour force and unpaid work estimates.[16][17] We calculated the cost for each health resource from the frequency of resources consumed and their unit prices. We aggregated indirect and direct costs (Province of Ontario, Canada, Ministry of Health perspective) to determine the societal perspective. Because of the duration of the study, we did not discount costs.

*Eradication of H pylori*
We calculated the proportion of patients in whom *H pylori* was eradicated on the basis of the result of the urea breath test at 12 months or, in the case of a missing 12 month value, the result at three months.

### Determination of sample size
We based calculations on estimates of the difference in rates of treatment success between treatments. The assumed treatment success rate was 39% for the eradication arm and 20% for the placebo arm. In order to achieve a two tailed significance level of 0.05 and a power of 90%, we needed 120 evaluable patients in each arm. To allow for a maximum dropout rate of 25%, we needed 150 patients per arm.

### Statistical evaluation
The intention to treat analysis included all randomised patients. Patients who discontinued at any time were considered treatment failures. We undertook a more clinically applicable analysis—"all evaluable patients"—in those patients who had data on symptoms at the 6-12 month assessments (figure). We carried data forward from six months and beyond to replace missing 12 month data. We used the Cochran-Mantel-Haenszel test to compare proportions of success by treatment group.

The main objective of the economic analysis was to measure and describe the costs per patient over the year of the study. As costs were not normally distributed, we used corrected α percentile bootstrap methods to measure mean costs per patient.[18][19]

## Results

The disposition of patients enrolled and randomised into the study is shown in the figure. Of patients with positive Helisal test results, 152 (33%) had a negative $^{13}$C-urea breath test result. A total of 294 patients were randomised, and the two groups were well matched (table 1).

The proportion of patients who were considered a treatment success was significantly greater for the eradication arm than for the placebo arm, with comparable results in the intention to treat and all evaluable patients analyses (table 2). The number needed to treat to achieve at least one treatment success in the eradication arm was 7 (95% confidence interval 4 to 63). A significant benefit for the eradication arm was also seen when we used the most stringent endpoint of

**Table 1** Baseline demographic characteristics of randomised patients (intention to treat). Values are numbers (percentages) unless stated otherwise

| Characteristic | Eradication group (n=145) | Placebo group (n=149) |
|---|---|---|
| Male | 69 (48) | 79 (53) |
| White | 128 (88) | 139 (93) |
| Mean age in years (range) | 50 (18-82) | 49 (19-81) |
| Current smoker | 42 (29) | 50 (34) |
| Consumer of alcohol | 83 (57) | 93 (62) |
| Previous *Helicobacter pylori* eradication treatment | 4 (3) | 1 (1) |
| Mean (SD) global overall symptom score (GOS) at presentation | 4.8 (0.8) | 4.9 (0.9) |
| Mean (maximum) years since first onset of dyspepsia | 10 (66) | 11 (57) |
| Adherent to drugs (≥12 of 14 doses) | 138 (95) | 145 (97) |

**Table 2** Treatment outcomes at 12 months

| Treatment | No of patients responding | Response rate (% (95% CI)) |
|---|---|---|
| **Treatment success (GOS 1 or 2)—intention to treat** | | |
| Eradication group (n=145) | 72 | 50 (42 to 58) |
| Placebo group (n=149) | 54 | 36 (28 to 44) |
| Difference | | 14 (2 to 25), P=0.02* |
| **Treatment success (GOS 1 or 2)—all evaluable patients** | | |
| Eradication group (n=133) | 72 | 54 (46 to 63) |
| Placebo group (n=134) | 54 | 40 (32 to 49) |
| Difference | | 14 (1 to 26), P=0.03* |
| **Patients completely asymptomatic (GOS=1)—intention to treat** | | |
| Eradication group (n=145) | 41 | 28 (21 to 36) |
| Placebo group (n=149) | 22 | 15 (9 to 20) |
| Difference | | 13 (4 to 24), P=0.008* |
| **Treatment success of reflux predominant dyspepsia subgroup—intention to treat** | | |
| Eradication group (n=54) | 23 | 43 (29 to 56) |
| Placebo group (n=53) | 17 | 32 (20 to 45) |
| Difference | | 11 (NT) |
| **Treatment success of non-reflux predominant dyspepsia subgroup—intention to treat** | | |
| Eradication group (n=91) | 49 | 54 (44 to 64) |
| Placebo group (n=96) | 37 | 39 (29 to 48) |
| Difference | | 15 (NT) |

GOS=global overall symptom score; NT=not tested.
*Statistical comparison by Cochran-Mantel-Haenszel test.

defining only completely asymptomatic patients as responders. The treatment responses in patients with reflux predominant dyspepsia and non-reflux predominant dyspepsia were of the same order of magnitude as for the overall groups (table 2).

The distribution of ulcer-like, dysmotility-like, and reflux-like dyspepsia subgroups was similar in both groups: 131 (90%), 76 (52%), and 122 (84%) in the eradication group (n=145) and 134 (90%), 93 (62%), and 129 (87%) in the placebo group (n=149). The subgroups showed considerable overlap, and only 29 (<10%) patients were in one category only. All dyspepsia subgroups showed a trend towards greater treatment success in the eradication arm than in the placebo arm (49% (64/131) v 36% (48/134) for ulcer-like dyspepsia, 39% (30/76) v 29% (27/93) for dysmotility-like dyspepsia, and 49% (60/122) v 36% (46/129) for reflux-like dyspepsia).

In multiple logistic regression analysis including age, sex, and treatment as predictors, only eradication treatment was significantly (P=0.009) associated with treatment success.

### Results according to *H pylori* status
*H pylori* was eradicated in 75% (109/145) of the patients in the eradication arm and in 14% (21/149) of those in the placebo arm in the intention to treat

**Table 3** Change in quality of life measured with quality of life in reflux and dyspepsia instrument (QOLRAD)

| Domain | Mean difference in change in quality of life (eradication arm–placebo arm)* | Range | P value |
|---|---|---|---|
| Emotional distress | 0.34 | 0.04-0.65 | 0.03 |
| Sleep disturbance | 0.18 | –0.10-0.46 | 0.21 |
| Problems with eating or drinking | 0.20 | –0.10-0.50 | 0.20 |
| Physical and social functioning | 0.25 | 0.01-0.48 | 0.04 |
| Vitality | 0.39 | 0.08-0.70 | 0.02 |

*A positive value indicates greater symptom improvement in the eradication arm.

population. During follow up, a second course of *H pylori* eradication treatment resulted in eradication in only 2 of 11 treated patients in the eradication arm compared with 15 of 23 treated patients in the placebo arm. The evaluable eradication rate in patients who received only the initial course of study treatment was 80% (107/134) in the eradication arm and 4.4% (6/136) in the placebo arm. In secondary analysis, patients who had *H pylori* eradicated had a treatment success rate of 54% (69/127; 95% confidence interval 45% to 63%) compared with 39% (54/137; 31% to 48%) in those who remained *H pylori* positive. For individual symptoms, eradication of *H pylori* also relieved epigastric pain or discomfort and belching symptoms but not heartburn, regurgitation, bloating, nausea, early satiety, or postprandial fullness (data not shown).

**Quality of life assessments**

Table 3 shows the impact of eradication treatment on disease specific measures of quality of life. The

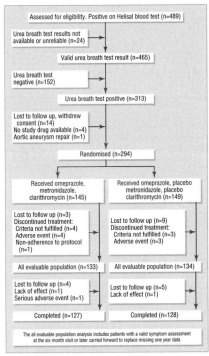

Assessed for eligibility. Positive on Helisal blood test (n=489)

Urea breath test results not available or unreliable (n=24)

Valid urea breath test result (n=465)

Urea breath test negative (n=152)

Urea breath test positive (n=313)

Lost to follow up, withdrew consent (n=14)
No study drug available (n=4)
Aortic aneurysm repair (n=1)

Randomised (n=294)

Received omeprazole, metronidazole, clarithromycin (n=145)

Received omeprazole, placebo metronidazole, placebo clarithromycin (n=149)

Lost to follow up (n=3)
Discontinued treatment:
Criteria not fulfilled (n=4)
Adverse event (n=4)
Non-adherence to protocol (n=1)

Lost to follow up (n=9)
Discontinued treatment:
Criteria not fulfilled (n=3)
Adverse event (n=3)

All evaluable population (n=133)

All evaluable population (n=134)

Lost to follow up (n=4)
Lack of effect (n=1)
Serious adverse event (n=1)

Lost to follow up (n=5)
Lack of effect (n=1)

Completed (n=127)

Completed (n=128)

The all evaluable population analysis includes patients with a valid symptom assessment at the six month visit or later carried forward to replace missing one year data

Flow of participants through the study

difference in the change in scores from pretreatment to study end showed significantly greater improvement in three of the five domains for the eradication arm. The gastrointestinal symptom rating scale assessment showed a significant change at 12 months in the eradication arm for the constipation dimension only (data not shown).

**Health resource utilisation**

Table 4 shows selected values for direct and indirect costs. The mean total annual costs from the perspectives of society and the Ontario Ministry of Health were lower for the eradication arm than the placebo arm, although the differences were not significant (table 5). Few patients had endoscopy or upper gastrointestinal barium examination in the follow up year (table 6). The increased costs for patients randomised to placebo were primarily incurred through increased visits to the physician and drugs for dyspepsia (table 6). The proportion of patients needing additional prescriptions was 50% (73/145) in the eradication arm and 58% (87/149) in the placebo arm. The total number of prescriptions for dyspepsia was also higher in the placebo arm than in the eradication arm (75 *v* 67 for proton pump inhibitors, 117 *v* 56 for $H_2$ antagonists, 19 *v* 12 for prokinetic agents).

**Adverse events**

The population consisted of all 294 randomised patients. Sixty one (42%) patients in the eradication arm and 62 (42%) patients in the placebo arm reported at least one adverse event. Diarrhoea, headache, increased abdominal pain, nausea, flatulence, and taste perversion were the most common events reported. One patient in the eradication arm stopped treatment owing to a skin rash. In the placebo arm, two patients stopped their pills because of adverse events: one had crampy abdominal pain and loose bowel movements, and the other had epigastric pain. Minor elevations of liver enzymes (aspartate aminotransferase, alanine aminotransferase, and alkaline phosphatase) occurred more often in the eradication group than in the placebo group, and all resolved within two to four weeks after the end of treatment.

Two deaths occurred during the study, both in the eradication arm. The first patient was diagnosed with metastatic brain cancer (primary tumour unknown) 10 months into the follow up phase and died before the 12 month visit. The second patient was a 69 year old man who was admitted to hospital with worsening dysphagia three months into follow up. He had no alarm symptoms at entry to the study. Investigations revealed inoperable oesophageal cancer, and the patient died one month later.

## Discussion

*H pylori* is known to cause duodenal ulcers and gastric ulcers and is linked to gastric cancer[20] and MALToma (mucosal associated lymphoid tumour),[21] but its association with dyspepsia remains unclear. Most studies of *H pylori* and dyspepsia have been done in patients with functional (that is, investigated) dyspepsia. Meta-analyses of these trials have shown either no benefit from eradication of *H pylori*[5] or at best a small benefit with a number needed to treat of 15.[4]

Patients do not present to the family physician with an identified cause for their dyspepsia, as they are uninvestigated at first presentation. They may have functional dyspepsia or diseases such as peptic ulcer or gastro-oesophageal reflux disease. Unfortunately, symptoms do not reliably predict endoscopic findings or allow reliable diagnosis.[3] The Rome definition of dyspepsia considers the symptoms of heartburn and acid regurgitation to be synonymous with gastro-oesophageal reflux disease and not part of the symptom complex of dyspepsia,[22] but it is well known that most patients have multiple, overlapping symptoms,[1][23] as we confirmed in this study. Even among patients with proved peptic ulcers, 28% can have heartburn or acid reflux as the predominant presenting symptom.[24] Therefore, a definition of dyspepsia that excludes reflux symptoms does not fit the conceptual framework of family physicians, and we believe that these symptoms form part of the symptom complex of dyspepsia.[2][8]

### Effect of *H pylori* eradication on symptoms of dyspepsia

Our study showed consistent results in favour of eradication of *H pylori* for most outcome measures, including global improvement (to mild or no symptoms) and complete resolution of dyspepsia, improvement in several specific symptoms (epigastric pain or discomfort, belching), and improvement in some aspects of quality of life. The number needed to treat to achieve one treatment success was 7 (4 to 63). The 14% clinical gain observed in this study may be attributable to the expected proportion of 5-15% of *H pylori* positive patients with a true ulcer diathesis.[25] This is speculative, as we did not perform endoscopy at the beginning of the study. Patients in whom *H pylori* was eradicated had better relief of symptoms than those in whom infection persisted, which is consistent with the hypothesis that *H pylori* is responsible for dyspepsia in some patients.

Although extensive overlap of symptoms makes it impossible to completely exclude patients with gastro-oesophageal reflux disease, we excluded patients with reflux disease previously diagnosed by endoscopy or 24 hour oesophageal pH study and patients with symptoms of only heartburn or acid regurgitation without epigastric pain or discomfort. Studies in

**Table 4** Selected values for direct and indirect costs

| Item | Costs ($C)* |
|---|---|
| Drugs†: | |
| Omeprazole 20 mg | 2.20 per tablet |
| Clarithromycin 250 mg | 1.48 per tablet |
| Metronidazole 500 mg | 0.056 per tablet |
| Hospital cost‡ | 432.05 per day |
| Visits to doctor§: | |
| Gastroenterologist | First visit 106.95, subsequent 23.45 |
| Surgeon | First visit 55.90, subsequent 19.20 |
| Visit to nurse¶ | 37.27 per visit |
| Endoscopy§ (physician charge) | 94.60 |
| Upper gastrointestinal barium meal§ (physician charge) | 84.85 |
| $^{13}$C-urea breath test** | 80.00 |
| Laboratory tests (selected)††: | |
| Full blood count | 8.77 per test |
| Creatinine | 2.74 per test |
| Blood sugar | 1.88 per test |
| Helisal rapid whole blood test | 22.00 per test |
| Lost productivity[17]: | |
| Men aged 20-65 | 79.39 per day |
| Men aged >65 | 19.27 per day |
| Women aged 20-65 | 73.84 per day |
| Women aged >65 | 21.61 per day |

*1 $C=0.60 US$=£0.43.
†Ontario Drug Benefit Formulary/Comparative Drug Index. Ontario Ministry of Health 35, Toronto, Canada, 1999. (Non-prescription drug costs were determined from the Medis Health and Pharmaceutical Services Inc Distributing Catalogue, Montreal, Canada, 1999.)
‡Canadian Coordinating Office for Health Technology Assessment (CCOHTA). A Manual of Standard Costs for Pharmacoeconomic Studies in Canada: Feasibility Study. Ottawa, Canada, 1995. (www.ccohta.ca)
§OHIP Schedule of Benefits: Physician Services Under the Health Insurance Act, 1999. Toronto, Canada.
¶Ontario Ministry of Health. System-Linked Research Unit. Approach to the Measurement of Costs (Expenditures) when Evaluating Health and Social Programmes. 1995. McMaster University, Hamilton, ON, Canada.
**MDS Laboratories charge, Ontario, Canada.
††Ontario Ministry of Health. OHIP Schedule of Laboratory Services. 1999. Ontario, Canada.

**Table 5** Mean (range) total costs to society and the Ministry of Health in $C by treatment arm (intention to treat population)

| Treatment arm | No of patients | Societal cost* | Ministry of Health cost† |
|---|---|---|---|
| Eradication | 142 | 477 (27-3069) | 136 (0-1066) |
| Placebo | 146 | 530 (31-3315) | 181 (0-1860) |

1 $C=0.60 US$=£0.43.
*Difference in cost $C53 (95% CI −$C86 to $C180).
†Difference in cost $45 (−$20 to $114).

patients with reflux disease who test positive for *H pylori* show that eradication of *H pylori* either does not affect the subsequent clinical course of gastro-

**Table 6** Main events counted to estimate use of resources over the one year follow up

| | Eradication group (No of events) | Eradication costs ($C) | Placebo group (No of events) | Placebo costs ($C) |
|---|---|---|---|---|
| Admissions to hospital for stomach problems | 1 | 432 | 6 | 2 592 |
| Visits to family practitioner | 120 | 2 186 | 150 | 2 787 |
| Visits to specialist (surgeon or gastroenterologist) | 24 | 1 631 | 32 | 2 033 |
| Upper gastrointestinal barium study | 13 | 1 103 | 14 | 1 188 |
| Upper gastrointestinal endoscopy | 11 | 1 041 | 16 | 1 514 |
| Cost of prescription drugs for dyspepsia* | 179 prescriptions (73 patients) | 25 816 | 299 prescriptions (87 patients) | 38 974 |
| Cost of non-prescription drugs for dyspepsia† | – | 3 527 | – | 4 486 |
| Laboratory tests | 24 | 714 | 36 | 1 254 |
| Days of work missed | 263 (30 patients) | 16 910 | 226 (24 patients) | 13 200 |
| Other‡ | 53 | 2 138 | 61 | 2 663 |

1 $C=0.60 US$=£0.43.
*Costs include drug treatment at start of study; costs taken from a log of gastrointestinal medications; includes antibiotics given for repeat *Helicobacter pylori* eradication treatment during the study.
†Cost of non-prescription drugs paid by the patient as reported in the questionnaire; the number and types of drugs taken were not captured.
‡Includes visits to a nurse, imaging studies (abdominal and chest radiography, ultrasonography of abdomen and pelvis, computed tomography of abdomen, barium enema), sigmoidoscopy, one colonoscopy, and transportation costs.

oesophageal reflux disease[26] or may worsen it. Inclusion of such patients in our study would have biased the results towards no effect. In this study, we saw a trend towards improvement and not worsening of dyspepsia in patients with predominant reflux symptoms (not statistically powered for these comparisons). These results are in keeping with a study in patients with peptic ulcers and concomitant reflux oesophagitis, in which symptoms improved after eradication of *H pylori*.[24] Our data thus suggest that a proportion of patients with uninvestigated dyspepsia with predominant reflux symptoms and epigastric pain or discomfort benefit from treatment to eradicate *H pylori*, and our results are robust and generalisable to primary care.

### Diagnosis and eradication of *H pylori*

Thirty three per cent of patients who were positive for *H pylori* by whole blood screening had a negative $^{13}$C-urea breath test. Thus whole blood testing is unreliable for use in a "test and treat" strategy, and we recommend the more accurate $^{13}$C-urea breath test as the diagnostic method of choice.[27]

The 80% *H pylori* eradication rate in this study is consistent with eradication rates achieved with omeprazole-metronidazole-clarithromycin in the community.[28] The treatment was well tolerated, and adherence was high. The frequency of adverse events was similar in both arms of the study, and most were minor. In this study, one patient (age 69) was diagnosed with oesophageal cancer three months after inclusion. At the time of randomisation, alarm symptoms (particularly dysphagia) were absent. We believe it is unlikely that earlier endoscopy could have prevented this patient's death.

### Treatment guidelines

Most dyspepsia guidelines recommend investigations in patients over 50.[6 8 29] We agree that endoscopy should be considered in patients at an earlier age in areas with high prevalence of gastric cancer.[30] However, in Canada, gastric cancer has steadily declined over the past 40 years. Our study and the recently reported Canadian adult dyspepsia empiric treatment—prompt endoscopy (CADET-PE) study were not restricted in age. No cases of gastric cancer occurred in 1040 patients with uninvestigated dyspepsia in the prompt endoscopy study.[31] Although these findings are suggestive, adequately powered studies are needed to determine whether an age limit of over 50 is safe in patients with uninvestigated dyspepsia.

### Economic analysis

The cost analysis shows benefits in favour of eradication of *H pylori*, although the differences were not statistically significant. The study was not powered to detect economic differences. The cost data do, however, provide another justification to advocate the "test for *H pylori* and treat" strategy. As the time horizon for this study was only one year, economic benefits would be expected to increase over time for patients cured of their dyspepsia. Nevertheless, it is important to keep in mind that at least half of patients will need further prescriptions for dyspepsia after anti-*H pylori* treatment. We have done further economic modelling and analyses, which support the view that treatment to eradicate *H pylori* is cost effective.[32]

## What is already known on this topic

Dyspepsia is a common problem in primary health care, although controversy exists about its definition

Studies of *H pylori* eradication in patients with uninvestigated dyspepsia have shown reduced need for endoscopy and thus significant cost savings compared with a strategy of prompt endoscopy

The "test for *H pylori* and treat" strategy has been recommended for uninvestigated dyspepsia, but there have been no randomised controlled trials showing improvement in symptoms

## What this study adds

When given eradication treatment in primary care, *H pylori* positive patients with uninvestigated dyspepsia show improvement in overall dyspepsia symptoms at 12 months

This supports the "test for *H pylori* and treat" strategy

### Conclusion

This primary care study has shown that the "test with $^{13}$C-urea breath test and treat to eradicate *H pylori*" strategy in patients with uninvestigated dyspepsia provides long term relief from symptoms and may reduce healthcare costs.

We thank Joanna Lee, AstraZeneca Canada, for statistical work. We also acknowledge the assistance of the other members of the CADET Summary Group: Alan Thomson, Alan Barkun, and David Armstrong. The CADET-*Hp* Study Group of principal investigators are G Achyuthan, Regina; D Barr, London; K Bayly, Saskatoon; W Booth, Antigonish; M Cameron, Regina; S Cameron, Halifax; H S Conter, Halifax; S J Coyle, Winnipeg; B N Craig, Saint John; R K Dunkerley, London; J Hii, Vancouver; W P House, Vancouver; E Howlett, Saskatoon; F F Jardine, Manuels; D Johnson, Winnipeg; K Kausky, Whistler; H Langley, Kingston; K R Loader, Brandon; P V Mayer, Kingston; D M McCarty, Edmonton; S Moulavi, Montreal; M Murty, Orleans; W O'Mahony, Corunna; P O'Shea, St John's; G Pannozzo, Waterloo; J Price, Portage La Prairie; P Sackman, Calgary; C L Sanderson-Guy, Nepean; K Saunders, Winnipeg; D Shu, Coquitlam; RJ Smith, Mount Pearl; T Tobin, Guelph; G R Webb, Grand Bay; P Whitsitt, Oshawa; W Winzer, Orleans; and P Wozniak, Cambridge.

Contributors: NC, SJOVvanZ, and PS were responsible for conception and design of the study, analysis and interpretation of data, drafting the article and revising it critically for important intellectual content, and approval of the version to be published. RAF, SE, and EG were responsible for analysis and interpretation of data, drafting the article and revising it critically for important intellectual content, and approval of the version to be published. NC and SJOVvanZ act as guarantors of this paper.

Funding: The study was financially supported by Astra-Zeneca Canada Inc.

Competing interests: NC and SJOVvanZ have acted as consultants and have received research support and honorariums for giving talks on this subject by the sponsor, AstraZeneca Canada, who manufacture omeprazole. PS and RAF are former employees of AstraZeneca Canada, and SE and EG are current employees of AstraZeneca Canada (sponsors of the study).

1  Tougas G, Chen Y, Hwang P, Liu MM, Eggleston A. Prevalence and impact of upper gastrointestinal symptoms in the Canadian population: findings from the DIGEST study. *Am J Gastroenterol* 1999;94:2845-54.

2  Chiba N, Bernard L, O'Brien BJ, Goeree R, Hunt RH. A Canadian physician survey of dyspepsia management. *Can J Gastroenterol* 1998;12:83-90.

3  Talley NJ, Silverstein MD, Agreus L, Nyren O, Sonnenberg A, Holtmann G. AGA technical review: evaluation of dyspepsia. *Gastroenterology* 1998;114:582-95.

4  Moayyedi P, Soo S, Deeks J, Forman D, Mason J, Innes M, et al. Systematic review and economic evaluation of Helicobacter pylori eradication treatment for non-ulcer dyspepsia. *BMJ* 2000;321:659-64.

5  Laine L, Schoenfeld P, Fennerty MB. Therapy for Helicobacter pylori in patients with nonulcer dyspepsia. A meta-analysis of randomized, controlled trials. *Ann Intern Med* 2001;134:361-9.

6  Hunt RH, Fallone CA, Thomson AB. Canadian Helicobacter pylori consensus conference update: infections in adults. *Can J Gastroenterol* 1999;13:213-7.

7  Chiba N. Definitions of dyspepsia: time for a reappraisal. *Eur J Surg* 1998;164(suppl):14-23.

8  Veldhuyzen van Zanten SJ, Flook N, Chiba N, Armstrong D, Barkun A, Bradette M, et al. An evidence-based approach to the management of uninvestigated dyspepsia in the era of Helicobacter pylori. *CMAJ* 2000;162(suppl 12):S3-23.

9  Moayyedi P, Carter AM, Catto A, Heppell RM, Grant PJ, Axon AT. Validation of a rapid whole blood test for diagnosing Helicobacter pylori infection. *BMJ* 1997;314:119.

10  Mock T, Yatscoff R, Foster R, Hyun JH, Chung IS, Shim CS, et al. Clinical validation of the Helikit: a 13C urea breath test used for the diagnosis of Helicobacter pylori infection. *Clin Biochem* 1999;32:59-63.

11  Veldhuyzen van Zanten SJO, Tytgat KMAJ, Pollak PT, Goldie J, Goodacre RL, Riddell RH, et al. Can severity of symptoms be used as an outcome measure in trials of non-ulcer dyspepsia and Helicobacter pylori associated gastritis? *J Clin Epidemiol* 1993;46:273-9.

12  Junghard O, Lauritsen K, Talley NJ, Wiklund IK. Validation of 7-graded diary cards for severity of dyspeptic symptoms in patients with non-ulcer dyspepsia. *Eur J Surg* 1998;164(suppl 583):106-11.

13  Jaeschke R, Singer J, Guyatt GH. Measurement of health status. Ascertaining the minimal clinically important difference. *Controlled Clin Trials* 1989;10:407-15.

14  Wiklund IK, Junghard O, Grace E, Talley NJ, Kamm M, Veldhuyzen van Zanten SJO, et al. Quality of life in reflux and dyspepsia patients. Psychometric documentation of a new disease-specific questionnaire (QOL-RAD). *Eur J Surg* 1998;164(suppl 583):41-9.

15  Svedlund J, Sjodin I, Dotevall G. GSRS—a clinical rating scale for gastrointestinal symptoms in patients with irritable bowel syndrome and peptic ulcer disease. *Dig Dis Sci* 1988;33:129-34.

16  Drummond MF, O'Brien B, Stoddart GL, Torrance GW. Cost benefit analysis. In: *Methods for the economic evaluation of health care programmes.* New York: Oxford Medical Publications, Oxford University Press, 1997:205-31.

17  *Labour force and unpaid work of Canadians: selected labour force, demographic, cultural, educational and income characteristics by sex (based on the 1991 standard occupational classification) for Canadian provinces, territories and CMAs, 1996 census (20% sample data).* Ottawa, Canada: Statistics Canada, 1996.

18  Thompson SG, Barber JA. How should cost data in pragmatic randomised trials be analysed? *BMJ* 2000;320:1197-200.

19  Vinod HD. Bootstrap methods: applications in econometrics. In: Maddala GS, Rao CR, Vinod HD, eds. *Handbook of statistics 11: econometrics.* Elsevier Science, Chapman and Hall, CRC, 1993:629-61.

20  Huang JQ, Sridhar S, Chen Y, Hunt RH. Meta-analysis of the relationship between Helicobacter pylori seropositivity and gastric cancer. *Gastroenterology* 1998;114:1169-79.

21  Bayerdörffer E, Neubauer A, Rudolph B, Thiede C, Lehn N, Eidt S, et al. Regression of primary gastric lymphoma of mucosa-associated lymphoid tissue type after cure of Helicobacter pylori infection [see comments]. *Lancet* 1995;345:1591-4.

22  Talley NJ, Stanghellini V, Heading RC, Koch KL, Malagelada JR, Tytgat GN. Functional gastroduodenal disorders. *Gut* 1999;45(suppl 2):II37-42.

23  Tytgat GN. GERD remains an intriguing enigma. *Gastroenterology* 2001;120:787.

24  McColl KEL, Dickson A, El-Nujumi A, el-Omar E, Kelman A. Symptomatic benefit 1-3 years after H. pylori eradication in ulcer patients: impact of gastroesophageal reflux disease. *Am J Gastroenterol* 2000;95:101-5.

25  Laine L. Helicobacter pylori and complicated ulcer disease. *Am J Med* 1996;100:52-9S.

26  Moayyedi P, Bardhan C, Young L, Dixon MF, Brown L, Axon AT. Helicobacter pylori eradication does not exacerbate reflux symptoms in gastroesophageal reflux disease. *Gastroenterology* 2001;121:1120-6.

27  Chiba N, Veldhuyzen van Zanten SJ. 13C-Urea breath tests are the noninvasive method of choice for Helicobacter pylori detection. *Can J Gastroenterol* 1999;13:681-3.

28  Chiba N, Marshall CP. Omeprazole once or twice daily with clarithromycin and metronidazole for Helicobacter pylori. *Can J Gastroenterol* 2000;14:27-31.

29  Malfertheiner P on behalf of the European Helicobacter Pylori Study Group (EHPSG). Current European concepts in the management of Helicobacter pylori infection. The Mäastricht consensus report. *Gut* 1997;41:8-13.

30  Lam SK, Talley NJ. Report of the 1997 Asia Pacific consensus conference on the management of Helicobacter pylori infection. *J Gastroenterol Hepatol* 1998;13:1-12.

31  Thomson ABR, Armstrong D, Barkun AN, Chiba N, Veldhuyzen van Zanten SJO, Daniels S, et al. Is prompt endoscopy necessary in uninvestigated dyspeptics? Prevalence of upper gastrointestinal abnormalities—the CADET-PE study. *Gastroenterology* 2001;120;(suppl 1):A50-1.

32  Chiba N, Veldhuyzen van Zanten SJO, Grace EM, Sinclair P, Simons WR, Lee JSM. The cost-effectiveness of a test and treat approach in primary care patients. *Gut* 2000;47(suppl 1):A113-4.

*(Accepted 25 January 2002)*

# Research on Predatory Algae: From Environmental Event to Environmental Policy

The United States was stunned in the mid 1980s by reports of massive unexplained fish kills in estuaries along its southeastern coast. In 1988, a team headed by Dr. JoAnn Burkholder, an aquatic botanist at North Carolina State University, discovered a new genus of "predatory" algae present at the site of several of these disasters: a dinoflagellate that kills fish by releasing a lethal neurotoxin into the water. This startling discovery was first reported to the scientific community in a letter to *Nature* (1992) and quickly gained the attention of the press. The letter and an early article from *Discover* magazine by journalist Patrick Huyghe (1993) are included as the first two documents in this chapter. The third entry is an RFP from the National Sea Grant College Program, followed by a 1994 proposal to this program in which Burkholder and co-investigator Parke Rublee proposed a plan for developing gene probes to help detect the dinoflagellate in estuarine waters. (You'll notice that the dinoflagellate was referred to as *Pfiesteria piscimorte* in early reports but was later formally named *Pfiesteria piscicida*.) The *Nature* letter is cited in the next document, a full-length report in the *Journal of Plankton Research* in 1995 in which Burkholder and her colleagues reported a later study exploring the possibility of controlling the algae with a plankton predator.

The documents that follow these studies are drawn from government agencies attempting to develop guidelines and procedures based on what the research on *Pfiesteria piscicida* has shown. The North Carolina Department of Environment and Natural Resources (NCDENR) began posting *Pfiesteria*-related safety guidelines to their website in the late 1990s; we've included two of these documents, a set of recommended safety practices for workers on the waterways developed by the North Carolina Department of Health and Human Services in 1998, and a set of instructions for lab technicians handling fish pathology samples, posted in 2001. During this time, the state of Maryland issued guidelines for closing and reopening rivers potentially affected by *Pfiesteria*, also included in this chapter.

While these three documents were issued for state agencies that regulate and patrol the state's waterways, the final text in this chapter, a "fact sheet," was issued by the EPA for the general public, who regularly use the rivers for commerce or recreation. Originally published in hard copy in 1998, this document was later posted to EPA's website (updated 2002). In this research case we thus see basic scientific research transformed into public information, procedures, and policy.

# New 'phantom' dinoflagellate is the causative agent of major estuarine fish kills

**JoAnn M. Burkholder, Edward J. Noga\*, Cecil H. Hobbs & Howard B. Glasgow Jr**

Department of Botany, Box 7612, North Carolina State University, Raleigh, North Carolina 27695-7612, USA
\* Department of Companion Animal and Special Species Medicine, North Carolina State University, 4700 Hillsborough Street, Raleigh, North Carolina 27606, USA

A WORLDWIDE increase in toxic phytoplankton blooms over the past 20 years[1,2] has coincided with increasing reports of fish diseases and deaths of unknown cause[3]. Among estuaries that have been repeatedly associated with unexplained fish kills on the western Atlantic Coast are the Pamlico and Neuse Estuaries of the southeastern United States[4]. Here we describe a new toxic dinoflagellate with 'phantom-like' behaviour that has been identified as the causative agent of a significant portion of the fish kills in these estuaries, and which may also be active in other geographic regions. The alga requires live finfish or their fresh excreta for excystment and release of a potent toxin. Low cell densities cause neurotoxic signs and fish death, followed by rapid algal encystment and dormancy unless live fish are added. This dinoflagellate was abundant in the water during major fish kills in local estuaries, but only while fish were dying; within several hours of death where carcasses were still present, the flagellated vegetative algal population had encysted and settled back to the sediments. Isolates from each event were highly lethal to finfish and shellfish in laboratory bioassays. Given its broad temperature and salinity tolerance, and its stimulation by phosphate enrichment, this toxic phytoplankter may be a widespread but undetected source of fish mortality in nutrient-enriched estuaries.

The alga, which represents a new family, genus and species within the order Dinamoebales (K. Steidinger, personal communication), was inadvertently discovered by fish pathologists[5] who observed sudden death of cultured tilapia (*Oreochromis aureus* and *O. mossambica*) several days after their exposure to freshly collected water from the Pamlico River. A small dinoflagellate increased in abundance before fish death, followed by a sharp decline unless additional live fish were introduced. Within 1-2 h of fish death, the sudden algal decline occurred because the cells produced resting cysts or non-toxic amoeboid stages and settled to the bottom of culture vessels (J.M.B., unpublished results).

The behaviour of the culture contaminant provided clues about its potential activity in estuarine habitat. We suspected that if low cell densities were lethal to fish in the wild, as in culture, then the alga would probably not be detected by routine monitoring efforts but might be found while a kill was in progress. We observed the dinoflagellate swarming in Pamlico water collected during a kill of nearly one million Atlantic menhaden (*Brevoortia tyrannus*; Table 1). Less than one day after the kill, few toxic vegetative cells remained suspended in the water although many fish carcasses still were present. In standard bioassays[6,7] maximum algal growth coincided with fish death, followed by rapid encystment (Fig. 1). This toxic dinoflagellate was found in abundance during several other major fish kills in the Pamlico and Neuse estuaries and in local aquaculture facilities (Table 1). Isolates from these events have been confirmed in laboratory bioassays as lethal to 11 species of finfish including commercially valuable striped bass (*Morone saxatilis*), southern flounder (*Paralichthys lenthostigma*), menhaden and eel (*Anguilla rostrata*).

The algal isolates all exhibit similar behaviour. Flagellated, photosynthetic toxic vegetative cells excyst after live fish or their fresh excreta are added to aquaria cultures under low light (Fig. 2a, b). The lag period for excystment ranges from minutes to days and increases with dormancy period or cyst age. The cells appear athecate but have thin cellulosic deposits (thecal plates, Fig. 2c)[8] beneath outer membranes that obscure the plate arrangement required for formal speciation[9]. This dinoflagellate completes its sexual cycle while killing fish. Vegetative cells produce poorly pigmented, anisogamous gametes; the male is smaller, and its extended longitudinal flagellum is five- to sixfold longer than the cell (Fig. 2d). The vegetative cells can also form large amoeboid stages (pigmented or colourless, 80–250 μm long with extended pseudopodia) with unknown role in the life cycle.

Gamete fusion results in planozygote formation followed by production of additional toxic vegetative cells if live fish are present. Without live fish, the planozygotes lose their flagella and form thick-walled cysts while remaining gametes continue to multiply especially under phosphate enrichment (batch cultures enriched with 50–400 μg $PO_4^{3-}$ $l^{-1}$ yielded significantly more gametes after 6 days than cultures with $PO_4^{3-} \leq 10$ μg $l^{-1}$; Student's $t$ test, $P < 0.01$)[10]. Such stimulatory effects have not been observed with either nitrate or ammonium. A ciliated protozoan, *Stylonichia* sp.[11], is common in local estuarine waters and consumes both toxic vegetative cells and cysts. But a pigmented, toxic amoeboid stage of the alga, in turn, attacks the protozoan predator.

Each flagellated vegetative algal cell forms a peduncle or pseudopodium shortly after excystment[12,13]. The peduncle becomes fully extended during toxic activity (Fig. 3a), particularly at the optimal salinity of 15‰ (among tested salinities 5, 10, 15, 25 and 35‰; significantly higher vegetative cell production at 15‰, Students $t$ test, $P < 0.01$)[10]. The lethal agent is an excreted neurotoxin (under analysis by D. Baden, personal communication). Water from which cells had been removed by gentle drop-filtration (0.22 μm-pore filters) induces neurotoxic signs by fish including sudden sporadic movement, disorientation, lethargy and apparent suffocation followed by death. The alga has not been observed to attack fish directly. It rapidly increases its swimming velocity to reach flecks of sloughed tissue from dying fish, however, using its peduncle to attach to and digest the tissue debris. Within several hours of fish death, toxic vegetative cells encyst in the presence or absence of fish carcasses (Fig. 3b–d). The cysts can be destroyed by dilute bleach. But after treatment with concentrated sulphuric acid or ammonium hydroxide, 35 days of desiccation, or nearly 2 years of dormancy, small percentages of the cysts have yielded viable toxic cells when placed in saline water (10–15‰) with live fish (J.M.B., unpublished results).

TABLE 1    Fish kills associated with the new toxic dinoflagellate, documented in North Carolina during 1991–92

| Date | Location | Temperature (°C) | Salinity (‰) | Fish killed | Alga (cells ml$^{-1}$) |
|---|---|---|---|---|---|
| **1991** | | | | | |
| May | Pamlico River | 24 | 3 | Menhaden | 1,300 |
| June | Pamlico River | 29 | 10 | Menhaden, others | 1,100 |
| Aug. | Pamlico River | 31 | 12 | Menhaden | 600 |
| Aug. | Pamlico River | 30 | 8 | Flounder walk | 800 |
| Sept.–Oct. | Neuse River | 26 | 11 | Menhaden, Blue crabs | 1,200 |
| Dec. | Taylor's Creek | 15 | 30 | Fish walk (flounder, eel, mullet, others) | ~35,000 |
| **1992** | | | | | |
| Jan. | Aquaculture | 9 | 0 | Striped bass | |
| Feb. | NC maritime museum (Newport River) | 21 | 25 | Fish spp. | |
| Feb. | National Marine Fisheries Service (Newport River) | 15–21 | 25 | Menhaden, others | |

Locations on the Pamlico River followed a salinity gradient from 3 to 12‰ over an 8-km distance. Low dissolved oxygen (DO, 2.8 mg l$^{-1}$) was measured from the bottom water in the vicinity of the kill during June 1991. On all other dates, DO was ≥4.5 mg l$^{-1}$ throughout the water column (R. Carpenter, K. Lynch and K. Miller, NC Dept. of Environmental Health and Natural Resources, personal communication). In each case from 1991, we observed abundant pigmented, flagellated vegetative cells in water samples taken during the kill. (Note, 'Walk' refers to cases in which stressed fish attempted to leave the water and beach before they died.) Water from fish kills in 1992 was collected within 1–4 days after fish death; hence, it was not possible to obtain data for cell abundances while the kills were in progress. Samples from a kill at a freshwater aquaculture facility in Aurora, NC contained ~40 vegetative cells ml$^{-1}$ as well as cysts. Water from both kills in February 1992 (original source, the Newport River, 11 °C, 25‰) yielded toxic vegetative cells as well as gametes, planozygotes, amoeboid stages and cysts. We confirmed virtually identical morphology of these with previous isolates using scanning electron microscopy, and verified toxicity of each population using aquarium bioassays with tilapia and striped bass as test species.

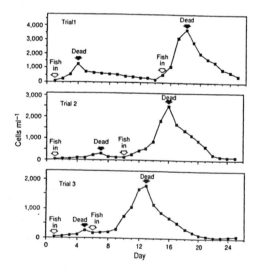

FIG. 1 Response of the dinoflagellate (toxic flagellated, vegetative stage) to *Oreachromis aureus* (length 3 cm, approximate age 40 days) in repeat-trial experiments. The trials were conducted as batch-culture aquarium bioassays (18° C, 10‰, 30 μEinst m$^{-2}$ s$^{-1}$) which varied in the time interval between death of the first fish and addition of a second live fish (fish length 5 cm). The toxic vegetative stage attained greatest abundance just before fish death, followed by rapid decrease in abundance as the cells encysted or formed non-toxic, colourless amoebae and settled out. As we reduced the time interval without live fish, the second fish died more rapidly.

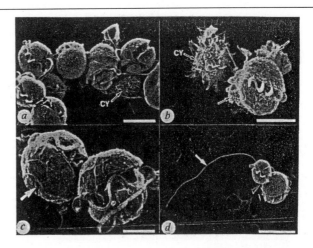

FIG. 2 Scanning electron micrographs of the toxic dinoflagellate. *a, b, c,* Several hours after a live tilapia was added to aquarium cultures (with environmental characteristics described in Fig. 1) after a 2-day period without fish. Most cysts (cy) had already produced flagellated vegetative cells (*a,* scale bar 10 μm; *b,* arrows indicate remains of cyst wall; scale bar, 8 μm); *c,* Cell showing the outline of plates or 'armour' (for example, arrow indicating an apical plate) beneath surficial membranes (scale bar, 5 μm); *d,* Male and female gametes (left and right, respectively). The presumed gametes were abundant ~1 day after addition of a live fish, when the fish began to exhibit neurotoxic signs of lethargy, inability to maintain balance, and apparent respiratory distress. In viewing live samples under light microscopy, we have observed such cells in fusion; note the extended length of the longitudinal flagellum on the male (arrow)(scale bar, 15 μm). Photos by C. Hobbs.

Laboratory bioassays and field collections have confirmed that this dinoflagellate is lethal to finfish across a temperature gradient from 4 to 28 °C[14-17], and across a salinity range of 2–35‰. It has also killed hybrid striped bass *Morone saxatilis* × *M. chrysops* in fresh water (0‰ salinity, in aquaculture facilities and in bioassays) with moderate concentrations of divalent cations (Table 1). Both native and exotic fishes are affected, suggesting stimulation by amino acids or other common, labile substance in fish excreta[18,19] (Table 1). Shellfish such as blue crabs (*Callinectes sapidus,* carapace width 8–10 cm) and bay scallops (*Aequipecten irradians,* shell width 4–5 cm) do not stimulate excystment or toxic activity. These shellfish remained viable for a 9-day test period while filtering low concentrations of the dinoflagellate (~50 cells ml[−1]), although the scallop closing reflex slowed perceptibly. When placed in aquaria with dying finfish, however, blue crabs were killed within hours to several days, and scallops died within minutes.

Long-term data sets from Europe, Asia, and North America

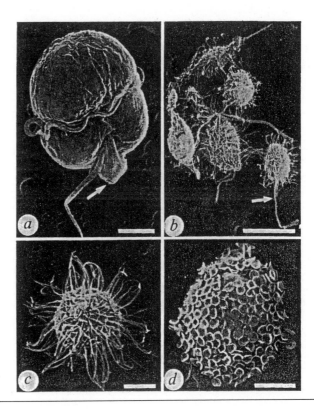

FIG. 3 Scanning electron micrographs of the dinoflagellate from aquarium cultures. *a,* While tilapia were dying. At its optimal salinity of 15‰, the flagellated vegetative cells assume a swollen appearance and the peduncle (arrow) becomes fully extended for use in saprotrophic feeding on bits of fish tissue (scale bar, 3 μm); *b,* Two hours after the tilapia died and were not replaced with live fish. Most vegetative cells had begun to form cysts; note that the longitudinal flagellum was still attached to one of the cells (arrow)(scale bar, 10 μm); *c,* One day after fish death, many completed cysts presumed to contain dinoflagellate cells were present. Note that the outer covering consists of scales with long bristles (scale bar, 3 μm); *d,* Nearly 1 month after fish death, the outer scales on most cysts had lost the bristle extensions (scale bar, 5 μm).

strongly correlate toxic phytoplankton blooms with increasing nutrient enrichment[20]. The Pamlico and Neuse Estuaries receive high anthropogenic loading of phosphorus and nitrogen[21,22]. Over the past century, cultural eutrophication has probably shifted the habitat to more favourable conditions for algal growth and toxic activity. Unlike other toxic phytoplankton, however, this dinoflagellate is ephemeral in the water column. The lethal flagellated vegetative cells move from the sediment surface to the water in response to lingering finfish which become lethargic from the toxin. The alga kills the organisms which stimulate it and then rapidly descends. Increasing reports worldwide describe unexplained kills involving fish that die quickly after exhibiting neurotoxic signs[23,24]. It is unlikely that this eurythermal, euryhaline dinoflagellate occurs only in these estuaries. We predict that with well timed sampling, this alga will be discovered at the scene of many fish kills in shallow, turbid, eutrophic coastal waters extending to geographic regions well beyond the Pamlico and the Neuse.    □

Received 6 April; accepted 5 June 1992.

1. Hallegraeff, G. M., Steffensen, D. A. & Wetherbee, R. *J. Plankt. Res.* **10**, 533–541 (1988).
2. Robineau, B., Gagne, J. A., Fortier, L. & Cembella, A. D. *Mar. Biol.* **108**, 293–301 (1991).
3. Shumway, S. E. *J. World Aquaculture Soc.* **21**, 65–104 (1990).
4. Miller, K. H. *et al. Pamlico Environmental Response Team Report* (North Carolina Department of Environment, Health and Natural Resources, Wilmington, 1990).
5. Smith, S., Noga, E. J. & Bullis, R. A. *Proc. 3rd Int. Colloquium Path. Mar. Aquaculture* 167–168 (1988).
6. Ryther, J. H. *Ecology* **35**, 522–533 (1954).
7. Roberts, R. J., Bullock, A. M., Turner, M. K., Jones, K. & Tett, P. *J. mar. Biol. Ass. UK* **63**, 741–743 (1983).
8. Taylor, F. J. R. in *The Biology of Dinoflagellates* (ed. Taylor, F. J. R.) 24–92 (Blackwell, Boston, 1987).
9. Taylor, F. J. R. in *The Biology of Dinoflagellates* (ed. Taylor, F. J. R.) 723–731 (Blackwell, Boston, 1987).
10. Sokal, R. R. & Rohlf, F. J. *Biometry* 2nd edn (Freeman, San Francisco, 1981).
11. Thorp, J. H. & Covich, A. P. *Ecology and Classification of North American Freshwater Invertebrates*, 77 (Academic, New York, 1991).
12. Spero, H. J. *J. Phycol.* **18**, 357–360 (1982).
13. Gaines, G. & Elbrachter, M. in *The Biology of Dinoflagellates* (ed. Taylor, F. J. R.) 224–268 (Blackwell, Boston, 1987).
14. North Carolina Department of Environment, Health and Natural Resources *1990 Algal Bloom Reports* (North Carolina Department of Environmental Management, Raleigh, 1990).
15. North Carolina Department of Natural Resources and Community Development *1986 Algal Bloom Reports* (North Carolina Division of Environmental Management, Water Quality Section, Raleigh, 1987).
16. North Carolina Department of Natural Resources and Community Development *1987 Algal Bloom Reports* (North Carolina Division of Environmental Management, Water Quality Section, Raleigh, 1988).
17. North Carolina Department of Natural Resources and Community Development *1988 Algal Bloom Reports* (North Carolina Division of Environmental management, Water Quality Section, Raleigh, 1989).
18. Palenik, B., Kieber, D. J. & Morel, F. M. M. *Biol. Oceanogr.* **6**, 347–354 (1988).
19. Lagler, K. F., Bardach, J. E. & Miller, R. R. *Ichthyology* 268–272 (New York, 1962).
20. Smayda, T. J. in *Novel Phytoplankton Blooms* (ed. Cosper, E. M. *et al*) 449–484 (Springer, New York, 1989).
21. Stanley, D. H. in *Proc. Symp. Coastal Water Resources* 155–164 (American Water Works Association, Wilmington, 1988).
22. Rudek, J., Paerl, H. W., Mallin, M. A. & Bates, P. W. *Mar. Ecol. Prog. Ser.* **75**, 133–142 (1991).
23. White, A. W. *Mar. Biol.* **65**, 255–260 (1981).
24. White, A. W. in *Proc. Int. Conf. Impact of Toxic Algae on Mariculture* 9–14 (Aqua-Nor 1987 Int. Fish Farming Exhib., Trondheim 1988).

ACKNOWLEDGEMENTS. We thank K. Lynch, K. Miller and R. Carpenter (NC Dept. of Environment, Health and Natural Resources), and N. McNeill and W. Hettler (National Marine Fisheries Service) for their help in obtaining water samples and V. Coleman, J. Compton, C. Harrington, and C. Johnson for laboratory assistance. This work was supported by the Albemarle-Pamlico Estuarine Study through the UNC Water Resources Research Institute, and the Department of Botany and the College of Veterinary Medicine at North Carolina State University.

## CORRECTIONS

### New 'phantom' dinoflagellate is the causative agent of major estuarine fish kills

**JoAnn M. Burkholder, Edward J. Noga, Cecil H. Hobbs & Howard B. Glasgow Jr**

*Nature* 358, 407–410 (1992)

In this letter in the 30 July issue, Stephen A. Smith (Virginia-Maryland Regional College of Veterinary Medicine, Virginia Polytechnic Institute, Blacksburg, Virginia 24061-0442, USA), a former collaborator of the second author, was inadvertently omitted and should be added as a fifth author. The Office of Sea Grant (NOAA, US Department of Commerce, and the UNC Sea Grant College), and the North Carolina Agricultural Research Foundation should be acknowledged for their contributions to funding support.

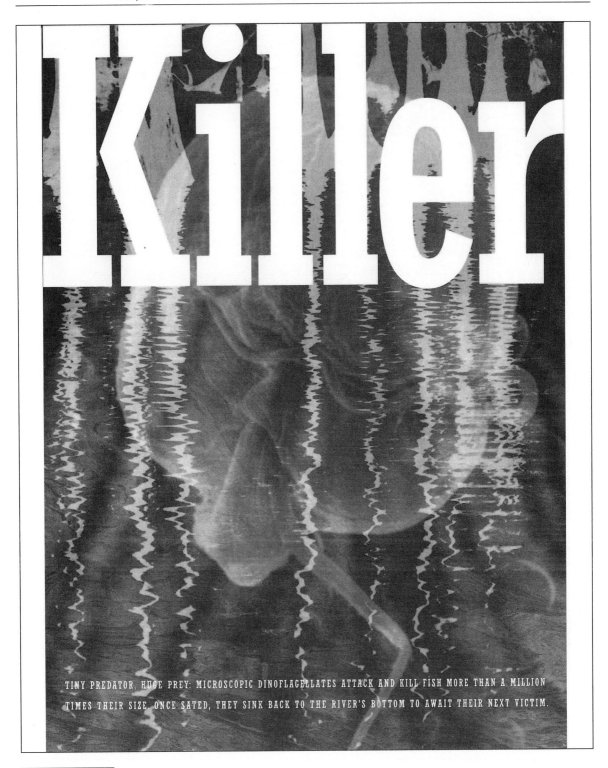

TINY PREDATOR, HUGE PREY: MICROSCOPIC DINOFLAGELLATES ATTACK AND KILL FISH MORE THAN A MILLION TIMES THEIR SIZE. ONCE SATED, THEY SINK BACK TO THE RIVER'S BOTTOM TO AWAIT THEIR NEXT VICTIM.

*Discover*, April 1993

BY PATRICK HUYGHE

Photographs by Andrew Moore

A horrific, predatory little plant is beating up on fish around the world.

# Algae

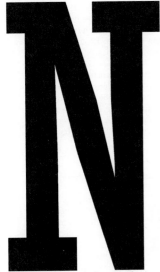ot long ago, one of the most bizarre creatures this world has ever seen showed up in North Carolina. Efforts to describe it have evoked the strangest comparisons: like grass feeding on sheep, said one scientist. And that's not stretching things much. The creature is in fact a tiny plant with a Jekyll and Hyde personality, one that preys on animal life a million times its size. It has caused massive kills of fish and other marine animals around the world. Though hordes of the little monsters could easily fit on the head of a pin, keeping a culture of them active, say its discoverers, requires about 15 fish a day. "It's really not like a normal culture," they say. No joke.

The discovery of this deadly little squirt has caused quite a commotion among marine ecologists; though they have long known the monster's relatives, no other members of the family have ever exhibited this upstart's horrific predatory behavior. The fish killer extraordinaire is a measly dinoflagellate, a kind of alga, or single-celled aquatic plant. Dinoflagellates are twilight-zone creatures: half-plant, half-animal. They produce chlorophyll, but they also move about, using their two flagella, or whiplike tails, to swim rapidly through the water. Dinoflagellates, along with diatoms and a dozen other forms of microalgae, serve as the very bottom of the food chain and, at 450 million to a billion years old, are among the most primitive forms of life on this planet.

Dinoflagellates have long been regarded as the bad boys of the phytoplankton community, though perhaps a little unjustly—so far, only 42 of some 2,000 marine species are known to be toxic. The oldest mention of their nasty behavior occurs in the Old Testament: the first plague visited upon Egypt is a blood red tide that kills the fish and fouls the water. Red tides—which can actually also be green, yellow, or brown—are caused by algae "blooms," excessive concentrations of dinoflagellates and other microalgae. The Red Sea itself may have been named for such a bloom.

More recently an increase of toxic blooms worldwide has alarmed marine ecologists. Not only do blooms appear to have proliferated during the last two decades, but species formerly thought harmless are now proving toxic. Their deadly appearances have led to immense revenue losses for the fishing industry and posed a threat to human health. A huge bloom in the Adriatic in 1989, for example, devastated the local tourist industry and the fishing community and caused $800 million in losses. An earlier toxic bloom, off the coast of Guatemala in 1987, caused one of the largest outbreaks of shellfish poisoning among humans in recent times. Although algal toxins can accumulate in shellfish without causing harm, when those shellfish are eaten by humans, serious illness and even death may occur. Almost 200 people fell ill during the Guatemalan outbreak; 26 died.

Researchers don't know for sure what is causing the increase in blooms. Some believe they may be a manifestation of a long-term cyclical trend. Others think that what we're seeing is the blooming of algal species that were always there in deep ocean waters but had never bloomed before. The new growth spurts, these researchers say, may be caused by natural factors, such as changes in currents or climate. Still others think the boom in blooms, especially in shallow waters and estuaries, is being triggered by an increase in nutrients via human sewage and agricultural runoff and that it signals the global deterioration of the marine environment. Whatever the cause, the discovery of the newest toxic dinoflagellate has done nothing but add a ghastly new twist to an old and increasingly perplexing problem.

The story properly begins one night early in 1988, with a massacre in Edward Noga's laboratory at North Carolina State University in Raleigh. Noga is a fish pathologist at the university's College of Veterinary Medicine. The scene was an aquarium filled with brackish water, and the victims were about 300 tilapias, African fish about two to three inches long; Noga was planning to use them for an immunology project. "All the fish were okay on Friday," recalls Noga, "and on Saturday they were all either dying or dead."

Noga and graduate student Stephen Smith immediately tried to trace the cause of the massive mortality. But they could find no pathogens on the skin or gills of the fish, and the cultures Smith took tested negative for bacteria. "The only thing that seemed unusual to me," recalls Noga, "was the presence in the water of an awful lot of dinoflagellates that seemed to have the same appearance. This suggested some kind of a bloom."

To test whether the dinoflagellates were responsible for the fish deaths, Noga and Smith set up five aquariums and stocked them with six fish apiece. Three of the tanks received a dose of a thousand dinoflagellates each; the two controls received none. Within 15 days blooms occurred in the three test tanks without discoloring the water, and 48 hours later all the fish in these tanks had died. It was a first: no dinoflagellate had ever been known to kill fish in an aquarium system before. The researchers then took water from a tank where fish were dying and filtered it, removing all bacteria, viruses, and dinoflagellates. When the water still proved lethal to 60 percent of the fish, they knew, says Smith, that "the organism was producing something very toxic to fish."

Noga, wanting the killer identified, began sending out cultures. But the dinoflagellates were so small (just 10 to 20 microns, or less than 4 to 8 ten-thousandths of an inch, in diameter), so nondescript, and so difficult to culture by ordinary

# Quickly the creature filled the water with a lethal neurotoxin that paralyzed the fish, causing slow suffocation.

techniques that few scientists were interested in the little critters, and those who were couldn't come up with an answer. JoAnn Burkholder, an aquatic botanist then new to the North Carolina State campus, was no exception.

When Noga first contacted Burkholder, she expressed little interest in the creature, citing no special knowledge of dinoflagellates. But he insisted, and she finally agreed to take a look. The critters, she told him, "were just like three or four other nondescript little dinoflagellates common to North Carolina estuaries." She suggested that one of her part-time graduate students, Cecil Hobbs, might be interested in working on the ecology and life cycle of the mysterious "dino." Hobbs, a former high school biology teacher, had a personal interest in—and perhaps something of a vendetta against—dinoflagellates. His family's oyster beds in Sneads Ferry, North Carolina, had been wiped out by a red tide that hit the coast in 1987. The Hobbses' oysters weren't the only ones hit, of course. That bloom caused $25 million in losses to the local shellfish industry.

Making skillful use of a scanning electron microscope, Hobbs and Burkholder managed to get the first decent shots of the killer. Then they began to characterize the changes taking place in the dinoflagellate before, during, and after its kills. What they found had never been seen before. With no fish present, the creature simply sat in the sediment, encrusted in a hard, scaly, eggshell-like cyst. But when one or more fish began lingering overhead to feed, the creature shed its cyst, often within minutes. What emerged was a flagellated cell—the dinoflagellate stage by which this creature has become known. Quickly it filled the water with a lethal neurotoxin that paralyzed the fish, causing slow suffocation. In the face of impending death, the stunned fish leaned against the side of the aquarium, thrashing about as they struggled to get to the top. Then they dropped to the bottom, bumping their heads or falling on their tails.

Unlike most dinoflagellates, which move in a leisurely, winding sort of way, these made a beeline for their target—flecks of fish tissue stripped off by the toxin. They used a tonguelike absorption tube called a peduncle to attach themselves. As the frenzy increased, the dinos nearly doubled in size, the peduncle itself becoming swollen and assuming different shapes. "In one picture," says Burkholder, "it looks like a huge hand with tentacles and a clawlike thing at the end. We call it our Darth Vader shot." In the presence of this tiny terror, fish are not long for this world. Although the time frame varies, death can occur in as little as 20 minutes.

Burkholder and Hobbs think the attack is actually signaled by something excreted by the fish, though precisely what, they don't yet know. The massacre ends, they think, as the amount of excreta in the water diminishes. The dinos then do one of two things. In less than a minute they can create and encase themselves in a new cyst and drop to the bottom

JoAnn Burkholder found that massive fish kills in North Carolina's Pamlico River were being caused by a new genus of dinoflagellate

to await more prey. This is apparently a protective measure; the cyst marks a dormant stage spurred by a lack of food or some other modification in their environment—a sudden change in water temperature, or turbulence from a storm. But alternatively, if there is no environmental stress and food (fish tissue) still remains, the dinos bizarrely enter instead a stage in which they transform themselves into amoebas. They shed their flagella, lose some of their toxicity, stop photosynthesizing, and become more animallike; once their rapid transformation is complete, they continue to feed on the fish bits at a more leisurely pace.

Burkholder has discovered that even after lying dormant for two years the dinoflagellate can still kill its prey, although it may then take about six weeks for the toxic cells to emerge from their cysts (whereupon, if no fish are present, they simply return to the cyst stage to await another passage of fish). While in the cyst stage, though, the dinos look absolutely harmless. Indeed, they fooled everyone—the fish, obviously, but also the scientists who had seen these cysts before but had never seen in them the face of a killer. Because these cysts so resembled those of a completely different kind of alga, a harmless chrysophyte, to anyone examining the water after a fish kill it looked as if no trace of a villain remained.

Hobbs and Burkholder had been working with the dinoflagellates for about a year when another wave of die-offs struck

Noga's aquariums, killing a thousand fish. "I'm not sure how his tanks got contaminated," says Hobbs, "but it was probably because we were all using the same refractometer" (an instrument used to measure the water's salinity). "We thought we could contain this thing in one room," adds Noga, "but we were very much mistaken. This is worse than any other infectious agent I've ever dealt with."

Hobbs and Burkholder went on to discover just how deadly their creature was. They found that the dinoflagellates could kill fish in anything from fresh water to full-strength seawater, though optimal growth occurred in water of midrange salinity. The researchers also learned that no fish was immune to the dinos' toxin, although some, like striped bass, were more susceptible than others.

Burkholder and Noga now knew enough about the dino to search for it in the wild. But since the killer was too elusive to be caught by haphazard sampling, they enlisted help from North Carolina state biologists in hopes of nailing the phantom while a kill was in progress. "They'd had a lot of sudden-death kills in the past decade or so with no explanations," recalls Burkholder. "During 1988, the best-monitored year, they found 88 fish kills in the local Pamlico River between May and October. Their reports noted the fish sometimes exhibited neurotoxic symptoms and acted panicky. They told me that in some of the kills the fish were actually trying to get out of the water and onto the beach before they died; they called it a flounder, or crab, walk."

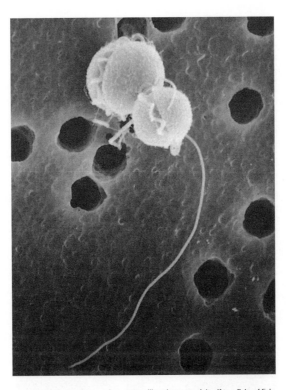

During its feeding frenzy, a dino can nearly double in size; it uses a long, tonguelike tube to attach itself to a flake of fish.

nos left in the water. Their retreat had been as rapid as their blitzkrieg attack.

Burkholder and Hobbs now definitely had their hands on the killer. But they had yet to tag it with an identity. The following October, however, they managed to take a photograph of it that revealed the faint outline of two cellulose structures sitting beneath surface membranes on the dinoflagellate cells. They suspected that these could be "armored plates" that serve as a protective covering for the cell when it is active. The presence or absence of such plates is one of the ways botanists classify a dinoflagellate species. Until this photo, the researchers had assumed the killer belonged among the "naked" species, which lack such armor.

At the end of the month, at an international toxic-phytoplankton conference in Rhode Island, Burkholder announced her findings: that this particular dino had plates, that the species went into a cyst stage, and that it hunted fish as prey. The audience, Burkholder recalls, was stunned. Most of the researchers there had assumed that phytoplankton toxins played a more passive role, as chemical deterrents to predators. But if Burkholder was right, this "plant" was a hunter armed with the equivalent of poison-tipped arrows. Burkholder's findings started a scramble; though somewhat doubtful, researchers from the United States and 14 other countries with similar sudden-death fish kills returned home to hunt for the phantom alga in their local waters.

A few months later, in January 1992, the supervillain struck again—but this time its target was human. Howard Glasgow Jr., a research associate of Burkholder's, was kneeling next to a drain on the floor of Noga's lab, where the dinoflagellate experiments were being conducted. He was carefully rinsing out a ten-gallon aquarium that he planned to use for his upcoming project on the phantom dinos when he began to realize that he was not thinking properly.

Thoughts began rolling through his mind at a terrific pace; but when he reached to pull himself up, his hand seemed to take forever to move. Sensing trouble, Glasgow decided to leave the chamber, but his steps turned into something of a moon walk. "I don't know if it was really slow, or if I was thinking fast," he recalls, "but something was drastically wrong." Glasgow recov-

In May 1991 the phantom killer was finally caught with its peduncle on the goods. Kevin Miller, a state environmental technician, was sampling menhaden (a marine fish of the herring family) from an estuary off the Pamlico when a squall blew up. Forty minutes later, after seeking shelter in a protected bay, he returned to the sampling site to find the fish behaving abnormally. Then they started dying—by the millions. He took water samples and sent them off to Burkholder and Hobbs, who confirmed their toxicity and determined that the dinoflagellate responsible was the same one they'd had in culture for three years. And as expected, samples taken at the site the next day showed almost no diered 15 minutes later and now remembers the experience as more euphoric than frightening. But the aerosol effects of toxin produced by dinoflagellates can be serious. For example, whenever the red tide produced by another dino, *Gymnodinium breve*, hits the coast of Florida, a dangerous aerosol is released for several days. During that period health warnings are posted telling people with heart and respiratory conditions to stay off the beach. "The alga will come in on wave action," explains Burkholder, "and when the cell fragments become airborne in sea spray, they can elicit sneezing, coughing, and severe asthmatic reactions."

Meanwhile the search for the killer's identity went on. Burkholder, unsuccessful in her attempts to strip the outer membranes from the cells, turned for help to a colleague, Karen Steidinger, chief of research at the Marine Research Institute of Florida's Department of Natural Resources and an expert in dinoflagellates. After several unsuccessful attempts on her own, in June 1992 Steidinger, using 100-proof alcohol, finally succeeded in stripping the stubborn membranes off the dino to reveal the armored plates beneath. With the plates' existence finally confirmed, Steidinger and Burkholder were able to verify that their killer represented a new genus and a new species. They have proposed calling it *Pfiesteria piscimorte* (the genus name was picked to honor the late dino specialist Lois Pfiester of the University of Oklahoma and also because Burkholder liked the name's echo of both *feast* and *cafeteria*; the species name means "fish killer").

Over the next few months Burkholder and Glasgow continued to discover just how bizarre the new dinoflagellate really is. Its life cycle consists of more than 15 stages—it's a veritable metamorphosing monster. The dino lives most of its life as an amoeba; in one toxic "giant" amoeboid stage it grows to be nearly 20 times the size of the flagellated toxic cells. The stage when the dino leaves its cyst to attack fish is actually quite ephemeral, appearing only when fish are present. The researchers also found that this stage is the only time the creature sexually reproduces; they again suspect that something in fish excreta triggers the reproductive urge.

Never before had marine dinoflagellates shown such stages. "I've been working with marine dinoflagellates for 29 years now," Steidinger says, "and I've never seen anything like it before. It has the most diverse life cycle I've ever seen. And this can't be the only species out there. There have to be others."

Since the 1991 kill of menhaden off the Pamlico River, Burkholder and her students have confirmed the presence of the dinoflagellate in 11 other fish kills in North Carolina. Outside the state, the phantom killer has been found in Delaware in the Indian River. Anecdotal evidence suggests it was also responsible for a kill in Maryland in the Wye River, a tributary of the Chesapeake. They expect that many other kills previously labeled "unknown" may now be explained.

While researchers the world over are looking for the elusive dino in their own waters, others are trying to identify

Howard Glasgow, seen through an aquarium used to hold the killer algae, was briefly affected by too close an encounter with their toxin.

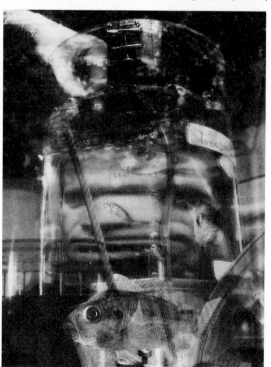

the toxin. Glasgow, for his part, is searching for the algal aphrodisiac—what it is in fish excreta that stimulates dinoflagellates to burst out of their cysts, feed on fish, and reproduce sexually. Burkholder is working on discovering the kind and concentration of organic and inorganic nutrients that stimulate the phantom dino's growth. She's already found that waters with excessive amounts of phosphorus and nitrogen seem to favor the organism. Last but not least, she hopes to find natural predators as a way to eventually control the dinoflagellate. One candidate, a microscopic animal called a rotifer that is a natural predator in estuaries, looks promising. In dino-infested waters, Burkholder has seen the guts of rotifers packed with dinoflagellates.

But finding an able opponent for the dino may not be easy. Burkholder laughs nervously as she recalls the day an undergraduate came running into her office with the news that all the dinoflagellates in a fish tank were dying. When Burkholder looked under the microscope she was surprised to find a protozoan, which had somehow gotten into the tank, busily consuming both the dinos and their cysts as the dinos themselves were attacking and killing the fish. But once the fish had died, the remaining dinoflagellates started to circle the protozoan. Then, like something from a grade B science fiction movie, a few of those dinoflagellates formed into giant toxic amoebas, which—you guessed it—completely engulfed the protozoan.

# NATIONAL SEA GRANT COLLEGE PROGRAM

# STATEMENT OF OPPORTUNITY FOR FUNDING: MARINE BIOTECHNOLOGY

### DEADLINE: FEBRUARY 15, 1994

### FIVE COPIES OF PROPOSAL AND SUMMARY REQUIRED

UNC SEA GRANT COLLEGE PROGRAM
BOX 8605, N.C. STATE UNIVERSITY
RALEIGH, NC 27696
(919) 515-2454

**PLEASE CIRCULATE**

National Sea Grant College Program, 1994

# NATIONAL SEA GRANT COLLEGE PROGRAM

# STATEMENT OF OPPORTUNITY FOR FUNDING: MARINE BIOTECHNOLOGY

## [Excerpts]

## INTRODUCTION

The National Sea Grant College Program is accepting proposals in marine biotechnology for projects of one, two, or three years duration (with annual funding) to begin on or about August 1, 1994.

The primary emphasis will be on research opportunities although advisory and educational proposals also are eligible to compete for funds from an appropriation of $3.2 million. The maximum annual project budget that will be considered for single and multiple investigators is $75,000 and $500,000, respectively, in federal funds. As required by law, at least one third of the total cost of all projects must come from nonfederal matching funds. Industrial matching funds and other close interactions with industry are highly desirable. The National Sea Grant College Program has identified the following preferred areas for proposals:

Aquaculture
Biomaterials/Biosensors
Bioprocessing/Environmental Remediation
Policy/Economics
Education/Technology Transfer

Because of limited funding and widespread interest in marine biotechnology, proposals will be evaluated rigorously. Clear and complete proposals that meet the objectives of the National Sea Grant College Program as outlined in the enclosed materials will have the best chance of success.

## BACKGROUND

The United States is the leader in research expertise in marine biotechnology. However, it faces strong competition in other countries that are moving ahead with national investment and planning in this field. Focused research in marine biotechnology in concert with commercial development offers the opportunity to provide scientific, economic, and social advancements. It will lead to new industries and new jobs and will help higher education to meet U.S. needs for scientists and technicians in an increasingly technical and competitive world. It will assist in reversing our trade deficit, which is $2.7 billion a year in seafood alone. It will lead to economic development and increased exports.

A national commitment to research and development in marine biotechnology will help respond to societal needs by (1) increasing the food supply through aquaculture, (2) developing new types and sources of industrial materials and processes, (3) opening new avenues to monitor health and treat disease, (4) providing innovative techniques to restore and protect aquatic ecosystems, (5) enhancing seafood safety and quality, and (6) expanding knowledge of processes in the world ocean.

A Sea Grant report, "Marine Biotechnology: Competing in the 21st Century," outlines a framework for research in support of marine biotechnology. Its three broad themes—Molecular Frontiers in the Ocean Sciences, Applications of Marine Biotechnology, and Marine Biotechnology and Society—encompass a range of research, education, and outreach needed to develop this field. (The report is available from the National Sea Grant Office and the thirty state and regional Sea Grant programs.)

The U.S. House of Representatives recently passed the "Marine Biotechnology Investment Act" which is designed to enhance the National Sea Grant College Program efforts in biotechnology to (1) expand the range and increase the utility of products from the oceans, (2) improve condition of marine ecosystems by developing substitute products that decrease the harvest pressure on living resources, (3) improve production of aquaculture, (4) provide new tools for understanding ecological and evolutionary processes, and (5) improve techniques for remediation of environmental damage.

The U.S. Senate is considering a companion bill to (1) expand the range and increase the utility of products from the oceans, (2) understand and treat human illness, (3) enhance the quality and quantity of seafood, (4) improve the stewardship of marine resources by developing and applying methods to restore and protect marine ecosystems, manage fisheries, and monitor biological and geochemical processes, and (5) to contribute to business and manufacturing innovations, create new jobs, and stimulate private sector investment.

In increasing the National Sea Grant College Program's budget by $3.2 million for fiscal year 1994, the U.S. Congress specified that the increase was to be used to expand research, education, and outreach in marine biotechnology. Focused research, education, and outreach in concert with commercial development offers promise of economic and social benefits. They will lead to new industries and new jobs and advance higher education to meet U.S. needs in an increasingly technical and competitive world.

Biotechnology may be defined as the application of scientific and engineering principles to provide goods and services through mediation of biological agents. The Federal Coordinating Council for Science, Engineering, and Technology, which identified marine biotechnology as an important field, defined biotechnology as "any technique that uses living organisms (or parts of organisms) to make or modify products, to improve plants or animals, or to develop microorganisms for specific use. The development of materials that mimic molecular structures or functions of living organisms is also included . . . [and] research involving recombinant DNA, DNA transfer techniques, macromolecular structure, cell fusion, bioprocessing, etc., and the infrastructure that supports that research." These broad definitions encompass more than DNA technology. Exclusive of agriculture, application of biotechnology in sewage treatment and water purification now comprises the largest sector in volume. Production of beer and spirits, cheese and other dairy products, baker's yeast, organic acids, and antibiotics follow in order of decreasing value. These traditional applications of biotechnology, which are based primarily on use of terrestrial organisms, are enormously important to the economy as well as human health and nutrition. For example, worldwide production of antibiotics by fermentation generates $10s of billions in sales annually.

While biotechnology is not new, developments in modern molecular biology indicate that it is still in its emerging phase. Many authorities expect biotechnology to be a primary basis of America's economic development and strength in the twenty-first century. Oceanic organisms harbor a major portion of the Earth's genetic resources (biodiversity), yet the large majority of marine organisms are not known well enough for their gene pool and biological processes to be accessible to those who develop and practice biotechnology in industry and academe. However, exploratory research shows the rich potential for exploiting the biochemical capabilities of marine organisms to provide models for new classes of pharmaceuticals, polymers, other chemical products, and new industrial processes as well as vaccines, diagnostic and analytical reagents, and genetically altered organisms for commercial use. Sea Grant has interests in the development of modern tools and technologies for aquaculture, resource management, seafood processing, bioprocessing, bioremediation, production of biomaterials and analogues, and controlling biofouling and biocorrosion.

The opportunities for advancing science and providing results directly supporting the development of marine biotechnology are broad. Because many of these developments will be realized only in the long-term it is essential to recognize that results will be enhanced through dissemination of information, through personnel mobility, and through cooperation and collaboration between universities and the private sector.

# IMPROVED DETECTION OF AN ICHTHYOTOXIC DINOFLAGELLATE IN ESTUARIES AND AQUACULTURE FACILITIES

**A Research Proposal Submitted to
The National Sea Grant College Program
Marine Biotechnology Initiative**

by
JoAnn M. Burkholder and Parke A. Rublee

JoAnn M. Burkholder
Department of Botany
North Carolina State University

Parke A. Rublee
Department of Biology
University of North Carolina–Greensboro

Ernest D. Seneca, Head
Department of Botany
North Carolina State University

North Carolina State University

Burkholder and Rublee Sea Grant proposal, 1994

# MARINE BIOTECHNOLOGY PROPOSAL: PROJECT SUMMARY

**Title:** Improved Detection of an Ichthyotoxic Dinoflagellate in Estuaries and Aquaculture Facilities

## Principal Investigators:

JoAnn M. Burkholder, Associate Professor     *and*     Parke A. Rublee, Assistant Professor
Department of Botany, Box 7612                         Department of Biology
North Carolina State University                        UNC–Greensboro
Raleigh, NC 27695                                      Greensboro, NC 27412
Telephone:                                             Telephone:
Time Devoted to Project: 15%                           Time Devoted to Project: 25%

**Project Period:** 080194–073196          **Budget Period:** 080194–073195

**Amount:** $96,243 federal / $48,135 matching (NC Agricultural Research Service)

## Objectives:

1) To refine rDNA probes, using cultures of the toxic dinoflagellate *Pfiesteria piscimorte,* to ensure high specificity in detecting flagellated, amoeboid, and encysted forms of this fish pathogen.
2) To test the probes to detect *P. piscimorte* in estuarine waters / sediments and aquaculture facilities.
3) To develop a "family" of probes that enable rapid, routine detection / quantification of intraspecific genetic strains of this toxic dinoflagellate at fish kill / ulcerative disease sites.
4) To apply fluorescent gene probes in screening for targeted chemosensory areas (portals of entry, attachment sites) for the dinoflagellate pathogen on/in fish prey.

## Methodology:

State-of-the-art PCR amplification techniques for targeted sequence(s) of the dinoflagellate's rRNA genes from algal cultures will enable refinement of gene probes for use in routine detection of *P. piscimorte* in water / sediment samples from estuaries and aquaculture facilities. A fluorescent anti-sense oligoprobe will be developed and tested for rapid visual identification and localization of the alga in water, sediments, and fish tissues. Culture isolates from North Carolina estuaries and other fish kill sites in the mid-Atlantic and Southeast will be tested for intraspecific sequence variation, needed to develop a "family" of probes that vary in specificity depending on the targeted strain of *P. piscimorte*. These probes, in combination with fish bioassay experimental data, will also enable us to detect the presence of other toxic species that may closely resemble this dinoflagellate.

## Rationale:

Unexplained "sudden-death" fish kills and ulcerative diseases have increased in North Carolina's estuaries over the past decade. The dinoflagellate *P. piscimorte* recently has been implicated as the causative agent of at least 30% of the major fish kills in these estuaries, and also has been tracked to fish kill sites on the mid-Atlantic Coast and Southeast. Despite its importance, the ephemeral toxicity, multiple cryptic flagellated and amoeboid stages, substantial size range (5–250 μm), and close resemblance of this pathogen to other non-toxic algae make reliable detection difficult in light microscopy analysis. Development of sensitive, highly specific gene probes for *P. piscimorte* will answer a critical need for its routine, early detection by regulatory staff and finfish / shellfish aquaculturists.

## Some Suggested Reviewers:

Dr. Edward DeLong, Biol. Dept., Univ. Calif., Santa Barbara, CA 93106

Dr. Lucie Maranda, Dept. of Pharmacognosy & Env., URI, Kingston, RI 02881

Dr. Norman Pace, Inst. Molec. & Cell. Biol., Indiana Univ,, Bloomington, IN 47405

Dr. John Paul, Dept. Mar. Sci., Univ. S. Florida, St. Petersburg, FL 33701

Dr. Karen Steidinger, FL DEP, FL Mar. Res. Inst., St. Petersburg, FL 33701

Dr. Susan Weiler, Dept. of Biol., Whitman College, Walla Walla, WA 99362

# TABLE OF CONTENTS*

*The complete table of contents has been reprinted here to convey the full structure of the proposal package, but only the main body of the proposal argument has been reproduced.

**TITLE:** Improved Detection of an Ichthyotoxic Dinoflagellate in Estuaries and Aquaculture Facilities

**PRINCIPAL INVESTIGATORS:** JoAnn M. Burkholder and Parke A. Rublee

## INTRODUCTION AND BACKGROUND

### A. The Established Linkage between Sudden-Death Fish Kills, Unexplained Fish Disease, and a Toxic Ambush-Predator Dinoflagellate

A recent cosmopolitan rise in the frequency and spatial extent of toxic phytoplankton blooms suggests that these noxious species can significantly reduce our estuarine and marine fishery resources (White 1988, Shumway 1990, Robineau *et al.* 1991). Among the most notorious of toxic phytoplankton are the dinoflagellates that cause "red tides," resulting in the death or "poisoning" of many finfish and shellfish as well as humans (Steidinger & Baden 1984). The United States and other industrialized nations spend billions of dollars annually in research aimed toward prediction, mitigation, and control of toxic outbreaks (Shumway 1990, Culotta 1992). Despite these efforts, accumulating evidence suggests that toxic dinoflagellates may be increasing their activity and geographic range (Steidinger & Baden 1984, Smayda 1989, Hallegraeff 1993). Moreover, the past decade has yielded discoveries of previously unknown toxic species that have suddenly appeared in bloom concentrations, causing millions of dollars of damage to coastal fisheries and aquaculture industries in many U.S. coastal waters (Steidinger & Baden 1984, Smayda 1989, Shumway 1990).

Among the many estuaries in the United States with increasing incidence of unexplained ulcerative fish disease and "sudden-death" fish kills are the Pamlico, Neuse and New Estuaries in North Carolina (Miller *et al.* 1990, North Carolina Division of Marine Fisheries 1992). In the past three years a phytoplankter contaminant from these estuaries was implicated as the causative agent of approximately 30% of North Carolina's major estuarine fish kills (involving $10^3$–$10^9$ fish; Burkholder *et al.* 1992a, 1993; Fig. 1). The dinoflagellate, *Pfiesteria piscimorte* (gen. et sp. nov.) Steidinger & Burkholder (Steidinger *et al.* 1994), represents a new family, genus and species. Unlike red tide dinoflagellates, its lethal forms are ephemeral in the water column and require live fish for toxic activity. When a school of finfish comes within detectable range, the dinoflagellate exhibits remarkable "phantom-like" behavior. The population swims up into the water from dormant cysts or amoebae on bottom sediments; the cells excrete a lethal toxin, complete sexual reproduction while killing the fish, and then rapidly descend back to the sediments (Burkholder *et al.* 1993). The toxin strips skin tissue from the fish and creates open bleeding sores; it also attacks the renal system, induces hemmorrhaging, suppresses the immune system, throws the nervous system into dysfunction, and apparently causes death by suffocation from muscle paralysis (Burkholder *et al.* 1993, E. Noga unpubl. data).

Toxic outbreaks of this "ambush-predator" dinoflagellate have been documented at salinities ranging from freshwater (0%/oo salinity) to full-strength seawater (35%/oo), and across temperatures of 4–33°C (Burkholder 1993; Burkholder *et al.* 1992a, 1993). Further, the alga has proven lethal to all 28 species of native and exotic finfish and shellfish tested thus far (Burkholder *et al.* 1993; Table 1).

### B. The Need for Highly Specific Molecular Probes to Detect This Fish Pathogen

Detection of *Pfiesteria piscimorte* (gen. et sp. nov.) demands well-timed sampling because the algal population displays rapid encystment without live fish, typically "disappearing" or settling out of the water

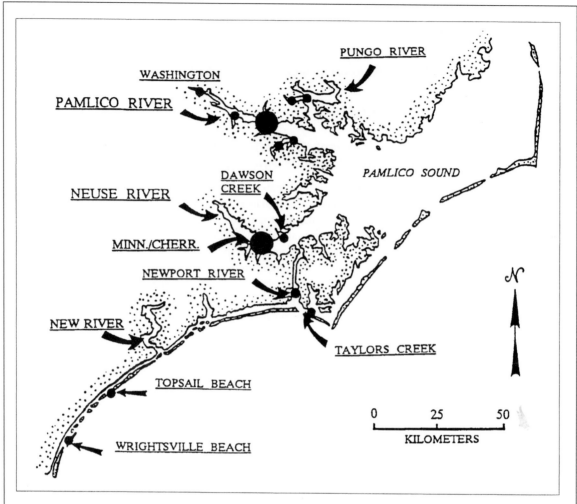

**FIGURE 1.** Locations where lethal stages of *Pfiesteria piscimorte* (gen. et sp. nov.) has been verified in North Carolina. Kill sites (documented in association with this toxic dinoflagellate in all labeled locations except the New River Estuary) are designated by blackened circles, with large circles representing sites where fish kills are most frequent with substantial area affected. The large circle on the Neuse River Estuary at Minnesott Beach / Cherry Point (MINN./CHERR.) designates the site with highest known fish loss; more than 1 billion Atlantic menhaden were killed, requiring bulldozers to clear the beaches over a 6-week period when the menhaden schools were moving out to sea.

column within only a few hours after fish death (Burkholder *et al.* 1992a). Efforts to establish the presence of *P. piscimorte* (gen. et sp. nov.) in fish kills also have been hampered because this small dinoflagellate resembles several other common nontoxic species so closely that it is difficult to reliably identify it in routine light-microscope analyses. The confusion is further compounded because the alga has a complex life cycle that includes rapid transformations (sometimes within minutes) among at least 15 different flagellated and amoeboid forms, many of which are amorphous and colorless and, hence, easily missed or mistaken for

**TABLE 1.** Species of finfish and shellfish that are known to be killed by *Pfiesteria piscimorte* (nov. gen et sp.)*

NATIVE ESTUARINE / MARINE SPECIES

American eel *(Anguilla rostrata)*
Atlantic croaker *(Micropogonias undulatus)*
Atlantic menhaden *(Brevoortia tyrannus)*
Bay scallop *(Aequipecten irradians)*
Black grouper *(Mycteroperca bonaci)*
Blue crab *(Callinectes sapidus)*
Channel catfish *(Icatalurus punctatus)*
Hogchoker *(Trinectes masculatus)*
Killifish (mummichog) *(Fundulus heteroclitus)*
Largemouth bass *(Micropterus salmoides)*
Littleneck clam *(Mercenaria mercenaria)*
Pinfish *(Lagodon rhomboides)*
Red drum *(Scianops ocellatus)*
Redear sunfish *(Lepomis microlophys)*
Southern flounder *(Paralichthys lethostigma)*
Spot *(Leiostomus xanthuris)*
Spotted sea trout *(Cynoscion nebulosus)*
Striped bass *(Morone saxatilis)*
Striped mullet *(Mugil cephalus)*
White perch *(Morone americana)*

EXOTIC (INTRODUCED) SPECIES

Clownfish *(Amphiprion percula)*
Goldfish *(Carrasius auratus)*
Guppie *(Poecilia reticulata)*
Hybrid Striped Bass *(Morone saxatilis × Morone chrysops)*
Mosquitofish *(Gambusia affinis)*
Tilapia *(Oreochromis aureus, Oreochromis mossambicus, Tilapia nilotica)*

* All 28 species of finfish and shellfish species tested thus far have proven susceptible to lethal effects of this ambush predator dinoflagellate (Burkholder *et al.* accepted <u>b</u>).

debris (Burkholder *et al.* 1993; research supported by a UNC Sea Grant mini-grant to PI JMB; Fig. 2, Plate 1). These considerations have led many U.S. coastal regulatory agencies to express the need for development of a species-specific "marker" that would enable rapid detection of this dinoflagellate, in its many stages, at sites of fish disease outbreaks or kills in their coastal waters and aquaculture facilities.

In the short time since its discovery in the Pamlico Estuary in 1992, *Pfiesteria piscimorte* (gen. et sp. nov.) has come to be regarded as a potentially widespread but typically undetected source of major fish mortality in nutrient-enriched estuaries (Burkholder *et al.* 1992a, 1993). Within the past two years, JMB and colleagues have tracked this organism to fish kill sites in eutrophic estuaries from the Delaware Bay south to Florida on the western Atlantic Coast, and as far west as Mobile Bay, Alabama on the Gulf Coast (Burkholder *et al.* 1993, Burkholder *et al.* submitted, Lewitus *et al.* submitted; Fig. 3). Scientists and regulatory staff on

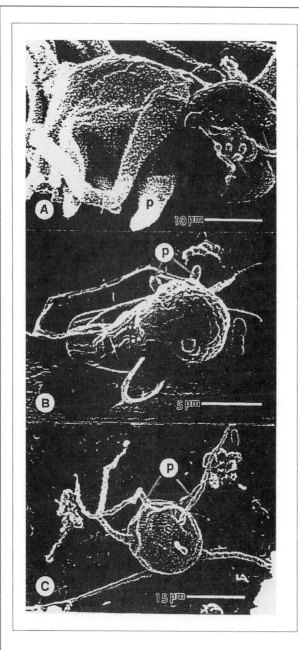

**PLATE 1.** Scanning electron micrographs of the ambush predator *Pfiesteria piscimorte* (gen. et sp. nov.), including (A) a large lobose amoeba [AM] with pseudopodia [p] and a TFVC [DI] beginning to form pseudopodia [p], but with the transverse flagellum [tf] still intact (× 3520); (B) More advanced state of TFVC conversion to an amoeba [ep = epitheca, hy = hypotheca] (× 2570); and (C) "Star" or filipodial amoeba stage [S] (formerly a TFVC) with long pseudopodia (× 1520).

the Atlantic, Gulf and Pacific Coasts have begun to hunt for the dinoflagellate at sites of ulcerative fish disease and sudden-death kill areas (Anonymous 1992; Kay 1992; Rensberger 1992a,b; Greer 1993; Huyghe 1993). **Methodology for rapid, reliable detection of this pathogen in advance of fish disease outbreaks**

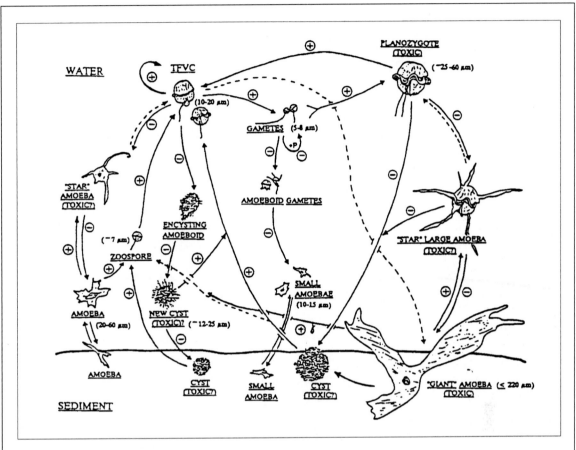

**FIGURE 2.** The life cycle of *Pfiesteria piscimorte* (gen. et sp. nov.), a representative toxic fish ambush-predator dinoflagellate (TFVC = toxic flagellated vegetative cell, the most toxic stage). Symbols indicate how the dinoflagellate appears in the presence (+) versus absence (–) of live finfish. Solid lines represent verified pathways in the complex life cycle; dashed lines indicate hypothesized additional pathways. "P" designates stimulation of amoeboid gamete production by phosphate enrichment. Stages known or suspected (?) to be toxic are also shown (from Burkholder *et al.* 1993).

**and kills is critically needed in order to develop effective management strategies to predict its toxic activity and mitigate its acute and chronic effects on our estuarine fisheries.**

## C. Progress to Date on Gene Probe Development for *Pfiesteria piscimorte*

Within the past three years, both national and international workshops on toxic marine phytoplankton have identified an "imperative" for improved techniques to enable accurate, rapid detection and identification of harmful algae by regulatory staff on a routine basis (UNESCO 1991, Anderson *et al.* 1993). From a practical standpoint, the need for a sensitive highly species-specific technique to detect harmful species such as *Pfiesteria piscimorte* (gen. et sp. nov.) with multiple cryptic stages would be especially critical. Under a cur-

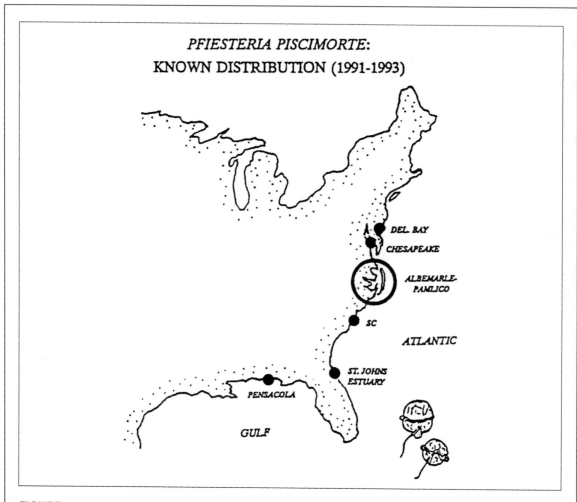

**FIGURE 3.** Locations where water samples have been confirmed to contain toxic stages of *Pfiesteria pisci-morte* (gen. et sp. nov.) (data from PI JMB with H. Glasgow, B. Anderson, R. Lewis, K. Steidinger, and J. Landsberg); Mobile Bay, AL recently has been confirmed as an additional distribution site.

rent mini-grant from the UNC Sea Grant and the UNC Water Resources Research Institute (9/93–7/94), we are developing a gene probe to cultures of *P. piscimorte* (gen. et sp. nov.) that was isolated from the most concentrated known region of toxic activity by this organism—the Albemarle–Pamlico Estuarine System of North Carolina. At present we have successfully extracted DNA from *P. piscimorte* (gen. et sp. nov.), and have amplified a 1.0–1.1 kb fragment (Fig. 4) using "universal" primers to the small subunit ribosomal-DNA gene (16S-rDNA; Pace *et al.* 1986, Lynn & Sogin 1988, Britschgi & Giovannoni 1991, Manhart & McCourt 1992, Weisburg *et al.* 1991). The fragment has been ligated into the *BamHI-XhoI* restriction sites of the pBluescript II SK plasmid, which was then used to transform competent *Escherichia coli* DH5α cells with standard techniques (Sambrook *et al.* 1989, Ausubel *et al.* 1992). We now have available a total of 40 bacte-

**FIGURE 4.** Gell electrophoresis of PCR reaction of *Pfiesteria piscimorte* (gen. et sp. nov.) from the large lobose amoeba stage. Lanes 1 & 8, Lambda, *BstE*II DNA marker. Lane 2, positive control. Lanes 5 & 6, PCR products from the dinoflagellate (length ca 1.1 kb).

rial clones from 2 isolates of the dinoflagellate. We are set to begin sequencing these clones by the dideoxy chain termination method (Sanger 1977), while continuing to establish additional clones from other isolates. After the sequence has been determined, we will look for 17–34 base pair consensus regions within the 16S rDNA that are unique to this toxic alga, for use as species-specific gene probes.

The consensus sequence will be tested as a probe against samples of *P. piscimorte* (gen. et sp. nov.) and other algae. A radioactively end-labeled oligonucleotide probe corresponding to the sequence will be tested at high-stringency with Southern blots and later with dot/slot blots using preparations of the dinoflagellate's DNA in varying amounts. Salt and temperature concentrations will be adjusted so that the probe hybridizes specifically with samples containing the alga's DNA (high-stringency). Its specificity will be further tested by mixing samples of the dinoflagellate's DNA and nontarget DNA in known proportions, to determine whether the probe evokes a signal proportional to the amount of DNA from *P. piscimorte* (gen. et sp. nov.) in the mixed samples. We actually plan to test several probes under these conditions to identify optimal sequences for sensitivity and specificity among culture isolates representing different strains of this dinoflagellate. We are confident that we will have a radiolabeled probe that can be used on extracted DNA samples of cultured *P. piscimorte* (gen. et sp. nov.) by July 1994.

## D. Goals and Hypotheses of the Proposed Research

**The major goal of the proposed research is to refine state-of-the-art techniques for application of gene probes to enable routine detection of *Pfiesteria piscimorte* (gen. et sp. nov.) in natural estuarine habitat and aquaculture facilities.** In this study we will extend application of gene probes to this toxic dinoflagellate for use on water and sediment field samples. We plan to determine the distribution of *P. piscimorte* (gen. et sp. nov.) in coastal systems and assess (a) genetic variability among populations; (b) whether this species is unique or representative of additional, as-yet-unknown toxic dinoflagellates, and (c) the distribution of "cyst banks" as it relates to nutrient conditions, especially to sites of anthropogenic nutrient input. We will also examine physical relationships (portals of entry / attachment) of this fish pathogen to its fish and invertebrate hosts, as indicative of sites of chemosensory attraction. Our research will focus on the following hypotheses:

1. Diverse races or strains within the species *Pfiesteria piscimorte* (gen. et sp. nov.) are abundant and widespread in eutrophic estuaries of North Carolina. Intraspecific genetic variability increases among populations in similar habitat of major estuaries from the mid-Atlantic to the Gulf Coast in the southeastern United States.

2. The unique primers that allow for PCR (polymerase chain reaction) amplification of targeted sequence(s) of rRNA genes in this toxic dinoflagellate can be used as a template from which other gene probes can be developed to enable detection of various strains within *P. piscimorte* (gen. et sp. nov.).

3. Sensitive, highly specific radiolabeled or fluorescent gene probes for *P. piscimorte* (gen. et sp. nov.), developed from cultured material, can be applied successfully for rapid detection of flagellated, amoeboid, and encysted stages of this fish pathogen—even in low abundance—in estuarine water column / sediment samples, and in aquaculture facilities.

## OBJECTIVES

The overall objective of this research is to develop, test and refine application of gene probes for rapid, routine detection / quantification of the toxic dinoflagellate *Pfiesteria piscimorte* (gen. et sp. nov.), in all its stages, from natural estuarine waters and sediments and from aquaculture facilities. Specific objectives to accomplish this goal are as follows:

1. **Extend application of the gene probe(s) for this toxic dinoflagellate so that we can (a) detect it quickly during fish kills or disease outbreaks; (b) assay estuarine sediments for detection of ambient active (amoeboid) and dormant (encysted) populations; and (c) apply the probes as a basic tool** for further research on the ecology, toxicology, and developmental biology of this fish pathogen. [Years I–II]. Specifically, this work will require development in two directions:

   (i) **Gene probe** development for use as an extremely sensitive "field" detection tool to determine genetic variability in *P. piscimorte* (gen. et sp. nov.) populations from both the water column and the sediments, and for use as a basis for additional gene probes to related species. This gene probe will consist of the unique primers that allow for PCR amplification of the target sequence of rRNA genes, even if the rRNA genes are present in low copy number. The high quantity of DNA per dinoflagellate cell (ca 3–200 pg cell$^{-1}$; Triplett *et al.* 1993), relative to DNA content in other algae (average of 0.54 pg DNA cell$^{-1}$; Rizzo 1987) assures that substantial material can be obtained from relatively few cells of this organism.

(ii) **Fluorescent probe** development as an *"in situ"* probe for rapid visual identification and localization of the target species. This second probe will actually be an anti-sense **mRNA oligoprobe** with a fluorescent tag that binds to target mRNA, found in high copy number in the dinoflagellate's cells, to facilitate microscopic visualization (DeLong *et al.* 1989).

2. **Apply the probes for detection / quantification of this dinoflagellate from the water column and sediments of known sudden-death fish kill sites, and at major fish kill events** [Year I (pilot); Years II–III]. This will involve examining the intraspecific genetic variability in *P. piscimorte* (gen. et sp. nov.), a process that will also lead to detection of closely related species with similar behavior and appearance.

3. **Screen and identify targeted areas (portals of entry, sites of attachment) of the pathogen on/in fish prey,** of use in related research to identify the substance(s) from fish tissues that trigger toxin production, and to aid in understanding the pathology of this organism. The fluorescent probe will facilitate light-microscopy inspection of specific tissues. This application will also be of value in providing evidence for the presence of the organism in "individual" fish kill events.

4. **Determine utility of gene probes for this pathogen to the aquaculture industry** [Years II, III]. Our probes will enable identification of problematic periods for use of estuarine waters, and rapid determination of the degree of water source contamination by the toxic dinoflagellate. The probes will also be of use in discerning contaminated tanks or fish.

## METHODOLOGY AND RATIONALE

### A. Culturing This Unique Toxic Dinoflagellate (Burkholder; Objective 1; Years I–III)

Accomplishment of the central goal of this research—refinement of gene probes for *Pfiesteria piscimorte* (gen. et sp. nov.)—depends on successful maintenance of culture isolates for quality assurance in efficiency of extraction of the dinoflagellate's DNA from water and sediment samples. This organism is unique among toxic dinoflagellates in two major characteristics: (i) Unlike nearly all other known toxic dinoflagellates, it is animal-like rather than photosynthetic. As mentioned, it is capable of assuming a photosynthetic habit only by consuming other algae and digesting all but their chloroplasts, which it retains as cleptochloroplasts for extended periods (Steidinger *et al.* 1994); and (ii) *P. piscimorte* (gen. et sp. nov.) requires an as-yet-uncharacterized substance(s) from live fish at least daily to maintain toxic activity and complete its life cycle. These features mandate aeration of cultures, with best dinoflagellate growth when aeration is accomplished using box filters (Burkholder *et al.* 1993).

The challenge of safely culturing this alga demands considerable expense and effort. Because of the potential for impairment to human health from neurotoxic aerosols, all cultures are maintained in an isolated, quarantined modular facility with restricted access (Burkholder *et al.* 1993). Each aquarium must be separately housed within a completely contained, custom-designed isolation chamber that is vented (through HEPA filters) to the air outside the facility. As we collect isolates from sudden-death fish kill sites in North Carolina and other areas for testing with our gene probes, these isolation chambers will protect cultures from cross-contamination while affording safe working conditions. We have successfully maintained high concentrations ($\geq$ 1,000 to ca 90,000 cells mL$^{-1}$) of toxic flagellated stages and cysts using tilapia prey (*Oreochromis mossambica;* 3 fish, each 5–7 cm in length, added 1x or 2x daily per 9-L aquarium). The culture medium consists of 15$^0$/oo-salinity water obtained by adding Instant Ocean salts to well water (NCSU School of Veterinary Medicine). To obtain unialgal cultures, fish are physically separated from dinoflagellate cells that have been isolated by hand (with micromanipulator at 400x) and inoculated into 0.45–0.80 μm-porosity dialysis tubing that allows passage of the toxin and finfish secreta / excreta.

Our modular culture facility includes a separate "cold" (uncontaminated) room for final cleaning of all aquaria and glassware with dilute bleach, a procedure which destroys even the encysted dinoflagellates without leaving toxic residues from the alga or the fixative. A larger adjacent room serves as the culture area, with facilities at one end for soaking contaminated labware and aquaria in strong bleach as initial steps in our cleaning procedures. We routinely use respirators (with HEPA + organic acid filters), disposable gloves, disposable boots, and clothing that is removed and treated with dilute bleach (0.05%) after use, as standard operating procedure. To de-contaminate aquaria, all reusable labware is soaked in dilute bleach for 12 hr to destroy the cells and cysts, and then is rinsed thoroughly with deionized water. All work in the culture facility further mandates a "buddy system" for additional safety assurance, with at least two people present at all times.

## B. Rationale for Use of Genetic Probes

Recent advances in molecular biology have shown great promise in unraveling many basic and applied ecological questions. Brand (1988) reviewed the topic of genetic variation in phytoplankton and concluded that such variation, while not well understood, likely is important in determining the ecological "success" of algal species. Manhart & McCourt (1992) noted the potential value of molecular tools in clarifying phytoplankton speciation and species identifications. Within the past five years, an extensive literature on molecular biology of phytoplankton has become available, including studies of structural characteristics of algal nuclear, chloroplast, and mitochondrial genomes (e.g., Coleman & Goff 1991, Li & Cattolico 1992, Anderson *et al.* 1992, Reynolds *et al.* 1993), as well as information on ribosomal RNA genes (e.g., Rowan & Powers 1992, Scholen *et al.* 1993) and specific functional genes such as large subunit ribulose-1,5-bisphosphate carboxylase (*rbcL;* e.g., Pichard *et al.* 1993).

The potential value of such knowledge has been recognized by recent workshops to identify essential research needs for understanding and mitigating harmful algal blooms (UNESCO 1991, Anderson *et al.* 1993), where strong recommendations have called for development of immunologic and molecular probes to economically important toxic algal species. Such probes could provide rapid sample analysis while avoiding subjective visual identifications, made with light microscopy, that can be challenging even to scientists who are experienced in working with this dinoflagellate (Burkholder *et al.* 1993). Further, many of the difficulties associated with probe development (e.g., limited knowledge base, intensive laboratory effort, expense) have been resolved or reduced in recent years (e.g., significant advances in PCR technology and applications).

Our choice of gene probes over immunologic probes is based on several considerations. First, gene probes offer a high degree of specificity, and technology has improved to the point that development of such probes is realistic within the projected time frame. Second, gene probes have all the advantages of immunologic probes (e.g., specificity, sensitivity, capacity for linkage with radiochemical, fluorescent, or enzymatic markers), with the additional advantage of providing direct insights about genetic variability among isolates via the nucleotide sequence. Selection of rDNA probes over alternative gene probes to specific functional genes (e.g., enzymes) derives from their common use (e.g., DeRijk *et al.* 1992) as well as the state of our knowledge about the genetic structure of *Pfiesteria piscimorte* (gen. et sp. nov.). The rDNA probes have been developed and used extensively over the past decade as both intra- and interspecific markers, since ribosomal RNA is abundant in cells of all living organisms. Further, the DNA that codes for the RNA contains regions which are conserved across all kingdoms as well as variable regions that are unique at species / subspecies levels (Pace *et al.* 1986, Britschgi & Giovannoni 1991, Scholin *et al.* 1993). As a result, a large information base is available that includes both applications and rDNA sequences (e.g., GenBank, EMBL), and which spans a wide array of prokaryotic and eukaryotic organisms including dinoflagellates (e.g., Lenaers *et al.* 1989, Geraud *et al.* 1991, Fritz *et al.* 1991, Rowan & Powers 1992, DeRijk *et al.* 1992).

The use of probes to specific functional genes can offer insights about organism abundance and activity, as Pichard & Paul (1993) and Pichard *et al.* (1993) have shown with *rcbL* genes in photosynthetic algae. Our limited knowledge of *Pfiesteria piscimorte* (gen. et sp. nov.), however, makes it difficult to identify functional genes of optimal use as targets for such probes. For example, the organism "borrows" or retains chloroplasts from algal prey; aside from these "cleptochloroplasts," this dinoflagellate has none of its own (Steidinger et al. 1994). Hence, use of *rcbL* probes would be misleading. Further, the ability of this fish pathogen to transform rapidly among stages that are distinct in morphology and function suggests that targeting highly specific functional genes would be premature. Thus, at present we have focused our efforts toward developing "conservative" probes which we believe will have the most utility. In particular, with respect to aquaculture application of pretesting facilities for the presence of *P. piscimorte* (gen. et sp. nov.), this approach is less likely to lead to false negatives and subsequent economic loss. Ultimately, we hope to develop not only markers to specific strains of this dinoflagellate, but also to find markers that are unique to different stages in the life cycle and/or different functions of the organism—that is, unique signatures of mRNA expression to each stage (e.g., markers specific to flagellated, encysted, or amoeboid structures).

## C. Research Plan for Extending Probe Applications

*1. Fluorescent Probe Development* (Burkholder/Rublee; required for Objectives 1-ii, 2–5; Years I–III; Table 2)—Immediately upon confirming suitability of the gene probe currently under development from cultured isolates, we will generate an anti-sense oligonucleotide probe tagged with a fluorescent marker (procedure of DeLong *et al.* 1989), using aminoethyl phosphate linkers and fluorescein dyes. This probe will first be tested for specificity and sensitivity against cultured dinoflagellate cells, followed by tests against water samples from fish kills or fish kill areas in which *Pfiesteria piscimorte* (gen. et sp. nov.) has been implicated as the causative agent.

*2. Selection of Sediment Extraction Techniques* (Rublee/Burkholder; required for Objective 3; Year I)—In sandy sediments, fluorescent probes should facilitate detection of the dinoflagellate's cysts as well as filipodial and lobopodial amoeboid stages, which tend to wrap their pseudopodia around particulate matter and tightly adhere when disturbed so that they are exceedingly difficult to detect by light microscopy. We anticipate that the amoebae will behave in this manner when they are sampled during or following distribution into aquarium cultures for bioassay of toxic activity in the presence of fish. The probes will be applied to water-column samples from the aquarium bioassays, of use as an aid in "ground truthing" our direct microscope counts.

We expect that detection of *Pfiesteria piscimorte* (gen. et. sp. nov.) in sediments with low cell abundance or high organic content will require use of more sensitive assays. We will probe these types of samples by extracting sediment DNA and using PCR amplification with our species-specific primers (e.g., Tsai & Olsen 1991, 1992; Moran *et al.* 1993). Since estuarine sediments vary in mineral and organic content (Rublee 1982; which may affect DNA extraction efficiency), we will first test defined sediments that have been seeded

**TABLE 2.** Timetable for completing the proposed research

| PROJECT ACTIVITY | YEAR I | YEAR II | YEAR III |
|---|---|---|---|
| **1.** Culturing the dinoflagellate | | ———— | |
| **2.** Extending probe applications: | | | |
| A. Fluorescent probe development | — | | |
| B. Development of sediment extraction methods | —— | | |
| C. Development of probes with varying sensitivity | ———— | – – – – – – – – | |
| **3.** Field sampling effort: | | | |
| A. Analysis of natural "bloom" samples | – – – | —— | —— |
| B. Collection trips—North Carolina | – – —— | —— | —— |
| C. Sampling from other locations (including network) | —— | —— | —— |
| **4.** Probe application to detect the dinoflagellate on fish prey | – – ———— | — – – – | |
| **5.** Testing aquaculture sites | | ———— | ———— |
| **6.** Completion of reports and publications | —— | —— | —— |

with the dinoflagellate's cells, followed by tests of natural sediments using added cells as an "internal standard" for DNA extractions. The lysozyme-SDS-freeze-thaw extractions procedure of Tsai & Olsen (1992) will be followed by purification by electrophoresis (cf. Erb & Wagner-Döbler 1993, Herrick *et al.* 1993), to the lysozyme-hot phenol extraction method of Moran *et al.* (1993, omitting the DNAse digestion step). An alternative procedure, to be considered if the rDNA probes prove unsuitable, will be to assay directly for rRNA with anti-sense oligonucleotide primers using reverse transcriptase PCR (Moran *et al.* 1993).

### 3. Development of Probes that Differ in Sensitivity  (Rublee/Burkholder; required for Objective 2; Year I [pilot]; Years II–III)—Within the past year, the discovery of a second Pfiesteria-like species (Landsberg *et al.* 1994) suggests that *Pfiesteria piscimorte* (gen. et sp. nov.) likely is one of a number of similar ambush-predator dinoflagellates (Burkholder 1993) that may play important roles in coastal ecosystems due to their varied life forms. An approach to screen for other such species which, like *P. piscimorte* (gen. et sp. nov.), may be cryptic with ephemeral stages, is to extend the utility of our probes by altering their specificity and developing a "family" of probes (*sensu* Anderson 1993). This will be accomplished either by (1) varying the conditions under which hybridization is allowed to occur (altering stringency), or (2) altering the nucleotide sequence in the probes themselves. We plan to test isolates as well as natural samples from kill sites under varying stringency conditions.

Improved selection of probe sequence will be achieved over time as we strengthen our understanding of the intraspecific genetic variability in *P. piscimorte* (gen. et sp. nov.) and related taxa. In particular, for all estuarine / aquaculture sampling efforts (see D, below), where we find dinoflagellates that appear visually similar to *P. piscimorte* (gen. et sp. nov.) but do not hybridize with probes under high-stringency conditions, we will reprobe using lower-stringency conditions as a test for other strains or closely related species (cf. Barkay *et al.* 1990). In either case, if we are able to isolate dinoflagellate cells from such samples, we will then use "universal" PCR primers to repeat the amplification, cloning, and sequencing process that was initially used to develop the *P. piscimorte* (gen. et sp. nov.) gene probes (see "Introduction and Background Information"). These steps will enable us to confirm new strains or species, and to develop additional probes of appropriate specificity.

### D. Analysis to Verify Ability to Track the Pathogen in Natural Habitat
(Rublee/Burkholder; required for Objective 2; Year I [pilot], Years II–III)—

As a component of a related research effort (with separate funding support), over the next four years PI JMB will be intensively sampling the Minnesott Beach / Cherry Point region of the Neuse Estuary, which represents a major repeat-kill site for this dinoflagellate (Burkholder *et al.* 1993). In the proposed research, we plan to take advantage of that sampling effort in order to assess the utility of our probes in tracking toxic *Pfiesteria piscimorte* (gen. et sp. nov.) blooms. The water column will be sampled at biweekly to monthly intervals from April–October, which is the period associated with maximal growth and toxic outbreaks of this organism. We will check for increasing abundance of toxic stages using the Utermöhl technique (Lund *et al.* 1958) as in Burkholder & Wetzel (1989), and will verify our identifications using SEM. The data will be compared with / without application of our fluorescent probe to ground truth our direct counts by light microscopy. Fish kill events will be sampled more intensively (2-hr intervals on short-term kills; daily intervals during kills of longer duration [≥ 1 week]), extending for 1 week after the kill to track the dinoflagellate's decline. Analysis of the results will provide insights about the level of spatial / temporal resolution required to adequately sample such events. Live water samples from fish kills (parts D,E) will be screened for the presence of toxic stages using aquarium bioassays with our standard tilapia test species (Burkholder *et al.* 1992a). Tilapia is not endemic to North Carolina estuaries; it offers the advantages of constant availability, wide salinity tolerance, and certainty of no prior contamination by the alga.

### E. Analysis of Natural Samples for Geographic Distribution
(Burkholder/Rublee; required for Objective 2; Year I [pilot-NC]; more intensive during Years II–III, including surveys of selected estuaries in DE, MD, GA, FL)—

An understanding of regional as well as local genetic variability among strains of *Pfiesteria piscimorte* (gen. et sp. nov.) is critical to development of highly specific gene probes that can be used by regulatory staff to detect this dinoflagellate in water samples on a routine basis. Our knowledge of intraspecific variability will be strengthened by the field component of the proposed research. Throughout this study we plan to collect natural water and sediment samples from sudden-death fish kills, known fish kill sites (from surveys of the Pamlico, Neuse and New River Estuaries in North Carolina; the Indian River in Delaware; the Choptank River in Maryland; the Savannah River in Georgia; and the St. Johns (Atlantic) and Pensacola (Gulf) areas in Florida that are suspected to harbor *Pfiesteria piscimorte* (gen. et sp. nov.). Samples will be obtained as follows:

1. **Bloom events during kills—We plan to continue to coordinate a network of state personnel and volunteer citizens** (established in 1991, with help from NC Division of Environmental Management and the NC Division of Marine Fisheries, and from concerned citizens groups such as the Neuse River Foundation and the Tar-Pamlico River Foundation) to help sample estuarine fish kills while in progress. We will also sample waters and sediments during fish kills ourselves whenever possible (i.e., whenever we are able to mobilize and complete the 3-hr drive to the North Carolina coast to sample the sudden-death fish kills while in progress). Water samples will be taken with an integrated water-column sampler (Neuse mean depth in the study area = 3.5 m); the upper 2 cm of sediment will be quantitatively sampled in two randomly positioned 0.1 $m^2$ quadrats per site with gentle siphoning guided by SCUBA divers (technique tested well at 26 sites in the Cherry Point area, October 1993).

   Our network has been highly successful in helping to establish the presence of the toxic dinoflagellate at kills during previous years. Based on data from May–Oct. 1991–1993, we expect that 8–12 major fish kills $yr^{-1}$ (involving $\geq 10^3$ fish) will be linked to this organism (Burkholder *et al.* 1993). Following a kill in which *P. piscimorte* (gen. et sp. nov.) is found swarming in the water, with toxicity verified using aquarium bioassays, we will sample biweekly for 3 months to follow the dinoflagellate's population dynamics over time in both the water and sediments. Environmental variables to be sampled (as part of the separate funding support) will include temperature, salinity, pH, dissolved oxygen, total phosphorous, phosphate, total Kjeldahl nitrogen, nitrate, and ammonium, using techniques described in Burkholder *et al.* (1992b).

2. **Local surveys—We will collect replicate water and sediment samples from ca 50 representative estuarine sites in North Carolina** that previously have experienced fish kills (Neuse and Pamlico Rivers) or high incidence of unexplained ulcerative disease (Neuse, Pamlico, and New Rivers; Burkholder *et al.* 1993). Control sites will also be tested, using areas without incidence of fish kill / disease that are regarded as prime habitat for commercially important fisheries. These sites will span salinity gradients from rivers to sounds, and will include areas with/without wastewater discharge. Sampling of surveys sites in NC will begin during Year I, with major field effort planned for July–October (maximal kill / disease incidence) of Years II–III. These samples will be quickly assayed by our gene probe(s) for comparison with direct microscope counts, to confirm utility of the probe and strengthen our database about the distribution of *P. piscimorte* (gen. et sp. nov.) and abundance of flagellated / amoeboid stages in North Carolina estuaries.

   **As part of this effort, we plan to examine correlations between the abundance of this pathogen and nutrient enrichment (P,N), to determine whether the dinoflagellate may be a useful biosensor of cultural eutrophication** (as indicated by nutrient bioassays on some stages in culture [Burkholder *et al.* 1992], and by surveys of the New River Estuary that yielded significantly higher abundance of the toxic precursor stage in sewage outfall sites than in control areas without wastewater discharge [Burkholder & Glasgow unpubl. data; Fig. 5]). Accompanying environmental data will include the variables listed for analysis in the Minnesott Beach / Cherry Point area (1, above).

3. **Geographic distribution (Southeast region)—We will complete a sampling survey of sites in the selected estuaries of Delaware, Maryland, Georgia and Florida,** known or suspected to support populations of this dinoflagellate and close relatives. Culture isolates will be established from each location for tests against our gene probes. [Note that any research with the Pfiesteria-like species discovered by Landsberg *et al.* will be pursued in collaborative efforts led by our Florida Marine Research Institute colleagues, Drs. K. Steidinger and J. Landsberg.]

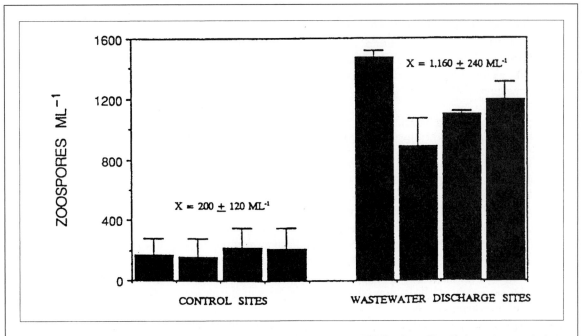

**FIGURE 5.** Abundance of the zoospore precursor stage to lethal TFVCs (see Fig. 2) from 4 control sites (upper Stones Bay, Swan Point, Ellis Cove, and Sneads Ferry Marina, in order of presentation) and 4 wastewater discharge sites (sampled within ca 100 m from the discharge point: Morgan Bay, mouth of French's Creek, lower Northeast Creek, and lower Stones Bay) at the end of the growing season in the New River Estuary, NC (duplicate samples per site; means ± 1 standard deviation). The data suggest that *Pfiesteria piscimorte* (gen. et sp. nov.) may be of use as a biosensor of cultural eutrophication.

## F. Localization of Sites with Maximal Chemosensory Attraction on Finfish
### (Burkholder; required for Objective 3; Year II)—

In a related project, we have begun to screen fish tissues (e.g., gill, cloacal area, blood, mouth) for "swarming" activity or high incidence of attachment of dinoflagellate cells indicating obvious chemosensory attraction. This task has been approached using both light microscopy and scanning electron microscopy (SEM), but with only limited success. Light microscopy does not afford the high resolution needed to readily discern colorless amoeboid stages and small flagellates among fish debris. SEM greatly improves resolution, but analysis is restricted to extremely small tissue areas. We anticipate that fluorescently labeled probe(s) will greatly facilitate screening of larger quantities of tissue by light microscopy.

Juveniles or young adults of three fish species (hybrid striped bass, tilapia, and mosquitofish *[Gambusia affinis]*), known to span a range of sensitivity to the toxins of this dinoflagellate (Burkholder & Glasgow unpubl. data), will be examined with the fluorescent probe(s) to discern sites / tissues that attract high concentrations of toxic cells. In separate bioassays, each fish species will be exposed to an actively growing culture of *Pfiesteria piscimorte* (gen. et sp. nov.). At intervals between introduction and death of prey, individual animals (each considered a "repeat trial" for the given species) will be sacrificed and gills, cloacal area, blood, mouth, and epidermis in localized regions of interest (e.g., for hybrid striped bass, the lymphatic canal region below the dorsal fin which typically fills with blood in response to the dinoflagellate's toxin) will be exposed to our fluorescent probe in the manner of DeLong *et al.* (1989). Time course analysis (1 hr, 2 hr, and 4 hr) of incubation with the fluorescent probe will be completed to determine the optimum exposure period for labeling dinoflagellate cells. Samples can then be screened for detection of the pathogen's flagellated / amoeboid stages among fish debris.

### G. Aquaculture Site Testing
### (Rublee and Burkholder; required for Objective 4; Year III)

This project component will be addressed using two approaches. First, we plan to obtain water samples and bottom scrapings in culture tanks used in finfish aquaculture at NMFS–Beaufort. For the past three years, PI JMB has counseled colleagues at this facility (Dr. A. Powell and Mr. W. Hettler, who are culturing southern flounder, Atlantic menhaden, spot *[Leiostomus xanthuris],* bay scallops and other species) at this facility. The seawater supply is Taylors Creek, which is a known "hot spot" and kill site for *Pfiesteria piscimorte* (gen. et sp. nov.; Burkholder *et al.* 1993). In collaborative efforts we have learned that menhaden eggs do not hatch when lethal stages of the alga are abundant in the culture tank water (Burkholder *et al.* 1993). Moreover, southern flounder delay reproduction and develop large open, bleeding ventral sores when populations of toxic amoebae begin to grow on the tank floor. We will provide gene probe versus light microscope analyses of the samples as our probes come on line especially during years II–III of this effort. Secondly, through UNC–Sea Grant aquaculture extension specialists, we will obtain and similarly test samples from large-scale commercial operations for hybrid striped bass *(Morone saxatilis × Morone chyrsops),* a species which is known to be extremely susceptible to the dinoflagellate (Burkholder *et al.* 1993). Such sampling will enable additional ground truthing or quality assurance of our probes for an extended period of time, and also will provide a means to introduce this early detection technology into the aquaculture industry.

### EXPECTED RESULTS

With bacterial clones containing rDNA fragments now established from two *Pfiesteria piscimorte* (gen. et sp. nov.) isolates, we have made significant progress toward developing a sensitive and reliable detection method for this toxic dinoflagellate. We expect to complete development of this probe for routine monitoring and detection by the fishing industry, as well as for use in basic ecological studies. Insights into the ecology, life history, and genetic character of this organism is a necessary prelude to the development of applications in medical research, and to developing management strategies to mitigate impacts of *P. piscimorte* (gen. et sp. nov.) on estuarine waters which serve as nursery areas and habitat for many commercially and recreationally important finfish and shellfish species.

Within the past year, one Ph.D. candidate has undertaken thesis research to examine nutritional controls on the population dynamics and transformations of *Pfiesteria piscimorte* (gen. et sp. nov.), and will be trained in development of gene probes. This graduate assistant will be continuing his dissertation experiments as a member of the project team for the proposed research. A Master of Science graduate research assistant will be conducting field population studies of this dinoflagellate as thesis research in this project, and she will also be trained in probe development. A second M.S. student and an undergraduate will be involved in gene probe development for the organism at UNC–Greensboro. At least four publications are planned per year from this research in internationally refereed journals. In other education-related activities, PI JMB has presented an average of more than 80 lectures and seminars per year in 1991–1993 about this toxic dinoflagellate and its effects on estuaries and aquaculture, including presentations to undergraduate students, graduate students, concerned citizens groups, regulatory staff, legislators, and scientists.

Research on this organism by PI JMB and her staff (1991–1993) has been translated into 37 languages and disseminated by the British Broadcasting Corporation, the Canadian Broadcasting Corporation, CNN, "Good Morning America," *Discover* Magazine, the *New York Times,* the *Washington Post,* and *Time* Magazine. There is strong public support in North Carolina, the United States, and other nations for rapid progress to be made in understanding and mitigating the effects of this organism on our fishery resources. To acquire this knowledge, techniques are critically needed to enable rapid, reliable, early detection of *P. piscimorte* (gen. et sp. nov.) prior to major ulcerative disease outbreaks or kill events. The central goal of the proposed research will be to meet this essential need by developing sensitive, highly specific gene probes for fish pathogen.

Education, graduate student training, and lecture / seminar presentations are important components of information transfer. To further aid in disseminating information about the availability of our gene probes for *Pfiesteria piscimorte* (gen. et sp. nov.), we plan to conduct demonstration / training workshops among regulatory agencies in the North Carolina Department of Environment, Health & Natural Resources (notably the Division of Environmental Management, Shellfish Sanitation, and Division of Marine Fisheries); the National Marine Fisheries Service–Beaufort, NC; and aquaculture specialists in UNC–Sea Grant extension program to inform them of the availability and power of this new technology as it is refined and tested. Staff members from these agencies have already expressed interest in participating in such workshops.

## APPLICATION

This work, addressing several key targets identified by a national Sea Grant initiative for research in Marine Biotechnology, will yield applications of both immediate and long-term value. Development of a sensitive, highly specific diagnostic technique for this ichthyotoxic dinoflagellate pathogen will enable routine *in situ* monitoring of water quality in aquaculture. First, the probes can immediately be used to test for the presence of *Pfiesteria piscimorte* (gen. et sp. nov.) as a contaminant of aquaculture facilities, thereby preventing introduction of valuable fish stocks into ponds with high probability of later fish kills. The probes will enable early detection of the organism's presence so that steps can be taken (e.g., removal of fish so that ponds can be drained and treated with chlorox pellets, known to destroy all stages of the dinoflagellate including resistant resting cysts) so that toxic outbreaks in the ponds or tanks can be mitigated or avoided. In addition, the gene probes will facilitate detection of contaminated fish, leading to improved quarantine measures and improved quality assurance when fish and fish products are transported to/from other states—also an important consideration for human health, since this alga is known to be toxic to humans as well as fish (Burkholder *et al.* 1993; Burkholder & Glasgow unpubl. data and medical records; D. Baden, NIEHS Center Director & Rosenstiel Institute of U. Miami, unpubl. data on the neurotoxins of *P. piscimorte* [gen. et sp. nov.]).

In more long-term benefits, our research will enable more accurate assessment of environmental quality, as related to the presence of this dinoflagellate, in natural estuarine ecosystems. As we improve our knowledge of the distribution of the organism and its relation to nutrient enrichment, we will gain insights about controlling *Pfiesteria piscimorte* (gen. et sp. nov.) and mitigating its acute and chronic effects on commercially important fisheries. Additonal long-term value of this research relates to medical and biological research applications. Medical concerns include the potential for toxin accumulation in humans and, conversely, use of the toxin(s) in medical applications (cf. Carmichael 1994). In related efforts we are working with a colleague, Dr. Daniel Baden (also president of Chiral Corporation, Miami, FL, a firm that specializes in developing test kits for routine analysis of dinoflagellate toxins), to characterize the toxins produced by this alga. Once this is accomplished, we plan to join forces to develop rapid detection / quantification procedures for both the organism and its toxins in natural estuarine samples and aquaculture ponds. *P. piscimorte* (gen. et sp. nov.) also offers interesting features from a developmental biology standpoint; in response to environmental cues, its ability to transform rapidly among stages that differ dramatically in both form and function (Burkholder *et al.* 1993) suggest well-regulated developmental controls.

From an ecosystem standpoint, greater understanding of this abundant, widespread organism—with multiple amoeboid stages that previously were not even recognized as dinoflagellates—has already begun to change existing paradigms about the role of dinoflagellates in estuarine food webs (Mallin *et al.* submitted). Moreover, the data on dinoflagellate-related fish kills gained from our coordinated monitoring effort will contribute to a long-term data base that will be of value in correlating toxic outbreaks with environmental conditions, and evaluating the impacts of this toxic alga on our fisheries resources. Development of sensitive, highly specific gene probes for *P. piscimorte* (gen. et sp. nov.) will answer a critical need for routine, early detection of this toxic dinoflagellate by regulatory staff and finfish / shellfish aquacultrists.

## REFERENCES

Anderson, D.A. & L. Hall. 1992. *State of Maryland Water Quality Monitoring Report for Fish Kills in 1990–1991.* Wye River Center, State of Maryland Department of Environmental Management.

Anderson, D.M. 1993. Identification of harmful algal species using molecular probes. In: *Proceedings, Sixth International Conference on Toxic Marine Phytoplankton.* Nantes, France, p. 22.

Anderson, D.M., S. Galloway & J.D. Joseph. 1993. *Marine Biotoxins and Harmful Algae: A National Plan.* Woods Hole Technical Report WHOI 93-02. NOAA, National Marine Fisheries Service.

Anderson, D.M., A.W. White and D.G. Baden. 1985. *Toxic Dinoflagellates.* New York, Elsevier Science Publishing Company.

Anderson, D.M., A. Grabher & M. Herzog. 1992. Separation of coding sequences from structural DNA in the dinoflagellate *Crypthecodinium cohnii. Molec. Mar. Biol.* 1:89–96.

Anonymous. 1992. "Phantom" algae are linked to mass fish deaths. *The New York Times,* 30 July, p.1.

Ausubel, F.M., R. Brent, R.E. Kingston, D.D. Moore, J.G. Siedman, J.A. Smith & K.H. Struhl. 1992. *Short Protocols in Molecular Biology.* 2nd ed. New York, John Wiley & Sons.

Barkay, T., M. Gillman & C. Liebert. 1990. Genes encoding mercuric reductases from selected gram-negative aquatic bacteria have a low degree of homology with *merA* of transposon Tn*501. Appl. Environ. Microbiol.* 56:1695–1701.

Brand, L.E. 1988. Review of genetic variation in marine phytoplankton species and the ecological implications. *Biol. Oceanogr.* 6:397–409.

Britschgi, T.B. & S.J. Giovannoni. 1991. Phylogenetic analysis of a natural marine bacterio-plankton population by rRNA gene cloning and sequencing. *Appl. Environ. Microbiol.* 57:1707–1713.

Burkholder, J.M. 1992. Phytoplankton and episodic suspended sediment loading: Phosphate partitioning and mechanisms for survival. *Limnol. Oceanogr.* 37:974–988.

Burkholder, J.M. 1993. Tracking a "phantom:" The many disguises of the new ichthyotoxic dinoflagellate. In: *Proceedings, Fifth International Conference on Modern and Fossil Dinoflagellates.* Zeist (The Netherlands), p. 22.

Burkholder, J.M. & R.G. Wetzel. 1989. Epiphytic microalgae on natural substrata in a hardwater lake: seasonal dynamics of community structure, biomass and ATP content. *Arch. Hydrobiol./Suppl.* 83:1–56.

Burkholder, J.M., R.G. Wetzel & K.L. Klomparens. 1994. Track SEM-autoradiography of adnate microalgae. In: *Periphyton Methods Manual,* by R.G. Wetzel (Ed.). Boston, Dr. W. Junk Publishers (in press).

Burkholder, J.M., H.B. Glasgow Jr. & C.W. Hobbs. Response of a ichtyotoxic estuarine dinoflagellate to gradients of salinity, light and nutrients. *Mar. Ecol. Prog. Ser.* (submitted).

Burkholder, J.M., H.B. Glasgow, E.J. Noga & C.W. Hobbs. 1993. *The Role of a New Toxic Dinoflagellate in Finfish and Shellfish Kills in the Neuse and Pamlico Estuaries.* Report No. 93-08, Albemarle–Pamlico Estuarine Study. Raleigh, North Carolina Department of Environment, Health & Natural Resources and U.S. Environmental Protection Agency–National Estuary Program, 58 pp. (3rd printing).

Burkholder, J.M., E.J. Noga, C.W. Hobbs, H.B. Glasgow, Jr. & S.A. Smith. 1992a. New "phantom" dinoflagellate is the causative agent of major estuarine fish kills. *Nature* 358:407–410; *Nature* 360:768.

Burkholder, J.M., K.M. Mason & H.B. Glasgow Jr. 1992b. Water-column nitrate enrichment promotes decline of eelgrass *Zostera marina* L.: Evidence from seasonal mesocosm experiments. *Mar. Ecol. Prog. Ser.* 81:163–178.

Carmichael, W.W. 1994. The toxins of cyanobacteria. *Sci. Amer.* 270:78–86.

Coleman, A.W. & J.L. Goff. 1991. DNA analysis of eukaryotic algal species. *J. Phycol.* 27:463–473.

Culotta, E. 1992. Red menace in the world's oceans. *Science* 257:1476–1477.

DeLong, E.F., G.S. Wickham & N.R. Pace. 1989. Phylogenetic strains: Ribosomal RNA-based probes for the identification of single cells. *Science* 243:1360–1363.

DeRijk, P., J-M. Neefs, Y. Vn de Peer & R. De Wachter. 1992. Compilation of small ribosomal subunit RNA sequences. *Nucleic Acids REs.* 20(Supp.):2075–2089.

Droop, M.R. 1974. Heterotrophy of carbon, pp. 530–559. In: *Algal Physiology and Biochemistry,* by W.D.P. Stewart (Ed.). Botanical Monograph Vol. 10. Berkeley, University of California Press, 989 pp.

Erb, R.W. & I. Wagner-Döbler. 1993. Detection of polychlorinated biphenyl degradation genes in polluted sediments by direct DNA extraction and polymerase chain reaction. *Appl. Environ. Microbiol.* 59:4065–4073.

Fritz, L., P. Milos, D. Morse & J.W. Hastings. 1991. *In situ* hybridization of luciferin-binding protein anti-sense RNA to thin sections of the bioluminescent dinoflagellate *Gonyaulax polyedra. J. Phycol.* 27:436–441.

Gaines, G. & M. Elbrachter. 1987. Heterotrophic nutrition, pp. 224–268. In: *The Biology of Dinoflagellates,* by F.J.R. Taylor (Ed.). Boston, Blackwell Scientific Publications, 785 pp.

Geraud, M.-L., M. Sala-Roviera, M. Herzog & M.-L. Soyer-Gobillard. 1991. Immunochemical localization of the DNA-binding protein HCs during cell cycle of the histone-less dinoflagellate protocista *Crypthecodinium cohnii* B. *Biol. Cell* 71:123–134.

Greer, J. 1993. Alien in our midst? Phantom algae suspected in Bay. *Marine Notes.* University of Maryland Sea Grant, March.

Hallegraeff, G.M. 1992. A review of harmful algal blooms and their apparent global increase. *Phycologia* 32:79–99.

Hallegraeff, G.M., D.A. Steffenson and R. Wetherbee. 1988. Three estuarine Australian dinoflagellates that can produce paralytic shellfish toxins. *J. Plankt. Res.* 10:533–541.

Hauser, D.C.R., M. Levandowsky, S.H. Hutner, L. Chunosoff & J.S. Hollwitz. 1975. Chemosensory responses by the heterotrophic marine dinoflagellate *Crypthecodinium cohnii. Microb. Ecol.* 1:246–254.

Herrick, J.B., E.L. Madsen, C.A. Batt & W.C. Ghiorse. 1993. Polymerase chain reaction amplification of naphthalene-catabolic and 16S rRNA gene sequences from indigenous sediment bacteria. *Appl. Environ. Microbiol.* 59:687–694.

Huyghe, P. 1993. A horrific predatory little plant is beating up on fish around the world. *Discover* Magazine, April.

Kay, J. 1992. Eerie killer algae may be stalking Bay fish. *San Francisco Examiner,* 17 Aug., p.1.

Landsberg, J.H., K.A. Steidinger & B. Blakesley. 1994. Fish-Killing dinoflagellates in a tropical aquarium. In: *Proceedings, Sixth International Conference on Toxic Marine Phytoplankton.* Amsterdam, Elsevier (in press).

Lenaers, G., C. Scholin, Y. Bhaud, D. Saint-Hilaire & M. Herzog. 1991. A molecular phylogeny of dinoflagellate protists (Pyrrhophyta) inferred from the sequence of 24S rRNA divergent domains D1 and D8. *J. Mol. Evol.* 32:53–63.

Levine, J.F., J.H. Hawkins, M.J. Dykstra, E.J. Noga, D.W. Moye & R.S. Cone. 1990. Species distribution of ulcerative lesions on finfish in the Tar-Pamlico Estuary, North Carolina. *Dis. Aquat. Org.* 8:1–5.

Lewitus, A.J., R.V. Jesien, T.M. Kana, J.M. Burkholder & E. May. Discovery of the "phantom" dinoflagellate in Chesapeake Bay. *Estuaries* (submitted).

Li, N. & R.A. Cattolico. 1992. *Ochromonas danica* (Chrysophyceae) chloroplast genome organization. *Mol. Mar. Biol. Biotech.* 1:165–174.

Lund, J.W.G., C. Kipling and E.D. LeCren. 1958. The inverted microscope method of estimating algal numbers and the statistical basis of estimates by counting. *Hydrobiologia* 11:143–170.

Lynn, D.H. & M.L. Sogin. 1988. Assessment of phylogenetic relationships among ciliated protists using ribosomal RNA sequences derived from reverse transcripts. *BioSystems* 21:249–254.

Mallin, M.A., J.M. Burkholder, L.M. Larsen & H.B. Glasgow Jr. Response of two zooplankton grazers to an ichthyotoxic estuarine dinoflagellate. *Mar. Ecol. Prog. Ser.* (submitted). [published 1995 in *J. Plankton Research.* 17:351–363]

Manhart, J.R. & R.M. McCourt. 1992. Molecular data and species concepts in the algae. *J. Phycol.* 28:730–737.

Miller, K.H., J. Camp, R.W. Bland, J.H. Hawkins, III, C.R. Tyndall and B.L. Adams. 1990. *Pamlico Environmental Response Team Report* (June–December 1988). North Carolina Dept. of Environment, Health, and Natural Resources. Wilmington, North Carolina, 50 pp.

Moran, M.A., V.L. Torsvik, T. Torsvik & R.E. Hodson. 1993. Direct extraction and purification of rRNA for ecological studies. *Appl. Environ. Microbiol.* 59:915–918.

North Carolina Division of Marine Fisheries. 1992. *Description of North Carolina's Coastal Fishery Resources.* Draft Report. Morehead City, North Carolina Department of Environment, Health and Natural Resources–Division of Marine Fisheries, 215 pp.

Pace, N.R., D.A. Stahl, D.J. Lane & G.J. Olsen. 1986. The analysis of natural microbial populations by ribosomal RNA sequences. *Adv. Microb. Ecol.* 9:1–55.

Pederson, B.H. & K.P. Anderson. 1992. Induction of trypsinogen secretion in herring larvae *(Clupea harengus). Mar. Biol.* 112:559–565.

Pichard, S.L. & J.H. Paul. 1993. Gene expresion per gene dose: a specific measure of gene expression in aquatic microorganisms. *Appl. Environ. Microbiol.* 59:451–457.

Pichard, S.L., M.E. Fisher & J.H. Paul. 1993. Ribulose bisphosphate carboxylase gene expression in subtropical marine phytoplankton populations. *Mar. Ecol. Prog. Ser.* 101:55–65.

Rensberger, B. 1992a. Huge fish kills linked to slumbering algae. *The Washington Post,* 30 July, p.1.

Rensberger, B. 1992b. Look! In the water! It's a plant, it's an animal, it's a toxic creature. *The Washington Post,* 17 August, Science Feature.

Reynolds, A.E., B.L. McConaughy & R.A. Cattolico. 1993. Chloroplast genes of the marine alga *Heterosigma carterae* are transcriptionally regulated during a light/dark cycle. *Molec. Mar. Biol. Biotech.* 2:121–128.

Rizzo, P.J. 1987. Biochemistry of the dinoflagellate nucleus, pp. 143–171. In: *The Biochemistry of Dinoflagellates,* by F.J.R. Taylor (Ed.). Oxford, Blackwell Scientific.

Robineau, B., J.A. Gagne, L. Fortier and A.D. Cembella. 1991. Potential impact of a toxic dinoflagellate *(Alexandrium excavatum)* bloom on survival of fish and crustacean larvae. *Mar. Biol.* 108:293–301.

Rowan, R. & D.A. Powers. 1992. Ribosomal RNA sequences and the diversity of symbiotic dinoflagellates (zooxanthellae). *Proc. Natl. Acad. Sci.* 89:3639–3643.

Rublee, P.A. 1982. Seasonal distribution of bacteria in salt marsh sediments of North Carolina. *Est. coastal Shelf Sci.* 15:67–74.

Rublee, P.A. 1992. Community structure and bottom-up regulation of heterotrophic microplankton in arctic LTER lakes. *Hydrobiologia* 240:133–141.

Sambrook, J., E.F. Fritsch & T. Maniatis. 1989. *Molecular Cloning: A Laboratory Manual.* 2nd ed. New York, Cold Spring Harbor.

Sanger, F., S. Nicklen & A.R. Coulson. 1977. DNA sequencing with chain-terminating inhibitors. *Proc. Nat. Acad. Sci.* 74:5463–5467.

Scholen, C.A., D.M. Anderson & M.L. Sogin. 1993. Two distinct small-subunit ribosomal RNA genes in the North American toxic dinoflagellate *Alexandrium fundyense* (Dinophyceae). *J. Phycol.* 29:209–216.

Shimizu, Y. 1991. Dinoflagellates as sources of bioactive molecules, p.71. In: *Program and Abstracts of the Second International Marine Biotechnology Conference.* Baltimore, Society for Industrial Microbiology, October (Abstract).

Shumway, S.E. 1990. A review of the effects of algal blooms on shellfish and aquaculture. *Journal of the World Aquaculture Society* 21:65–104.

Smayda, T.J. 1989. Primary production and the global epidemic of phytoplankton blooms in the sea: A linkage?, pp. 449–484. In: *Novel Phytoplankton Blooms,* by E.M. Cosper, V.M. Bricelj and E.J. Carpenter (Eds.). Coastal and Estuarine Studies No. 35. New York, Springer-Verlag, 799 pp.

Steidinger, K.A. & D.G. Baden. 1984. Toxic marine dinoflagellates, pp. 201–262. In: *Dinoflagellates,* by D.L. Spector (Ed.). New York, Academic Press, 545 pp.

Steidinger, K.A., E.W. Truby, J.K. Garrett & J.M. Burkholder. 1994. The morphology and cytology of a newly discovered toxic dinoflagellate, 6 pp. In: *Proceedings, 6th International Conference on Toxic Marine Phytoplankton.* Amsterdam, Elsevier (in press).

Tsai, Y.-L. & B.H. Olsen. 1991. Rapid method for direct extraction of DNA from soil and sediments. *Appl. Environ. Microbiol.* 57:1070–1074.

Tsai, Y.-L. & B.H. Olsen. 1992. Rapid method for separation of bacterial DNA from humic substances in sediments for polymerase chain reaction. *Appl. Environ. Microbiol.* 58:2292–2295.

Triplett, E.L., N.S. Govind, S.J. Roman, R.V.M. Jovine & B.B. Prezelin. 1993. Characterization of the sequence organization of DNA from the dinoflagellate *Heterocapsa pygmaea* (*Glenodinium* sp.). *Molec. Mar. Biol. Biotech.* 2:239–245.

UNESCO. 1991. *Programme on Harmful Algal Blooms.* Workshop Report No. 80. Newport (RI), Intergovernmental Oceanographic Commission.

Weisberg, W.G., S.M. Barns, D.A. Pelletier & D.J. Lane. 1991. 16S ribosomal DNA amplification for phylogenetic study. *J. Bact.* 173:697–703.

White, A.W. 1988. Blooms of toxic algae worldwide: Their effects on fish farming and shellfish resources. In: *Proceedings of the International Conference on Impact of Toxic Algae on Mariculture.* Aqua-Nor '87 International Fish Farming Exhibition, August 1987, Trondeim, Norway, pp. 9–14.

# Response of two zooplankton grazers to an ichthyotoxic estuarine dinoflagellate

Michael A.Mallin, JoAnn M.Burkholder[1], L.Michael Larsen[1] and Howard B.Glasgow,Jr[1]

*Center for Marine Science Research, University of North Carolina at Wilmington, 7205 Wrightsville Avenue, Wilmington, NC 28403 and [1]Department of Botany, Box 7612, North Carolina State University, Raleigh, NC 27695-7612, USA*

**Abstract.** The dinoflagellate *Pfiesteria piscicida* (gen. et sp. nov.), a toxic 'ambush predator', has been implicated as a causative agent of major fish kills in estuarine ecosystems of the southeastern USA. Here we report the first experimental tests of interactions between *P.piscicida* and estuarine zooplankton predators, specifically the rotifer *Brachionus plicatilis* and the calanoid copepod *Acartia tonsa*. Short-term (10 day) exposure of adult *B.plicatilis* to *P.piscicida* as a food resource, alone or in combination with the non-toxic green algae *Nannochloris* and *Tetraselmis*, did not increase rotifer mortality relative to animals that were given only non-toxic greens. Similarly, short-term (3 day) feeding trials using adult *A.tonsa* indicated that the copepods survived equally well on either *P.piscicida* or the non-toxic diatom *Thalassiosira pseudonana*. Copepods given toxic dinoflagellates exhibited erratic behavior, however, relative to animals given diatom prey. The fecundity of *B.plicatilis* when fed the toxic dinoflagellate was comparable to or higher than that of rotifers fed only non-toxic greens. We conclude that, on a short-term basis, toxic stages of *P.piscicida* can be readily utilized as a nutritional resource by these common estuarine zooplankters. More long-term effects of *P.piscicida* on zooplankton, the potential for toxin bioaccumulation across trophic levels, and the utility of zooplankton as biological control agents for this toxic dinoflagellate, remain important unanswered questions.

## Introduction

The toxic estuarine dinoflagellate *Pfiesteria piscicida* (Dinophyceae; gen. et sp. nov.; Steidinger *et al.*, in preparation) was first observed in 1988 as a culture contaminant of unknown origin (Smith *et al.*, 1988; Noga *et al.*, 1993). Its lethal effects on a wide array of finfish and shellfish have since been documented at major fish kills and in laboratory trials (Burkholder *et al.*, 1992, 1995a). *Pfiesteria*-like species act as 'ambush predators' with chemosensory stimulation by live fish or their fresh secretions and tissues (Burkholder, 1993; Glasgow and Burkholder, 1993). Since its discovery in 1991 during a fish kill in the Pamlico Estuary, North Carolina, USA (Burkholder *et al.*, 1992), this dinoflagellate has been implicated as a causative agent of major fish kills in estuarine and coastal waters (Burkholder *et al.*, 1995a). *Pfiesteria*-like species apparently are widespread; recent investigations have demonstrated their presence at 'sudden-death' estuarine fish kill sites from the mid-Atlantic (Delaware and Chesapeake Bays) to both the Atlantic and Gulf Coasts of Florida, extending to the Alabama coast (Burkholder *et al.*, 1995a; Lewitus *et al.*, 1995).

*Pfiesteria piscicida* is the first known estuarine or marine dinoflagellate to have a complex life cycle with at least 19 flagellated, amoeboid, and encysted stages (Burkholder and Glasgow, 1995). 'Phantom-like' toxic flagellated vegetative cells (TFVCs, diameter ~9–20 μm), the most lethal known stage, develop from zoospores that emerge from resting cysts or are produced by other non-toxic stages when live fish are detected, e.g. when a school of fish enters an estuarine tributary to feed. The TFVCs swim up from the sediment in chemosensory response to fish secreta; they excrete a potent neurotoxin which creates open bleeding sores and causes sloughing of tissue in affected fish. The active TFVCs consume small flagellated algae, protozoans and the sloughed bits of fish tissue (Burkholder *et al.*, 1992, 1993). The fish are also narcotized by the toxin so that they linger in the area. Consumption of small flagellated algae by TFVCs (Steidinger *et al.*, 1995) while attacking fish results in a temporary, specialized form of mixotrophy in which the chloroplasts are sequestered and retained for short-term functional capacity (days). Hence, the TFVCs can utilize the small algal flagellates' cleptochloroplasts as a 'stolen' means of photosynthesis (e.g. Schnepf *et al.*, 1989; Fields and Rhodes, 1991; Schnepf and Elbrächter, 1992).

TFVCs produce gametes (diameter 5–8 μm) that are stimulated to complete sexual reproduction (formation of planozygotes, diameter 10–60 μm) when the prey begin to die. Upon fish death, the population may follow several possible courses: (i) the cells transform into multiple, active, heterotrophic amoeboid stages (length 5–250 μm; from TFVCs, gametes and planozygotes) which consume the fish remains and prey upon other microorganisms; (ii) they revert from sexual to non-toxic asexual forms (zoospores, diameter 5–9 μm), especially in nutrient-enriched waters with abundant flagellated algal prey (TFVCs and gametes); or (iii) they encyst under stressed conditions (TFVCs, planozygotes, amoebae, and possibly gametes and zoospores; Burkholder *et al.*, 1992, 1995b).

Even extremely lethal cultures, fed live fish repeatedly, retain their toxicity for <24 h when denied further access to fish prey (Burkholder *et al.*, 1992, 1993). Although *P.piscicida* is often ephemeral in the water column (hours)—killing its prey and then either encysting or transforming to amoeboid stages and following the carcasses down to the sediments or into shore—its toxic outbreaks have been known to extend for days to weeks when schools of fish are available, such as during autumn migrations of Atlantic menhaden (*Brevoortia tyrannus* Latrobe) from shallow estuaries out to sea (Burkholder *et al.*, 1992).

The ciliated protozoan *Stylonichia* cf. *putrina* has been observed to consume the TFVCs of *P.piscicida* without apparent adverse toxic effects (Burkholder *et al.*, 1992). During prolonged feeding events, however, remaining planozygote stages of the dinoflagellate can transform into large amoebae which in turn, engulf the ciliate as predator becomes prey (Burkholder *et al.*, 1992, 1993). Interactions between toxic stages of *P.piscicida* and other potential predators are currently unknown, such as whether mesozooplankton or microzooplankton aside from *Stylonichia* can consume or control it, or whether they are adversely affected by its toxin. Other toxic dinoflagellates are used as food resources by some zooplankters (Watras *et al.*, 1985; Turner and Tester, 1989), but they reduce fecundity and survival in other zooplankton species or adversely affect higher trophic levels through toxin bioaccumulation (White, 1980, 1981; Huntley *et al.*, 1986).

The objective of this research was to determine the effects of *P.piscicida* on the survival and fecundity of the common estuarine parthenogenic rotifer

*Brachionus plicatilis* Mueller, and the survival of an obligate sexually reproducing calanoid copepod, *Acartia tonsa* Dana. We expected that if the zooplankters were adversely affected by the toxic dinoflagelate, survival and/or fecundity would be depressed when TFVCs were presented as the available algal food supply, relative to the response of animals given non-toxic algae alone or in combination with *P.piscicida*.

### Method

Culture isolates of *P.piscicida* were collected on 23 May 1991 from the Pamlico River Estuary near Channel Marker no. 9 at the mouth of Blount Bay in Beaufort County, North Carolina, during an active bloom of TFVCs while ~1 million Atlantic menhaden, southern flounder (*Paralichthys lethostigma* Jordan & Gilbert), hogchokers (*Trinectes maculatus* Block & Schneider) and spot (*Leiostomus xanthuris* Lacepede) were dying (Burkholder *et al.*, 1993). The cultures were maintained in a walk-in culture facility under 50 $\mu$E m$^{-2}$ s$^{-1}$ illumination (cool white fluorescent lamps) at 20°C with a 12:12 h light:dark (L:D) cylce in 40 l aerated aquaria filled with artificial seawater at 15‰ salinity (seawater derived by adding Instant Ocean salts to water from a well-water source on the grounds of North Carolina State University).

The dinoflagellate's TFVCs require an unidentified substance in fresh fish excreta (Burkholder *et al.*, 1992); hence, it was necessary to maintain cultures using live fish. We routinely fed the dinoflagellate tilapia (*Oreochromica mossambica* Peters, each 5–7 cm in length and washed thoroughly with deionized water) at a density of 15–20 fish day$^{-1}$, and removed all dead fish as live replacements were added. The cultures also contained the small blue–green algae, *Lyngbya* sp. (Cyanophyceae; mean cell diameter 2 $\mu$m, mean filament length 9 $\mu$m) and *Gloeothece* spp. (gelatinous colonial forms; mean cell biovolume 3 $\mu$m$^3$). Contact with culture water and aerosols has been associated with serious human health effects (Glasgow and Burkholder, 1994); therefore, all work was completed using full-face respirators with organic acid filters; disposable gloves, boots and hair covers; and protective clothing that was bleached after use to kill all dinoflagellate stages ($\geq$30% bleach).

Since some TFVCs in a given population are induced to transform to amoeba shortly after disturbance from transport or gentle pouring (more so when the TFVCs are pipetted), we minimized amoeboid transformations by completing all zooplankton feeding trials in the dinoflagellate culture room within close proximity of the stock culture tanks. In all treatments containing *P.piscicida*, the dinoflagellates were added by slow pouring to containers marked for the appropriate volume. Treatments were randomly distributed daily within four covered containers to reduce potential contamination from aerosol components of toxin from the stock dinoflagellate culture tanks, which were also covered but could not be tightly sealed.

The common estuarine rotifer, *B.plicatilis* (Lee *et al.*, 1985), was obtained from W.F.Hettler (NOAA Southeast Fisheries Center, Beaufort, NC). *Acartia tonsa*, an abundant estuarine zooplankter in the Albemarle–Pamlico estuarine system (Peters, 1968; Mallin, 1991), was cultured from samples that were taken from the Newport River Estuary near Beaufort, NC, mixed with cultures supplied by H.Millsaps (Chesapeake Biological Laboratory, Solomon, MD) where *P.piscicida* is also known to occur (Burkholder *et al.*, 1995a). Stock cultures of these organisms were maintained in a S/P Cryo-Fridge environmental chamber at 20°C, 15–20‰ salinity and a 12:12 L:D cycle. The calanoid copepod was grown on the centric diatom *Thalassiosira pseudonana* Hasle (Bacillario-phyceae; mean diameter and mean biovolume 10 μm and 490 μm$^3$, respectively; $n = 25$ cells), and the rotifers were maintained on a green algal culture containing *Nannochloris* sp. and *Tetraselmis* sp. [Chlorophyceae (Bold and Wynne, 1985); mean diameter and biovolume of *Nannochloris* = 3 μm and 9 μm$^3$, respectively; mean length and biovolume of *Tetraselmis* = 9 μm and 450 μm$^3$, respectively; $n = 25$ cells each; supplied by W.F.Hettler, NOAA]. Size ranges of all algal prey were well within that which is known to be grazed by each predator (for the copepod, adult and copepodite stages; Berggreen *et al.*, 1988). Algal cell densities and biovolumes were determined from samples preserved in acidic Lugol's solution (Vollenweider, 1974) with the Utermöhl method (Lund *et al.*, 1958), following the procedure of Burkholder and Wetzel (1989). Cells were quantified under phase contrast at 600× using an Olympus IMT2 inverted microscope.

*Brachionus plicatilis* was tested for survival and fecundity over a 9 day period as follows. Single rotifers (non-egg bearing) were placed in 15 ml of filtered test water (salinity 15‰, 21–22°C, taken from a fish culture without dinoflagellate contamination and filtered through Whatman 934AH filters with pore size ~1.5 μm) within 50 ml polyethylene cups. Water that had previously contained fish was used to facilitate comparison with dinoflagellate treatments, which required addition of *P.piscicida* in fish culture water to enhance toxic activity as long as possible, and to reduce the transformation of TFVCs to less toxic, inedible amoebae or inactive cysts. Each treatment included 10 replicate animals maintained separately. A designated 'unfed' treatment consisted of filtered fish water that contained only small blue–green algae [*Lyngbya* sp. and *Gloeothece* spp., ~2.40 × 10$^5$ μm$^3$ biovolume ml$^{-1}$; both non-toxic (Gorham and Carmichael, 1988)] and bacteria in the filtrate as potential prey items. *Brachionus plicatilis* is known to consume bacteria (Turner and Tester, 1992) and, hence, would not have been completely starved. In all treatments, both animals would also have had access to the blue–greens (~20% of the total available algal resources by biovolume), which are considered a poor food resource, especially for copepods (Fulton and Paerl, 1988; Gorham and Carmichael, 1988).

The first of three 'fed' treatments for the rotifer trials contained the dinoflagellate as a food resource, added at the fish culture density of ~2500 cells ml$^{-1}$, which is typical of phytoplankton densities in local estuaries during the summer growing season (Carpenter, 1971; Thayer, 1971; Mallin *et al.*, 1991). The toxic dinoflagellate culture was capable of killing three test tilapia at 3 h intervals; it was comprised of 88% TFVCs [mean diameter 10 μm, biovolume 510 ± 18 μm$^3$ (mean ±1 SE)], 10% zoospores and gametes (mean diameter

$7 \pm 2$ μm, biovolume $195 \pm 26$ μm$^3$), and 2% amoebae (biovolume $410 \pm 58$ μm$^3$; $n = 25$ cells for each stage), for a total initial dinoflagellate biovolume of $9.33 \times 10^5$ μm$^3$ ml$^{-1}$. The *Nannochloris* component of the non-toxic green algal prey consisted of much smaller cells than the dinoflagellate; hence, we provided approximately equal biovolumes of toxic and non-toxic prey types in the designated treatments. The second 'fed' treatment consisted of the non-toxic green algal culture added to filtered fish water ($3.38 \times 10^5$ μm$^3$ *Nannochloris* sp. + $4.92 \times 10^5$ μm$^3$ *Tetraselmis* sp. ml$^{-1}$, for a total of $8.30 \times 10^5$ μm$^3$ non-toxic greens ml$^{-1}$). The third 'fed' treatment was a 1:1 mixture by biovolume of *P.piscicida* (1250 cells ml$^{-1}$) and non-toxic greens. Treatments with the dinoflagellate alone or mixed with green algal food thus included full volume or half volume, respectively, as 'contaminated' filtered fish water from actively killing cultures—an important point since the lethal agents of *P.piscicida* is an exotoxin, with little residual toxic materials retained within the cells (H.B. Glasgow and J.M.Burkholder, unpublished data; D.Baden, U.Miami, unpublished data). Toxic filtrate from dinoflagellate cultures with fish (0.22 μm porosity filters) is known to be lethal to tilapia for 12–24 h, depending on the TFVC concentration (Burkholder *et al.*, 1992, 1993). On a daily basis, we used Palmer cells (Wetzel and Likens, 1991) to quantify algal cells from stock cultures 2 h before transferring animals to fresh food supply, and adjusted dilutions to ensure that the targeted algal densities were added throughout the experiment.

Rotifer tests were run for 9 days, with daily removal of each animal into a fresh test water/phytoplankton preparation. Mortality and the presence of eggs and live young were recorded daily at 17:00–20:00 h. After scanning a given replicate and removing the animal, the water and remaining phytoplankton were preserved with acidic Lugol's solution for quantification. From each treatment we randomly selected 4–6 replicate samples (excluding those that had contained dead animals) for analysis by the Utermöhl method, to quantify changes in the phytoplankton food supply during 24 h grazing periods.

*Acartia tonsa* was tested for survival in a similar manner over three repeat 3-day trials ($n = 10$ replicate animals maintained separately), with the following alterations. Test organisms were late-stage copepodites (CIV–CVI) of either sex, placed randomly into polyethylene cups as they were encountered. Non-toxic algal food was supplied as the centric diatom, *T.pseudonana*, of similar dimensions and biovolume as TFVCs of *P.piscicida*. The toxic dinoflagellate culture used in the copepod trials was capable of killing three small tilapia at 30 to 60 min intervals; the culture consisted of 93% TFVCs, 6% zoospores and gametes, and ~1% amoebae. Treatments for trials 1 and 2 included (i) 'unfed', without algae aside from small blue–greens; (ii) *T.pseudonana* at 2500 cells ml$^{-1}$); (iii) *P.piscicida* at 2500 cells ml$^{-1}$; and (iv) a mix of 1250 cells each of the diatom and the dinoflagellate. Trial 3 was conducted similarly, except that total cell densities were ~2000 ml$^{-1}$. Media volumes were maintained at 30 ml rather than the 15 ml volume used for the rotifer tests. Standard cladoceran toxicity bioassays use one animal per 15 ml water (Mount and Norberg, 1984; Horning and Weber, 1985). We increased the volume to 30 ml for the copepod to minimize containment effects. Volumes >15 ml for *B.plicatilis*, and >30 ml for

*A.tonsa*, would have imposed visual and manipulation problems when attempting to capture/transfer individual animals to fresh algal supply.

As for the rotifer feeding trials, the copepods were transferred daily into freshly prepared algal treatments. We also quantified changes in the phytoplankton food supply per 24 h grazing period similarly as for the rotifer bioassays. The selected diatom cell densities and biovolumes exceeded quantities necessary to achieve maximal feeding rates for adult *A.tonsa* (Houde and Roman, 1987; Paffenhöfer, 1988). After terminating bioassays, the guts of both *B.plicatilis* and *A.tonsa* were densely packed with both non-toxic algae and/or flagellated stages of *P.piscicida* when viewed in Lugol's-preserved material under light microscopy. Although we did not observe dinoflagellates consuming non-toxic algae in the mixed treatment for *B.plicatilis* or *A.tonsa*, it is likely that TFVCs also accounted for a portion of the non-toxic algal losses (Burkholder *et al.*, 1993; Burkholder and Glasgow, 1995).

Fisher's exact test (Horning and Weber, 1985) was used to test for significant differences in animal mortality among treatments within each trial. The mean number of young produced among all *N* animals exposed (regardless of time of mortality) is considered an approximate measure of an overall ecological effect of a toxicant upon a test organism (Hamilton, 1986). Thus, to analyze rotifer fecundity, we tested for differences in live young produced among treatments by ANOVA followed by the LSD. One-way ANOVA was also used to compare ingestion rates among treatments for both test animals (Statistical Analysis Systems Inc., 1985; significance level for all tests set at $\alpha = 0.05$).

## Results

Bioassay trials with the rotifer, *B.plicatilis*, showed no mortality among any of the three algal food treatments (Table I). In the unfed treatment, 7 of 10 replicate animals died during the 9 day experimental period, and survival of unfed rotifers was significantly lower than in the other treatments (Fisher's exact test, $P = 0.0015$). Moreover, significantly fewer young were produced in the unfed treatment as compared with all algal treatments (Table I, Figure 1). There was no significant difference in rotifer fecundity when given *P.piscicida* or the mixed dinoflagellate/green algal food treatment, nor a significant difference in fecundity with non-toxic greens alone versus *P.piscicida* alone. The fecundity of *B.plicatilis* when given non-toxic greens + *P.piscicida* was marginally significantly greater than with non-toxic greens alone as a food source ($P = 0.056$). These data were supported by significantly greater ingestion rates after 1 day for animals given mixed food as opposed to either non-toxic or toxic algae, alone ($P = 0.05$; Table II). After day 6, however, rotifer fecundity in the non-toxic green algal treatment showed an increasing trend, whereas after day 7 fecundity in the dinoflagellate treatment decreased (Figure 1).

*Brachionus plicatilis* consumed $61 \pm 10\%$ (mean $\pm$ 1 SE) of the *Nannochloris* biovolume and $70 \pm 17\%$ of the *Tetraselmis* daily when fed non-toxic green algae, with consumption increasing to $98 \pm 2\%$ and $85 \pm 15\%$ of each green, respectively, in the mixed green/dinoflagellate treatment (Table III).

**Table I.** Mean survival and fecundity of *B.plicatilis* when exposed for 9 days to non-toxic green algae (greens) and/or *P.piscicida* (dino), including the SD and the LSD ranking of mean live young produced per rotifer ($n = 10$; $\alpha = 0.05$ for all treatments)

| Treatment | Surviving adults | Total young | Young/adult | SD | LSD[a] |
|---|---|---|---|---|---|
| Unfed | 3 | 4 | 0.4 | 1.0 | A |
| Greens | 10 | 88 | 8.8 | 5.1 | B |
| Dino | 10 | 104 | 10.4 | 2.8 | BC |
| Greens + dino | 10 | 131 | 13.0 | 3.2 | C |

[a]Means with the same letter were not significantly different.

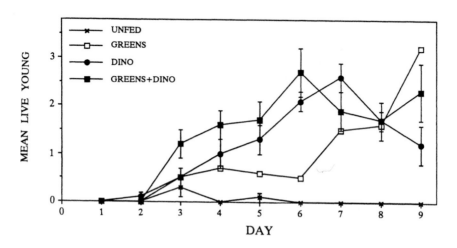

**Fig. 1.** Live young produced daily by *B.plicatilis* in each feeding treatment (means ± 1 SE).

Rotifer consumption of edible stages (TFVCs, gametes and zoospores) of *P.piscicida* was also higher in the mixed treatment, increasing from 53 ± 15% of the cells when given dinoflagellates alone to 84 ± 4% with mixed non-toxic/toxic algal food. After 24 h the percentage of inedible amoebae, transformed from TFVCs and gametes, was 20 ± 6% in the dinoflagellate-alone treatment and 47 ± 19% in the mixed algal treatment. Approximately one-third of the dinoflagellate cells available at the end of a daily feeding period were TFVCs, however, and more than half consisted of edible TFVCs, zoospores or gametes.

Trial 1 for *A.tonsa* showed no significant differences in survival among copepods fed any of the three algal treatments (Figure 2). Survival in the unfed treatment was significantly less than in all algal treatments ($P < 0.05$). In trial 2, there was no significant difference in survival between diatom-fed copepods and those fed *P.piscicida*, alone or in mixed toxic/non-toxic food supply (Figure 2). Copepods in all three 'fed' treatments survived significantly better than those in the unfed treatment ($P = 0.01$). In trial 3, the unfed copepods had significantly poorer survival than those given either dinoflagellate treatment (Figure 2).

**Table II.** Ingestion rates after the first 24 h (algal cells animal$^{-1}$ h$^{-1}$, considering all trials) by the rotifer. *B.plicatilis*. and the copepod. *A.tonsa*. in bioassays with non-toxic algae. *P.piscicida* (flagellated stages) and mixed food supply ($n = 4$–6 randomly selected animals; data given as means $\pm$ 1 SE). Asterisks indicate *B.plicatilis* ingestion rates that were significantly higher in the mixed treatment than with one algal food source ($\alpha = 0.05$)

| Treatment | Non-toxic algae | Dinoflagellate |
|---|---|---|
| Rotifer + dino | – | 750 $\pm$ 210 |
| Rotifer + greens | | |
|    *Nannochloris* | 19 880 $\pm$ 3260 | – |
|    *Tetraselmis* | 600 $\pm$ 150 | – |
| Rotifer + mixed | | |
|    *Nannochloris* | 31 900 $\pm$ 600 | 1180 $\pm$ 60* |
|    *Tetraselmis* | 730 $\pm$ 130 | – |
| Copepod + dino | – | 1240 $\pm$ 240 |
| Copepod + diatom | 1350 $\pm$ 320 | – |
| Copepod + mixed | 1240 $\pm$ 390 | 1070 $\pm$ 380 |

**Table III.** Consumption of algal prey after the first 24 h (as a percentage of the total biovolume initially available, considering all trials) by the rotifer. *B.plicatilis*. and the copepod. *A.tonsa* ($n = 4$–6 randomly selected animals; data given as means $\pm$ 1 SD)

| Treatment | Non-toxic algae (%) | Dinoflagellate (flagellated; %) |
|---|---|---|
| Rotifer + dino | – | 53 $\pm$ 15 |
| Rotifer + greens | 61 $\pm$ 10 *Nannochlorois*; 70 $\pm$ 17 *Tetraselmis* | – |
| Rotifer + mixed | 98 $\pm$ 2 *Nannochloris*; 85 $\pm$ 15 *Tetraselmis* | 84 $\pm$ 4 |
| Copepod + dino | – | 79 $\pm$ 16 |
| Copepod + diatom | 86 $\pm$ 20 | – |
| Copepod + mixed | 79 $\pm$ 25 | 68 $\pm$ 24 |

Copepods fed *P.piscicida* alone maintained significantly higher survival than those fed only diatoms ($P < 0.01$), whereas ingestion rates and survival were not significantly different among the non-toxic, toxic and mixed algal treatments ($P > 0.1$; Table II).

*Acartia tonsa* consumed ~80% of non-toxic diatom cells or toxic dino-flagellate cells when given either food source separately (Table III). Further, consumption of both non-toxic and toxic algae was comparable in the 'alone' or mixed algal treatment. After 24 h the inedible amoebae remained at <20% of the *P.piscicida* population in both dinoflagellate-alone and mixed algal treatments, with the remainder comprised of edible flagellated stages. Only 18% of the dinoflagellate cells were TFVCs, however, whereas >60% of the dinoflagellates were smaller and were probably either zoospores or gametes.

Although copepod survival was not impaired by the dinoflagellate during our short-term feeding trials, the presence of *P.piscicida* appeared to adversely affect the animals' behavior. In both dinoflagellate treatments, we observed that movement of *A.tonsa* was often extremely rapid and erratic, to the extent that some animals repeatedly collided head-first with the culture vessel walls. During trial 3, we attempted to quantify movement by counting the number of moves

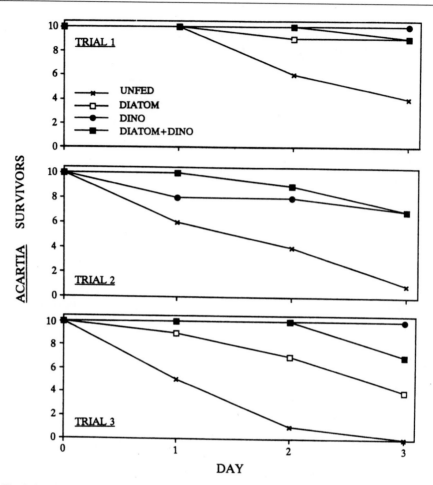

**Fig. 2.** Survival of *A.tonsa* for each feeding treatment over each of three experimental feeding trials. In trial 2, the copepod response in the diatom-only treatment was identical to that in the diatom + dinoflagellate treatment.

generated by use of swimming legs or antennae within a 30 s interval. For active animals, we estimated 29 ± 14 movements in treatments with the dinoflagellate. By comparison, copepods within the diatom-only treatment were 5- to 6-fold less active, and none were observed to collide with the vessel walls. Analysis of samples throughout the duration of the feeding trials did not indicate feeding impairment for *A.tonsa*, however, with similar ingestion rates and dinoflagellate losses daily within each trial. In trial 3, we extended the observations for an additional 4 days and found similar ingestion rates for *A.tonsa* among algal food treatments until day 7, when most animals probably had become stressed from handling and senescence.

## Discussion

The lack of mortality among *B.plicatilis* in the *P.piscicida* treatments indicated that the toxin did not affect the rotifer. Most dinoflagellates had not encysted, as visual observations supported by cell counts confirmed that they were actively swimming in the test media after 24 h prior to water exchange. The rotifers actively consumed the dinoflagellate and benefited from this food resource, based on cell counts, observations of *P.piscicida* cells in the guts, and the high fecundity in the dinoflagellate-only treatment as compared to the low fecundity of unfed animals. In analogous behavior, the congeneric species *Brachionus calyciflorus* Pallas is capable of resisting the toxins produced by the blue–green alga *Microcystis aeruginosa* Kütz., and can utilize the blue–green as a supplementary nutritional source (Fulton and Paerl, 1987).

Since the fecundity of *B.plicatilis* in the two *P.piscicida* treatments was slightly elevated overall, relative to fecundity when given green algae alone, this dinoflagellate is probably a nutritious food source for the rotifer. The abundance of dinoflagellates in general has been positively correlated with zooplankton grazing rates in the Neuse River Estuary (Mallin and Paerl, 1994). Other researchers have noted the food value of, or grazing preference toward, various dinoflagellate species (Burkhill *et al.*, 1987; Uye and Takamatsu, 1990; Sellner *et al.*, 1991). For example, the copepods *A.tonsa*, *Oncaea venusta* Philippi and *Labidocera aestiva* Wheeler ingested *Gymnodinium breve* Davis cells in proportion to their availability over a broad range of concentrations without mortality or physiological incapacitation (Turner and Tester, 1989). *Centropages typicus* did not ingest *G.breve*. When offered a choice, however, *A.tonsa*, *L.aestiva* and *C.typicus* avoided *G.breve*, and instead consumed the non-toxic diatom, *Skeletonema costatum* (Greve.) Cleve.

Over the long term (weeks), the positive influence of *P.piscicida* on the fecundity of *B.plicatilis* may change, considering the decreasing trend shown in the fecundity data from days 7–9. This trend may be irrelevant in many field situations, however, since most *P.piscicida*-induced fish kills are <3 days in duration, and since the dinoflagellate often represents only 10–20% of the total available phytoplankton during toxic outbreaks (Burkholder *et al.*, 1995a).

The calanoid copepod *A.tonsa* also apparently found toxic *P.piscicida* cells a nutritious food source. Their survival was comparable to or higher than that of copepods fed a diatom diet, and their ingestion rates of diatoms and dinoflagellates were comparable when given either food resource alone or in combination. However, the animals' ability to detect and avoid the walls of the culture vessels was impaired in both the dinoflagellate-only and the dino-flagellate/diatom treatments, as evidenced by their extremely rapid and erratic movements and frequent, repeated collisions against the wall surfaces. In a natural habitat, such erratic behavior would be expected to increase the animals' susceptibility to visual predators, and might also impair their ability to detect and successfully mate with other individuals.

The utility of zooplankton as biocontrol agents for the suppression of toxic *P.piscicida* blooms remains an open question. The common estuarine zooplankters tested were able to consume the dinoflagellate's most lethal stage, and the rotifer was capable of reproducing successfully when TFVCs were its only food resource. Most zooplankton species, however, cannot reproduce quickly enough to control toxic or non-toxic short-term blooms. An additional question concerns whether the toxin, as yet uncharacterized, remains viable in zooplankton guts after consumption of the cells. If this is the case, then feeding by higher organisms on zooplankton in bloom areas may lead to bioaccumulation, with possible lethal or more insidious sublethal effects on their finfish and shellfish predators.

## Acknowledgements

Funding support for this research was provided by the Office of Sea Grant, NOAA, US Department of Commerce (grant NA90AA-D-SG062) and the UNC Sea Grant College (project R/MER-17), the National Science Foundation (grant OCE-9403920), the UNC Water Resources Research Institute (grant 70136), the US Marine Corps–Cherry Point Air Station, the NCSU College of Agriculture & Life Sciences, the NC Agricultural Research Foundation (project 006034), the NC Agricultural Research Service (project 06188) and the Department of Botany at NCSU. Research assistance was contributed by J.Compton and F.Johnson. *Thalassiosira* cultures were provided by D. Kamykowski and R.Reed; *Acartia* cultures were provided, in part, by H.Millsaps, recommended by K.Sellner; and *Brachionus* and *Nannochloris* cultures were provided by W.Hettler. D.Rainer and B.MacDonald contributed sound counsel on laboratory safety precautions. We thank B.J.Copeland, J.Wynne and E.D.Seneca for their counsel and administrative support.

## References

Berggreen,U., Hansen,B. and Kiørboe,T. (1988) Food size spectra, ingestion and growth of the copepod *Acartia tonsa* during development: implications for determination of copepod production. *Mar. Biol.*, **99**, 341–352.

Bold,H.C. and Wynne,M.J. (1985) *Introduction to Phycology.* Prentice-Hall, New York, pp. 476–512.

Burkhill,P.H., Mantoura,R.F., Llewellyn,C.A. and Owens,N.J.P. (1987) Microzooplankton grazing and selectivity of phytoplankton in coastal waters. *Mar. Biol.*, **93**, 581–590.

Burkholder,J.M. and Glasgow,H.B.,Jr (1995) Interactions of a toxic estuarine dinoflagellate with microbial predators and prey. *Archiv. für Protistenka*, in press.

Burkholder,J.M. and Wetzel,R.G. (1989) Epiphytic microalgae on natural substrata in a hardwater lake: seasonal dynamics of community structure, biomass and ATP content. *Arch. Hydrobiol.*, **83** (Suppl.), 1–56.

Burkholder,J.M., Noga,E.J., Hobbs,C.W., Glasgow,H.B.,Jr and Smith,S.A. (1992) New 'phantom' dinoflagellate is the causative agent of major estuarine fish kills. *Nature*, **358**, 407–410; *Nature*, **360**, 768.

Burkholder,J.M., Glasgow,H.B.,Jr, Noga,E.J. and Hobbs,C.W. (1993) *The Role of a New Toxic Dinoflagellate in Finfish and Shellfish Kills in the Neuse and Pamlico Estuaries.* Research Institute of the University of North Carolina, Raleigh, pp. 1–58.

Burkholder,J.M., Glasgow,H.B.Jr. and Hobbs,C.W. (1995a) Distribution and environmental conditions for fish kills linked to a toxic ambush-predator dinoflagellate. *Mar. Ecol. Proc. Ser.*, in press.

Burkholder,J.M., Glasgow,H.B.,Jr and Steidinger,K.A. (1995b) Stage transformations in the complex life cycle of an ichthyotoxic 'ambush predator' dinoflagellate. In Lassus,P., Arzul,G.,

Erard,E., Gentien,P. and Marcaillou,C. (eds), *Proceedings of the Sixth International Conference on Toxic Marine Phytoplankton*. Elsevier, Amsterdam, in press.

Carpenter,E.J. (1971) Annual phytoplankton cycle of the Cape Fear River estuary, North Carolina. *Chesapeake Sci.*, **12**, 95–104.

Fields,S.D. and Rhodes,R.G. (1991) Ingestion and retention of *Chroomonas* spp. (Cryptophyceae) by *Gymnodinium acidotum* (Dinophyceae). *J. Phycol.*, **27**, 525–529.

Fulton,R.S.,III and Paerl,H.W. (1987) Toxic and inhibitory effects of the blue-green alga *Microcystis aeruginosa* on herbivorous zooplankton. *J. Plankton Res.*, **9**, 837–855.

Fulton,R.S.,III and Paerl,H.W. (1988) Effects of the blue-green alga *Microcystis aeruginosa* on zooplankton competitive relations. *Oecologia (Berlin)*, **76**, 383–389.

Glasgow,H.B.,Jr and Burkholder,J.M. (1993) Comparative saprotrophy by flagellated and amoeboid stages of an ichthyotoxic estuarine dinoflagellate. In Lassus,P. (ed.), *Proceedings, Sixth International Conference on Toxic Marine Phytoplankton*. Institut Francais de Recherche Pour L'Exploitation de la Mer, Nantes, p. 78 (Abstract).

Glasgow,H.B.,Jr and Burkholder,J.M. (1994) Insidious effects on human health and fish survival by the toxic dinoflagellate *Pfiesteria piscimorte* from eutrophic estuaries. In *Proceedings of the 1st International Symposium on Ecosystem Health & Medicine*, Ottawa (Abstract).

Gorham,P.R. and Carmichael,W.W. (1988) Hazards of freshwater blue-green algae (cyanobacteria). In Lembi,C.A. and Waaland,J.R. (eds), *Algae and Human Affairs*. Cambridge University Press, New York, pp. 403–431.

Hamilton,M.A. (1986) Statistical analysis of the cladoceran reproductivity test. *Environ. Tox. Chem.*, **5**, 205–212.

Horning,W.B.,II and Weber,C.I. (eds) (1985) *Short-term Methods for Estimating the Chronic Toxicity of Effluents and Receiving Waters to Freshwater Organisms*. Report EPA/600/4-85/014. US Environmental Protection Agency, Cincinnati.

Houde,S.E.L. and Roman,M.R. (1987) Effects of food quality on the functional ingestion response of the copepod *Acartia tonsa*. *Mar. Ecol. Prog. Ser.*, **40**, 69–77.

Huntley,M., Sykes,P., Rohan,S. and Marin,V. (1986) Chemically mediated rejection of dinoflagellate prey by the copepods *Calanus pacificus* and *Paracalanus parvus*: Mechanism, occurrence and significance. *Mar. Ecol. Prog. Ser.*, **28**, 105–220.

Lee,J.J., Hutner,S.H. and Bovee,E.C. (1985) *An Illustrated Guide to the Protozoa*. Society of Protozoologists. Allen Press, Lawrence, KS.

Lewitus,A.J., Jesien,R.V., Kann,T.M., Burkholder,J.M., Glasgow,H.B. and May,E. (1995) Discovery of the "phantom" dinoflagellate in Chesapeake Bay. *Estuaries*, in press.

Lund,J.W.G., Kipling,C. and LeCren,E.D. (1958) The inverted microscope method of estimating algal numbers and the statistical basis of estimates by counting. *Hydrobiologia*, **11**, 143–170.

Mallin,M.A. (1991) Zooplankton abundance and community structure in a mesohaline North Carolina estuary. *Estuaries*, **14**, 481–488.

Mallin,M.A. and Paerl,H.W. (1994) Planktonic trophic transfer in an estuary: Seasonal, diel, and community structure effects. *Ecology*, **75**, 2168–2185.

Mallin,M.A., Paerl,H.W. and Rudek,J. (1991) Seasonal phytoplankton composition, productivity and biomass in the Neuse River Estuary, North Carolina. *Estuarine Coastal Shelf Sci.*, **32**, 609–623.

Mount,D.I. and Norberg,T.J. (1984) A seven-day life-cycle cladoceran toxicity test. *Environ. Tox. Chem.*, **3**, 425–434.

Noga,E.J., Smith,J.A., Burkholder,J.M., Hobbs,C.W. and Bullis,R.A. (1993) A new icthyotoxic dinoflagellate: cause of acute mortality in aquarium fishes. *Veterinary Record*, **1339**, 96–97.

Paffenhöfer,G.-A. (1988) Feeding rates and behavior of zooplankton. *Bull. Mar. Sci.*, **43**, 430–445.

Peters,D.S. (1968) A study of the relationships between zooplankton abundance and selected environmental variables in the Pamlico River Estuary of Eastern North Carolina. MS Thesis, North Carolina State University, Raleigh, NC, pp. 1–38.

Schnepf,E. and Elbrächter,M. (1992) Nutritional strategies in dinoflagellates. A review with emphasis on cell biological aspects. *Eur. J. Protistol.*, **28**, 3–24.

Schnepf,E., Winter,S. and Mollehauer,D. (1989) *Gymnodinium aeruginosum* (Dinophyta): A blue-green dinoflagellate with a vestigial, cryptophycean symbiont. *Plant Syst. Evol.*, **164**, 75–91.

Sellner,K.G., Lacoutre,R.V., Cibik,S.J., Brindley,A. and Brownlee,S.G. (1991) Importance of a winter dinoflagellate–microflagellate bloom in the Patuxent River Estuary. *Estuarine Coastal Shelf. Sci.*, **32**, 27–42.

Smith,S.A., Noga,E.J. and Bullis,R.A. (1988) Mortality in *Tilapia aurea* due to a toxic dinoflagellate bloom. In *Proceedings of the Third International Colloquium on the Pathology of Marine Aquaculture*, Glouster Point, VA, pp. 167–168 (Abstract).

Statistical Analysis Systems Inc. (1985) *SAS/STAT Guide for Personal Computers, Version 6*. SAS Institute, Inc., Cary, NC, pp. 1–378.

Steidinger,K.A., Truby,E.W., Garrett,J.K. and Burkholder,J.M. (1995) The morphology and cytology of a newly discovered toxic dinoflagellate. In Lassus,P., Arzul,G., Erard,E., Gentien,P. and Marcaillou,C. (ed.), *Proceedings of the Sixth International Conference on Toxic Marine Phytoplankton*. Elsevier, Amsterdam, in press.

Steidinger,K.A., Burkholder,J.M., Glasgow,H.B.,Jr, Truby,E.W., Garrett,J.K., Noga,E.J. and Smith,S.A. *Pfiesteria piscicida*, a new toxic dinoflagellate genus and species of the order Dinamoebales. *J. Phycol.*, in preparation.

Thayer,G.W. (1971) Phytoplankton production and the distribution of nutrients in a shallow unstratified estuarine system near Beaufort, N.C. *Chesapeake Sci.*, **12**, 240–253.

Turner,J.T. and Tester,P.A. (1989) Zooplankton feeding ecology: Copepod grazing during an expatriate red tide. In Cosper,E.M., Bricelj,V.M. and Carpenter,E.J. (eds), *Novel Phytoplankton Blooms: Causes and Impacts of Recurrent Brown Tides and Other Unusual Blooms*. Springer-Verlag, Berlin, pp. 359–374.

Turner,J.T. and Tester,P.A. (1992) Zooplankton feeding ecology: bactivory by metazoan microzooplankton. *J. Exp. Mar. Biol. Ecol.*, **160**, 149–167.

Uye,S. and Takamatsu,K. (1990) Feeding interactions between planktonic copepods and red-tide flagellates from Japanese coastal waters. *Mar. Ecol. Prog. Ser.*, **59**, 97–107.

Vollenweider,R.A. (ed.) (1974) *A Manual on Methods for Measuring Primary Production in Aquatic Environments, 2nd edn International Biology Program Handbook 12*. Blackwell Scientific Publications, Oxford, pp. 1–213.

Watras,C.J., Garcon,V.C., Olson,R.J., Chisholm,S.W. and Anderson,D.M. (1985) The effect of zooplankton grazing on estuarine blooms of the toxic dinoflagellate *Gonyaulax tamarensis*. *J. Plankton Res.*, **7**, 891–908.

Wetzel,R.G. and Likens,G.E. (1991) *Limnological Analyses*, 2nd edn. Springer-Verlag, New York.

White,A.W. (1980) Recurrence of kills of Atlantic herring (*Clupea harengus*) caused by dinoflagellate toxins transferred through herbivorous zooplankton. *Can. J. Fish. Aquat. Sci.*, **37**, 2262–2265.

White,A.W. (1981) Marine zooplankton can accumulate and retain dinoflagellate toxins and cause fish kills. *Limnol. Oceanogr.*, **26**, 103–109.

*Received on March 14, 1994; accepted on September 22, 1994*

# Safety and Personal Protection During Fish Kills

**Recommended Work Practices and Use of Personal Protective Equipment for Reducing Algal Toxin Exposure and Human Health Risk in Estuarine Waters**

It is unclear whether toxins generated by *Pfiesteria piscicida* and morphologically related organisms (MROs) have produced human health effects such as dermatitis, neuropsychological changes or upper respiratory irritation when exposure occurs in the natural environment. Some anecdotal reports, published rodent studies[1] and medical evaluation data from Maryland researchers[2,3], suggest that environmental health risk is plausible. However, a careful review of the evidence to date also suggests that environmentally-mediated adverse human health impacts are not widespread, are minor, and are of short duration if they have occurred at all. Nonetheless, while health effects research is ongoing, definitive results, if found, are many months to possibly years away. In the absence of such results, risk management must be accomplished using the information that we currently have for those workers whose jobs may place them in harm's way. The objective of this document is to make recommendations for work practices and the use of personal protective equipment to reduce potential exposure to toxins generated by *Pfiesteria* and MROs.

## Avoiding Exposure

Minimizing risk should be accomplished by avoiding exposure. Fishermen and other workers should not enter or pass through risky waters. Not enough is known about *Pfiesteria* and MROs, the toxins they produce, nor their effect on human health to define with certainty which waters are risky and which are not. However, for the purposes of this document, risky waters will be considered those that are declared, by officials of the State of North Carolina, to be risky because of recent fish kills or because of the presence of potentially harmful algal toxins. Estuarine waters are assumed to be risky in the area of a fish kill, defined as "a significant number of fish - more than a few dozen - dead, dying, behaving abnormally, exhibiting lesions, or showing other signs of disease". [9] If the cause of the fish kill is known to result from a nontoxic event (such as catch discards, low dissolved oxygen, etc.), exposure to this water is not considered to pose a human health hazard and these work practices do not apply to such situations.

Workers should not enter risky waters unless there is a critical need (e.g., rescue, collection of water samples, or collection of diseased or dead fish). The amount of time spent on or in risky waters should be minimized as much as possible to reduce exposure dose. Current information about fish kills and the presence of toxic algae in NC estuaries can be obtained by contacting the NC Occupational and Environmental Epidemiology Section at (919) 733-3410 or on the Internet at www.ehnr.state.nc.us/EHNR/SCHS/

## Work Practices

Because of the wide variety of job functions that might be conducted on or in risky waters, only general work practice recommendations are offered here:

Ø Skin contact with risky water should be avoided. Tasks that involve splashing, handling of fish or water samples, or overhead placement of wet nets or lines should be minimized or modified to avoid skin contact with risky water.

Ø Exposure could conceivably occur by inhalation of aerosolized water droplets containing toxin or by inhalation of a gas or vapor phase toxin or toxins. Agitation of contaminated water could generate a water mist (airborne water particles) and could increase the rate of volatilization of a gas or vapor phase toxin. Operation of power boats and the

North Carolina Department of Health and Human Services, 1998 placement of nets or sampling equipment should be conducted so that water agitation is minimized.

Ø Hand washing facilities (potable water, soap, and disposable towels) should be available to workers while on the water. In addition, there should be potable water available for flushing or washing skin that might become irritated. There should be at least 10 gallons of potable water (in addition to drinking water) on board per worker that may have direct contact with estuary water. Hands should be washed immediately after working in or around risky water or after handling fish from risky waters. Skin that comes in contact with risky water should be flushed with potable water immediately.

Ø If 10 gallons of water per person cannot be provided on-board because of small boat size, 2 ½ gallons of water per person in a pressurized sprayer (garden sprayers) can be provided. There should be 10 gallons per person available on the larger boat or vehicle from which the small boat was launched for additional rinsing.

Ø Whenever possible, a fish kill site should be approached from the upwind side. If working along the edge of a fish kill, work should be conducted on the upwind edge of the site, if possible.

In general workers should be aware that the potential routes of exposure to the suspect toxins are inhalation and dermal exposure, and they should know how to modify their job tasks to minimize exposure.

Protective Clothing

Personal protective equipment (PPE) should be used only when exposure to risky water cannot be avoided. Personal protective equipment can reduce worker exposure to suspected toxins if used properly. However, when used in hot environments, PPE will put workers at risk of heat stress related illness or death. Precautions for managing heat stress risk are described later in this document.

When direct contact with risky water cannot be avoided, workers should wear disposable elbow length, puncture resistant gloves, shin length aprons, and boots that extend above the bottom of the apron. This protective clothing should remain impermeable to water throughout the work shift. All protective clothing should be removed when the worker leaves the work area. Gloves should be disposed of and boots and aprons should be thoroughly rinsed with potable water before being reused. Workers should use aprons and boots only when necessary to prevent their feet, legs or torso from wetting.

Workers should take a change of clothing including shoes, socks, and underwear when entering risky waters. If clothing becomes wet, those items should be changed as quickly as possible and the wetted skin should be rinsed with potable water.

Respiratory Protection

It is not known whether *Pfiesteria* and MRO toxins in gas phase or in aerosolized water droplets can occur in the breathing zone of watermen. Respiratory protection equipment, when used in risky waters, should be capable of reducing gas phase contaminant and aerosol (airborne particle) exposure.

When working in risky waters, workers should wear half face-piece, air purifying respirators or powered air purifying respirators with combination P-100 filter and organic vapor cartridges. All equipment should be NIOSH approved and compatible. Because the nature of the toxins generated by *Pfiesteria* and MROs is not known, it is possible that the toxins will not be efficiently captured by the respirator cartridges recommended. The respirator cartridges recommended do offer the broadest range of protection from mists, organic vapors, and gases.

Before any employer requires an employee to wear a respirator, the requirements of OSHA Standard 1910.134 "Respiratory Protection"[4] must be met. The requirements in this standard include: a written respiratory protection program, medical evaluations of employees required to wear respirators, fit testing procedures, and employee training. An Industrial Hygienist from DHHS, dedicated to the Harmful Algal Blooms Program, will be available for assistance, such as fit-testing, training, etc. Ensuring appropriate medical evaluations and providing a written program are the responsibilities of the employer.

Divers

Divers should not dive in risky waters. If there is a critical need (e.g., rescue) to dive in risky waters, divers should follow their organization's procedures for diving in polluted waters. At a minimum, this would include the use of dry suits and rinsing with clean water prior to removal of the dry suit at the end of the dive. Support personnel and divers out of the water should follow other worker protection recommendations contained in this document.

Medical Monitoring

In addition to a physical examination to determine physical ability to use PPE, any North Carolina state government employee that may be required to enter risky waters should receive baseline medical monitoring, including neuropsychological testing through Duke or UNC providers, prior to entering risky waters. Ongoing medical monitoring may be recommended for these employees. The nature of this medical monitoring and the frequency of ongoing monitoring will be determined by the Occupational and Environmental Epidemiology Section of the NC Department of Health and Human Services.

Heat Stress

The estuarine fish kill season in North Carolina generally occurs during the hottest months of the year. If the risk is not properly managed, there is risk of illness or death from heat stress for workers that are required to wear protective clothing and respirators while working outdoors.[5,6,7]

The use of PPE may reduce the natural ability of the body to cool itself by evaporative cooling or the evaporation of sweat. Negative pressure air purifying respirators require increased effort for breathing which causes the body to generate more heat. Powered air purifying respirators may impair the body's ability to reduce heat load by sweat evaporation. If heat cannot be dissipated, a worker's body temperature can rise, leading to heat-related illness. This includes decreased alertness and coordination (resulting in increased risk of accidents and injury), heat exhaustion, and heat stroke which can result in permanent brain injury and death.

The following guidelines should be followed to reduce the risk of heat induced illness when PPE is used outdoors during summer months:

Ø Work requiring the use of PPE should be conducted during early morning and late afternoon to avoid the highest daily temperatures. Workers should not be required nor allowed to wear PPE as specified above at air temperatures of 95 degrees F or above. As temperatures approach 90 degrees F, the duration and strenuousness of job tasks should be lessened. Workers should be allowed more time to accomplish tasks so that breaks can be taken more frequently.

Ø Shade must be available for workers while on the water. If possible, the worker should be sheltered from the sun while performing job tasks. At a minimum, shade must be available during work breaks.

Ø Only workers that have been acclimatized to the heat should be allowed to wear PPE outdoors during hot weather. Five to seven days of exposure to the heat should be allowed for acclimatization.

Ø Physically fit and healthier workers become acclimatized to heat more easily and are more capable of dealing with heat stress. Workers that are obese, hypertensive, have heart disease, respiratory disease, or diabetes (among other medical conditions) may be at greater risk for heat-related illness.

Ø Cooled (10 to 15 degrees C) drinking water should be available to workers at all times. During periods of rapid fluid loss, thirst does not adequately warn of the need for fluid replenishment. Workers should be encouraged to drink 5 to 7 ounces of water every 15 to 20 minutes.

Ø Workers and their supervisors should receive training that outlines the risk presented by the use of PPE in hot environments and the means of managing that risk. Procedures for risk control should be written and in possession of the manager. Workers should not be allowed to wear PPE in hot outdoor environments until the workers and their supervisors have received the training. The written heat stress management program and training will include: heat avoidance, use of a buddy system, recognition of and first aid for heat stress, monitoring of environmental heat (WBGT values), and work practice modification.

Monitoring of heat stress will be conducted during the summer of 1998 for North Carolina state government employees using PPE on the water. A written monitoring plan will be developed and carried out by the NC Division of Epidemiology. The objective of the monitoring plan will be to define environmental heat conditions that pose unacceptable risk. Workers will be instructed not to enter risky waters during times when those conditions exist.

Summary

Fishermen and other workers should not pass through or enter risky waters. If exposure cannot be avoided, personal protective equipment, if used properly, can reduce exposure to airborne or waterborne toxins that may be generated by *Pfiesteria* and MROs. It is not clear that taking these precautions provides any health benefit to the worker relative to exposure control. The use of PPE as described above in hot outdoor environments could put the worker at risk of heat-related illness or death if the risk is not properly managed. Workers and their supervisors must be made aware of the relative risks involved with potential exposure to marine toxins and the use of PPE to reduce exposure.

References

1. Levin E et al: "Persistent Learning Deficits in Rats After Exposure to *Pfiesteria Piscicida.*" Environmental Health Perspectives. Vol. 105, No. 12. December 1997.

2. Morris G et al: Unpublished data.

3. Matuszak D et al: "Toxic *Pfiesteria* and Human Health." Maryland Medical Journal November/December 1997.

personal protection                                                  http://www.esb.enr.state.nc.us/Fishkill/persprotect.html

4. North Carolina Department of Labor. Division of Occupational Safety and Health. 29 CFR 1910. "Occupational Safety and Health Standards for the General Industry." 1910.134, "Respiratory Protection." January 8, 1998.

5. National Institute for Occupational Safety and Health. "Working in Hot Environments". DHHS (NIOSH) Publication No. 86-112. 1986.

6. Gasper A, Benatar M: "Abatement Equipment Report; Focus: Heat Stress and Protective Clothing". Asbestos Issues '88. October, 1988.

7. Iowa State University, Environmental Health and Safety. "Heat Stress in the Workplace". EH&S Home Page. www.ehs.iastate.edu/ih/heat.htm. 1998.

8. American Conference of Governmental Industrial Hygienists. "Threshold Limit Values for Chemical Substances and Physical Agents". 1997.

9. USEPA, NOAA, USDA, USGS, DHHS, ASIWPCI. "What You Should Know About *Pfiesteria piscicida"* Fact Sheet draft 5/98.

NCDHHS

Occupational and Environmental Epidemiology Section

William Service, MSPH

Industrial Hygiene Consultant

(919)715-6431

will_service@mail.ehnr.state.nc.us

PPE16.WPD June 2, 1998

Return to Top

 Return toFish Kill Event Homepage

Restoration | Bay | Harmful | Bay | Bay Life | Bay
& Protection | Grasses | Algae | Monitoring | Guide | Education

## MARYLAND GUIDELINES FOR CLOSING AND REOPENING RIVERS POTENTIALLY AFFECTED BY *PFIESTERIA* OR *PFIESTERIA*-LIKE EVENTS

### November 1, 1999

### OVERVIEW

It is the goal of the State to reduce the risk to public health posed by toxic outbreaks of *Pfiesteria* or *Pfiesteria*-like organisms. To date, however, there is no rapid method to definitively identify a toxic *Pfiesteria* or *Pfiesteria*-like event. Therefore, State environmental and health officials exercise their best professional judgment and apply the latest scientific knowledge and analytical methods to determine if river closures are necessary.

The following guidelines generally describe the conditions considered for evaluating *Pfiesteria* or *Pfiesteria*-like events, as well as procedures followed for issuing closures. Since many factors may cause lesions in fish, the guidelines do not reference lesions as a sole basis for closing rivers.

### A CLOSURE WILL BE RECOMMENDED WHEN:

sufficient evidence exists that a toxic outbreak of *Pfiesteria* or *Pfiesteria*-like organisms is likely to be in progress and poses a potential threat to public health. Because no definitive test currently exists to identify a toxic outbreak, the decision to close a river will be based on the best professional judgment of the State environmental and health officials, taking into account the full suite of conditions present. Factors to be considered in this decision include the following:

- a significant fish kill is confirmed and there is no apparent explanation for the kill other than a toxic outbreak of *Pfiesteria* or *Pfiesteria*-like organisms;

- a significant number of fish are confirmed to be acting erratically and no other explanation for the behavior is apparent, such as low dissolved oxygen;

- 20% or more of one species (from a minimum of 50 fish of that species) are exhibiting fresh sores of a kind typically associated with *Pfiesteria* outbreaks;

http://www.dnr.state.md.us/bay/cblife/algae/dino/pfiesteria/river_closure.html          3/29/03

Maryland Department of Natural Resources, 1999

- there is evidence of increased *Pfiesteria* or *Pfiesteria*-like activity, as reflected by an increase in the number of fish with sores typically associated with *Pfiesteria* outbreaks;

- environmental conditions (temperature, salinity, etc) are within ranges that may allow for a toxic outbreak; and

- an evaluation using the best available rapid technologies (such as light microscopy and molecular/toxin probes) reveals the presence of possible toxic *Pfiesteria* or *Pfiesteria*-like cells.

## Procedures:

- Based on an evaluation of the above factors, the Secretary of the Department of Natural Resources (DNR) will recommend closure of the affected area to the Secretary of the Department of Health and Mental Hygiene (DHMH) who will consult with and coordinate closure implementation with local health officials and Maryland Department of the Environment (MDE).

- Closure boundaries will be determined through evaluation of habitat conditions, assessment of potential human exposure, and the best analytical methods available. Closure areas will be monitored and assessed for fish health and *Pfiesteria* or *Pfiesteria*-like activity throughout the duration of the closure.

- The public will be notified of river closures through postings and other appropriate communications.

Note: In some cases it may be necessary to collect additional fish health and water quality information before a decision to close a river can be made. During such investigations, Natural Resources Police will regularly patrol the area in question and will advise individuals who fish or recreate on the waterway that an investigation is underway. DNR will also notify DHMH, MDE, area communities, and local officials of such investigations, and the Governor may issue a public caution.

**IN THE EVENT OF A CLOSURE, REOPENING WILL BE RECOMMENDED WHEN:**

- the conditions that initiated the closure have ceased for 14 days.

## Procedure:

- The DNR Secretary will recommend reopening to the DHMH Secretary,

Chesapeake Bay Front Page                                                                    Page 3 of 3

who will consult and coordinate reopening with local health officials and
MDE.

**COMMENTS:**

- Much work is ongoing to identify the nature and activity of *Pfiesteria*
  toxins. These guidelines are based on evolving scientific knowledge and
  may be modified as necessary.

- Current scientific knowledge indicates that the behavior of *Pfiesteria* and
  *Pfiesteria*-like organisms is seasonal and episodic in nature. Therefore, it
  may be necessary to repeatedly close a river as toxic outbreaks reoccur.

*Pfiesteria* Links:  **FAQ | Monitoring | Publications | Contact Us**
***Pfiesteria piscicida | Pfiesteria shumwayae***

**Search Maryland DNR**

⦿  Search www.dnr.state.md.us

Restoration and Protection | Bay Grasses | Harmful Algae | Bay Monitoring
Bay Life Guide | Bay Education

Return to the *Maryland DNR Home Page*.
Your opinion counts! Take a survey!

http://www.dnr.state.md.us/bay/cblife/algae/dino/pfiesteria/river_closure.html                3/29/03

# Collection and Handling of Fish Pathology Samples

The following is a set of instructions for handling fish specimens destined for postmortem examination, including euthanasia, proper fixation, and transport.

The best specimen to examine is a fish collected alive which was exhibiting the clinical signs most common to the population of diseased fish from which the specimen came. Whenever possible, a fish should be examined immediately after removal from the water. If not, the fish should be kept cool and examined as soon as possible. After euthanasia, either put the fish in a 10% solution of buffered formalin (*see below*), or store fish in individual plastic bags and place in a cooler or refrigerator. Do not freeze, as freezing creates artifacts, making evaluation of lesion morphology difficult, if not impossible.

Post mortem changes commence immediately after a fish is dead. These changes occur in both fish tissues and in parasites, making the task of assessing pathological conditions more difficult. Moreover, migrations of helminths after the death of the fish often occur, so that the parasites' true location becomes uncertain. For these reasons, fish collected dead may be of little use for diagnostic purposes.

Fish Collection, Sample Numbers

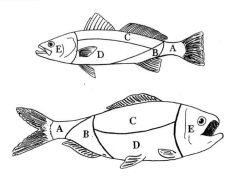

Record species, fork length, lesion prevalence, and lesion location on **all** fish caught, when possible (see diagram below). Example: 50 menhaden are caught; 20 have focal ulcers in area B, and 10 have lesions in area "BD". Record N=50, 40% B, 20% BD. Also record location, date, and time of day, since these data will be correlated with water quality data. Note method of capture; e.g., whether a school of fish was sought by cast net. Even the smallest observation could turn out to be critical to determining the pathogenesis of these lesions.

When possible, **20 menhaden per area sampled**, bimonthly, should be <u>randomly</u> chosen for pathology. Preliminary data suggest that sites with low dissolved oxygen will have few to no fish. Thus, a smaller "N" will have to suffice for that place and time. Obviously, water quality data will be critical in these areas. Fish of other species **with lesions** should also be collected for pathology. If they are too large for the container, cut out the lesion area along with a non-lesion area (*see next section below*).

Euthanasia and Fixation of Fish

If the fish is still alive, it will be necessary to humanely euthanize it before preparing it for transport. This may be done by placing the live fish into a solution of the anesthetic MS-222 at 500 mg/liter (about 2 teaspoons per gallon of water). Pre-measured MS-222 will be provided. This solution may be re-used for multiple fish. (FYI: other methods of euthanasia are covered in Dr. Noga's *Fish Disease* book). After gill flaring has ceased, make a ventral cut from the gills to right above the anal pore (so as to open the body cavity for proper fixation). **HOWEVER, avoid cutting into an obvious lesion area** (e.g., do not

North Carolina Department of Environment and Natural Resources, 2001 cut into an ulcer around the anal pore). Place the fish into a 10% solution of neutral buffered formalin in a sealable container. There should be about ten parts formalin to one part tissue. If this is not possible, or formalin is unavailable, place each fish in an individual, sealed plastic bag and place on ice.

?Some fish will be too large for the container (e.g. flounder). In this case, cut out the lesion(s) with wide (at least 2 cm) margins of normal surrounding tissue and place in formalin. Also, include a 2 cm wide piece of a non-lesioned skin with its underlying body wall/muscle. Include the viscera when possible.

Please ship to the Vet School at NC State as soon as possible (address on following page). We want these fish to fix in formalin for no longer than about 48 hours, so that we can use the tissues for later immunohistochemistry procedures. This will greatly facilitate our understanding of the pathogenesis of these lesions.

## SUMMARY OF FISH COLLECTION:

1. Collect fish as fresh as possible.

2. Euthanize live fish humanely.

3. Record fork length, lesion prevalence, and lesion location.

4. Make a ventral slit to open body cavity.

5. Place in 10% neutral buffered formalin in a sealable container


Return to Top

 Return to Fish Kill Event Homepage

# WHAT YOU SHOULD KNOW ABOUT
# *PFIESTERIA PISCICIDA*

## WHAT IS *PFIESTERIA* ?

*Pfiesteria piscicida* (fee-STEER-ee-uh pis-kuh-SEED-uh) is a toxic dinoflagellate that has been associated with fish lesions and fish kills in coastal waters from Delaware to North Carolina. A natural part of the marine environment, dinoflagellates are microscopic, free-swimming, single-celled organisms, usually classified as a type of alga. The vast majority of dinoflagellates are not toxic. Although many dinoflagellates are plant-like and obtain energy by photosynthesis, others, including *Pfiesteria*, are more animal-like and acquire some or all of their energy by eating other organisms.

Discovered in 1988 by researchers at North Carolina State University, *Pfiesteria piscicida* is now known to have a highly complex life-cycle with 24 reported forms, a few of which can produce toxins. Three typical forms are shown on the right. A few other toxic dinoflagellate species with characteristics similar to *Pfiesteria* have been identified but not yet named. These are referred to as "*Pfiesteria*-like organisms," and they occur from Delaware to the Gulf of Mexico.

**amoeboid form**

**flagellated form**

**encysted form**

Three forms of *Pfiesteria piscicida*. Photos courtesy of the Aquatic Botany Laboratory, North Carolina State University.

## HOW DOES *PFIESTERIA* AFFECT FISH ?

*Pfiesteria* normally exists in non-toxic forms, feeding on algae and bacteria in the water and in sediments of tidal rivers and estuaries. Scientists believe that *Pfiesteria* only becomes toxic in the presence of fish, particularly schooling fish like Atlantic menhaden, triggered by their secretions or excrement in the water. At that point, *Pfiesteria* cells shift forms and begin emitting a powerful toxin that stuns the fish, making them lethargic. Other toxins are believed to break down fish skin tissue, opening bleeding sores or lesions. The toxins or subsequent lesions are frequently fatal to the fish. Fish may also die without developing lesions. As fish are incapacitated, the *Pfiesteria* cells feed on their tissues and blood. *Pfiesteria* is NOT an infectious agent like some bacteria, viruses, and fungi. Thus, fish are NOT killed by an infection of *Pfiesteria*, but rather by the toxins it releases, or by secondary infections that attack the fish once the toxins have caused lesions to develop.

US Environmental Protection Agency, 1998

## IS *PFIESTERIA* THE ONLY CAUSE OF FISH LESIONS AND FISH KILLS ?

A lesion is any sore, wound, or area of diseased tissue. There are many possible causes for fish lesions other than *Pfiesteria* and *Pfiesteria*-like organisms. These include physical injury in nets or traps, bites by other fish or birds, chemical pollutants, generally poor water quality, and infectious disease agents such as certain viruses, bacteria, and fungi. A fish kill is a situation in which many fish -- more than a few dozen -- die over a short period of time -- hours or days. *Pfiesteria* and *Pfiesteria*-like organisms are only one cause of fish kills on the southeast and Gulf coasts. Other causes include a lack of dissolved oxygen in the water, sudden changes in factors such as salinity or temperature, sewage or chemical spills, blooms of other kinds of harmful or toxic algae, infectious disease agents, and other environmental changes.

## HOW LONG DO TOXIC *PFIESTERIA* OUTBREAKS LAST ?

Toxic outbreaks of *Pfiesteria* are typically very short, no more than a few hours. After such an event, *Pfiesteria* cells change back into non-toxic forms very quickly, and the *Pfiesteria* toxins in the water break down within a few hours. However, once fish are weakened by the toxins, *Pfiesteria*-related fish lesions or fish kills may persist for days or possibly weeks.

## WHERE HAS *PFIESTERIA* BEEN FOUND ?

*Pfiesteria piscicida* is known to occur in brackish coastal waters from the Delaware Bay to North Carolina. Other *Pfiesteria*-like organisms occur along the southeast coast from Delaware to the Gulf of Mexico. These organisms are believed to be native, not introduced species, and are probably common inhabitants of estuarine waters within their range. These microbes have not been found in freshwater lakes, streams, or other inland waters.

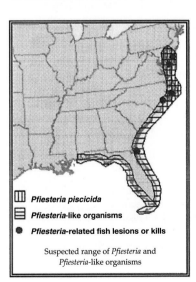

▥ *Pfiesteria piscicida*

▤ *Pfiesteria*-like organisms

● *Pfiesteria*-related fish lesions or kills

Suspected range of *Pfiesteria* and *Pfiesteria*-like organisms

*Pfiesteria piscicida* has been implicated as a cause of major fish kills at many sites along the North Carolina coast, particularly the New River and the Albemarle-Pamlico estuarine system, which includes the Neuse and Tar-Pamlico Rivers. Millions of fish have died from *Pfiesteria* in North Carolina. In 1997, *Pfiesteria* or *Pfiesteria*-like organisms killed thousands of fish in several Eastern Shore tributaries of the Chesapeake Bay, including the Chicamacomico and Manokin Rivers and King's Creek in Maryland, and the lower Pocomoke River in Maryland and Virginia. *Pfiesteria piscicida* is the probable cause for a 1987 fish kill in Delaware's Indian River. Fish kills in coastal aquaculture operations in Maryland and North Carolina have also been linked to *Pfiesteria* and *Pfiesteria*-like organisms. Lesioned fish found in association with *Pfiesteria* or *Pfiesteria*-like organisms have been documented in several Maryland and Virginia tributaries of the Chesapeake Bay, in many coastal areas of North Carolina, and in the St. John's River in Florida.

### CAN *PFIESTERIA* CAUSE HUMAN HEALTH PROBLEMS ?

*Pfiesteria* is not a virus, fungus, or bacterium. It is not contagious or infectious, and cannot be "caught" like a cold or flu. Any human health problems associated with the microbe stem from its release of toxins into river and estuarine waters.

Preliminary evidence suggests that exposure to *Pfiesteria* toxins in the air, water, or fish at the site of an outbreak can cause skin irritation as well as short-term memory loss, confusion, and other cognitive impairments in people. It has been shown that similar human health effects can be caused by exposure to *Pfiesteria* toxins in a laboratory setting. However, there is no evidence that illnesses related to *Pfiesteria* are associated with eating fish or shellfish. To date, only *Pfiesteria piscicida* has been linked to human health problems; other *Pfiesteria*-like organisms have not been shown to cause human illness.

As of October, 1997, 146 people had reported possible *Pfiesteria*-related health problems, including researchers working with the toxins in the laboratory, commercial fishermen, a water-skier, and officials working in the field during a fish kill. Symptoms reported by these individuals include skin irritation; memory loss and other cognitive impairments; nausea and vomiting; and respiratory, kidney, liver, vision, and immune system problems. Recent studies suggest that some of these symptoms may be temporary. Establishing a definite link between generalized symptoms and the microbe is difficult, but health officials are studying the situation carefully.

### IS IT SAFE TO EAT SEAFOOD ?

**YES**. In general, it IS safe to eat seafood.
- There has never been a case of illness from eating fish or shellfish exposed to *Pfiesteria*.
- There is no evidence of *Pfiesteria*-contaminated fish or shellfish on the market.
- All seafood products and processing facilities are required by law to have programs to ensure the safety of the fish and shellfish they sell. Seafood from restaurants, supermarkets, and other retailers is considered safe.
- There is no evidence that illnesses related to *Pfiesteria* are associated with eating fish or shellfish.

To be on the safe side, the following common-sense precautions are recommended:
- Comply with state closures of water bodies and public health advisories. Do not harvest or consume fish or shellfish from areas that are closed by the state.
- Do not handle or consume fish that you have harvested that are dead or dying; that exhibit sores, peeling, lesions, or other signs of disease; or that were acting abnormally when caught.
- If you notice significant numbers of fish that are dead or that exhibit lesions or other signs of disease, avoid contact with the fish and water, and promptly report the incident to your state's environment or natural resource agency.

Lesions found on fish in association with *Pfiesteria* or *Pfiesteria*-like organisms. Photo courtesy of the Aquatic Botany Laboratory, North Carolina State University.

### Is it safe to swim and boat in coastal waters ?

Swimming, boating, and other recreational activities in coastal waters are generally safe. To be on the safe side, the following common-sense precautions are recommended:

- Comply with state closures of water bodies and public health advisories. Do not go into or near the water in areas that are closed by the state.

- If you notice significant numbers of fish that are dead or that exhibit lesions or other signs of disease, avoid contact with the fish and water, and promptly report the incident to your state's environment or natural resource agency.

### Is *Pfiesteria* a "harmful algal bloom" ?
### How is it related to red and brown tides ?

Most species of algae are not harmful. Algae are the energy producers at the base of the ocean's food web, upon which all other marine organisms depend. However, a few species of algae and other microbes can become harmful to marine life and to people under certain conditions. Scientists call such events "harmful algal blooms." Brown tides, toxic *Pfiesteria* outbreaks, and some kinds of red tides are all considered types of harmful algal blooms. Some harmful algal blooms, like toxic *Pfiesteria* outbreaks, can cause detrimental effects when the microbes are at low concentrations in the water and cannot be visibly detected. In other cases, like certain red and brown tides, harmful effects occur when the algae reach high concentrations that discolor the water. However, not all algal blooms that discolor the water are harmful -- many red tides appear to have no negative effects on marine life, people, or the environment.

Some kinds of algal blooms are harmful because the algae produce one or more toxins that poison fish or shellfish, and can pose human health risks when people come in contact with affected waters. These toxic algal blooms may also kill seabirds and other animals indirectly as the toxins are passed up the food chain. Certain kinds of these toxic algal blooms can cause human health problems via contaminated seafood, like Ciguatera Fish Poisoning, Amnesic Shellfish Poisoning, and Paralytic Shellfish Poisoning. However, there is no evidence that *Pfiesteria*-related illnesses are associated with eating fish or shellfish.

Most algal blooms are not toxic, but they are still considered harmful if they reduce the amount of light or oxygen in the water, consequently killing sea grasses, fish or other marine life. Blooms of macro algae -- seaweed -- can also be harmful if they damage underwater habitats such as coral reefs or sea grass beds.

## WHAT CAUSES TOXIC *PFIESTERIA* OUTBREAKS ?

The exact conditions that cause toxic outbreaks of *Pfiesteria* to develop are not fully understood. Scientists generally agree that a high density of fish must be present to trigger the shift of *Pfiesteria* cells into toxic forms. However, other factors may contribute to toxic *Pfiesteria* outbreaks by promoting the growth of *Pfiesteria* populations in coastal waters. These factors include warm, brackish, poorly flushed waters and high levels of nutrients.

Nutrients such as nitrogen and phosphorus are thought to encourage the growth of *Pfiesteria* populations by stimulating the growth of algae that *Pfiesteria* feeds on when in its non-toxic forms. Some evidence suggests that nutrients may also directly stimulate the growth of *Pfiesteria*, but more research is needed to show this conclusively. At this time, the precise role that nutrients and other factors may play in promoting toxic outbreaks of *Pfiesteria* is not clear, and is an area of active research.

Excess nutrients are common pollutants in coastal waters. Chief sources of nutrient pollution in coastal areas are sewage treatment plants, septic tanks, polluted runoff from suburban landscapes and agricultural operations, and air pollutants that settle on the land and water.

## WHAT IS BEING DONE ABOUT *PFIESTERIA* ?

State and federal agencies are working closely with local governments and academic institutions to address the problems posed by *Pfiesteria*. Federal agencies involved in the effort include the U.S. Environmental Protection Agency, the National Oceanic and Atmospheric Administration, the Centers for Disease Control and Prevention, the National Institute of Environmental Health Sciences, the Food and Drug Administration, the U.S. Geological Survey, and the U.S. Department of Agriculture. Together with state departments of health and natural resources, these agencies are working to:

- Manage the risk of human health effects by monitoring and rapid response through river closures and public health advisories.
- Direct funding and technical expertise to *Pfiesteria*-related research and monitoring.
- Make current and accurate information widely available to the public.
- Understand and address the causes of *Pfiesteria* outbreaks, especially the possible role of excess nutrients.

## WHOM SHOULD I CONTACT TO REPORT FISH LESIONS, FISH KILLS, OR POSSIBLE HUMAN EXPOSURE TO *PFIESTERIA* ?

A few fish with lesions or even a few dead fish are not cause for alarm. However, if you notice a significant number of fish -- more than a few dozen -- that are dead, dying, behaving abnormally, exhibiting lesions, or showing other signs of disease, please contact your state's department of environment or natural resources. If you experience health problems after being exposed to fish, water, or air at the site of a fish kill or suspected toxic *Pfiesteria* outbreak, contact your physician and your state or local public health agency at once. Several states have set up *Pfiesteria* hotlines, listed on the following page.

WHERE CAN I GET MORE INFORMATION ABOUT *PFIESTERIA*?

**State *Pfiesteria*, Fish Kill, or Related Health Effects Hotlines:**

| | | | |
|---|---|---|---|
| Delaware | 1-800-523-3336 | North Carolina | 1-888-823-6915 |
| Maryland | 1-888-584-3110 | Florida | 1-800-636-0511 |
| Virginia | 1-888-238-6154 | | |

## On the Internet:

Federal:
U.S. Environmental Protection Agency
    http://www.epa.gov/owow/estuaries/pfiesteria/
National Oceanic and Atmospheric Administration, Coastal Ocean Program
    http://www.cop.noaa.gov/pfiesteria
USDA National Agricultural Library
    http://www.nal.usda.gov/wqic/pfiest.html
National Office for Marine Biotoxins and Harmful Algal Blooms
    http://www.redtide.whoi.edu/hab/
U.S. Geological Survey
    http://www.usgs.gov/outreach/fishlesions/

State:
Delaware Department of Natural Resources and Environmental Control
    http://www.dnrec.state.de.us/tpff1.htm
Maryland Department of Natural Resources
    http://www.dnr.state.md.us/fishhealth.html
Virginia Department of Health
    'http://www.vdh.state.va.us/misc/alert.htm
North Carolina Department of Environment and Natural Resources
    http://www.ehnr.state.nc.us/EHNR/files/pfies.htm
North Carolina Department of Health and Human Services
    http://www.dhr.state.nc.us/DHR/docs/pfanswer.htm
Florida Department of Environmental Protection
    http://www.ces.fau.edu/library/info/*Pfiesteria/Pfiesteria*.html
Coastal States Organization
    http://www.sso.org/cso/
Association of State and Interstate Water Pollution Control Administrators
    http://www.asiwpca.org/

Academic Institutions:
University of Maryland Sea Grant- Fish Health in the Chesapeake Bay
    http://www.mdsg.umd.edu:80/fish-health/
Virginia Institute of Marine Science
    http://www.vims.edu/welcome/news/pfiesteria/
North Carolina State University Aquatic Botany Laboratory
    http://www2.ncsu.edu/unity/lockers/project/aquatic_botany/pfiest.html
North Carolina Sea Grant- *Pfiesteria* Research Results
    http://www2.ncsu.edu/ncsu/CIL/sea_grant/pfiest.html

For sale by the U.S. Government Printing Office
Superintendent of Documents, Mail Stop: SSOP
Washington, DC 20402-9328
ISBN 0-16-049645-4

Free copies available from the National Center for
Environmental Publications and Information
1-800-490-9198
Document # EPA 842-F-98-011

# Research on Supernova Remnants:
# From Proposal to Publication

This case illustrates the research life cycle by tracing one project from initial grant proposal to publication of results. In presenting this picture of the research process, it must be acknowledged that we are oversimplifying that process, for rarely do scientific explorations have distinctly identifiable starting and ending points. The work represented here began long before the call for proposals that opens the chapter, and it continues today. We've focused on one line of study within an active complex of research agendas, in order to illustrate the types of interactions a successful grant proposal puts in motion.

A distinctive feature of research in astrophysics is the complementarity of two types of work: astronomical observations and theoretical modeling. In August 1993, NASA issued a call for proposals for theoretical research to support the agency's ongoing Space Astrophysics program. Issued under the Astrophysics Theory Program, this NRA (NASA Research Announcement) was designed to encourage the development of theoretical models and tools to be used in interpreting data generated in Space Astrophysics missions, that is, the astronomical observations carried out by the ROSAT, ASCA, and other satellites. The first item in this chapter is a brief excerpt from this NRA, describing the purpose of the program. It is followed by a proposal submitted by astrophysicist Stephen Reynolds in response to NASA's call. Reynolds and co-investigators Kasimierz Borkowski and John Blondin proposed an array of studies designed "to advance the current understanding of SNRs [supernova remnants] and to provide a more physical model with which to interpret new high-quality X-ray observations" (page 369). Their successful proposal provided funding for three years of research, to be undertaken by the primary authors and a number of graduate and undergraduate students working with them. We have included the yearly progress reports required by NASA, in which the team described their accomplishments to date and their plans for subsequent phases of the project.

The other three entries in the chapter are published reports from this line of research. We've included a 1996 research letter published in *The Astrophysical Journal*, in which Reynolds outlines a theoretical model that is shown to generate X-ray images consistent with X-ray observations from the ASCA satellite of the supernova remnant SN 1006. This is one of many publications produced under this initial grant. The Reynolds, Borkowski, and Blondin team have since received two further grants from the Astrophysics Theory Program for this line of work, which as of this writing is funded through March 2004. We've included two versions of a report from a later stage of the research, coauthored by Reynolds with Sean Hendrick, a graduate student who joined the project during the second grant period. In this later work, Hendrick, Reynolds, and Borkowski apply their developing theoretical models to archived satellite data to examine SNR shock acceleration of electrons, an important step toward understanding the origins and properties of cosmic rays. These results were first presented at the Eleventh Astrophysics Conference, on Young Supernova Remnants, in October 2000, appearing in an edited volume of conference proceedings in 2001. The work also was published as a more developed research article by Hendrick and Reynolds in *The Astrophysical Journal* in 2001, the final item in this collection. In these reports we see the promise of NASA's research call fulfilled as theory and observation are brought together to advance the field's understanding of the mysteries of the universe.

# ASTROPHYSICS THEORY PROGRAM

**NASA Research Announcement**

**Soliciting Proposals
for Theory in Space Astrophysics
for period ending
November 29, 1993**

**NRA 93-OSSA-06
Issued: August 27, 1993**

**Office of Space Science
National Aeronautics and Space Administration
Washington, DC 20546**

**[Excerpt]**

This NASA Research Announcement (NRA) solicits proposals for theoretical research with ultimate relevance to the interpretation of data from Space Astrophysics observations. The program, which was initiated 9 years ago, consists of research projects with a duration of 3 years.

The NASA Astrophysics Theory Program is intended to support efforts to develop the basic theory needed for NASA's Space Astrophysics program. This program seeks to address theoretical problems in Space Astrophysics of a wider scope than those typically studied within any particular subdiscipline of Space Astrophysics, or studied as a part of a data analysis program. Participation in this program is open to all categories of organizations including educational institutions, profit and nonprofit organizations, NASA Centers, and other Government agencies. Proposals may be submitted at any time during the period ending November 29, 1993. The proposal deadline will be adhered to strictly, and proposals received after November 29, 1993 will be held for the next review cycle which will commence in 1996. Proposals will be evaluated by scientific peer review panels with the goal of announcing selection by the middle of December 1993, and with availability of funds tentatively scheduled for January 1994 (subject to the NASA budget cycle).

Funds for awards under this NRA are expected to be available subject to the annual NASA budget cycle. The Government's obligation to make awards is contingent upon the availability of appropriated funds from which payment for award purposes can be made and the receipt of proposals which the Government determines are acceptable for award under this NRA.

National Aeronautics and Space Administration, 1993

# X-RAY EMISSION AND DYNAMICS OF SUPERNOVA REMNANTS

### A PROPOSAL TO THE
### NATIONAL AERONAUTICS AND SPACE ADMINISTRATION'S
### ASTROPHYSICS THEORY PROGRAM

### In response to NRA 93-OSS-06

**Proposing Organization: North Carolina State University**

**Principal Investigator: Stephen P. Reynolds, Department of Physics**

**December 7, 1994**

————————————————————

Stephen P. Reynolds
Associate Professor of Physics, North Carolina State University
Principal Investigator

————————————————————

Kazimierz J.B. Borkowski
Research Associate, Department of Astronomy, University of Maryland
Co-Investigator

————————————————————

John M. Blondin
Assistant Professor of Physics, North Carolina State University
Co-Investigator

————————————

Reynolds et al. NASA proposal, 1994

# TABLE OF CONTENTS[*]

---

[*]The complete table of contents has been reprinted here to convey the full structure of the proposal package, but only the main body of the proposal argument has been reproduced.

## I. INTRODUCTION

Supernovae and their remnants are called upon to perform a wide range of duties in the galactic ecosystem. They seed the interstellar medium (ISM) with heavy elements; dump kinetic energy into turbulence; accelerate cosmic rays, which affect ionization levels in molecular clouds with profound consequences for astrochemistry and star formation; heat the "hot phase" of the interstellar medium; and perhaps trigger star formation by compressing clouds on the point of instability. Understanding these diverse responsibilities demands a detailed knowledge of the physics of supernova remnants (SNRs), from the gross hydrodynamics of the blast wave (and reverse shock, at early times) to the nonthermal physics of the shock waves.

X-ray emission provides perhaps the most direct and fundamental knowledge about the physics of SNRs. Its diagnostic values are many: X-ray intensity depends quadratically on the gas density, X-ray spectra are sensitive to electron temperatures, gas abundances and SNR ages, X-ray morphology presents us with information on the SNR structure (e.g., see Canizares 1990 for a brief introduction to X-ray emission from SNRs). X-ray evidence has provided some of the most fundamental information we have on supernovae and remnants: total energy of SN events, amount and composition of ejected material, supernova subtype: information essential to many areas of astrophysics beyond the study of SNRs for their own sake. This information is extracted from observations, however, in a highly model-dependent fashion. The current analysis of X-ray spectra relies heavily on one-dimensional self-similar solutions such as a point blast explosion (Sedov) model or (more infrequently) on one-dimensional hydrodynamical simulations, coupled with nonequilibrium ionization calculations. An account of early work on X-ray emission from young SNRs is given by Chevalier (1988), with references to papers which describe the basic strategy for modeling X-ray observations in use today.

The major problem with using X-ray observations as the diagnostic of hot shocked plasmas has been the modest spectral and spatial resolution of X-ray satellites. This problem is being addressed by new observatories such as ROSAT and ASCA

These new high-quality X-ray observations mark a beginning of a new era in SNR research, to be developed further with the anticipated launch of AXAF in 1999. ROSAT observations of SNRs are just beginning to be analyzed, and spectral observations are being gathered by the ASCA satellite. Initial results indicate that these new data are providing qualitatively new information which in many cases cannot be explained within the framework of existing theoretical models. Once these data are analyzed and interpreted, a new, more complete picture will emerge of SNRs, their progenitors, and their interactions with the circumstellar and interstellar medium.

To take advantage of this observational breakthrough, intensive theoretical efforts in X-ray emission modeling are necessary. The spectral imaging data provided by ASCA will require multidimensional hydrodynamic simulations coupled with non-equilibrium ionization calculations. The authors of this proposal have developed just such a tool, which allows the calculation of X-ray spectra from multidimensional hydrodynamic simulations. One of this proposal's main objectives is to use this unique tool to generate a new class of detailed theoretical predictions of the X-ray morphology and spectrum of SNRs at various evolutionary phases, and to interpret new observations of young, X-ray bright SNRs with the help of these predictions.

However, X-ray emission alone is not a perfect tracer of SNR dynamics. Absorption effects and uncertainties about electron temperatures, or even the possibility of strongly non-Maxwellian electron distributions, can impair the diagnostic power of X-rays. Many of these problems can be addressed by combining radio information with X-ray information. While many remnants show radio emission beyond the confines of X-rays, as in W49 B (Moffett & Reynolds 1994) and other middle-aged remnants, there are no examples of X-rays found in the absence of radio emission; radio emission thus more reliably indicates at least the outer shock wave. Turbulence in remnant interiors probably amplifies magnetic field, so that one might expect synchrotron enhancements at the location of such turbulence which would allow dynamically unstable regions to be identified in observed objects. In addition, nonthermal effects, for instance collisionless heating of electrons at the shock, may influence

X-ray morphology and spectra. Such collisionless heating may depend on the obliquity angle $\Theta_{Bn}$ between the shock normal and upstream magnetic field. The remnant of SN1006, with apparently lineless power-law X-ray spectra from some areas, and a bipolar morphology resembling the radio, provides an excellent example. The efficacy of processes such as electron heat conduction may depend on the degree of small-scale order in the magnetic field, which can be investigated with radio polarization information. Full use of X-ray spectra and images of SNRs requires concomitant study of radio emission.

Much is unknown about the physics responsible for producing radio emission, for instance the relative importance of shock acceleration and second-order ("stochastic") acceleration in producing the relativistic electron populations we infer, the detailed mechanisms involved in injection of thermal electrons into the shock acceleration mechanism, and the possible importance of magnetic-field amplification in determining SNR radio morphology. To use the radio information as efficiently as possible requires expertise in both radio observations and in the modeling of strong shocks and of turbulence.

Current models for analyzing X-ray spectra of SNRs often rest on extremely simple dynamical substrates which cannot accurately describe all facets of real SNR evolution. For example, it is still common practice to use Sedov-Taylor models when interpreting the X-ray emission of SNRs, despite the fact that many, if not most, SNRs are not well described by such a model. Other similarity solutions exist to describe the global dynamics of SNRs in the early two-shock phase (the "self-similar driven wave", or SSDW; Chevalier 1982), and such solutions have been used in the interpretation of X-ray emission. Furthermore, many young SNRs may actually be in a transition phase between these self-similar solutions, for which time-dependent hydrodynamic simulations must be used (e.g., Band & Liang 1988). More recently, advances have been made using computational hydrodynamics to study the evolution of SNRs in multidimensions, addressing the role of dynamical instabilities (Chevalier, Blondin, & Emmering 1992), and the interaction of SNRs with a non-spherical circumstellar medium (Borkowski, Blondin, &

Sarazin 1992). However, a similar problem exists in the study of SNR dynamics; to advance the understanding of SNR dynamics, one tends to make simplistic assumptions about the emissivity of the gas in order to use, say, SNR X-ray images for comparison with hydrodynamic models.

It is time to develop a combined approach, putting the most advanced techniques of each of these three areas—X-ray emission, synchrotron emission, and global dynamics—into one research effort. We shall propose below to combine our well-developed programs in each of these areas to make such progress. We shall bring together sophisticated, multidimensional hydrodynamic simulations and advanced modeling of nonequilibrium ionization and X-ray emissivity to produce model X-ray images, to confront published and new SNR observations at X-ray wavelengths. Our model calculations will allow the production of spectra integrated over all or part of an image, and images integrated over arbitrary bandpasses, to compare with any satellite imaging or spectral information. Where necessary, we shall also produce model radio images, which can further constrain the process of abstracting astrophysically interesting information from observations. Without such a combined approach of individually state-of-the-art techniques, the next generation of understanding of SNRs, and all the galactic processes that depend on them, cannot be achieved.

The proposed research is timely—it is new observations of SNRs with the ASCA satellite, and to a lesser degree with the ROSAT satellite, which have been the primary cause of the current renaissance of SNR studies. We bring to this field the only tool capable of addressing the spectral imaging data being collected by the ASCA satellite, and one well able to meet the theoretical challenge of the AXAF mission.

We summarize the thrust of our proposed work in the following questions:

1. What role do instabilities play in mixing ejecta and interstellar material, forming clumps in the ejecta, generating turbulence, amplifying magnetic fields, and in creating the observed X-ray and radio morphology of various SNRs?
2. What are the chemical abundances in the shocked ambient medium and in the shocked

SN ejecta as a function of position within the remnant? The latter question is of paramount importance in view of the SN role in enriching the interstellar medium (ISM) in heavy elements, and must be understood for the correct interpretation of spatially resolved spectra.

3. What is the nature of supernova progenitors and of their surroundings, as inferred from the geometry and dynamics of individual SNRs?

4. What are the major outlines of the hydrodynamical evolution of a supernova remnant as it interacts first with circumstellar material, then with possibly inhomogeneous interstellar medium? How do multidimensional treatments differ from one-dimensional (1D) analytic results? Are there distinctive X-ray, optical, or radio signatures of these evolutionary stages which will allow us unambiguously to identify the stage of any observed SNR?

5. How is the heating of electrons (producing X-ray emission) related to the acceleration of electrons (producing radio emission)? What information can radio observations provide on collisionless heating of electrons, an important issue for X-ray modeling? Where are energetic particles accelerated, and with what efficiency?

These questions must be addressed to attack such fundamental issues as the total energy input from SNRs into the ISM, and the proportions introduced as turbulent motions, thermal energy, fast particles, and magnetic field: the origin of cosmic rays; the microphysics of collisionless shock waves; the statistics of SN and SNRs, and the use of remnants to diagnose the state of the circumstellar medium (CSM) and ISM. Thus they reach beyond the particulars of understanding SNRs for their own sake, and are important for a wide range of astronomical applications; studying the Galactic energy budget, the nature of the ISM, the properties of SN progenitors, and the properties of cosmic rays.

## II. PROJECT DESCRIPTION

### A. BACKGROUND

A complete understanding of SNR dynamics must include a self-consistent account of both the thermal plasma and relevant aspects of the physics of strong shock waves, both because of the influence of those aspects on the dynamical evolution, and because proper interpretation of X-ray and radio observations of SNRs, and hence testing of theory against those observations, cannot be accomplished without such understanding. We describe below the current state of research in X-ray and radio emission from SNRs and in the global dynamics of SNRs.

### 1. X-Ray Emission of SNRs

X-ray observations of tenuous hot plasmas have been used extensively to study physical conditions and chemical abundances in a variety of astrophysical objects, including SNRs and clusters of galaxies. This is possible because X–ray spectra of such objects are dominated by numerous emission lines, whose analysis can yield a wealth of information about the X-ray emitting gas. When such systems are in an equilibrium state (kinetic equilibrium and collisional ionization equilibrium), their X-ray emission properties depend only on the current physical state of the system, which may be deduced from the observations. The ages of young SNRs are much too short to allow them to achieve full ionization or kinetic equilibrium by the present time, implying that the present-day X-ray emission from these systems depends on their history, and coupling their emission and dynamics.

The previous decade has seen the development of methods to account for departures from equilibrium ionization and equipartition in simple, one-dimensional hydrodynamical models. X-ray spectra of young SNRs are now routinely analyzed with the help of one-dimensional nonequilibrium ionization models (Itoh 1977; Gronenschild and Mewe 1982; Shull 1982; Hamilton, Sarazin, and Chevalier 1983; Nugent et al. 1984; Hughes and Helfand 1985; Hamilton, Sarazin, and Szymkowiak 1986a, b; Itoh, Masai, and Nomoto 1988; see Kaastra & Jansen 1993 for recent developments in numerical techniques). Morphological analysis of SNR images is at a much less developed stage, though some work has been done (e.g., Pye et al. 1981; Reid, Becker, and Long 1982; Seward, Gorenstein, and Tucker 1983; Petre et al. 1982; Matsui et al. 1984; Hughes 1987; Jansen et al. 1988; Matsui, Long, and Tuohy 1988). It is straightforward to compute the radial brightness distribution in one-dimensional models (Kaastra & Jansen 1993 illustrate this procedure for Cas A SNR). While one-dimensional nonequilibrium ionization models have been adequate to interpret older X-ray observations of young SNRs in most cases, this certainly does not hold true for new high-quality observations gathered with the ASCA and ROSAT satellites. A good example is provided by Cas A where X-ray emitting material expands asymmetrically (Mushotzky 1994), suggesting a ring-like distribution of the SN ejecta and/or of the ambient CSM. A preliminary analysis of these ASCA data on Cas A reinforces the conclusion about the inadequacy of the current modeling efforts—it is impossible to model these data in the framework of existing nonequilibrium ionization models (Mushotzky 1994). This failure is most likely related to the use of vastly oversimplified, one-dimensional models for Cas A SNR. It seems improbable that these models can properly describe the complex geometry of this remnant. One-dimensional models are hardly adequate for an initially spherically-symmetric distribution of the ambient medium and the SN ejecta because the gas in SNRs is often subject to various hydrodynamical instabilities. This is one reason why one-dimensional models fail completely to account for the morphology of young SNRs: they produce thin X-ray emitting shells which have never been seen in any SNR. In general, one must calculate the ionization and kinetic state of the plasma in two- or three-dimensional hydrodynamical simulations in order to determine the X-ray emission. X-ray properties depend sensitively on the details of the flow, because of the quadratic dependence of the X-ray emissivity on gas density.

While multidimensional ionization calculations have been computationally prohibitive in the past, the situation is changing rapidly due to the availability of fast computers, vast data storage, and efficient algorithms. It is now possible to match rapid advances in X-ray astronomy with multidimensional hydrodynamics and ionization calculations. Our new technique for accomplishing this task, described in § IIB, is fully developed, and has already been applied to X-ray data on Kepler's SNR (Borkowski, Sarazin, & Blondin 1994). This novel technique will allow us to investigate a number of fundamental problems which had been intractable until now because of their multidimensional nature. The solution of these problems is necessary in order to interpret new X-ray observations of SNRs.

X-ray emission is dependent on the extent and location of electron heating. Indeed, the very observation of X-rays from SNRs directly implies the thermalization of electrons in strong shock waves. However, this thermalization process is not well understood. (Electron heating in all X-ray emission models, including our numerical code, is usually parameterized in a simple fashion, and then deduced by comparison with observations.) First, the collisionless shocks in SNRs directly heat ions only, leaving electrons to gain energy only through relatively slow Coulomb collisions. For young SNRs the Coulomb collision time scale is long enough that electrons become heated relatively slowly, and may never reach the ion temperature (Itoh 1977; Cox and Anderson 1982; Masai et al. 1988). In fact, for shock velocities greater than 2000 km s$^{-1}$ as inferred for the historical SNRs, the ion temperature should be $3m_p v_s^2/16k \sim 10^8$ K, which is not unambiguously observed. Furthermore, the X-ray morphology of many SNRs is puzzling. While some work has been done on modeling the fairly common situation of a radio shell but filled-center X-ray morphology (White and Long 1991), invoking evaporation by saturated conduction of small dense clouds in the ISM, the odd bipolar morphology of SN 1006 in X-rays (Pye et al. 1981), mirroring the radio structure (Reynolds and Gilmore 1986), seems to demand an influence in electron heating also manifested in nonthermal electron acceleration. Such an influence may be the operation of plasma instabilities that depend on the shock obliquity (Cargill and Papadopoulos 1988). The understanding of the physics of electron heating is important for the interpretation of X-ray images of SNRs and their comparison with dynamical simulations, as well as an important problem in plasma astrophysics in its own right.

### 2. Radio Emission of SNRs

SNR shocks are commonly assumed to produce the Galactic cosmic rays (see, e.g., Blandford and Eichler 1987 for a review) by first-order Fermi acceleration, though many details of the process are not well understood (Jones and Ellison 1991). The process must be intrinsically nonlinear, as SNRs must put 5%–10% or more of their energy into cosmic rays to explain the observations, so the cosmic rays can influence the shock structure, through both broadening the velocity profile (pre-accelerating the upstream fluid, in an SNR frame of reference) and changing the bulk fluid's adiabatic index and hence compressibility. Furthermore, the particles probably generate the Alfvén waves from which they also scatter, and the waves can damp into the thermal fluid. Many authors have considered these issues, separately and in combination, in the context of SNRs.

Our chief concern will be with the use of radio synchrotron radiation from SNRs to trace locations of shocks and of interior turbulence, and to give information on shock properties which may be relevant for understanding electron heating. However, in spite of the central position held by electron acceleration both to provide synchrotron tracing of the SNR dynamics, and to provide evidence for shock acceleration of ions, firm conclusions about the nature and location of the process are scarce. The observed synchrotron emissivities of older SNRs known to have cooling, high-compression shocks do not require any further electron acceleration beyond the compression of ambient cosmic-ray electrons and magnetic field (van der Laan 1962), though such acceleration may occur (Blandford & Cowie 1982). However, the inferred emissivities of brighter, adiabatic-phase remnants almost certainly require generation of new electrons or amplification (beyond mere compression) of magnetic field. Two mechanisms and locations are available for electron acceleration: in addition to shock acceleration, so-called second-order Fermi (stochastic) acceleration in the remnant interior, wherever levels of MHD turbulence are high, for instance at the unstable contact discontinuity between shocked ejecta and shocked ISM (Cowsik and Sarkar 1984) or just in the region behind the blast wave (Schlickeiser and Fürst 1989). Arguments based on morphology, reviewed in Reynolds (1988), Dickel et al. (1989), and Fulbright and Reynolds 1990, suggest that shock acceleration of electrons provides most or all of the required syn-

chrotron emissivity, and various authors have invoked it, in varying degrees of detail (Bell 1978; Reynolds & Chevalier 1981; Heavens 1984a,b). However, shock acceleration at the blast wave is not unambiguously demanded by the observations, and has never been considered at the reverse shock.

Even if electron acceleration occurs primarily at shock waves, the synchrotron emissivity may still be dominated by magnetic-field effects, such as turbulent amplification at the contact discontinuity. It will be necessary to resolve such issues as the relative importance of electron and magnetic-field effects in SNR morphology in order to make use of radio data to constrain SNR physics. A start in this direction was accomplished by Jenkins, Blondin, and Reynolds (1993) who constructed predicted morphologies of synchrotron radiation from young remnants assuming a fixed fraction of turbulent energy density appearing as magnetic field, and with shock acceleration of electrons. The turbulence profiles were formed from suitable averages over 2-D hydrodynamic simulations (Chevalier, Blondin, and Emmering 1992). For a large range of values of the (unknown) aspect angle of the local magnetic field with respect to the line of sight, models showed a thin rim of emission at the shock and a thicker interior band of emission, resembling Tycho's supernova remnant (e.g., Dickel et al. 1991). More work along these lines should solidify our understanding of the non-thermal processes in SNRs, at least well enough to allow the use of synchrotron radiation as a powerful dynamical diagnostic.

### 3. Dynamics of SNRs

The dynamical evolution of young SNRs through various stages has been documented by 1D numerical and analytic models. Soon after the expanding shock wave breaks out of the progenitor star a reverse shock is formed where the remains of the star (the ejecta) collide supersonically with the CSM that has been shocked by the leading blast wave. At early times the density profile of the ejecta is well described by a steep power law in radius (e.g., Arnett 1988). If the ambient gas is also fit by a power law, as is the case for a progenitor stellar wind or a constant-density ISM, then the region between the reverse and forward shocks can be modelled with a self-similar solution (Chevalier 1982). This self-similar driven wave (SSDW) phase applies until either the shock runs into something (e.g., a wind-swept shell; Chevalier and Liang 1989) or the ejecta density profile flattens out (Band and Liang 1988). In the latter case the SNR will evolve into a Sedov-Taylor (ST) blast wave. In the former case the forward shock will penetrate the shell and break out into the (presumably) low-density interstellar medium. In either case the expansion velocity will eventually slow to the point where the recently shocked gas is cool enough to radiate a substantial fraction of its energy, and the SNR enters the radiative pressure-driven snow-plow (PDS) phase (Cioffi, McKee, and Bertschinger 1988). Eventually the remnant expansion becomes subsonic, and the SNR merges with the ISM.

This theoretical picture of SNR evolution is lacking in several respects. First, the dynamics of the transition phases (e.g., between the SSDW and the ST phases) have not been sufficiently explored. Secondly, all these results are for 1D, spherically symmetric models, thus ignoring the possibility of non-radial instabilities or evolution into a non-spherical CSM. Multidimensional calculations are needed to understand the role of instabilities in creating the X-ray and radio morphology of SNRs, in forming clumps in young SNRs, in mixing of ejecta with the surrounding medium, in the geometry and amplification of magnetic fields, and possibly in the generation of energetic particles. The ability to model non-spherical SNRs will also allow one to use SNRs as tracers of the ambient density structure, both in the immediate CSM and on the larger scale of several parsecs.

Virtually every phase of evolution, including the transition phases, can be shown to be dynamically or thermally unstable to non-radial perturbations. Gull (1973) has shown that the contact discontinuity separating shocked ejecta and shocked CSM in young SNRs is subject to a convective or Rayleigh-Taylor instability. Goodman (1990) and Ryu and Vishniac (1991) found that Sedov-Taylor blast waves are subject to a convective instability for sufficiently steep radial density profiles in the surrounding medium. The transition to the radiative PDS phase has been shown to be unstable to radial perturbations in numerical simulations (Falle

1981), and analytic work by Bertschinger (1986) suggests that this instability is also present in non-radial modes. The radiative PDS phase is itself subject to an oscillatory over-stability (Vishniac 1983).

These dynamical instabilities may have a profound effect on the observed morphology of SNRs. As an example, the X-ray image of Tycho's remnant apparently shows emission from clumps within a shell interior to the leading shock wave (Seward, Gorenstein, and Tucker 1983). The radio emission also shows a broad region of emission peaking inside the shock front (Dickel, van Bruegel, and Strom 1991), providing at least a suggestion that instabilities of the contact discontinuity are important in this remnant.

Non-radial instabilities may affect the global dynamics as well, changing the results of 1D solutions. If the instabilities can redistribute the entropy within the remnant, the global expansion rate may deviate from the spherically-symmetric solution. If this is the case, many commonly accepted inferences about SNR ages, energetics, and statistics may be inaccurate. A possible example of this situation may arise in the transition from the SSDW phase to the ST phase, where a large fraction of the interior may be unstable as it is swept up by the reverse shock after it reflects off itself near the center of the remnant (a process that itself is likely to be grossly distorted if spherical symmetry is assumed).

The studies of non-radial dynamical stability of SNRs mentioned above all rely on linear perturbation analysis. While such an analysis can provide valuable information on regions of instability and linear growth rates, it cannot address the question of nonlinear evolution and hence cannot explain the impact of the instabilities on the global evolution and structure of SNRs. To answer questions related to the non-linear evolution of dynamical instabilities one must turn to numerical simulation. Two recent examples of numerical studies of dynamical instabilities in SNRs are Chevalier, Blondin, and Emmering (1992; see Figure 1) and Mac Low and Norman (1993). Both of these works confirmed the linear perturbation analysis of the relevant instability and went on to quantify the non-linear evolution.

The use of 1D spherically symmetric models is also a limitation in that many SNRs are distinctly non-spherical. In the case of young SNRs with massive progenitors, such objects provide information on the CSM surrounding the supernova. The evidence for an asymmetric distribution of CSM is provided by very young SNRs such as SN 1988Z (Chevalier & Fransson 1993), SN 1986J (Bartel, Rupen, and Shapiro 1989) or SN 1987A (Crotts, Kunkel, & Heathcote 1994). Somewhat older remnants with the asymmetric CSM include Cas A, Kepler's SNR, or N132D in the Large Magellanic Cloud.

In the case of older SNRs, large asymmetries may be tracing density gradients in the ISM. An excellent example is 3C 391 (Reynolds & Moffett 1993), where two levels of brightness gradient may reflect a complex structure in the surrounding medium. The bright shell is only about two-thirds complete, with the brightest region opposite the faintest, while a broad, faint plateau to the SE extends far beyond the shell. The asymmetry in the bright structure can be simply modeled with a picture of shock acceleration and magnetic field varying with external density, so that a density gradient in the NW direction in the external medium can produce a partial shell like that observed (Reynolds & Fulbright 1990). The broad extension may mark a "breakout" of the SNR into a region of considerably lower density. These somewhat speculative conclusions could be placed on a much firmer foundation by incorporating HD simulations into the model imaging.

Finally, scant effort has been invested relating the theoretical models to observations in a critical manner. While the self-similar solutions can be used to derive global parameters such as the age of a remnant, there is no way to check that the assumed evolutionary stage is in fact correct. Robust observational signatures of each stage are not available, making more detailed inferences difficult or impossible. X-ray spectral analysis has had the most success in determining important physical properties of SNRs; radio spectra are difficult to understand and have hardly been analyzed, since a nonlinear theory of electron acceleration is required (but see Reynolds & Ellison 1992). Since most SNRs have complex, confused morphology, imaging observations have been interpreted even less, though some work has been done (see below). Here again, however, basic observational signatures of important features such as the reverse shock remain undetermined. We shall work to correct this major gap between theory and observations, so that the observations can in fact serve as a useful check on the theoretical work.

## B. PROPOSED METHODOLOGY

We propose a comprehensive investigation of the structure and evolution of SNRs through the various stages outlined above, emphasizing how dynamical instabilities affect the structure and evolution of SNRs, and how these various stages may be discerned from X-ray and radio observations. At the same time, we shall study the processes by which electrons are accelerated and heated in SNRs, as intrinsically important physics and as essential to relate dynamical calculations to observations. Our project contains two main ingredients: hydrodynamic simulations to study the dynamical evolution of SNRs, and model images and spectra in X-rays and radio to critically compare theoretical results with observations. One of us (SPR) will continue work (funded elsewhere) on radio and X-ray observations of SNRs which will complement this project very well.

*i. Hydrodynamic Simulations*

The first ingredient of our investigation is numerical hydrodynamic simulations. The primary work horse for the proposed computations will be VH-1, a hydrodynamics code developed by two of us (JMB and KJB) and colleagues at the University of Virginia. This code has been thoroughly tested and successfully applied to a variety of problems in astrophysics. VH-1 is based on the piece-wise parabolic method of Colella and Woodward (1984) and includes radiative cooling, external forces, and several orthogonal grid geometries. The third-order accuracy in the spatial representation of the fluid variables allows this code to resolve flow patterns on significantly cruder grids than would otherwise be required with other types of numerical methods. The use of a Riemann solver at zone interfaces insures accurate shock jump conditions and alleviates the need for artificial viscosity, which broadens shocks and may dampen the growth of the dynamical instabilities under study. These properties of VH-1 make it an ideal code for the study of dynamical instabilities in supersonic flows, as demonstrated in Chevalier, Blondin, and Emmering (1992).

Computational hydrodynamics will allow us to study the full evolution of dynamical instabilities as well as the evolution of asymmetric SNRs. Because of the minimal amount of dissipation in the PPM algorithm, we can accurately study the growth of instabilities in the linear regime where we can make contact with linear analysis and quantify linear growth rates. We can also study the non-linear evolution of instabilities with computational hydrodynamics, and thereby assess their impact on the structure and evolution of SNRs. Multidimensional simulations can also address the question of mixing between ejecta and the ISM with these techniques.

*ii. Model Images and Spectra*

We bring a well-developed formalism to bear here in both the X-ray and radio wavebands. To calculate X-ray spectra from hydrodynamical simulations, Borkowski, Sarazin & Blondin (1994) have devised a method for calculating the nonequilibrium ionization and kinetic state for the gas in two-dimensional hydrodynamical flows. This is the only fully developed code that we know of that can do this. Its construction took a considerable amount of effort because of a number of difficult technical problems associated with the complexity and the scale of the problem. Briefly, the idea is to perform hydrodynamical calculations first, storing pressure, density, and Lagrangian "labels" at intermediate times in "history" files. These labels are set to initial coordinates of fluid particles at the start of simulations, and then advected with the fluid across the computational grid with the help of the hydrodynamics code VH-1. The information stored at intermediate times is sufficient to reconstruct how pressure and density varied for each fluid element located at each computational cell at the final time. This reconstruction uses an efficient method of interpolation on irregular grids. It is then possible to find electron temperatures and nonequilibrium ionization for each fluid element, the latter with an efficient algorithm for computing nonequilibrium ionization developed by Hughes & Helfand (1985). In this form, this technique is applicable to nonradiative flows where gas cooling through X-ray emission does not affect the dynamics of the flow, allowing for separate hydrodynamical and ionization calculations. The nonradiative condition is satisfied in young SNRs. The final step involves calculations of X-ray spectra and images, which again must be computationally efficient because of a large number of grid cells required in a typical hydrodynamical simulation of a SNR.

We are hopeful that strong constraints on the structure of the CSM, on ejecta abundances, and on blast-wave properties can emerge from the comparison of these sophisticated model X-ray images and spectra with observations from *Einstein, ROSAT, ASCA, AXAF,* and other future X-ray imaging instruments. We will have the ability to produce predicted images in any instrumental bandpass, including all important spectral effects, at a level far beyond what has been accomplished in the past. This will allow us to make effective use of previously obtained images, as well as to anticipate results of upcoming observations. It will always be possible, of course, to integrate over images to obtain total spectra for comparison with observations of unresolved SNRs.

We have also been successful in developing model radio imaging techniques. Model radio synchrotron images of SNRs have been used to study obliquity-dependence of electron acceleration (Fulbright and Reynolds 1990), appearance of SNRs in a density gradient (Reynolds and Fulbright 1990), and preshock electron diffusion and synchrotron "halos" (Achterberg, Blandford, and Reynolds 1994; Reynolds 1994). In addition, work has begun on the contribution of turbulent amplification of magnetic fields to observed morphology of SNRs (Jenkins, Blondin, and Reynolds 1993). Our fully developed code can be immediately adapted for different dynamical substrates, and for different prescriptions for the sources and evolution of relativistic electrons and magnetic field. Additional near-term projects include allowing for additional electron acceleration at locations of strong turbulence, and examining the degree of order in the magnetic field with model polarization images, including internal Faraday depolarization effects as well as intrinsic disorder in the magnetic field. We hope by these means to set limits on the location, and fractional turbulent energy, in relativistic electron acceleration and magnetic-field amplification, and to constrain the possible extent of electron heat conduction.

*iii. Comparison with Observations*

Model X-ray images and spectra for an individual SNR must be compared with X-ray observations of this remnant. This task can be accomplished with the recently established High Energy Astrophysics Science Archive Center (HEARSAC) Facility at NASA/GSFC. Archived X-ray data can be easily accessed through the HEARSAC On-line Service. Many ROSAT observations are already available to the general astronomical community, and ASCA observations from the performance verification phase, including many bright SNRs, will become available in 1995 January. This date nearly coincides with the anticipated starting date for this project. We will compare our models with observations with the help of the XANADU (XSPEC, XIMAGE) software package supported by the HEARSAC, as we had already done for Kepler's SNR (Borkowski et al. 1994).

In order to ensure the most timely and efficient interpretation of X-ray observations of individual SNRs in terms of our two-dimensional models, we plan on close collaboration with X-ray observers. This would address instrumental and calibration problems associated with X-ray data, would allow us to concentrate our attention on the most important issues, and would keep us in close contact with X-ray observers. For example, existing data on Cas A SNR are quite extensive and the most recent observations with the ASCA satellite are likely to be subject to poorly-understood instrumental problems so early in the satellite's lifetime. We intend to work on this remnant with A. Szymkowiak from the NASA/GSFC, who was himself involved in recent BBXRT observations of this remnant. Moderately frequent visits to the NASA/GSFC are anticipated in order to foster this collaboration.

## C. PROPOSED WORK

In spite of a great deal of observational and theoretical effort, the connection between the theoretically calculated hydrodynamic evolution of SNRs, and the spectral and morphological properties of objects observed in X-rays, radio, and other wavebands is far from clear. Without a combined approach, the hydrodynamical modeling cannot be validated (or refuted), much less used to infer secondary properties such as SN energetics and the nature of surrounding material. We propose to use the twofold approach described above to advance the current understanding of SNRs and to provide a more physical model with which to interpret new high-quality X-ray observations. This work will take on two direct lines of attack: modelling individual SNRs in X-ray and radio to infer specific properties of those systems, and a systematic modelling of the continuum of evolutionary stages of SNRs. The first tack allows us to extract as much information as possible from a given set of observations. The second tack will build a new theoretical framework for the future interpretation of X-ray and radio observations. In all this work, the combination of state-of-the-art hydrodynamical modeling and sophisticated X-ray and radio visualization of the dynamical results is essential for success. The best hydrodynamical calculations are useless without the ability to predict the resulting appearance of objects to the X-ray or radio observer.

## 1. Modelling Individual SNRs

One of the main objectives of this proposal is to simulate X-ray spectra and X-ray images of young, X-ray bright SNRs, and compare them with most recent high-quality X-ray observations, such as imaging spectroscopy with the ASCA satellite. It is best to illustrate our strategy for addressing these issues by discussing a couple of well-known SNRs. The youngest (about 320 yr old) galactic SNR, Cas A, is a good example because of the availability of good-quality data from radio to $\gamma$-rays. Another historical SNR, SN 1006, is an ideal test case for studying electron heating and particle acceleration processes in view of its striking bipolar morphology in both X-rays and radio. Still another extremely interesting SNR is SN 1987A, not discussed below, but which will become a showcase at the end of this century. One of us (JMB) has already theoretically investigated the formation of its ring of circumstellar material—the next logical step is to study the interaction of the SN ejecta with the CSM, the outcome of which is of considerable interest to the astronomical community.

*i. Cassiopeia A*

There is overwhelming evidence that the progenitor of Cas A was a massive star, the most compelling being the detection of the radioactive $^{44}$Ti by the

Gamma-Ray Observatory (Iyudin et al. 1994). Recent spectroscopic data (Fesen & Becker 1991) indicate that the SN progenitor was a massive stellar core composed mainly of He, with a thin hydrogen layer at the surface. Most of the H-rich envelope was expelled from the star shortly before the SN explosion. This envelope is now seen in the remnant through its interaction with the SN ejecta.

The shell-like X-ray morphology of Cas A consists of a bright clumpy shell and outer, weaker shell-like emission (Fabian et al. 1980; Jansen et al. 1988). The remnant's morphology is strongly asymmetric—optical and X-ray spectroscopy reveal that the clumpy shell is actually a ring inclined to the line of sight (Markert et al. 1983). New data from the ASCA satellite provided images in various emission lines, high-quality spectra in various locations within the remnant, with over a dozen of emission lines detected at each location. Position-dependent Doppler shifts have been found, attesting to the asymmetric nature of the remnant (Mushotzky 1994).

X-ray modeling of Cas A has been limited to very simple models; for example, Jansen et al. (1988) interpreted EXOSAT data in terms of a simple point-blast explosion (Sedov) model. In view of the complex remnant morphology, it is doubtful that such models are adequate to describe even pre-ASCA observations. It is obvious that a major theoretical effort is required to interpret new X-ray data on Cas A, surpassing anything that has been done before.

Our approach is straightforward. We propose to start with a supernova ejecta structure expected for an exploded stellar core with a thin H envelope. We would then impose an appropriate density structure for the ambient medium, which would involve an asymmetric distribution of the circumstellar medium (CSM), possibly in a form of a ring such as seen in SN 1987A. We would then follow the interaction of SN ejecta with the ambient CSM and ISM with the help of the 2-D nonequilibrium-ionization hydrodynamical code described in § IIB. At the current remnant's age, we would calculate X-ray images and X-ray spectra, and compare them with observations. By matching the observed X-ray morphology of the remnant in various spectral fea-

tures, it should be possible to determine the geometry of the remnant, locate the shocked ejecta, the shocked CSM shell and find the primary shock (blast wave). Chemical abundances can be determined by fitting model spectra to observed spectra at each location within the remnant. The amount of electron heating in collisionless shocks can be found by analyzing the slope and intensity of high-energy X-ray continuum throughout the remnant.

Hydrodynamical instabilities might be very important in determining the remnant's morphology. This is suggested by the work of Chevalier, Blondin & Emmering (1992) who studied the development of a Rayleigh-Taylor instability for the reverse-shock solutions in spherical geometry. They found that apparent thickness and clumpiness of various large-scale emission features can be modeled in the framework of hydrodynamical instabilities. A similar numerical study can be also done for Cas A SNR with a two-dimensional hydrodynamical code. It could then be possible to interpret the observed morphology (such as the shell thickness) in terms of hydrodynamical instabilities.

### ii. SN 1006

SN 1006 was the brightest supernova recorded in history (Clark and Stephenson 1977), and has produced an extremely interesting remnant. Its height above the Galactic plane of almost over 400 pc probably accounts for both its faintness at radio and X-ray wavelengths and its extremely regular, bipolar morphology in both spectral regions (Reynolds and Gilmore 1986; Pye et al. 1981), because of a presumably lower-density, more orderly ambient medium than is true immediately in the Galactic plane. SN 1006 is customarily classified as Type Ia, because of the large $z$-height and the implied large peak brightness of the SN. Extensive searches have failed to turn up any evidence of a collapsed object at the center, and spectral work (e.g., Hamilton, Sarazin, and Szymkowiak 1986a) seems consistent with a Type Ia origin. Optical emission is confined to a few faint H$\alpha$ filaments in one region of the periphery, classing SN 1006 with Kepler, Tycho, and several Magellanic Clouds remnants as "Balmer-dominated" SNRs (Smith et al. 1991).

The shell X-ray morphology, with its two bright limbs and fainter regions between, strikingly resembles the radio, which has been explained as due to preferential shock acceleration of electrons in regions of particular obliquity $\Theta_{Bn}$. This acceleration may also provide preferential electron heating to explain the coincidence of the radio and X-ray bipolar structures. The soft X-ray emission is less concentrated in a shell, with significant emission from the remnant interior (Pye et al. 1981). The X-ray spectrum has always been problematic: early work (Becker et al. 1980) suggested a lineless power-law as the best fit, but later observations (summarized in Hamilton, Sarazin, and Szymkowiak 1986a; HSS) disagreed with the slope of this power-law and provided evidence for weak lines below 1 keV. Very recent ASCA data (summarized by J. Hughes, January 1994 AAS Meeting, Washington, DC) indicate that spectra from the fainter regions of the remnant show perfectly normal line spectra, while those from the brightest limbs persist in exhibiting a lineless power-law of energy index about $-1.8$.

The puzzling X-ray spectrum was modeled by Reynolds and Chevalier (1981) as the extension of the radio synchrotron spectrum, steepened by losses. This picture requires accelerating electrons to $10^{12}$ eV or higher near the shock, after which they diffuse into a region of higher magnetic field somewhat further behind the shock. The new spectral slope reported by ASCA causes serious problems for this simple picture. An elaborate non-equilibrium ionization thermal model was put forward by HSS, assuming the progenitor was a white dwarf. This picture used a particular 1-D analytic dynamical model of a forward and reverse shock, and explained the lack of lines by a strict segregation of ejecta by radius, so that the X-ray continuum was produced in an outermost carbon-rich layer of dense X-ray emitting gas. The model predicted a rather high expansion rate, marginally at odds with radio observations of the expansion (Moffett, Goss, and Reynolds 1993). In addition, while HSS did not make detailed morphological predictions, they mention the difficulty of explaining the interior soft X-ray emission with their model.

The resolution of these problems seems to be at hand with the new ASCA observations. Because of the anticipated pivotal role of SN 1006 in understanding of electron heating and particle accelera-

tion in SNRs, a thorough interpretation of X-ray observations of this remnant is desirable. Although SN 1006 appears more homogeneous on small scales than other SNRs, radio and X-ray observations still show a fair amount of clumpiness. On theoretical grounds, hydrodynamical instabilities must have been operating in this remnant during its lifetime (Chevalier, Blondin, & Emmering 1992). Our goal is to construct a hydrodynamical model which would include these instabilities, as well as electron heating at strong shocks, possibly dependent on the magnetic field inclination to the shock normal in order to reproduce the observed bipolar morphology. (We are also well equipped to study a hypothetical contribution to X-ray emission from relativistic electrons if necessary.) We will take the structure of expanding SN ejecta appropriate for an exploding white dwarf, and study its interaction with the uniform ambient medium. We will be able to study the development of instabilities, estimate the amount of energy contained in turbulent motions, calculate radio and X-ray maps and position-dependent X-ray spectra. SN 1006 is clearly a prime target for our combined approach.

### 2. Evolutionary Stages of SNRs

To complement the work described above in which observations of individual remnants are analysed, we propose to study the various evolutionary stages of an SNR using the same techniques. Specifically, we will use hydrodynamic simulations in 1D and 2D to study the SSDW phase, the transition from SSDW to ST, and the transition from ST to a radiative SNR. These hydrodynamic simulations can then be coupled with X-ray emission calculations to investigate differences between 1D and 2D models and to search for important clues in the X-ray spectra that may betray the evolutionary status of a given remnant.

The end product of this line of research will be an X-ray catalog of simulated images and spectra covering an appropriate range in the parameter space, somewhat similar to a catalog constructed by Hamilton, Sarazin, & Chevalier (1983) for the Sedov solution. Because of the availability of standard X-ray software packages such as XANADU, the emphasis will be on a creation of a small computer database compatible with the XSPEC and XIMAGE packages within XANADU. Our results could then be used by X-ray observers to interpret new and archival observations of SNRs. The final form of such computer catalog will be decided upon successful completion of the goals outlined below.

*i. Early Evolution: SSDW*

The work of Chevalier, Blondin, and Emmering (1992) represents an important step in the study of SNRs dynamics, and we propose to take advantage of this pioneering work by extending it in several directions. The first immediate step will be the calculation of the X-ray emission to study the effects of the dynamical instabilities on the resulting X-ray spectra and morphology. In particular, dynamical instabilities can have the effect of broadening the region of X-ray emission, producing a more extended region of X-ray emission compared to the relatively thin shell of X-rays implied by 1D models.

To complement the X-ray analysis, we will also create model radio images from the 2D hydrodynamic simulations. We will use these models to investigate the possibility that turbulent amplification of CSM magnetic field can enhance the synchrotron emission from SNRs. The dynamical instability in the SSDW phase converts roughly 5% of the post-shock energy into turbulent motions. If one accepted the postulate that this turbulent motion can efficiently amplify an ambient magnetic field, one would expect a fairly large local magnetic energy density in a thick shell interior to the forward shock wave.

We propose to extend these simulations of the SSDW phase to a study of young SNRs for which the reverse shock is radiatively cooling. This situation may be relevant to a class of SNRs that are believed to be propagating into very dense circumstellar material resulting in very high densities in the shocked ejecta, and hence short cooling times (Chevalier 1990). An interesting example of this class of young SNRs is SN 1986J, the brightest SN observed in the radio. This object has been mapped in the radio using VLBI, and shows strong deviations from spherical symmetry (Bartel, Rupen, and Shapiro 1989). These observations suggest either an initially very non-uniform beginning or a very strong non-radial instability. Could dynamical instabilities be sufficiently strong to distort the for-

ward shock wave? While the work on instabilities in the adiabatic SSDW phase suggests that the convective instability is confined to the interior of the forward shock, it may be possible that strong cooling at the reverse shock can enhance the instability by increasing the entropy contrast in the shocked shell. This scenario can be readily tested with the present HD code by including an appropriate energy loss rate for the shock-heated ejecta.

Another investigation of the SSDW phase is the structure of an SNR evolving into an asymmetric CSM. A direct question to be answered by this work is "What magnitude of asymmetry is required in the CSM to produce observed asymmetric SNRs?" Such a question is more directly applicable to "old" supernovae (or very young SNRs) both on an observational basis and on theoretical grounds that the immediate CSM of a SN may be much more asymmetric than the surrounding ISM. As an example, the interpretation of SN 1988Z given by Chevalier and Fransson (1993) requires an extremely large ambient density asymmetry, with the polar region being $10^3$ times less dense than the equatorial region. However, if the asymmetry leads to significant non-radial flow, as would be revealed by multidimensional HD simulations, then such a large asymmetry might not be required. This project represents a very straightforward integration of the simulations in Chevalier, Blondin, and Emmering (1992) and the simulations of stellar wind bubble evolution into an asymmetric CSM in Blondin and Lundqvist (1992). This work would apply to many astrophysical systems, including the SN explosion of a Be star with a dense equatorial wind, or the SN explosion of a red supergiant in a moderately close binary system.

*ii. Transition from SSDW to ST*

The transition from the SSDW phase to the ST phase has not been investigated in detail, and yet it is expected to be relevant to many observed SNRs because of the length of time a given SNR spends in this transitional phase. Band and Liang (1988) used 1D numerical HD to study this phase, but their computational techniques produced considerable numerical noise, they considered only 3 specific models, and they did not evolve their models long enough to see some of the more interesting effects. We have begun preliminary work using 1D HD simulations to quantify the behavior of SNRs in this phase. We have found that for SNRs propagating into a preexisting stellar wind this transition is very slow, lasting as much as a thousand times longer than the SSDW phase. For SNRs propagating into a constant density ambient medium, this transition is much quicker and more violent. The reverse shock is driven into the center by the high pressure in the shocked shell, reaching the center roughly 7 times the age at which the turn-over in the ejecta density profile reaches the reverse shock (defining the beginning of the transitional phase). After bouncing off itself at the center of the SNR, this reverse shock then travels outward, re-accelerating the shocked ejecta to a velocity such that it expands with the self-similar rate of the expanding forward shock. From then on the ejecta occupy a region of roughly one-tenth the volume of the entire SNR. This 1D result in and of itself is very exciting, but the evolution will be even more interesting in 2D when the effects of this reflected reverse shock on the hydrodynamic instabilities are included, and when the "bounce" of the reverse shock is treated realistically. Perhaps the interaction of the reflected shock and the clumpy ejecta could enhance the local X-ray emission leading to a center-brightened X-ray SNR.

We propose to expand this 1D investigation of the transitional phase to include predictions of the X-ray and optical morphology of such a SNR, and to extend the HD calculations to 2D. The study of this transitional phase in 2D will be a straight forward extension of the work described in Chevalier, Blondin, and Emmering (1992), with the exception that all of the interior of the blast wave must be included in the evolution. This will necessitate the use of a cylindrical grid in order to avoid prohibitively small time steps in the time-explicit HD calculations. For the constant ambient density case the instability properties may be substantially different from the SSDW phase. First of all, the rapid deceleration of the shocked shell may enhance the growth rate of the convective instability studied in the SSDW phase. Secondly, the passage of the reflected reverse shock may re-invigorate the instability growth through the action of the Richtmyer-Meshkov instability.

*iii. Transition from ST to Radiative PDS*

The onset of radiative cooling in an expanding SNR has been shown to induce large-amplitude radial perturbations (Falle 1981), which is likely to affect the global morphology of older SNRs. This is an example of the global thermal overstability of radiative shocks first studied analytically by Chevalier and Imamura (1982) and later examined with the aid of 1D numerical simulations by, among others, Imamura, Wolff, and Durisen (1984). It is possible that this overstability is enhanced with the addition of non-radial wave modes. Bertschinger (1986) used a 3D linear stability analysis to investigate the role of non-radial perturbations in the evolution of a radiative spherical shock wave. He found slight differences in 3D compared to the 1D results, but again, the linear analysis does not provide information on the end state of the overstability. Recently, Strickland & Blondin (1994) have applied hydrodynamic simulations to study this overstability in 1D and 2D. This work has shown that the back of the radiative shock—where the shock-heated gas has cooled down to the preshock temperature—develops strong non-radial structure with a wavelength of order the length of the radiative shock.

We propose to continue this multidimensional study of the thermal overstability of radiative shock waves, extending the recent non-linear simulations to spherical geometry with specific application to SNRs. In particular, we will study the effect of this overstability in multidimensions during the transition to the radiative phase, when the wavelength of instability is of order the radius of the remnant. As an example of the kinds of investigations that we can accomplish with relatively little effort, we will examine the evolution of a SNR entering this transition phase in a slightly non-uniform environment. If one side of the SNR begins to cool slightly ahead of the other side, a strong asymmetry—much larger than the asymmetry imposed by the surrounding medium—could develop in the SNR. One of the goals of this work will be to find observational signatures that can distinguish between the different possible sources of structure in older SNRs. For example, the filamentary structure observed in many SNRs has been attributed to shocked clouds in the ISM, local thermal instabilities in the radiative shock of a SNR (Blondin & Cioffi 1989), the global thermal instability to be studied here, and the dynamical thin-shell instability discovered by Vishniac (1983) and studied numerically by Mac Low & Norman (1993).

While this work is less relevant to X-ray emission (the bulk of the X-ray flux from these older SNRs will come from the interior of the remnant, away from the interesting hydrodynamics), it is directly applicable to high-resolution optical observations using the *Hubble Space Telescope.* Such observations will provide detailed information on the scale and structure of the shock front, relating directly to dynamical instabilities we intend to investigate.

### D. INSTITUTIONAL FACILITIES

Crucial to the success of the computational projects described in this proposal is continuing access to large allocations of supercomputer time. One of us (JMB) has maintained access to such large allocations through the North Carolina Supercomputing Center (NCSC—conveniently located only twenty miles from campus). The current annual allocation to JMB is over 900 cpu hours on a Cray YMP. This allotment is extended some 40% by running large jobs during non-prime hours. At the time of this writing NCSC is planning on upgrading to a next-generation supercomputer, possibly the new Cray Triton. Whatever the upgrade, the state of North Carolina appears committed to providing state-of-the-art supercomputing to its residents. NCSC is also a vital resource in the field of scientific visualization. This ready access to supercomputing and visualization at NCSC is an indispensable resource supporting the PI, co-I's and students in the activities here proposed.

Another essential element is local workstations that possess intermediate-level processing power and substantial amounts of disk space for downloading and post-processing data from supercomputer simulations. The astrophysics group at NCSU maintains a cluster of 5 UNIX workstations, 4 of which provide better than 15 Mflops of computing speed each. One of these workstations also possesses advanced 3D graphics hardware for visualization and analysis of 3D data. This will be particularly advantageous for the proposed research, in that we will be able to rapidly produce volume rendered images of the X-ray emission predicted from multidimensional simulations. Substantial disk space (over 9Gb) and archiving (Exabyte and DAT) are also available, although significantly more disk space will be added for the research proposed herein due to the extremely large datasets generated by such work. These local facilities represent an institutional commitment toward the development of a computational astrophysics laboratory at NCSU.

### III. BIBLIOGRAPHY

Achterberg, A., Blandford, R. D., & Reynolds, R. P. 1993, AA, submitted

Arnett, W. D. 1988, ApJ, 331, 337

Band, D. L., & Liang, E. P. 1988, ApJ, 334, 266

Bartel, N., Rupen, M., & Shapiro, I. I. 1989, ApJ, 337, L85

Becker, R. H., Szymkowiak, A. E., Boldt, E. A., Holt, S. S., & Serlemitsos, P. J. 1980, ApJ, 240, L33

Bell, A. R. 1978. MNRAS, 182, 443

Bertschinger, E. 1986, ApJ, 304, 154

Blandford, R. D., & Cowie, L. L. 1982, ApJ, 260, 625

Blandford, R. D., & Eichler, D. 1987, Physics Reports, 154, 1

Blondin, J. M., & Lundqvist, P. 1993, ApJ, 405, 337

Borkowski, K. J., Blondin, J. M., & Sarazin, C. L. 1992, ApJ, 400, 222

Borkowski, K. J., Sarazin, C. L., & Blondin, J. M. 1994, ApJ, 429, in press

Canizares, C. R. 1990, in Physical Processes in Hot Cosmic plasmas, ed. W. Brinkmann, A. C. Fabian & F. Giovanelli (Dordrecht: Kluwer), 17

Cargill, P. J., & Papadopoulos, K. 1988, ApJ, 329, L29

Chevalier, R. A., 1982, ApJ, 258, 790

Chevalier, R. A. 1988, in Supernova Remnants and the Interstellar Medium, ed. R. S. Roger & T. L. Landecker (Cambridge University Press: Cambridge), 31

Chevalier, R. A. 1990, in Supernovae, ed. A. G. Petschek (Berlin: Springer), p.91

Chevalier, R. A., Blondin, J. M., & Emmering, R. T. 1992, ApJ, 392, 118

Chevalier, R. A., & Fransson, C. 1993, preprint

Chevalier, R. A. & Imamura, J. N. 1982, ApJ, 261, 543

Chevalier, R. A., & Liang, E. P. 1989, ApJ, 344, 332

Cioffi, D. F., McKee, C. F., & Bertschinger, E. 1988, ApJ, 334, 252

Colella, P., & Woodward, P. R. 1984, J. Comp. Phys., 54, 174

Cowsik, R., & Sarkar, S. 1984, MNRAS, 207, 745

Cox, D. P., and Anderson, P. R. 1982, ApJ, 253, 268

Crotts, A. P. S., Kunkel, W. E., & Heathcote, S. R. 1994, ApJ, in press

Dickel, J. R., Eilek, J. A., Jones, E. M., & Reynolds, S. P. 1989, ApJS, 70, 497

Dickel, J. R., van Bruegel, W. J. M., & Strom, R. G. 1991, AJ, 101, 2151

Fabian, A. C., Willingale, R., Pye, J. P., Murray, S. S., & Fabbiano, G. 1980, MNRAS, 193, 175

Falle, S. A. E. G. 1981, MNRAS 195, 1011

Fesen, R. A., & Becker, P. 1991, ApJ, 371, 621

Fulbright, M. S., & Reynolds, S. P. 1990, ApJ, 357, 591

Goodman, J. 1990, ApJ, 358, 214

Gronenschild, E. H. B. M., and Mewe, R. 1982, A&AS, 48, 305

Gull, S. F. 1973, MNRAS, 161, 47

Hamilton, A. J. S., Sarazin, C. L., & Chevalier, R. A. 1983, ApJS, 51, 115

Hamilton, A. J. S., Sarazin, C. L., & Szymkowiak, A. E. 1986a, ApJ, 300, 698

Hamilton, A. J. S., Sarazin, C. L., & Szymkowiak, A. E. 1986b, ApJ, 300, 713

Heavens, A. F. 1984a, MNRAS, 210, 813

Heavens, A. F., 1984b, MNRAS, 211, 195

Hughes, J. P., & Helfand, D. J. 1985, ApJ, 291, 544

Hughes, J. P. 1987, ApJ, 314, 103

Imamura, J. N., Wolff, M. T., & Durisen, R. H. 1984, ApJ, 276, 667

Itoh, H. 1977, PASJ, 29, 813

Itoh, H., Masai, K., and Nomoto, K. 1988, ApJ, 334, 279

Iyudin, A. F., Diehl, R., Bloemen, H., Hermsen, W., Lichti, G. G., Morris, D., Ryan, J., Schönfelder, V., Steinle, H., Varendorff, M., deVries, C., & Winkler, C. 1994, A&A, 248, L1

Janesen, E., Smith, A., Bleeker, J. A., de Korte, P. A. J., Peacock, A., & White, N. E. 1988, ApJ, 331, 949

Jenkins, G. C., Blondin, J. M., & Reynolds, S. P. 1993, BAAS, 25, 1421

Jones, F. C., & Ellison, D. C. 1991, Ap&SS, 58, 259

Kaastra, J. S., & Jansen, F. A. 1993, AAS, 97, 873

Kesteven, M. J., & Caswell, J. L., 1987, A&A, 183, 118

Mac Low, M. M., & Norman, M. L. 1993, ApJ, 407, 207

Markert, T. H., Canizares, C. R., Clark, G. W., & Winkler, P. F. 1983, ApJ, 268, 134

Masai, K., Hayakawa, S., Inoue, H., Itoh, H., & Nomoto, K. 1988, Nature, 335, 804

Matsui, Y., Long, K. S., Dickel, J. R., and Greisen, E. W. 1984, ApJ, 287, 295

Matsui, Y., Long, K. S., and Tuohy, I. R. 1988, ApJ, 329, 838

Moffett, D. A., Goss, W. M., & Reynolds, S. P. 1993, AJ, 106, 1566

Moffett, D. A., & Reynolds, S. P. 1994, ApJ, in press

Mushotzky, R. 1994, in the proceedings of a symposium held at STScI in 1994 May, The Analysis of Emission Lines, in press

Nugent, J. J., Pravdo, S. H., Garmire, G. P., Becker, R. H., Tuohy, I. R., & Winkler, P. F. 1984, ApJ, 284, 612

Petre, R., Canizares, C. R., Kriss, G. A., and Winkler, P. F. 1982, ApJ, 258, 22

Pye, J. P., et al. 1981, MNRAS, 194, 569

Reid, P. B., Becker, R. H., and Long, K. S. 1982, ApJ, 261, 485

Reynolds, S. P.: 1988, in Galactic and Extragalactic Radio Astronomy, eds. G. L. Verschuur and K. I. Kellermann, Springer-Verlag, New York, p. 439.

Reynolds, S. P. 1994, ApJS, 90, 845

Reynolds, S. P., & Chevalier, R. A. 1981, ApJ, 245, 912

Reynolds, S. P., & Ellison, D. C. 1992, ApJ, 399, L75

Reynolds, S. P., & Fulbright, M. S. 1990, in Proc. 21st Int. Cosmic Ray Conf. (Adelaide), 4, 72

Reynolds, S. P., & Gilmore, D. M. 1986, AJ, 92, 1138

Reynolds, S. P., & Moffett, D. A. 1993, AJ, 105, 2226

Ryu, D. S., & Vishniac, E. J. 1991, ApJ, 368, 411

Schlickeiser, R., & Fürst, E. 1989, AA 219, 192

Seward, F. D., Gorenstein, P., & Tucker, W. 1983, ApJ, 266, 287

Shull, J. M. 1982, ApJ, 262, 308

Smith, R. C., Kirshner, R. P., Blair, W. P., & Winkler, P. F. 1991, ApJ, 375, 652

van der Laan, H. 1962, MNRAS, 124, 125

Vishniac, E. T. 1983, ApJ, 274, 152

White, R. L., & Long, K. S. 1991, ApJ, 373, 543

### X-ray Emission and Dynamics of Supernova Remnants

Stephen P. Reynolds, John M. Blondin, and Kazimierz J. Borkowski

#### First-Year Report

Our project involves the combination of several approaches to understanding the dynamics and radiation from supernova remnants: hydrodynamic simulations, detailed calculations of thermal X-ray emission, and possible nonthermal contributions to X-rays; and modeling of individual objects based on these general calculations. In our first year we have made progress in all three areas.

**General hydrodynamic simulations.** With graduate student Eric Wright, we have performed 1-D simulations of the transition from Sedov-Taylor to radiative evolution, with unprecedented resolution and with realistic cooling. It is possible that the rapid collapse of the post-shock cooling region to form a thin shell may result in hydrodynamic instabilities; these will be explored in Years 2 and 3.

We have done 3-D simulations of the convective instability in self-similar driven waves, following up the original work of Chevalier, Blondin, and Emmering (1992, ApJ, 392, 118). Results in 3-D show, relative to 2-D, more power in shorter wavelengths, longer reach of the Rayleigh-Taylor fingers, and lower power-law growth exponents.

**X-ray Modeling.** We have updated our detailed X-ray emission code, originally written by Hamilton & Sarazin at the University of Virginia. A lot of atomic data was revised – in particular, we used most recent calculations of collision strengths for the Fe L-shell emission lines, kindly provided to us by Duane Liedahl from the Lawrence Livermore National Laboratory. The X-ray emission code was installed in the XSPEC v9 software package, together with a number of models useful for interpretation of X-ray emission from SNRs. Further work on the X-ray emission code is in progress. We expect that this code will be useful for X-ray observers working on SNRs.

Undergraduate Jay Lyerly has been applying the upgraded spectral code to calculating complete non-equilibrium ionization spectra from Sedov-Taylor supernova remnants. We are currently studying the degree to which these complex models can be well-described by simpler models that can be calculated quickly in XSPEC. We find that models with high shock temperatures should be well-fit by a single-temperature, single ionization-timescale component, while lower temperatures are not well described even by a superposition of several such components.

In the project's first year, we developed an extensive program modeling nonthermal X-rays expected from young SNRs due to shock acceleration of electrons to a maximum energy limited by radiation losses or by the shock age. Both images and spectra of such radiation were calculated for a variety of parameters. Many remnants younger than 10,000 years or so could have substantial nonthermal components to their X-ray emission, which could dominate above 10 keV. These predictions were used in several proposals for ASCA and XTE observations of supernova remnants.

**Modeling individual objects: Cas A.** We have undertaken a comprehensive effort to model the Cassiopeia A SNR in the framework of the circumstellar medium (CSM) interaction picture, through one- and two-dimensional hydrodynamical simulations coupled with nonequilibrium X-ray emission calculations. A first attempt at understanding a spatially-integrated ASCA spectrum of this remnant was done under the assumption of spherical symmetry. This work (in collaboration with A. Szymkowiak (NASA/GSFC) and C. Sarazin (UVa)) revealed two distinct components in the Cas A X-ray spectrum: the dense circumstellar shell of material ejected prior to the SN explosion, and the Si- and S-rich SN ejecta. We found about $8 M_\odot$ of shocked gas in Cas A, half as much as

Reynolds et al. NASA First-Year Report, 1995

previous estimates. We are continuing our work on this remnant, with a particular emphasis on the determination of chemical abundances in the SN ejecta. We are also investigating the morphology of the remnant through two-dimensional hydrodynamical calculations. These calculations revealed the shocked CSM shell to be unstable because of the Richtmyer-Meshkov and Rayleigh-Taylor instabilities, in qualitative agreement with the remnant's ragged appearance.

In addition, we have performed 2-D simulations of Cas A, to follow up on these 1-D calculations. For a given set of initial parameters, the final shock velocity is some 10-20 percent larger in 2-D as a result of Rayleigh-Taylor and Richtmeyer-Meshkov instabilities (some of the shell gas gets left behind, and less momentum is transferred to shell). 2-D simulations will be used to compare the morphology of these models with that observed in Cas A.

**SNR 1987A.** In collaboration with R. McCray (University of Colorado), we have modeled the impact of the blast wave generated by the SN 1987A with its ring, which is expected to occur in several years. Our two-dimensional calculations show how the blast wave drives slow (several hundred km s$^{-1}$) shocks into the ring, and how the ring is eventually disrupted by hydrodynamical instabilities. We have made predictions about X-ray emission generated by this impact, through a combination of one- and two-dimensional hydrodynamical simulations coupled with nonequilibrium ionization calculations (this work is nearing completion). Future X-ray satellites such as AXAF and ASTRO-E should readily detect X-ray emission generated by this impact.

**SN 1006 AD.** The recently discovered nonthermal X-ray emission from this object was modeled with the shock acceleration code described above. Good spectral and morphological fits could be obtained for two classes of model: those with a low upstream magnetic field, in which the finite remnant age limited maximum electron energies, and those in which MHD waves with wavelengths above $10^{17}$ cm are assumed to be absent, so that electrons with gyroradii above this value can escape. Predictions were made for XTE observations that are now planned.

Bibliography

Blondin, J. M., Lundqvist, P., & Chevalier, R. A. "Axisymmetric Circumstellar Interaction in Supernovae." 1996, ApJ, submitted.

Borkowski, K. J., Szymkowiak, A. E., Blondin, J. M., & Sarazin, C. L. "A Circumstellar Shell Model for the Cassiopeia A Supernova Remnant". 1996, ApJ, 466, 866.

Borkowski, K. J., Szymkowiak, A. E., Blondin, J. M., & Sarazin, C. L. "The Cassiopeia A Supernova Remnant: Dynamics and Chemical Abundances". 1996, in proceedings of the October Maryland conference on "Cosmic Abundances", in press.

Reynolds, S. P. "Synchrotron Models for X-rays from the Supernova Remnant SN 1006." 1996, ApJ, 459, L13.

Reynolds, S. P., & Hornschemeier, A. "Synchrotron X-rays from Supernova Remnants." 1996, in "Röntgenstrahlung from the Universe," ed. H.U. Zimmermann, J. E. Trümper, & H. Yorke (MPE Report 263, Garching, Germany).

### X-ray Emission and Dynamics of Supernova Remnants

Stephen P. Reynolds, John M. Blondin, and Kazimierz J. Borkowski

Second-Year Report

Our project involves the combination of several approaches to understanding the dynamics and radiation from supernova remnants: hydrodynamic simulations, detailed calculations of thermal X-ray emission, and possible nonthermal contributions to X-rays; and modeling of individual objects based on these general calculations. In our second year we have continued work in all three areas, with significant progress.

**General hydrodynamic simulations.** With graduate student Eric Wright, we have performed 2-D simulations of the transition from Sedov-Taylor to radiative evolution, with unprecedented resolution and with realistic cooling. The comparison with our earlier 1-D simulations indicates that radial oscillations observed in 1-D appear less severe in 2-D.

**X-ray Modeling.** Undergraduate Jay Lyerly continued his project of using the upgraded spectral code to calculate complete non-equilibrium ionization spectra from Sedov-Taylor supernova remnants. It appears that no simple model can reproduce with reasonable fidelity a full Sedov-Taylor nonequilibrium-ionization calculation, for low shock temperatures, while a simple representation is possible at high shock temperatures. Mr. Lyerly presented this work last June at the Madison meeting of the American Astronomical Society. More recent work shows that for low temperatures and long ionization timescales, the models are very insensitive to the degree of electron heating. We are studying the extent to which this might be true at high shock temperatures.

In the project's first year, we developed an extensive program modeling nonthermal X-rays expected from young SNRs due to shock acceleration of electrons to a maximum energy limited by radiation losses or by the shock age. During the second year we have used the modeling code to perform a major parameter-space study of this process. We find that significant nonthermal X-ray emission can be expected for ages as large as 10,000 years. In many cases, reasonable spectral predictions can be obtained by simple scaling of a model template. A major publication describing this work will be submitted in the next month.

It is often not recognized that depletion of heavy elements onto dust grains affects X-ray spectra of SNRs. Dust grains contain elements such as Mg, Si, Ca, and Fe which, while in gaseous phase, produce strong X-ray lines frequently seen in X-ray spectra of SNRs. As noticed previously by several investigators, the depletion of these elements onto dust grains reduces the intensities of their X-ray lines, in an amount proportional to their depletion. However, X-ray lines can be produced in dust grains also. In collaboration with A. Szymkowiak (NASA/GSFC), we found that fluorescent Fe $K\alpha$ emission from Fe atoms in dust grains should be present in X-ray spectra emitted by dusty plasma, in addition to $K\alpha$ emission from highly ionized Fe ions. This affects the strength and shape of the Fe $K\alpha$ complex. We predict that these effects should be most pronounced in very young SNRs such as Tycho, Kepler, and Cassiopeia A SNRs, and detectable even with the spectral resolution of the ASCA satellite. This finding suggests that Fe X-ray lines in SNRs cannot be successfully modeled without taking into account depletion of Fe onto dust grains and without including fluorescent $K\alpha$ emission from dust grains. It also opens exciting prospects for studying dust in SNRs.

**Modeling individual objects: Cas A.** We have undertaken a comprehensive effort to model the Cassiopeia A SNR in the framework of the circumstellar medium (CSM) interaction picture, through one- and two-dimensional hydrodynamical simulations coupled with nonequilibrium X-ray emission calculations. During the first year, we focused our attention on modeling of X-ray spectra in one dimension, under the assumption of spherical symmetry. In the second year, we made progress on modeling this remnant in two dimensions. Undergraduate Scott Starin did a large number of two-dimensional hydrodynamical simulations, with the purpose of accounting for the asymmetric morphology of Cas A seen in the various wavelength bands from X-rays to radio. An off-center supernova explosion within a distorted CSM shell is likely responsible for this morphology. We are continuing our work on Cas A, with an emphasis on two-dimensional hydrodynamical simulations and on the determination of chemical abundances in the SN ejecta.

**SNR 1987A.**

In the second year we have completed modeling of the impact of the blast wave generated by the SN 1987A with its ring, which is expected to occur in several years. We then focused our attention on recent X-ray observations of this SNR by the ROSAT satellite. The soft X-ray emission seen from SN 1987A and the apparent deceleration of the radio source expansion suggest that the supernova blast wave has encountered a moderately dense H II region interior to the inner circumstellar ring. We simulated the hydrodynamics of this interaction in two dimensions and calculated the resulting X-ray and ultraviolet emission line spectrum and light curves. The soft X-ray spectrum is dominated by emission lines of hydrogenic and helium-like C, N, O, and Ne; it is consistent with the ROSAT observations if Fe is depleted on grains. NV $\lambda\lambda1240$ emission should be observable easily with the Hubble Space Telescope. We are clearly entering an exciting period in the SNR evolution and we will continue our work on this SNR. This work is done in collaboration with R. McCray (University of Colorado).

**SN 1006 AD.** The recently discovered nonthermal X-ray emission from this object was modeled with the shock acceleration code described above. Good spectral and morphological fits could be obtained for two classes of model: those with a low upstream magnetic field, in which the finite remnant age limited maximum electron energies, and those in which MHD waves with wavelengths above $10^{17}$ cm are assumed to be absent, so that electrons with gyroradii above this value can escape. This work was used to make predictions for what should be observable by XTE. These predictions were then used as the basis for a successful XTE proposal, which should allow a definitive test of this picture.

Bibliography

Blondin, J. M., Lundqvist, P., & Chevalier, R. A. "Axisymmetric Circumstellar Interaction in Supernovae." 1996, ApJ, in press.

Borkowski, A. E., Blondin, J. M., & McCray, R. "X-Ray and Ultraviolet Line Emission from Supernova Remnant 1987A." 1997, ApJ (Letters), in press.

Borkowski, A. E., Blondin, J. M., & McCray, R. "X-Rays from the Impact of SN 1987A with its Circumstellar Ring." 1997, ApJ, 477, in press.

Borkowski, A. E., Blondin, J. M., & McCray, R. "X-Ray Emission from Supernova Remnant 1987A." 1996, in proceedings of a workshop on High Throughput X-Ray Spectroscopy Mission, in press.

Borkowski, K. J., & Szymkowiak, A. E. "X-ray Emission from Dust Grains in Hot Plasmas." 1996, submitted to ApJ (Letters).

Borkowski, K. J., & Szymkowiak, A. E. "X-ray Emission from Dust Grains in Young Supernova Remnants." 1996, Bull. Am. Astr. Soc., 28, 949.

Borkowski, K. J., Szymkowiak, A. E., Blondin, J. M., & Sarazin, C. L. "A Circumstellar Shell Model for the Cassiopeia A Supernova Remnant". 1996, ApJ, 466, 866.

Borkowski, K. J., Szymkowiak, A. E., Blondin, J. M., & Sarazin, C. L. "The Cassiopeia A Supernova Remnant: Dynamics and Chemical Abundances". 1996, in proceedings of the October Maryland conference on "Cosmic Abundances."

Lyerly, W. J., Reynolds, S. P., Borkowski, K. J., & Blondin, J. "Nonequilibrium-Ionization X-ray Spectra from a Sedov-Taylor Blast Wave Model for a Supernova Remnant." 1996, Bull. Am. Astr. Soc., 28, 949.

Reynolds, S. P. "Synchrotron Models for X-rays from the Supernova Remnant SN 1006." 1996, ApJ, 459, L13.

Reynolds, S. P., & Hornschemeier, A. "Synchrotron X-rays from Supernova Remnants." 1996, in "Röntgenstrahlung from the Universe," ed. H.U. Zimmermann, J. E. Trümper, & H. Yorke (MPE Report 263, Garching, Germany).

## X-ray Emission and Dynamics of Supernova Remnants

Stephen P. Reynolds, John M. Blondin, and Kazimierz J. Borkowski

Final Report

Our project involved the combination of several approaches to understanding the dynamics and radiation from supernova remnants: hydrodynamic simulations, detailed calculations of thermal X-ray emission, and possible nonthermal contributions to X-rays; and modeling of individual objects based on these general calculations. We continued to make progress in all three areas since our second-year report. The project continues in a new contract under the same title.

**General hydrodynamic simulations.** With graduate student Eric Wright, we completed 2-D simulations of the transition from Sedov-Taylor to radiative evolution, with unprecedented resolution and with realistic cooling. As anticipated from the 1D results, the thin, dense shell formed by the onset of radiative cooling is subject to the nonlinear thin-shell instability, particularly if the ambient density is relatively high. This instability quickly shreds the thin shell, suggesting that SNRs evolving in a dense environment will have a very ragged appearance. This work appeared in 1998 (Blondin, Wright, Borkowski, & Reynolds 1998).

With Guilford College undergraduate Vladimir Rekovic, we studied the evolution of a SN blast-wave propagating through the circumstellar medium left behind by the evolution of the massive progenitor star. A realistic CSM was created following the methods described in Garcia-Segura et al. using a model for a 60 solar mass star. A key feature of this simulation is the relative darkness of the SNR until it reaches the shell of ISM swept up by the progenitor during its main sequence evolution. At that point the SNR becomes extremely bright in optical and UV emission.

**X-ray modeling.** During the project period, we have devoted considerable effort to our X-ray nonequilibrium-ionization spectral code. While we can model X-ray emission for any 1-D or 2-D hydrodynamical simulation, such an approach is laborious, time-consuming and impractical for most X-ray observers. This problem was painfully apparent during a workshop on shell-type SNRs at the University of Minnesota. To remedy this problem, Jay Lyerly (an NCSU student working with us) used the upgraded spectral code to calculate complete non-equilibrium ionization spectra from Sedov-Taylor supernova remnants. It appears that no simple model can reproduce with reasonable fidelity a full Sedov-Taylor nonequilibrium-ionization calculation, for low shock temperatures, while a simple plane-parallel shock provides a good approximation at high shock temperatures. Mr. Lyerly presented his progress at two meetings of the American Astronomical Society. These findings indicate that **most** published work on analysis of X-ray spectra of SNRs suffers from serious deficiencies. Our subsequent work focused on an efficient implementation of the Sedov-Taylor models and simpler shock models into our version of the XSPEC program. After considerable effort we now have a necessary tool for analysis of X-ray spectra of SNRs. We will report our results to astronomical community in the near future, after gaining sufficient practical experience with these software tools.

A major part of the project was the thorough investigation of nonthermal X-ray emission expected from young SNRs due to shock acceleration of electrons to a maximum energy limited by radiation losses or by the shock age. A demonstration of the ability of such a model to fit X-ray data on SN1006 AD was made (Reynolds 1996), and a major parameter-space study of this process was performed for Type Ia remnants (those presumed to be encountering uniform external material) (Reynolds 1998). We find that significant nonthermal X-ray emission can be expected for ages as large as 10,000 years. In many cases, reasonable spectral predictions can be obtained by simple scaling of a model template.

Reynolds et al. NASA Final Report, 1997

It is often not recognized that depletion of heavy elements onto dust grains affects X-ray spectra of SNRs. Dust grains contain elements such as Mg, Si, Ca, and Fe which, while in the gaseous phase, produce strong X-ray lines frequently seen in X-ray spectra of SNRs. As noticed previously by several investigators, the depletion of these elements onto dust grains reduces the intensities of their X-ray lines, in an amount proportional to their depletion. However, X-ray lines can be produced in dust grains also. In collaboration with A. Szymkowiak (NASA/GSFC), we found that fluorescent Fe $K\alpha$ emission from Fe atoms in dust grains should be present in X-ray spectra emitted by dusty plasma, in addition to $K\alpha$ emission from highly ionized Fe ions (Borkowski & Szymkowiak 1997). This affects the strength and shape of the Fe $K\alpha$ complex. We predict that these effects should be most pronounced in very young SNRs such as Tycho, Kepler, and Cassiopeia A, and detectable even with the spectral resolution of the ASCA satellite. This finding suggests that Fe X-ray lines in SNRs cannot be successfully modeled without taking into account depletion of Fe onto dust grains and without including fluorescent $K\alpha$ emission from dust grains. It also opens exciting prospects for studying dust in SNRs.

**Modeling individual objects: Cas A.** We modeled the Cassiopeia A SNR in the framework of the circumstellar medium (CSM) interaction picture, through one- and two-dimensional hydrodynamical simulations coupled with nonequilibrium X-ray emission calculations. During the first year, we focused our attention on modeling of X-ray spectra in one dimension, under the assumption of spherical symmetry (Borkowski et al. 1996). In the second year, we made progress on modeling this remnant in two dimensions. Undergraduate Scott Starin did a large number of two-dimensional hydrodynamical simulations, with the purpose of accounting for the asymmetric morphology of Cas A seen in the various wavelength bands from X-rays to radio. An off-center supernova explosion within a distorted CSM shell is likely responsible for this morphology. During the third year, Cas A was one of the focal points of a workshop on shell-type SNRs at the University of Minnesota. This workshop revealed continuing disagreements about the basic structure of Cas A, suggesting that a more sophisticated hydrodynamical modeling is necessary. Because it is our belief that Cas A holds a crucial clue for understanding young SNR with core collapse progenitors, we are continuing our work on this remnant. As reported in the first year report on our project extension, the clumpiness of SN ejecta seems to be the most important factor missing in the hydrodynamical simulations discussed here.

**SNR 1987A.** In the first and second years we have completed modeling of the impact of the blast wave generated by the SN 1987A with its ring, which is expected to occur in several years. We then focused our attention on X-ray observations of this SNR by the ROSAT satellite. The soft X-ray emission seen from SN 1987A and the apparent deceleration of the radio source expansion suggest that the supernova blast wave has encountered a moderately dense H II region interior to the inner circumstellar ring. We simulated the hydrodynamics of this interaction in two dimensions and calculated the resulting X-ray and ultraviolet emission line spectrum and light curves. The soft X-ray spectrum is dominated by emission lines of hydrogenic and helium-like C, N, O, and Ne; it is consistent with the ROSAT observations if Fe is depleted on grains. We predicted that the interaction should also be seen with STIS onboard the Hubble Space Telescope. This prediction was confirmed immediately after STIS became operational, with the detection of a broad Ly$\alpha$ emission generated by the reverse shock driven into the supernova ejecta. We also predicted that the ISO observatory should see infrared emission from the collisionally heated dust in the shocked H II region. This prediction was also confirmed. An analysis and interpretation of STIS data, done in collaboration with researchers from Colorado and Maryland, revealed significant discrepancies with our hydrodynamical models. Subsequent investigations, reported in the first year report on our project extension, resolved these problems. Our work on this SNR will continue, with increasing emphasis on X-ray observations in view of the anticipated launch of AXAF, XMM, and Astro-E. This work is done in close collaboration with R. McCray (University of Colorado).

**SN 1006 AD.** The synchrotron model for the nonthermal X-ray emission from this object, described above, was used to predict inverse-Compton TeV gamma rays emission, due to upscattering of cosmic-microwave background photons by 100 TeV synchrotron-emitting electrons. The detection of these TeV gamma rays by the CANGAROO collaboration confirmed that the synchrotron process is producing the nonthermal X-rays. The predicted images should show the same bilateral symmetry as the X-ray image, and the spectral predictions were used to eliminate one of the two synchrotron X-ray models, by fixing the mean magnetic field at a few microgauss (Reynolds 1997b).

## Bibliography

Blondin, J. M., Lundqvist, P., & Chevalier, R. A. "Axisymmetric Circumstellar Interaction in Supernovae." 1996, ApJ, in press.

Blondin, J.M., Wright, E.B., Borkowski, K.J., & Reynolds, S.P. "Transition to the Radiative Phase in Supernova Remnants." 1998, ApJ, 500, 342

Borkowski, K. J. "New Light on Supernova Remnants: An Introduction." 1997, Bull. Am. Astr. Soc., 29, No. 2

Borkowski, K. J., Blondin, J. M., & McCray, R. "X-Ray and Ultraviolet Line Emission from Supernova Remnant 1987A." 1997, ApJ, 476, L31

Borkowski, K. J., Blondin, J. M., & McCray, R. "X-Rays from the Impact of SN 1987A with its Circumstellar Ring." 1997, ApJ, 477, 281

Borkowski, K. J., Blondin, J. M., & McCray, R. "X-Ray Emission from Supernova Remnant 1987A." 1996, in Proceedings of the High Throughput X-Ray Spectroscopy Workshop, ed. H. Tananbaum, N. White, and P. Sullivan, p. 191

Borkowski, K. J., de Kool, M., McCray, R., & Wooden, D. H. "Infrared Emission from SN 1987A." 1997, Bull. Am. Astr. Soc., 29, No. 5

Borkowski, K. J., & Szymkowiak, A. E. "X-ray Emission from Dust Grains in Hot Plasmas." 1997, ApJ, 477, L49

Borkowski, K. J., Szymkowiak, A. E., Blondin, J. M., & Sarazin, C. L. "A Circumstellar Shell Model for the Cassiopeia A Supernova Remnant". 1996, ApJ, 466, 866.

Borkowski, K. J., Szymkowiak, A. E., Blondin, J. M., & Sarazin, C. L. "A Circumstellar Shell Model for the Cassiopeia A Supernova Remnant". 1996, in the Clemson University Workshop "The Radioactive Galaxy", p. 165

Borkowski, K. J., Szymkowiak, A. E., Blondin, J. M., & Sarazin, C. L. "The Cassiopeia A Supernova Remnant: Dynamics and Chemical Abundances". 1996, in proceedings of the October Maryland conference on "Cosmic Abundances", ed. S. S. Holt and G. Sonneborn, p. 294

Jones, T. W., Rudnick, L., Jun, B.-I., Borkowski, K. J., Dubner, G., Frail, D. A., Kang, H., Kassim, N. E., & McCray, R. "$10^{51}$ Ergs: The Evolution of Shell Supernova Remnants." 1998, PASP, 110, 125

Lyerly, W. J., Reynolds, S. P., Borkowski, K. J., & Blondin, J. "Nonequilibrium-Ionization X-ray Spectra from a Sedov-Taylor Blast Wave Model for a Supernova Remnant." 1997, Bull. Am. Astr. Soc., 29, 839

Michael, E., McCray, R., Borkowski, K. J., Pun, J., & Sonneborn, G. "High Velocity Lyα Emission from SNR 1987A." 1998, ApJ, 492, L143

Rekovic, V., Blondin, J., & Reynolds, S.P. "SNR Evolution Through a Realistic Circumstellar Medium." 1997, Bull. Am. Astr. Soc., 29, 794

Reynolds, S. P. "Synchrotron Models for X-rays from the Supernova Remnant SN 1006." 1996, ApJ, 459, L13.

Reynolds, S. P., & Hornschemeier, A. "Synchrotron X-rays from Supernova Remnants." 1996, in "Röntgenstrahlung from the Universe," ed. H.U. Zimmermann, J. E. Trümper, & H. Yorke (MPE Report 263, Garching, Germany).

Reynolds, S.P. "Shock Acceleration and High-Energy Nonthermal Emission from Supernova Remnants." 1997a, BAAS, 29, 826

Reynolds, S.P. "Nonthermal X-rays and Gamma Rays from Supernova Remnants in Stellar-Wind Bubbles." 1997b, BAAS, 29, 1267

Reynolds, S.P. "Models of Synchrotron X-rays from Shell Supernova Remnants." 1998, ApJ, 493, 375

Starin, S. "Hydrodynamical Simulation of an Elliptical Shell Surrounding the Progenitor of SNR Cassiopeia A" 1997, Bull. Am. Astr. Soc., 29, 813

Wright, E.B., Blondin, J.M., Borkowski, K.J., & Reynolds, S.P. "Transition to the Radiative Phase in Supernova Remnants" 1997, Bull. Am. Astr. Soc., 29, 813

# SYNCHROTRON MODELS FOR X-RAYS FROM THE SUPERNOVA REMNANT SN 1006

STEPHEN P. REYNOLDS

Physics Department, North Carolina State University, P.O. Box 8202, Raleigh, NC 27695
*Received 1995 November 13; accepted 1995 December 21*

## ABSTRACT

Recent observations with the *ASCA* satellite (Koyama et al. 1995) have finally settled the question of the nature of the X-ray spectrum from the remnant of the supernova of 1006 AD. The bright rims have a featureless power-law spectrum while fainter parts of the remnant show a normal thermal spectrum with the expected lines. I describe model images and spectra that fit the data from the bright rims well, on the premise that the X-rays are synchrotron emission from electrons with energies up to 100 TeV accelerated in the remnant blast wave. The maximum energy to which electrons can be accelerated is limited by the requirement that the acceleration time be less than the smaller of the remnant age or the electrons' radiative loss time. In addition, absence of magnetohydrodynamic waves with sufficiently long wavelength to scatter electrons above some energy would allow electrons to escape freely above that energy, rather than being further accelerated. The maximum energy is thus a function of time and position. I assume that the electron mean free path is proportional to gyroradius, with proportionality constant $f$. With no internal magnetic field amplification beyond the original shock compression, the observed morphology, spectral shape, and X-ray flux at 4 keV can be fitted well with two models, one with escape, and the other with a perhaps unreasonably low upstream magnetic field. The former model has an external magnetic field strength of 3 $\mu$G, $f = 10$, and a maximum MHD wavelength of $10^{17}$ cm. The latter, no-escape model has an external magnetic field strength of 0.6 $\mu$G and $f = 1$. Both models predict upstream emission at a level of a few percent of postshock emission, but with differing morphologies. Models with an upstream magnetic field of 3 $\mu$G and without escape overpredict X-rays at 4 keV by over an order of magnitude.

*Subject headings:* acceleration of particles — shock waves — supernova remnants — X-rays: ISM

## 1. INTRODUCTION

The X-ray spectrum of the remnant of SN 1006 AD has been a puzzle since it was first observed with *OSO 8* and *Einstein* (Becker et al. 1980). Those observations did not find any spectral features, but rather a smooth power law with energy index $\alpha$ ($S_\nu \propto \nu^\alpha$) of about $-1.1$, roughly half a power steeper than the radio index (Green 1991). This led Reynolds & Chevalier (1981, hereafter RC81) to propose a synchrotron model, invoking shock acceleration of electrons to energies above 100 TeV. Later observations (Galas et al. 1982; Vartanian, Lum, & Ku 1985) seemed to show that lines, though weak, were present, and the continuum was reported to be much steeper than Becker et al. had claimed, with a spectral index of about $-2.1$. Hamilton, Sarazin, & Szymkowiak (1986) produced an elaborate thermal model accounting for the power-law continuum by thermal bremsstrahlung from a range of temperatures, and explained the lack of strong lines by underionization and strict elemental stratification of the ejecta. There matters rested until the recent *ASCA* study (Koyama et al. 1995) reporting spatially resolved spectra. The center and faint limbs of the remnant show a normal thermal spectrum, with obvious lines of O, Ne, Mg, S, and Fe and an electron temperature of order 2 keV. However, the brightest, opposing limbs show the featureless power law again, with a spectral index between 0.8 and 8 keV of $-1.95 \pm 0.20$, roughly the same slope as the overall spatially unresolved spectrum reported by later X-ray observers.

Several possibilities can be envisioned to explain this featureless power-law spectrum. Koyama et al. (1995) argue convincingly that thermal bremsstrahlung models are very unlikely. Asvarov et al. (1990) describe nonthermal-bremsstrahlung models for supernova remnant (SNR) X-rays, which add a power-law tail to a thermal-bremsstrahlung spectrum, produced by the electrons at the low-energy end of a shock-accelerated nonthermal spectrum. However, their models all append this tail to an obviously thermal peak, which for the limbs of SN 1006 would then have to be well below 0.2 keV (Vartanian et al. 1985), an extremely unlikely result in view of the measured shock velocity of about 3000 km s$^{-1}$ (Smith et al. 1991), and given the 2 keV temperature obtained for the rest of the remnant by Koyama et al. (1995). Furthermore, the ratio of line excitation to bremsstrahlung cross sections for a single electron on a particular ion is independent of electron energy (up to logarithmic dependences in Gaunt factors), so lines should still be prominent, even with a non-Maxwellian electron energy distribution (K. Borkowski, private communication).

Inverse Compton emission could conceivably produce X-rays from an SNR, but the energy densities in radiation are low in SN 1006. The magnetic energy density is $\sim 4 \times 10^{-14}(B/1\ \mu\text{G})^2$ ergs cm$^{-3}$, which we compare to the radio energy density of about $10^{-15}$ ergs cm$^{-3}$ (from data in Green 1991), so synchrotron self-Compton emission is overwhelmed by synchrotron emission by the same electrons. Now the energy density may be much higher in infrared emission, as is true for many SNRs (Saken, Fesen, & Shull 1992), but SN 1006 is not seen in *IRAS* data (Arendt 1989) with an upper limit corresponding to an energy density of order $3 \times 10^{-14}$ ergs cm$^{-3}$, so that even the low-magnetic field model discussed below, with a mean postshock field of about 2 $\mu$G, would lead to synchrotron domination. Furthermore, electrons upscattering $10^{12}$ Hz photons to $10^{18}$ Hz would have energies of order 1 GeV. These electrons produce the well-observed radio spectrum between 30 and 1000 MHz (see data in Green 1991), which has a spectral index of about $-0.5$, and the inverse Compton spectrum should therefore have the same slope (e.g., Rybicki & Lightman 1979), totally at odds with the observations. I therefore conclude that the most reasonable model for the power-law X-ray spectrum is synchrotron emission from shock-accelerated electrons.

A more involved theoretical shock model for synchrotron X-rays based on generalization of the Reynolds & Chevalier model was produced by Ammosov et al. (1994), and was able to fit roughly the pre-*ASCA* overall spectrum. However, in the description of electron diffusion, the authors set the postshock diffusion coefficient to zero, chose an arbitrary momentum-dependence for the upstream diffusion coefficient, and ignored the differential diffusion of electrons along and across magnetic-field lines (Jokipii 1987). In addition, as noted by the authors, the broad spatial distribution of X-rays obtained seems at odds with the sharply bounded emission observed by *ROSAT* (Willingale et al. 1995). I report here a calculation which updates the RC81 calculation with much better shock physics, and which provides a spectrum and an image consistent with observations. The model constrains the microphysics of shock acceleration through parameterizations of the electron mean free path and free-escape energy. Most broad, however, is the confirmation of RC81's and Ammosov et al.'s basic result that it is not unreasonable to expect electron acceleration in SNR blast waves to X-ray emitting energies $\geq 1$ TeV, for remnants even considerably older than SN 1006, and that such possibilities should be taken into account in fitting the new generation of X-ray observations of supernova remnants.

## 2. CALCULATION

Details of the calculation are provided in Reynolds (1995), as well as in a larger work, in preparation, in which I describe the general parameter space occupied by models; here I shall provide a brief outline. The overall remnant dynamics are assumed to be well described by the Sedov similarity solution, though the spectrum and images are not strongly dependent on this assumption at X-ray energies. The remnant is taken to be expanding into a uniform medium with a uniform magnetic field, whose strength $B_1$ is a free parameter. Behind the shock the magnetic field strength increases by a factor between 1 and 4 as the tangential component is compressed by a factor of 4, so that the postshock field strength varies with obliquity angle $\theta_{Bn}$, the angle between the shock normal and the external magnetic field. As in RC81, the possible existence of a reverse shock is accounted for by setting the synchrotron emissivity to zero interior to the radius at which the Sedov extrapolation would have given an initial expansion velocity greater than 10,000 km s$^{-1}$ as appropriate for a Type Ia explosion. This rather crude approximation is again reasonable for X-rays, most of which come from fairly close behind the current blast-wave position.

The shock is assumed to endow each fluid element with a power-law distribution of electrons $N(E) = KE^{-s}$, with $s = 2.1$ consistent with radio observations (Green 1991), up to a maximum energy $E_m$ set by requiring that the acceleration time be less than the smallest of the remnant age $t$, the radiative loss time $\tau_{1/2}$, or a free-escape energy, corresponding to a maximum wavelength of magnetohydrodynamic (MHD) waves, above which the diffusion coefficient upstream becomes much larger (here by a factor of 10, though the spectra are insensitive to the exact value). The acceleration time is calculated from standard shock-acceleration theory (see, e.g., Blandford & Eichler 1987; Ellison, Jones, & Reynolds 1990), including the stronger magnetic field downstream (Reynolds 1995). In this calculation, the energy dependence of the electron diffusion coefficient must be specified, and I have assumed it proportional to electron energy. Specifically, I take the electron mean free path along the magnetic field $\lambda_\parallel = f r_g$ where $r_g$ is the gyroradius and I require $f \geq 1$ ($f = 1$ in the strong turbulence limit). The perpendicular mean free path is taken to be smaller by a factor $1 + f^2$, corresponding to the particle's diffusing one gyroradius perpendicular to the magnetic field with each scattering. This results in a considerable speedup in acceleration time where the shock is quasi perpendicular, by a factor $R_J < 1$ (Jokipii 1987).

The time for an electron of energy $E$ ergs to lose half its energy by synchrotron radiation in a magnetic field of $B$ G is given by

$$\tau_{1/2} = \frac{637}{B^2 E} \text{ s,}$$

after averaging over electron pitch angles (e.g., Pacholczyk 1970). For an electron diffusing between upstream and downstream regions where the magnetic field differs by a factor of between 1 and 4, a factor of order unity multiplies this expression (Reynolds 1995). The acceleration time to energy $E$ of an electron in a shock of speed $10^8 u_8$ cm s$^{-1}$ and preshock magnetic field strength $B_1$ is

$$\tau_{acc} = 6.25 \times 10^3 \left(\frac{fR_J}{B_1}\right) u_8^2 E$$

(Ellison et al. 1990), with another factor of order unity to take account of differing upstream and downstream magnetic field strengths (Reynolds 1995). The $\theta_{Bn}$ dependence of the acceleration time is primarily contained in $R_J$ which is near 1 for $\theta_{Bn} = 0$, and falls to $\sim f^{-2}$ for $\theta_{Bn} = \pi/2$. (The omitted factors of order unity also depend on the shock obliquity and give some $\theta_{Bn}$ dependence to the maximum energy, in addition to the strong dependence through $R_J$.) When the maximum energy is limited by the radiative loss time, equating the loss and acceleration timescales gives

$$E_{m1} \cong 0.1(fR_J B_1)^{-1/2} u_8 \text{ ergs.} \tag{1}$$

We see immediately that, for $f \sim 10$, $R_J = 1$, $B_1 \cong 3 \times 10^{-6}$ G, and $u_8 \sim 1$, maximum energies above 10 ergs are possible. In the case that the remnant lifetime is shorter than the radiative loss time, we have

$$E_{m2} \cong 8 \times 10^{-4}\left(\frac{B_1}{fR_J}\right) u_8^2 t \text{ ergs,} \tag{2}$$

which again, for an age of about 1000 yr $\sim 3 \times 10^{10}$ s, gives maximum energies of order 10 ergs. A third maximum energy $E_{m3}$, the free-escape energy, is essentially a free parameter, and is related to the maximum wavelength $\lambda_{max}$ of MHD turbulence:

$$E_{m3} = eB_1 \lambda_{max}/f. \tag{3}$$

We notice that if the shock velocity is dropping with time according to the Sedov relation $u \propto t^{-3/5}$, both $E_{m1}$ and $E_{m2}$ drop with time: $E_{m1} \propto t^{-3/5}$ and $E_{m2} \propto t^{-1/5}$, while $E_{m3}$ is of course constant. If the deceleration rate is less, as is true of earlier phases, $E_{m1}$ and $E_{m2}$ drop more slowly with time. For constant velocity, $E_{m1}$ is constant while $E_{m2} \propto t$. Thus, unless $\lambda_{max}$ is very short, the maximum energy to which electrons can have been accelerated was larger in the past, peaking when either the shock initially began its deceleration from the presumed initial value of about 10,000 km s$^{-1}$, or when $E_{m1} = E_{m2}$, and in the latter case, remaining constant until deceleration began. This result has been known for ions, whose maximum energies are always limited by the remnant age, for some time (Lagage & Cesarsky 1983). It is also useful to note that if expression $E_{m1}$ holds, the characteristic photon energy $h\nu_{\max}$ emitted by electrons with that energy is independent of the magnetic field, since it is proportional to $E^2 B$. One cannot explain higher X-ray energies as synchrotron emission by invoking stronger magnetic fields. (On the other hand, if the maximum energy is limited by the remnant lifetime or free escape, $h\nu_{\max}$ scales as $B^3$.)

Careful calculations (Webb, Drury, & Biermann 1984) show that the detailed form of the distribution function for particles undergoing shock acceleration and suffering radiative losses is well approximated by a power law and an exponential cutoff above a maximum energy given by an expression similar to equation (1). I have used such an exponential cutoff to describe the electron distribution in each just-shocked fluid element. I then evolve the distribution behind the shock, taking into account both adiabatic and radiative losses, using the Sedov description of the variation of density and velocity behind the shock. At each position in the current remnant, then, I obtain the full electron distribution, and I integrate the single-particle synchrotron emissivity (e.g., Pacholczyk 1970) over that distribution to obtain the synchrotron volume emissivity. The final images are obtained by integrating over a raster scan of lines of sight through the remnant, and fluxes are obtained by integrating over each image.

Particles must of course diffuse upstream to be accelerated, and while upstream they may contribute detectable radiation, termed a "synchrotron halo" in earlier work (Reynolds 1994). I allow the electrons to diffuse ahead of the shock with parallel and perpendicular coefficients of the diffusion tensor as described above, and a wave field assumed spatially constant so that the distribution is attenuated over that at the shock front by a factor $\exp(-\Delta r/r_d)$, where $r_d \equiv \kappa/u$ is the diffusion length and $\kappa$ the diffusion coefficient. For models with escape, above the escape energy particles diffuse away primarily along field lines, starting at the time their energy first reaches $E_{m3}$, with a diffusion tensor corresponding to $f = 100$. Achterberg, Blandford, & Reynolds (1994) used the absence of obvious radio halos in SNRs to deduce an upper limit to $f$ (or a lower limit to $\delta B/B \sim f^{-1/2}$): $f \lesssim 10$.

Since the models are optically thin at all wavelengths, the aspect angle from which they are observed, the angle $\phi$ between the line of sight and the ambient magnetic field, does not affect the total flux. It does, however, affect the observed morphology, and I have calculated models for varying values of $\phi$.

### 3. RESULTS

Figure 1 shows several model spectra from radio to X-rays, with a collection of radio (Ford & Reynolds 1996) and X-ray (references in Hamilton et al. 1986) data. The spectra do not have sharp breaks to steeper power laws, or to exponential cutoffs. Instead, they steepen gradually over many decades in frequency. One should not expect a break to a steeper power law; this picture is appropriate when the acceleration region is spatially distinct from the region in which losses take place, which is not the case here. A simple exponential cutoff is not to be expected either, for several reasons. First, an electron distribution with an exponential cutoff produces a considerably broader photon distribution after integration over the single-particle synchrotron emissivity. Second, the remnant is inhomogeneous in space and time, so that at any instant various regions contribute different cutoff energies. The superposition of all these broadened exponentials produces the very gradual steepening of Figure 1.

All models in Figure 1 assume dynamical parameters ap-

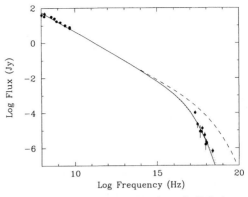

Fig. 1.—Model spectra from radio to X-ray frequencies. Radio data are collected in Ford & Reynolds (1995), and X-ray data are collected in Hamilton et al. (1986). The solid line shows the model with escape ($\lambda_{\max} = 10^{17}$ cm) and with $B_1 = 3\ \mu$G and $f = 10$. The dashed line has the same $B_1$ and $f$, but no escape. The dotted line invokes no escape, and has $B_1 = 0.6\ \mu$G and $f = 1$.

propriate for an assumed distance to SN 1006 of 1.7 kpc (Kirshner, Winkler, & Chevalier 1987), which implies a current shock radius $R$ of 7.4 pc and shock velocity $u_8 = 3.1$, taking the current expansion rate $d (\log R)/d (\log t) = 0.4$, consistent with the radio expansion rate measured by Moffett, Goss, & Reynolds (1993), as well as with optically measured shock velocities (Smith et al. 1991). An explosion energy of $10^{51}$ ergs then requires an external density $n_0 = 0.1$ cm$^{-3}$.

Two models can fit the spectrum well. One involves no escape, but invokes a very low upstream magnetic field, $B_1 = 0.6\ \mu$G. This low value would require about $10^{49}$ ergs in relativistic electrons to explain the observed flux density, at the edge of plausibility (since if ions are accelerated with about 40 times the energy density of electrons, as observed in cosmic rays, the remnant must have a total cosmic-ray acceleration efficiency of order 40%). The model uses $f = 1$; higher values can fit the spectrum, but require even lower magnetic fields. The other reasonable model uses conventional values of $B_1 = 3\ \mu$G and $f = 10$, but requires escape for electrons with gyroradii above $10^{17}$ cm (a value constrained to within a factor of 3 or so). Without escape, those values of $B_1$ and $f$ produce the dashed spectrum in Figure 1, obviously far too bright in X-rays. Clearly, different assumptions about the shock microphysics can produce drastic changes in model spectra. This model constrains $\lambda_{\max}$ fairly well, but requires only that $B_1 \gtrsim 1\ \mu$G and $f \lesssim 50$ so that $E_{m3} < \min(E_{m1}, E_{m2})$. In the larger work in preparation, I describe the full extent of the $(B_1, f, \lambda_{\max})$ parameter space (some results are also shown in Reynolds 1996).

The best-fitting models from Figure 1 are clearly broadly consistent with the observed mean X-ray slope at 4 keV of $-1.95$ seen by *ASCA*. However, the curvature should be measurable with some effort, for instance by *XTE*, which is ideally suited to test the predictions of this model up to 50 keV and beyond. Models fitting the data with one of these curved spectra in addition to thermal components should be compared in goodness-of-fit with straight power laws. While increasing $f$ decreases the X-ray flux at a given energy, its minimum value of 1, without escape, implies an upper bound to the predicted X-ray emission for a given value of $B_1$. Better knowledge of the spectrum can also allow better constraints on the parameters of the two models shown here.

Figure 2 (Plate L1) shows images and profiles of the models that fit, at a frequency of 1 GHz and at 1 keV ($2.4 \times 10^{17}$ Hz). The magnetic field is uniform upstream, with sky plane component vertical, and making an aspect angle of $60°$ with the line of sight. The images have 128 pixels across a diameter, and have been convolved with a Gaussian of FWHM = 2 pixels, corresponding in the case of SN 1006 to a spatial resolution of about 0.24 pc pixel$^{-1}$ or about $30''$ pixel$^{-1}$. (Only the low-field model is shown at 1 GHz; at that frequency the two images are identical.) Figure 2a is not a perfect fit to the radio imaging data, which would require adding Faraday depolarization, magnetic field randomization behind the shock, and other effects. Thus the details of these model images and profiles, in particular of the interior emission, are not unique predictions of the models, and, of course, thermal X-rays will make an unknown contribution to the X-ray images. The bipolar symmetry here is due only to the increased compression of magnetic field where $\theta_{Bn} \sim 90°$, though Fulbright & Reynolds (1990) showed that this effect alone could not account quantitatively for the radio observations, and that some extra enhancement of electron injection was required. Such an extra effect would affect both the radio and X-ray images similarly (modulating the source of both 10 GeV and 1 TeV electrons by equal factors). Simultaneous model fitting of radio and *ROSAT* HRI images is planned, to make sure that quantitative properties such as shell thickness and limb-to-center ratio can be reproduced. The constraint of a common origin for radio and X-rays is sufficiently strong that this joint fitting should be quite restrictive.

Between 1 GHz and 1 keV, the model images are broadly similar, differing most obviously in shell thickness, and in the significant exterior emission in X-rays. The 1 keV images in Figure 2 are not in gross conflict with the observations, especially since an unknown, but probably limb-brightened, contribution from thermal X-rays is also present in the X-ray image (Pye et al. 1981; Willingale et al. 1995). In particular, sharp exterior edges are evident in contrast to the work of Ammosov et al. (1994). However, the exterior emission in the escape model, from electrons diffusing away primarily along magnetic field lines, reaches a level of several percent of the peak brightness, and should be searched for in the *ROSAT* imaging data. That in the low $B_1$ model is considerably fainter. The shell emission in the low $B_1$ model is also thicker, and may be inconsistent with the structure of the nonthermal component, if the shell thermal component can be removed. Thus a discrimi-

nation between these two model classes may be possible with current imaging data.

### 4. CONCLUSIONS

I have demonstrated that a reasonably careful accounting of electron acceleration and energy-loss processes in a strong shock implies the presence of a strong synchrotron-emitting component in SN 1006, consistent with the observations. Such a component may be important in other young remnants such as Tycho, Kepler, and Cas A. Even if it is not seen at all, upper limits can provide useful constraints on shock acceleration, for instance requiring particle escape, with implications for the efficiency of acceleration of cosmic-ray ions. It is also possible to constrain $f$, the low value of which found here implies fairly strong turbulence. This turbulence, consistent with that deduced from sharp edges in the radio image, limits upstream diffusion by radio-emitting electrons (Achterberg et al. 1994).

These models have some implications for the acceleration of cosmic-ray ions as well. Electrons and ions of these energies are absolutely indistinguishable to the shock-acceleration mechanism (except in the helicities of the waves which they produce and scatter from, but for the high levels of turbulence I infer, waves of both helicities should certainly be present). Thus the production of the required $\gtrsim 1$ TeV electrons implies that ions of such energies should also be readily produced. The same reasoning implies that if electrons can escape above some energy, so can ions; and the required escape energy for the model shown here is considerably lower than the "knee" in the cosmic-ray spectrum at around $10^{15}$ eV, frequently invoked as the divide between cosmic-ray production in SNRs and other mechanisms, probably extragalactic (e.g., Blandford & Eichler 1987).

The prediction of an unbroken nonthermal spectrum from radio to X-rays also makes a marginally testable prediction at optical wavelengths. The models of Figure 1 imply a mean surface brightness in $V$ of about $32^m$ arcsec$^{-2}$, probably undetectable at the present but perhaps someday attainable, especially in the bright regions. In the bright radio rims, surface brightnesses as high as $28^m$ arcsec$^{-2}$ may occur, and fractional polarizations of at least 10% might mean that optical polarimetry might be considered.

I would like to acknowledge useful discussions with D. Ellison, R. Jokipii, and K. Borkowski, and with J. Hughes and R. Petre on the observations. This work was supported by NASA through grants NAG 5-2212 and NAG 5-2844.

REFERENCES

Achterberg, A., Blandford, R. D., & Reynolds, S. P. 1994, A&A, 281, 220
Ammosov, A. E., Ksenofontov, L. T., Nikolaev, V. S., & Petukov, S. I. 1994, Astron. Lett., 20, 157
Arendt, R. 1989, ApJS, 70, 181
Asvarov, A. I., Dogiel, V. A., Guseinov, O. H., & Kasumov, F. K. 1990, A&A, 229, 196
Becker, R. H., Szymkowiak, A. E., Boldt, E. A., Holt, S. S., & Serlemitsos, P. J. 1980, ApJ, 240, L33
Blandford, R. D., & Eichler, D. 1987, Phys. Rep., 154, 1
Ellison, D. C., Jones, F. C., & Reynolds, S. P. 1990, ApJ, 360, 702
Ford, A. J., & Reynolds, S. P. 1996, in preparation
Fulbright, M. S., & Reynolds, S. P. 1990, ApJ, 357, 591
Galas, C. M. F., Venkatesan, D., & Garmire, G. P. 1982, Astrophys. Lett., 22, 103
Green, D. A. 1991, PASP, 103, 209
Hamilton, A. J. S., Sarazin, C. L., & Szymkowiak, A. E. 1986, ApJ, 300, 698
Jokipii, J. R. 1987, ApJ, 313, 842
Kirshner, R. P., Winkler, P. F., & Chevalier, R. A. 1987, ApJ, 315, L135
Koyama, K., Petre, R., Gotthelf, E. V., Matsuura, M., Ozaki, M., & Holt, S. S. 1995, Nature, 378, 255

Lagage, P. O., & Cesarsky, C. J. 1983, A&A, 118, 223
Moffett, D. A., Goss, W. M., & Reynolds, S. P. 1993, AJ, 106, 1566
Pacholczyk, A. G. 1970, Radio Astrophysics (San Francisco: Freeman)
Pye, J. P., Pounds, K. A., Rolf, D. P., Seward, F. D., Smith, A., & Willingale, R. 1981, MNRAS, 194, 569
Reynolds, S. P. 1994, ApJS, 90, 845
———. 1995, Proc. 24th Int. Cosmic-Ray Conf. (Rome), 2, 17
———. 1996, in Proc. Int. Conf. Roentgenstrahlung from the Universe (Würzburg), in press
Reynolds, S. P. & Chevalier, R. A. 1981, ApJ, 245, 912, RC81
Rybicki, G. B., & Lightman, A. P. 1979, Radiative Processes in Astrophysics (New York: Wiley)
Saken, J. M., Fesen, R. A., & Shull, M. 1992, ApJS, 81, 715
Smith, R. C., Kirshner, R. P., Blair, W. P., & Winkler, P. F. 1991, ApJ, 375, 652
Vartanian, M. H., Lum, K. S. K., & Ku, W. H.-M. 1985, ApJ, 288, L5
Webb, G. M., Drury, L. O'C., & Biermann, P. 1984, A&A, 137, 185
Willingale, R., West, R. G., Pye, J. P., & Stewart, G. C. 1995, MNRAS, in press

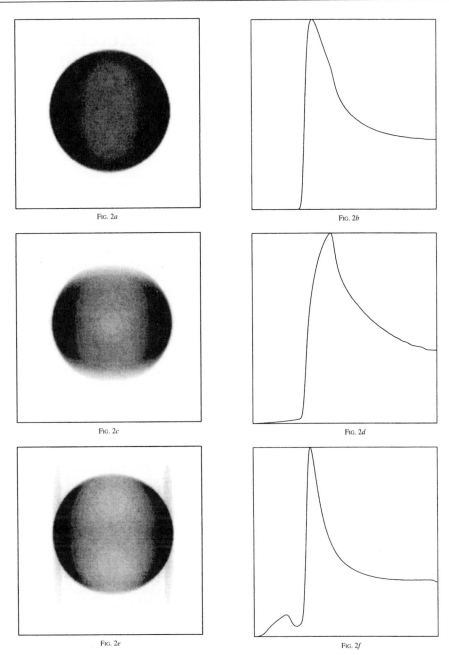

FIG. 2a

FIG. 2b

FIG. 2c

FIG. 2d

FIG. 2e

FIG. 2f

FIG. 2.—Model images corresponding to the solid and dotted lines in Figure 1. Both assume an aspect angle between the line of sight and ambient magnetic field of $\phi = 60°$. ($a$) Model with low $B_1$ and no escape, at $10^9$ Hz. ($b$) Profile of $a$ at a position angle of 63°. ($c$) Model with low $B_1$ and no escape, at 1 keV ($=2.4 \times 10^{17}$ Hz). ($d$) Profile of $b$ at a position angle of 63°. ($e$) Model with escape, at 1 keV. ($f$) Profile of $d$ at a position angle of 63°.

# Maximum Energies of Shock-Accelerated Electrons in Supernova Remnants in the Large Magellanic Cloud

S.P. Hendrick, S.P. Reynolds, & K.J. Borkowski

*North Carolina State University, Raleigh, NC 27695, USA*

**Abstract.**

Some supernova-remnant X-ray spectra show evidence for synchrotron emission from the extension of the electron spectrum producing radio synchrotron emission. For any remnant, if the extrapolated radio flux exceeds the observed X-ray flux (thermal or nonthermal), a rolloff of the relativistic-electron energy distribution must occur below X-ray emitting energies. We have studied the X-ray emission from a sample of 11 remnants in the Large Magellanic Cloud to constrain this rolloff energy. We assume that the electron distribution is a power law with an exponential cutoff at some $E_{max}$ and radiates in a uniform magnetic field. If the radio flux and spectral index are known, this simple model for the synchrotron contribution depends on only one parameter which relates directly to $E_{max}$. Here we have modeled the X-ray spectra by adding a component for thermal radiation of a Sedov blast wave to the synchrotron model. For all 11 supernova remnants in this sample, the limits for $E_{max}$ range between 10 and 70 TeV (for a mean magnetic field of 10 $\mu$G). We interpret $E_{max}$ in the context of shock acceleration theories.

## INTRODUCTION

Observations of the ion component of galactic cosmic rays show a straight power-law spectrum from a few GeV up to the "knee" at 1000 TeV. Shock acceleration of particles by supernova remnants is believed to create cosmic rays up to this energy. However, from radio synchrotron observations, we only learn about electron energies in the range 0.1 to 10 GeV. Both electrons and ions are accelerated by the shock, and are treated equally by the shock above 1 GeV, when all particles become relativistic. Between 5 GeV and 200 GeV, cosmic ray ions and electrons have power-law distributions that are consistent with the same slope at the cosmic ray source. Understanding the cosmic ray electrons should give us some information on the cosmic ray ions, which contain most of the energy in the cosmic ray spectrum.

The electron component of cosmic rays accelerated by a SNR shock can be cut off at high energies by any of several factors. Radiative (synchrotron and inverse-Compton) losses on electrons can cause the spectra to cut off. Finite shock age is also a factor since the time required for a particle to be accelerated to a given energy by diffusive propagation back and forth across the shock is a sharply increasing function of particle energy. Finally, there can be a change in scattering due to the lack of long wavelength MHD waves around the shock. The first cutoff cause affects only the electron component, but finite age and change in scattering properties should affect ions and electrons equally.

X-ray observations have shown several SNRs with spectral evidence for nonthermal emission [1]. This nonthermal emission could be explained by extrapolating the radio synchrotron emission across the entire electromagnetic spectrum to X-rays. Comparing the X-ray flux to the extrapolated radio flux, we see that the electron spectrum must have steepened between the radio and X-ray regimes. We can determine at which frequency this steepening begins using a synchrotron cutoff model. The cutoff frequency is related to the maximum energy to which the shock can accelerate particles. Therefore, X-ray observations can put limits on the maximum energy, $E_{max}$, to which a SNR can accelerate electrons. The remaining question is: Do electron limits correspond to limits of ion acceleration?

# DATA

The supernova remnants in this study (see Table 1) fall into two main categories: mature remnants, most likely in the Sedov evolutionary phase, and remnants from type Ia SNe, most of which exhibit Balmer lines in their optical spectrum, and which are probably in pre-Sedov phases. Remnants that appear to be in the Sedov phase include N23, N63A, N132D, and N49B [2]. Besides those objects, DEM 71 seems to be in a transition state from ejecta dominated spectra to a spectrum dominated by swept up ISM. 0534-69.9 has not been well studied, but its size and morphology (diffuse emission across the face of the shell) point toward Sedov phase. The other category is remnants of type Ia supernovae: 0453-68.5, N103B, DEM 71, 0509-67.5, 0519-69.0, and 0548-70.4, where the last four objects on this list also have an optical spectrum dominated by Balmer line emission [3]. 0453-68.5 and N103B are identified by characteristic lines in the X-ray spectra as type Ia. The youngest remnants in this survey are 0509-67.5, N103B, and 0519-69.0 [4]. Optical spectroscopy gives shock velocities which imply an age of 750-1500 years for 0519-69.0, and an upper limit of 1000 years for 0509-67.5 [3]. N132D has high velocity O-rich filaments within the shell, and both N103B and N63A contain optically bright knots.

Archival ACSA data were retrieved for each remnant, and processed with standard revision 2 criteria. The data for all four ASCA instruments, SIS0, SIS1, GIS2, and GIS3, were fit simultaneously. A complete discussion of the Sedov model used in this paper can be found in Borkowski, Lyerly, and Reynolds [5]. Details of the thermal fits will be presented in a later publication.

The possibility of synchrotron emission is modeled with a cutoff model (called *SRCUT* in XSPEC). This model assumes an exponential cutoff of the electron spectrum and is otherwise homogeneous. It should give the sharpest physically plausible cutoff in the synchrotron spectrum, allowing the most conservative upper limit on $E_{max}$ and on the frequency $\nu_{cutoff}$ radiated by electrons with $E_{max}$. This rolloff frequency, along with the radio spectral index and flux at 1 GHz, are the only input parameters for the *SRCUT* model. Values for the spectral index and flux at 1 GHz have been documented in the literature [6], and were frozen at their values during fitting. In some cases, the flux at 1 GHz was extrapolated from the value of the flux at 408 MHz. The cutoff frequency $\nu_{cutoff}$ is related to $E_{max}$ by

$$\nu_{cutoff} = 1.82 \times 10^{18} E_{max}^2 B \qquad (1)$$

## DISCUSSION

The results of this study can be found in Table 1, where we have assumed a typical SNR magnetic field of 10 $\mu$G to convert $\nu_{cutoff}$ to $E_{max}$. None of the SNRs in this LMC sample is able to accelerate electrons in an unbroken power-law beyond 70 TeV. Observations of the spectrum of Galactic cosmic ray electrons indicate that the spectrum has already begun to fall below the cosmic ray ion spectra at 1 TeV. It is cosmic ray ions that extend out to the "knee" at 1000 TeV; we have no direct evidence of cosmic ray electrons near that energy. In this regard, our results that SNRs cannot accelerate cosmic ray electrons beyond 100 TeV without a break are consistent with Galactic cosmic-ray observations.

Theoretical models of shock acceleration predict that ions and electrons ought to have similar power law spectra in their putative source, SNR shocks. If the cutoffs we infer in the electron spectrum are due solely to radiative losses, then the ion spectrum will not be affected. However, the finite age of the shock, and a change in scattering, will affect both ions and electrons, so that the ion spectrum would not be able to extend beyond the electron spectrum. A cutoff of the electron spectrum below 100 TeV then casts into doubt the ability of supernova remnants to accelerate cosmic ray protons up to the "knee" at 1000 TeV.

Although there are no data on the cosmic ray spectrum of the LMC, the supernova remnant sample spans a wide range of evolutionary stages. Properties of these remnants should be similar to the properties of galactic SNRs. As opposed to a galactic survey, the oldest (and, presumably the largest) remnants in the LMC can still be observed in their entirety within a single ASCA observation. Young remnants have had a short time to accelerate particles, and they have a high limit for radiative losses. Older remnants, while having more time to accelerate particles, have much lower limits for radiative losses. These SNRs could become loss limited after a few thousand years. Therefore, older remnants could have an ion spectrum that extends beyond the maximum energies we have reported.

| Object | $h\nu_{cutoff}$ (keV) | $E_{max}$ (TeV) | Object | $h\nu_{cutoff}$ (keV) | $E_{max}$ (TeV) |
|---|---|---|---|---|---|
| DEM 71 | 0.95 | 70 | N63A | 0.10 | 20 |
| N49B | 0.43 | 50 | 0534-69.9 | 0.06 | 20 |
| N103B | 0.26 | 40 | 0519-69.0 | 0.06 | 20 |
| N23 | 0.22 | 30 | 0509-67.5 | 0.05 | 20 |
| N132D | 0.18 | 30 | 0453-68.5 | 0.02 | 10 |
| 0548-70.4 | 0.11 | 20 | | | |

**TABLE 1.** Cutoff frequency and maximum electron energy upper limits for supernova remnants in the LMC. The $h\nu_{cutoff}$ values are the $3\sigma$ upper limits of our fitted results. Note that while 10 $\mu$G was assumed for the magnetic field, actual SNR magnetic fields are quite uncertain.

## CONCLUSION

This survey of LMC supernova remnants has found that the ability to accelerate electrons in the shock can be limited from the X-ray spectra. By determining the point at which the electron spectrum begins to steepen, we have found that 70 TeV is the upper limit for the maximum energy to which these LMC remnants can accelerate electrons in a straight power-law. With more detailed models and better X-ray data, the conservative upper limits on $E_{max}$ found in this study will become lower. The question of particle acceleration by supernova remnants being able to accelerate particles up to the "knee" at 1000 TeV remains uncertain. The ion component could continue beyond the electron component, but finite shock age and change in scattering properties should affect both equally. It is unknown if there is a similar "knee" in the LMC cosmic ray spectrum, but the properties of LMC supernova remnants should be consistent with remnants in the galaxy. The fact that both the youngest and the oldest remnants in the LMC fail to accelerate electrons to within even an order of magnitude of the "knee" suggests that some revision may be needed in the picture of galactic cosmic ray origin.

## REFERENCES

1. Koyama, K., et al. 1995, Nature, 378, 255; Koyama, K., et al. 1997, PASJ, 49, L7; Slane, P., et al. 1999, ApJ, 525, 357; The, L.-S., et al. 1996, AASupp, 120, C357
2. Hughes, J.P., Hayashi, I., & Koyama 1998, ApJ, 505, 732
3. Smith, R.C., Kirshner, R.P., Blair, W.P., & Winkler, P.F. 1991, ApJ, 375, 625
4. Hughes, J.P., et al 1995, ApJL, 444, L81
5. Borkowski, K.J, Lyerly, W.J., & Reynolds, S.P., 2000, ApJ, in press (astro-ph/0008066)
6. Mathewson, D.S, & Clarke, J.N., 1973, ApJ, 180, 725; Dickel, J.R., & Milne, D.K., 1995, AJ, 109, 200; Dickel, J.R., & Milne, D.K., 1998, AJ, 115, 1057; Filipovic, M.D., et al, 1998, A&A Supp, 127, 119

## MAXIMUM ENERGIES OF SHOCK-ACCELERATED ELECTRONS IN LARGE MAGELLANIC CLOUD SUPERNOVA REMNANTS

SEAN P. HENDRICK AND STEPHEN P. REYNOLDS

Physics Department, North Carolina State University, Raleigh, NC 27695-8202; sphendri@unity.ncsu.edu, steve_reynolds@ncsu.edu

*Received 2000 December 19; accepted 2001 June 5*

### ABSTRACT

Some supernova remnant X-ray spectra show evidence for synchrotron emission from the extension of the electron spectrum that produces radio synchrotron emission. For any remnant, if the extrapolated radio flux exceeds the observed X-ray flux, thermal or nonthermal, a roll-off of the relativistic electron energy distribution must occur below X-ray–emitting energies. We have studied the X-ray emission from 11 remnants in the Large Magellanic Cloud to constrain this roll-off energy. We assume that the electron distribution is a power law with an exponential cutoff at some $E_{max}$ and radiates in a uniform magnetic field. If the radio flux and spectral index are known, this simple model for the synchrotron contribution depends on only one parameter that relates directly to $E_{max}$. Here we have modeled the X-ray spectra by adding a component for thermal radiation of a Sedov blast wave to the synchrotron model. For all 11 supernova remnants in this sample, the limits for $E_{max}$ range between 10 and 80 TeV (for a mean magnetic field of 10 $\mu$G). This result is similar to a study of Galactic remnants in which 13 out of 14 objects had limits between 20 and 80 TeV. We interpret $E_{max}$ in the context of shock acceleration theories. Better data and models should allow either firm detections of nonthermal components or more restrictive limits on $E_{max}$.

*Subject headings:* acceleration of particles — Magellanic Clouds — shock waves — supernova remnants — techniques: spectroscopic — X-rays: ISM

### 1. INTRODUCTION

Observations of the ion component of Galactic cosmic rays show a straight power-law spectrum from a few GeV up to the "knee" around 1000 TeV (see references in Blandford & Eichler 1987 for these and other observational generalities about cosmic rays). Shock acceleration of particles by supernova remnants (SNRs) is believed to create cosmic rays up to this energy. However, from radio synchrotron observations, we learn only about electron energies in the range 0.1–10 GeV. Strong shocks should accelerate both electrons and ions and should treat them equally above a few GeV, when all particles become relativistic. Between 5 GeV and 200 GeV, cosmic-ray ions and electrons have different power-law distributions when viewed from Earth. However, these distributions are consistent with both ions and electrons having the same slope at the cosmic-ray source, but evolving differently in the course of propagating through the Galaxy. Understanding the cosmic-ray electrons should give us some information on the cosmic-ray ions, which contain most of the energy in the cosmic-ray spectrum.

The power-law distribution of cosmic-ray electrons accelerated by an SNR shock can be cut off at high energies by several factors. Synchrotron and inverse Compton radiative losses can cause the spectra to cut off. Finite shock age is also a factor since the time required for a particle to be accelerated to a given energy by diffusive propagation back and forth across the shock is a sharply increasing function of particle energy. Finally, there can be a change in scattering owing to the lack of long wavelength MHD waves ahead of the shock. The first cutoff cause affects only the electron component, but finite age and change in scattering properties should affect ions and electrons equally.

X-ray observations have shown several SNRs with spectral evidence for nonthermal emission (SN 1006, Koyama et al. 1995; G347.3−0.5, Koyama et al. 1997, Slane et al. 1999; Cas A, The et al. 1996). This nonthermal emission could be explained by extrapolating the radio synchrotron emission across the entire electromagnetic spectrum to X-rays. Comparing the X-ray flux to the extrapolated radio flux, we see that the electron spectrum must have steepened between the radio and X-ray regimes. We can bound from above the frequency at which this steepening begins using a synchrotron cutoff (SRCUT) model. The cutoff frequency is related to the maximum energy to which the shock can accelerate electrons, which can therefore also be bounded from above by the observations. We must still consider whether these electron limits also apply to ion acceleration.

A previous study by Reynolds & Keohane (1999, hereafter RK) used a sample of 14 Galactic SNRs to constrain the maximum energies of electron acceleration. In this paper, we have undertaken a parallel investigation, studying 11 SNRs in the Large Magellanic Cloud (LMC) using data from the *ASCA* Public Archive. Although there is little evidence concerning the cosmic-ray spectrum in the LMC, the SNRs in the LMC appear to be similar in most respects to those in the Galaxy. Since we are attempting to define the electron acceleration properties of SNRs in general, the LMC provides an excellent sample. It has many SNRs at many different stages of development, all at about the same distance. This sample has young remnants, where $E_{max}$ may still be rising owing to increasing age, and mature remnants, where $E_{max}$ may be dropping because of radiation losses competing with slowing rates of particle acceleration as the shock decelerates.

In § 2, we will review the sample of LMC remnants and discuss the data and the spectral models used in this study. Section 3 will summarize our results. In § 4, we will discuss the implications of the cutoffs we infer in the electron spectrum and what can be learned about the ion spectrum. Finally, § 5 will list our conclusions.

## 2. ANALYSIS TECHNIQUE

### 2.1. *The Sample*

Radio and X-ray properties of the SNRs appear in Table 1. The radio parameters quoted here are from several sources. The 1 GHz flux densities in column (2) were extrapolated from the flux at the observed frequency from 408 MHz (Mathewson et al. 1983) and 4.75 GHz (Filipovic et al. 1998) using the spectral indices, $\alpha$, in column (3) from these same references, except as noted in the table (we define $S_\nu \propto \nu^{-\alpha}$). SNR DEM L71 (0505−67.9) is a very dim radio source; flux densities are available only at 843 MHz (Mills et al. 1984), so a spectral index of 0.5 was assumed. The X-ray size and *ROSAT* Position Sensitive Proportional Counter (PSPC) count rate in columns (5) and (6), respectively, come from Williams et al. (1999). Columns (7) and (8) give the extrapolation of the synchrotron power law into the X-ray regime, assuming no change in slope. This 1 keV flux density is listed in both millijanskys and in photons s$^{-1}$ cm$^{-2}$ keV$^{-1}$ for the convenience of radio and X-ray observers.

The SNRs in this study fall into two main categories: mature remnants, most likely in the Sedov phase, and younger remnants from Type Ia supernovae, most of which exhibit Balmer lines in their optical spectrum. Remnants that appear to be in the Sedov phase include N23 (0506−68.0), N63A (0535−66.0), N132D (0525−69.6), and N49B (0525−66.0) (Hughes, Hayashi, & Koyama 1998, hereafter HHK). DEM L71 seems to be in a transition state from ejecta-dominated to dominated by swept-up ISM; 0534−69.9 has not been well studied, but its size and morphology (diffuse emission across the face of the shell; Williams et al. 1999) suggest the Sedov phase. Remnants suspected to have Type Ia progenitors include 0453−68.5, N103B (0509−68.7), DEM L71, 0509−67.5, 0519−69.0, and 0548−70.4. The last four objects on this list also have an optical spectrum dominated by Balmer line emission (Smith, Raymond, & Laming 1994). N103B is identified by characteristic lines in the X-ray spectra as Type Ia (Hughes et al. 1995), and 0453−68.5 is classified by Chu & Kennicutt (1988) as Type Ia. The youngest remnants in this survey are 0509−67.5, N103B, and 0519−69.0 (Hughes et al. 1995). Optical spectroscopy gives shock velocities implying an age of 750–1500 yr for 0519−69.0 and an upper limit of 1000 yr for 0509−67.5 (Smith et al. 1994). N132D has high-velocity oxygen-rich filaments within the shell (HHK), and both N103B (Hughes et al. 1995) and N63A (HHK) contain optically bright knots.

### 2.2. *The X-Ray Data*

The data for the 11 LMC SNRs were retrieved from the *ASCA* data archive on the High Energy Astrophysics Science Archive Research Center (HEASARC) Online Service.[1] The unscreened events files for each observation were downloaded and processed using the FTOOLS software package. *ASCA* contains four X-ray telescopes, each with a detector. Two of the detectors are Solid-State Imaging Spectrometers (SIS0 and SIS1), with each containing an array of four CCD camera chips. The other two detectors are Gas Imaging Spectrometers (GIS2 and GIS3), which are gas scintillation proportional counters. More details on *ASCA* and its components can be found in Tanaka, Inoue, & Holt (1994).

Table 2 summarizes the observations and the exposure times for both the SIS and GIS detectors. SIS0 and SIS1 data in 1-CCD mode were used. Data in faint data mode were converted to bright mode and corrected for dark frame error. These files were added to data in bright data mode to increase photon counts. The combined data were screened using the standard Revision 2 procedure. High bit rate data were used for each SIS observation. GIS2 and GIS3 data were collected in pulse height data mode, also with high bit rate. Standard Revision 2 screening criteria again were used. Details on the Revision 2 procedure can be found at the HEASARC web site.

Extracting the spectra involved drawing circular regions about the image of each remnant in SAOimage, making sure that each region included all the radiation from the source. To get the background spectra, all obvious sources

---

[1] Visit the Web site at http://heasarc.gsfc.nasa.gov.

TABLE 1

RADIO PARAMETERS FOR LMC SNRs

| Object (1) | 1 GHz FLUX (Jy) (2) | SPECTRAL INDEX ($\alpha$) (3) | RADIO REFERENCES (4) | X-RAY SIZE (arcmin) (5) | PSPC RATE (counts s$^{-1}$) (6) | EXTRAPOLATED 1 keV FLUX DENSITY — mJy (7) | EXTRAPOLATED 1 keV FLUX DENSITY — photons s$^{-1}$ cm$^{-2}$ keV (8) |
|---|---|---|---|---|---|---|---|
| N132D ......... | 5.6 | 0.58 | 3, 5 | 2.2 × 1.7 | 4.5 | 0.077 | 0.12 |
| N63A ........... | 1.9 | 0.53 | 1 | 1.35 × 1.35 | 2.8 | 0.068 | 0.10 |
| N103B ......... | 0.73 | 0.57 | 3, 5 | 0.65 × 0.6 | 0.72 | 0.012 | 0.018 |
| N49B .......... | 0.56 | 0.57 | 4 | 2.8 × 2.4 | 0.57 | 0.009 | 0.014 |
| N23 ............. | 0.40 | 0.60 | 4 | 1.55 × 1.4 | 0.41 | 0.004 | 0.006 |
| 0453−68.5...... | 0.24 | 0.40 | 1 | 2.4 × 2.3 | 0.20 | 0.108 | 0.16 |
| 0534−69.9...... | 0.11 | 0.48 | 5 | 2.4 × 2.3 | 0.086 | 0.001 | 0.002 |
| 0548−70.4...... | 0.11 | 0.55 | 5 | 1.9 × 1.8 | 0.059 | 0.003 | 0.004 |
| 0519−69.0...... | 0.11 | 0.47 | 3, 5 | 0.75 × 0.75 | 0.60 | 0.013 | 0.019 |
| 0509−67.5...... | 0.066 | 0.46 | 1 | 0.6 × 0.6 | 0.20 | 0.009 | 0.014 |
| DEM L71 ...... | 0.009 | (0.50) | 2 | 1.2 × 1.2 | 0.65 | 0.001 | 0.001 |

NOTE.—A summary of the radio parameters for SNRs in the LMC in order of decreasing flux density. The final column is the extrapolation of the synchrotron power law to 1 keV, listed in two units. For DEM L71, no spectral index has been published; we have assumed a typical value of 0.5.

REFERENCES.—References for the flux at 1 GHz and the spectral indices: (1) Mathewson et al. 1983, (2) Mills et al. 1984, (3) Dickel & Milne 1995, (4) Dickel & Milne 1998, and (5) Filipovic et al. 1998. X-ray information is from Williams et al. 1999.

TABLE 2

SUMMARY OF X-RAY OBSERVATIONS

| OBJECT | SEQUENCE NUMBER | OBSERVATION DATES | EXPOSURE TIME | |
|---|---|---|---|---|
| | | | SIS (ksec) | GIS (ksec) |
| N132D ....... | 50022000 | 1993 Nov 23 | 5 | 6 |
| 0519−69.0... | 51013000 | 1993 Nov 8–9 | 8 | 10 |
| N103B ....... | 51012000 | 1993 Nov 9 | 7 | 10 |
| N23 ......... | 51002000 | 1993 Nov 9–10 | 21 | 16 |
| 0509−67.5... | 51003000 | 1993 Nov 10 | 10 | 8 |
| N63A ........ | 51004000 | 1993 Nov 21 | 12 | 13 |
| N49B........ | 52010000 | 1994 Oct 1–2 | 10 | 13 |
| 0548−70.4... | 53027000 | 1995 Aug 24–25 | 16 | 16 |
| DEM L71 .... | 53026000 | 1995 Oct 8–9 | 13 | 11 |
| 0453−68.5... | 53018000 | 1995 Nov 25–26 | 10 | 7 |
| 0534−69.9... | 55050000 | 1997 May 12–14 | 20 | 20 |

NOTE.—*ASCA* observation dates and sequence numbers for the LMC SNRs. Exposure times for each instrument are listed as well.

were eliminated from the image before extraction. For the SIS data in 1-CCD mode, the chip has a field of view of $10' \times 10'$. The largest SNR in our survey is only slightly larger than $2' \times 2'$, so there was sufficient space on the chip to extract an uncorrupted background spectrum. For GIS data, the field of view is a circle $25'$ in radius, so there was less of a problem getting a source-free background spectrum.

Next, the appropriate response matrices were created using the FTOOLS software package. For SIS data, both the response matrix file and the ancillary response file had to be created for each observation. The response matrix files are the same for all GIS2 and GIS3 observations, and those files come with the data sets. Ancillary response files still needed to be created, using the FTOOLS package. All of the data were grouped with a minimum of 25 counts bin$^{-1}$.

The four spectra for each object were analyzed simultaneously in XSPEC, with relative normalizations of the fits to the spectra allowed to vary. This allowed us to freeze one data set and shift the others along the vertical axis to account for small differences in effective area. Bad channels were ignored, as were channels above 10.0 keV. For the SIS spectra, the range below 0.8 keV was ignored, and below 0.4 keV for the GIS spectra. A full study of each remnant is in preparation; this paper will concentrate on constraining the maximum electron energies. Each SNR was fitted with a Sedov model and the SRCUT model, as described below.

### 2.3. The Models

The details of the Sedov model used in this paper can be found in Borkowski, Lyerly, & Reynolds (2001). Calculations of the X-ray spectra include reliable atomic data for Fe L-shell lines and are based on the updated Hamilton-Sarazin spectral code (Hamilton, Sarazin, & Chevalier 1983). The code allows the electron temperature to differ from the (mean) shock temperature to allow for the possibility of incomplete temperature equilibration between electrons and ions. In order to speed up the fitting procedure, we used a precalculated table of Sedov model solutions with the fitting program interpolating between grid points. The Sedov grid of models was created by the FTOOL "kbmodels" (K. Arnaud 2000, private communication). The grid covers a range of shock temperatures from 0.1 to

3.0 keV, the same range in electron temperature, and a range of ionization timescales of $\tau \equiv n_e t = 10^9 – 10^{13}$ s cm$^{-3}$. For each parameter $kT_s$, $kT_e$, and $\tau$, 10 grid points were created. The models allow arbitrarily variable abundances, but since the Sedov model presumes emission dominated by the blast wave presumably encountering normal abundance material, we froze abundances to 0.4 solar as appropriate for the LMC.

Any nonthermal emission is modeled with the SRCUT model. This model assumes a spatially uniform electron spectrum, $N(E) = KE^{-s} \exp(-E/E_{max})$, radiating in a uniform magnetic field—that is, the synchrotron volume emissivity is just the convolution of $N(E)$ with the single-electron emissivity. The spectral shape is a power law with a cutoff dropping roughly as $\exp[-(\nu/\nu_{roll-off})^{1/2}]$, where $\nu_{roll-off} = 1.82 \times 10^{18} E_{max}^2 B$, all quantities in cgs units. For more details on this model, refer to Reynolds (1998). Since the model assumes continuity between the radio and X-ray emission, this roll-off frequency, the radio spectral index, and the flux at 1 GHz are the only parameters, the flux normalization being set by the observed radio flux. Values for the spectral index and flux at 1 GHz for our target objects have been reported in the literature (see references in Table 1) and were frozen during fitting. Thus, only a single adjustable parameter describes synchrotron emission for each remnant.

### 2.4. The Spectral Fitting Methods

The spectra of the 11 LMC SNRs were fitted using standard Wisconsin absorption (Morrison & McCammon 1983) applied to the Sedov and SRCUT models. Abundances were frozen at 0.4 solar. The radio parameters from Table 1 were frozen as well. The parameters that remained to be fitted were absorbing column density $N_H$, electron temperature $kT_e$, shock temperature $kT$, ionization timescale $\tau$, and cutoff frequency $\nu_{roll-off}$. In some cases, the column density was assumed to be $1.2 \times 10^{21}$ cm$^{-2}$ to the LMC.

Our purpose is to bound the maximum energy to which an SNR can accelerate electrons, from X-ray observations. We adopt the strategy of allowing the fitting program to employ a synchrotron component in addition to our best idea of the appropriate thermal model. The fitting process will then produce values for the one adjustable synchrotron parameter, $\nu_{roll-off}$. If we had total confidence in the appropriateness of these thermal models and if the fits were of high statistical quality, we could accept the results as indicating the actual presence of a synchrotron component. However, since neither of these is the case, we shall conservatively interpret the values of $\nu_{roll-off}$ as indicating the *maximum* possible synchrotron component allowable by the data. To be even more conservative, we shall use as our upper limit not the best-fit value of $\nu_{roll-off}$, but the $3\sigma$ upper bound to that fitted value, as determined by XSPEC. Remarkably, we obtain powerful constraints even with this highly conservative approach.

RK used a different method of data analysis. That study of Galactic remnants used only GIS data from the *ASCA* satellite, and the study limited itself to Galactic remnants that fit within the GIS field of view. Their spectral modeling used standard Wisconsin absorption and allowed the SRCUT model to account for the entire continuum. For some remnants it was necessary to add a bremsstrahlung component, and Gaussian lines were also added when

TABLE 3

ROLL-OFF FREQUENCIES AND ENERGY UPPER LIMITS

| OBJECT (1) | $\chi_v^2$ (2) | $h\nu_{\text{roll-off}}$ (keV) (3) | $3\sigma\ h\nu_{\text{roll-off}}$ (keV) | | $E_{\max}\sqrt{(10\mu G/B)}$ | |
|---|---|---|---|---|---|---|
| | | | This Work (4) | RK Method (5) | ergs (6) | TeV (7) |
| DEM L71 ...... | 1.9 | 1.03 | 1.32 | 9.4 | 130 | 80 |
| N49B............ | 3.4 | 0.28 | 0.39 | 0.71 | 72 | 40 |
| N103B .......... | 3.7 | 0.18 | 0.25 | 0.60 | 57 | 40 |
| N23 ............ | 1.5 | 0.17 | 0.20 | 0.41 | 52 | 30 |
| 0509−67.5...... | 6.9 | 0.12 | 0.16 | 0.18 | 45 | 30 |
| N132D .......... | 1.8 | 0.13 | 0.15 | 0.27 | 45 | 30 |
| 0548−70.4...... | 1.5 | 0.08 | 0.12 | 0.18 | 39 | 20 |
| N63A ........... | 1.9 | 0.08 | 0.10 | 0.21 | 36 | 20 |
| 0519−69.0...... | 4.2 | 0.006 | 0.07 | 0.25 | 30 | 20 |
| 0534−69.9...... | 2.2 | 0.01 | 0.03 | 0.07 | 19 | 10 |
| 0453−68.5...... | 1.2 | 0.004 | 0.02 | 0.03 | 17 | 10 |

NOTE.—Roll-off frequency fitted values, upper limits ($3\sigma$), and maximum electron energy upper limits, derived from the roll-off frequency upper limits. We rounded the energy limits in TeV to one significant figure. Note that while 10 $\mu$G was assumed for a standard SNR magnetic field, the actual magnetic fields are quite uncertain. The $\chi_v^2$-values indicate that some of the fits are not statistically acceptable, but the upper limits are insensitive to the quality of the fits.

needed. However, they made the same interpretation of the fitted values of $\nu_{\text{roll-off}}$ as upper limits.

To test the dependence of our results on the analysis method, we refit all the remnants using methods similar to RK. For all of the remnants in this LMC survey a bremsstrahlung component was needed. Only GIS data were used in this case for purposes of comparison to RK. The results of this comparison can be found in columns (4) and (5) of Table 3. We tested the RK method on the LMC remnants, but we also tested our methods on a Galactic remnant (3C 391) from the RK survey to compare with their results.

### 3. RESULTS

The results of this study can be found in Table 3. Column (2) lists the reduced $\chi_v^2$ from our fits with the Sedov model and the SRCUT model. Column (3) gives the best-fit value for $\nu_{\text{roll-off}}$, and column (4) gives the $3\sigma$ upper limit on $\nu_{\text{roll-off}}$ determined from the model fits, which we shall use to obtain limits on $E_{\max}$ as described above. Parallel results using the RK method fits on the LMC remnants (that is, $3\sigma$ upper bounds on the fitted values) are found in column (5). The final two columns are the calculated $E_{\max}$ upper limits in units of ergs (col. [6]) and TeV (col. [7]).

While the compilations of radio data did not quote errors for flux densities, the range of errors quoted for spectral index varied from $\pm 0.03$ for N132D to $\pm 0.19$ for N49B. Typical errors in radio parameters are 10% in flux density and $\pm 0.1$ in spectral index $\alpha$. This error in $\alpha$ dwarfs all other sources of uncertainty in radio data and in relative calibration of radio and X-ray flux scales, since a change in $\alpha$ of 0.1 results in a factor of 7 variation in extrapolating a spectrum from 1 GHz to 1 keV ( = $2.41 \times 10^{17}$ Hz). All other calibration errors are considerably smaller than a factor of 2. In order to test the dependence of our results on the radio parameters, we refitted the data varying $\alpha$ by 0.1 and $S_{1\ \text{GHz}}$ by 10%. Tables 4 and 5 show the results of this analysis for DEM L71 and N63A, respectively. The boldface values in these tables are the fitted results. For all the remnants, even in the worst-case conspiracy of extreme radio uncertainties, $\nu_{\text{roll-off}}$ rises by less than a factor of 6.3, implying an increase

of only $6.3^{1/2}$ = 2.5 in $E_{\max}$. The extreme insensitivity of our derived limits on $E_{\max}$ to all these sources of error and uncertainty is a major reason for our confidence in these results.

The error analysis for N63A found in Table 5 behaves as one would expect. When the spectral index increases by 0.1, the roll-off frequency increases (the inverse is also true). The effect of the 10% change on the flux shows an inverse relationship with the frequency. This is the expected result, if only the radio parameters were changed and the thermal

TABLE 4

ROLL-OFF FREQUENCY DEPENDENCE ON RADIO PARAMETERS FOR DEM L71

| $\alpha$ | $S_{1\ \text{GHz}}$ | | |
|---|---|---|---|
| | 0.00792 | 0.0088 | 0.00968 |
| 0.40...... | 1.03 | ... | 0.17 |
| 0.50...... | ... | 1.03[a] | ... |
| 0.60...... | 1.03 | ... | 1.50 |

NOTE.—Best-fit roll-off frequency values for DEM L71 (keV) as a function of radio flux density (Jy) and spectral index.
[a] Fitted result.

TABLE 5

ROLL-OFF FREQUENCY DEPENDENCE ON RADIO PARAMETERS FOR N63A

| $\alpha$ | $S_{1\ \text{GHz}}$ | | |
|---|---|---|---|
| | 1.701 | 1.89 | 2.079 |
| 0.43...... | 0.038 | ... | 0.037 |
| 0.53...... | ... | 0.083[a] | ... |
| 0.63...... | 0.26 | ... | 0.23 |

NOTE.—Best-fit roll-off frequency values for N63A (keV) as a function of radio flux density (Jy) and spectral index.
[a] Fitted result.

parameters were frozen to their values from the original fit. As mentioned above, we do not have total confidence in the thermal models. For this reason, the thermal parameters were allowed to vary as we refitted the remnants to create Tables 4 and 5. For N63A the thermal parameters remained relatively the same for each fit in Table 5, but that was not the case for DEM L71 in Table 4. The thermal parameters varied such that the roll-off frequency for each fit did not change for the lower flux values. For the higher flux values, we see the expected variation in the roll-off frequency.

Plots of the spectral fits for SNRs N63A and N132D are displayed as folded spectra in counts $s^{-1}$ $keV^{-1}$ in Figures 1 and 2, respectively. Both the Sedov blast-wave model and the SRCUT model component are shown on the plots, with the synchrotron component the lower curve at low energies.

From Table 3, we can see that no SNR in this survey is producing an unbroken power-law distribution of electrons beyond 100 TeV. This is the major result of our paper. Table 3 also includes the RK method results, which show that the Sedov model with the SRCUT component gives much lower cutoff frequencies than the values from the RK method. There are some changes in rank between the methods seen in the middle five remnants, but the top four

objects and the bottom two remained the same. The method of RK does seem to give us a higher (more conservative) upper limit on the cutoff frequency. The same result was seen in the analysis of 3C391, where the break frequency was slightly lower using our methods (0.049 keV) than the RK result (0.058 keV).

### 4. DISCUSSION

Observations of the spectrum of cosmic-ray electrons indicate that the spectrum has already begun to fall below the cosmic-ray ion spectrum at 1 TeV (see references in Blandford & Eichler 1987). It is cosmic-ray ions that extend out to the "knee" at 1000 TeV; we have no direct evidence of cosmic-ray electrons near that energy. In this regard, our results that SNRs cannot accelerate cosmic-ray electrons beyond 100 TeV are consistent with observations.

However, there is cause for concern. Cosmic-ray observations indicate that ions and electrons have similar power-law spectra at the cosmic-ray sources, while theory reassures us that strong shocks should treat electrons and ions equally at energies well above 1 GeV. The one obvious explanation for differences between electron and ion spectra is radiative losses on the electrons. If all our remnants have electron spectra limited by such losses, the ion spectra could extend unbroken to much higher energies. However, the finite age of the shock and a change in scattering will affect both ions and electrons. A cutoff of the electron spectrum below 100 TeV would then cast doubt on the ability of SNRs to accelerate cosmic-ray protons up to the "knee" at 1000 TeV.

The Galactic remnant study (RK) was limited to relatively small diameter, probably young SNRs, since the observations had to fit well inside the *ASCA* GIS field of view. But the oldest, largest remnants in the LMC are small enough in angular diameter to fit easily on a single SIS chip. Young remnants have had only a short time to accelerate particles, but since their shocks are strong and accelerate particles rapidly, they have a high limit for radiative losses. Older remnants, while having more time to accelerate particles, have much lower limits for radiative losses (Reynolds 1998). Such SNRs could become loss-limited after a few thousand years. Therefore, older remnants could have an ion spectrum that extends beyond the maximum energies we have reported.

In the previous study of Galactic remnants, the SRCUT model was allowed to represent the entire continuum (with Gaussian lines and a bremsstrahlung component, when necessary). We can see that the $E_{max}$ values decrease when the more realistic Sedov model is used. This result is expected owing to the thermal nature of most SNR X-ray spectra. By allowing the Sedov model to account for most of the continuum in each object, we have attempted to isolate any nonthermal component to the spectrum. The SRCUT model then fits the part of the spectrum not covered by our single-component Sedov model. In the younger remnants, emission from ejecta is probably also significant; this would require an additional thermal component. More detailed thermal models should continue to reduce the upper limits for electron acceleration found in this paper.

### 5. CONCLUSION

This survey of LMC SNRs has found that their blast waves' ability to accelerate electrons can be bounded from

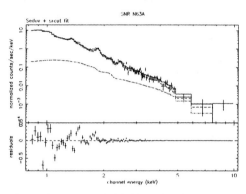

FIG. 1.—Fitted spectrum of N63A with combined SIS0 and SIS1 data. The synchrotron component is the curve with lower values at lower X-ray energies.

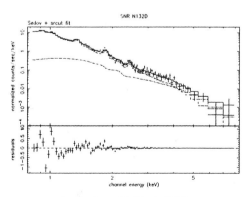

FIG. 2.—Same as Fig. 1, but for N132D

the X-ray spectra. By determining the point at which the electron spectrum begins to steepen, we have found that 80 TeV is the upper limit for the maximum energy to which these LMC remnants can accelerate electrons in an unbroken power law from radio-emitting energies. With more detailed models and better X-ray data, the conservative upper limits on $E_{max}$ found in this study will become lower. The question of SNR shock waves being able to accelerate particles up to the "knee" at 1000 TeV remains uncertain. The ion component could continue beyond the electron component, but finite shock age and change in scattering properties should affect both equally. It is unknown if there is a similar "knee" in the LMC cosmic-

ray spectrum, but the properties of LMC SNRs should be consistent with remnants in the Galaxy. The fact that both the youngest and the oldest remnants in the LMC fail to accelerate electrons to within even an order of magnitude of the "knee" suggests that some revision may be needed in the picture of Galactic cosmic-ray origin.

We are pleased to acknowledge helpful discussions with K. Borkowski, K. Arnaud, and K. Dyer. This research has made use of data obtained through the HEASARC Online Service, provided by the NASA/Goddard Space Flight Center.

REFERENCES

Blandford, R. D., & Eichler, D. 1987, Phys. Rep., 154, 1
Borkowski, K. J., Lyerly, W. J., & Reynolds, S. P. 2001, ApJ, 548, 820
Chu, Y., & Kennicutt, R. C. 1988, AJ, 96, 1874
Dickel, J. R., & Milne, D. K. 1995, AJ, 109, 200
———. 1998, AJ, 115, 1057
Filipovic, M. D., et al. 1998, A&AS, 127, 119
Hamilton, A. J. S., Sarazin, C. L., & Chevalier, R. A. 1983, ApJS, 51, 115
Hughes, J. P., et al. 1995, ApJ, 444, L81
Hughes, J. P., Hayashi, I., & Koyama, K. 1998, ApJ, 505, 732 (HHK)
Koyama, K., Kinugasa, K., Matsuzaki, K., Nishiuchi, M., Sugizaki, M., Torii, K., Yamauchi, S., & Aschenbach, B. 1997, PASJ, 49, L7
Koyama, K., Petre, R., Gotthelf, E. V., Hwang, U., Matsuura, M., Ozaki, M., & Holt, S. S. 1995, Nature, 378, 255
Mathewson, D. S., Ford, V. L., Dopita, M. A., Tuohy, I. R., Long, K. S., & Helfand, D. J. 1983, ApJS, 51, 345

Mills, B. Y., Turtle, A. J., Little, A. G., & Durdin, J. M. 1984, Australian J. Phys., 37, 321
Morrison, R., & McCammon, D. 1983, ApJ, 270, 119
Reynolds, S. P. 1998, ApJ, 493, 375
Reynolds, S. P., & Keohane, J. W. 1999, ApJ, 525, 368
Slane, P., Gaensler, B. M., Dame, T. M., Hughes, J. P., Plucinsky, P. P., & Green, A. 1999, ApJ, 525, 357
Smith, R. C., Raymond, J. C., & Laming, J. M. 1994, ApJ, 420, 286
Tanaka, Y., Inoue, H., & Holt, S. S. 1994, PASJ, 46, L37
The, L.-S., Leising, M. D., Kurfess, J. D., Johnson, W. N., Hartmann, D. H., Gehrels, N., Grove, J. E., & Purcell, W. R. 1996, A&AS, 120, C357
Williams, R. M., Chu, Y., Dickel, J. R., Petre, R., Smith, R. C., & Tavarez, M. 1999, ApJS, 123, 467

# Research on the Oracle at Delphi: From Ancient Myth to Modern Interdisciplinary Science

This case features the interdisciplinary collaboration of a geologist, an archeologist, a geochemist, and a clinical toxicologist, all centered on understanding the origin of prophecies given by the ancient priestesses of the Temple of Apollo in Delphi, Greece. The theory that the legendary trances and visions of the prophesying priestesses at the Delphic oracle were caused by their inhaling vapors rising from the floor of the subterranean vault into which the women were lowered was first postulated by writers in antiquity, most notably Plutarch. But this theory was cast into doubt in the late 19th century and again in the mid-20th century by French archeologists and geologists, who found no major fault in the region from which the fumes could rise. In 1981, however, Jelle Zeilinga de Boer, a geologist hired by the Greek government to determine the geological conditions for constructing nuclear reactors in the area, discovered a heretofore hidden fault that ran under the fissured floor of the Temple of Apollo. After discussing his discovery at a chance meeting with archeologist John Hale fourteen years later, the two explored the area and found a second fault that ran from a spring, and petrochemicals that can produce intoxicating gases in the bedrock. They applied to the Greek government for permission to take samples from the area for chemical analysis. Their proposal is included as the first document in this chapter, along with the Greek permit.

The analyses were performed back in the United States by geochemist Jeffrey Chanton, who then ventured to Greece where he found an unstable but even more intoxicating gas in the stream. In 2001, De Boer, Hale, and Chanton published a report on their findings in *Geology*, the next document in this chapter, in which they presented their evidence that the fault and the spring exist, that they were petro-gaseous, and that these gases were the likely cause of the priestesses' trances. Further pharmacological investigation by toxicologist Henry Spiller, director of the Kentucky Regional Poison Center, compared the effects of the gases found in the deposits with the known effects from 20th century uses of these gases in anesthesia; Spiller, Hale, and De Boer published a "A Multidisciplinary Defense of the Gaseous Vent Theory" the following year in *Clinical Toxicology*, the next piece in this chapter. In a 2002 article from the *New York Times*, William Broad reported the story for a wider audience, providing fuller context and details of this interdisciplinary collaboration: an account of chance and circumstance that led these four scientists to work together to solve an ancient riddle. The collaborative circle is closed in this case with the abstract of a 2002 conference presentation by all four scientists at the Annual Meeting of the Archeological Institute of America.

# PROPOSAL

**DATE:**

21 September 1996

**RESEARCH PROJECT:**

A study of the evidence for geological faults and gaseous emissions at the sanctuary of Delphi, in order to test the validity of ancient testimonia concerning the origins of the oracle of Apollo.

**INVESTIGATORS:**

Dr. Jelle Zeilinga de Boer, Professor of Earth and Environmental Sciences, Wesleyan University.
Dr. John R. Hale, Lecturer in Archaeology and Director of Liberal Studies, University of Louisville.

**ADDRESS:**

Department of Earth and Environmental Sciences
Wesleyan University
Middletown, Connecticut 06549-0139, U.S.A.
Phone: (860) 685-2254
Fax: (860) 685-3651
Email: jdeboer@wesleyan.edu

**REQUEST:**

Permission to take ten small samples of travertine rock from the vicinity of the Apollo temple at Delphi. Samples need to be no larger than 15 centimeters in diameter, and could consist of pieces formerly removed during excavation.

If permission is granted, the samples will be submitted to testing and isotopic analyses. Using a mass spectrometer the isotopic ratio C13 to C12 can be determined which will provide information on the source of the carbon (non-organic vs. organic) and the ratio 018 vs. 016 which will reveal the origin of the waters (rain or deep groundwater) that rose along the young fault that underlies the site. The latter ratio can also provide data on the original temperature of the spring. Using a gas chromatograph the dissolved travertine can be checked for gasses and/or minerals left behind during incorporation of gases in the travertine.

Analyses will be carried out by the Geochemical and Environmental Research Group at Texas A & M University which group has done extensive work on gas emissions along faults in the Gulf of Mexico that are similar to the fault exposed East and West of the Oracle site.

**RATIONALE:**

The travertine rock (often referred to as tufa) is very porous and spongy, indicating that the emerging waters of the springs in the sanctuary were rich in air and gases.

---

De Boer and Hale proposal to Government of Greece, 1996

New tufa rock was being formed on the slope around the Apollo temple throughout the historic period, as is shown by the heavy encrustation of stalactite-like material on the retaining wall just north of the temple. An elaborate system of conduits and drains suggests a massive discharge of water from the Kassotis spring during the period of the oracle.

Some of the pores in the spongy travertine may still contain trapped pockets of the gas that bubbled up from the spring in antiquity. Laboratory analysis could potentially determine both the temperature at which the water emerged from the ground, and the chemical composition of the gases.

## BACKGROUND:

Ancient sources referred to oracles coming "from the rock" at Delphi, and specified a chasm and vapor as important elements in the power of the oracle. This evidence accords with that for another oracular shrine, Didyma in Anatolia, where a prophetess customarily inhaled vapors rising from a spring in the temple of Apollo. As at Delphi, blocks of local tufa were used in the building of the temple at Didyma. A link between oracles and gas emissions is also attested in Italy and Persia.

In spite of this historic evidence, a number of modern archaeologists and geologists have denied the existence, or even the possibility of the existence, of a chasm and vapors at Delphi. Modern doubts are based on the failure of archaeological excavators to identify a chasm in the adyton of the Apollo temple, and on geological reports which concluded that the substrata of rock under the sanctuary solely consisted of limestones incapable of producing gaseous emissions.

Surveys in and near Delphi have on the contrary demonstrated that 1) beneath the travertine and shales at Delphi lies a stratum of bituminous limestone which could produce gases when sheared and superheated during seismic events; 2) fractures in the floor blocks of the temple, along with other evidence, suggest a history of seismic activity which could have stimulated emissions of gases; 3) the visible "bedrock" reached by the archaeologists appears to have once been covered by deposits of travertine, laid down by waters percolating through the site; and 4) the fault zone at Delphi is of a nature which is known to have produced gaseous emissions at similar tectonic settings elsewhere.

## LABORATORY:

In recent years, geochemists at Texas A&M University have tested gaseous emissions from submarine faults in the Gulf of Mexico. The region provides a close parallel with Delphi in terms of the system of faults and the underlying rock strata. Sediments formed along the lines of the faults were analyzed at the College of Geosciences and Maritime Studies.

The gases proved to be light hydrocarbons, predominantly methane with small proportions of ethane and propane. All these gases are colorless and odorless, and produce narcotic effects when inhaled in sufficient concentration.

We propose that the laboratory at Texas A&M University be permitted to analyze the travertine samples from Delphi so that a direct comparison can be made with the results already obtained from faults in the Gulf of Mexico.

ΕΛΛΗΝΙΚΗ ΔΗΜΟΚΡΑΤΙΑ
ΥΠΟΥΡΓΕΙΟ ΠΟΛΙΤΙΣΜΟΥ
ΓΕΝΙΚΗ ΔΙΕΥΘΥΝΣΗ ΑΡΧΑΙΟΤΗΤΩΝ
ΔΙΕΥΘΥΝΣΗ ΣΥΝΤΗΡΗΣΗΣ ΑΡΧΑΙΟΤΗΤΩΝ
ΤΜΗΜΑ ΕΡΕΥΝΑΣ

Αθήνα 21 Μαΐου 1997

Αριθ.Πρωτ. ΥΠΠΟ/ΣΥΝΤ/Φ44/1094

Ταχ.Δ/νση: Μπουμπουλίνας 20
Ταχ.Κώδικας: 106 82 Αθήνα
Πληροφορίες: Κ.Ασημενός
Τηλέφωνο: 3215548-3218475
FAX: 3310342

ΠΡΟΣ:

    Κλαυσικών Σπουδών
    Σουηδίας 54
    106 76 Αθήνα

ΚΟΙΝ:

    Μουσείο Δελφων
    330 54 Δελφοί

ΘΕΜΑ: Δειγματοληψία από ασβεστολιθικά
      πορώδη πετρώματα του Αρχ/κού χώρου Δελφών

*ΣΧΕΤ :α)1243/9-10-96/ΑΣΚΣ*
*       β)2/2/2708/30-4-97/Ι' ΕΠΚΑ*
*       γ)ΥΠΠΟ/ΑΡΧ/Α3/Φ64/21917/315/13-5-97*

      Εχοντας υπόψη τις διατάξεις του Κ.Ν. 5351/32 και την ΥΠ.ΠΟ/ΑΡΧ/Α3/2869/79/20-4-93 Υπουργική απόφαση χορηγούμε άδεια δειγματοληψίας και μεταφοράς των **δειγμάτων** στην Αμερική για ανάλυση στον κ.John Hale μέλος της ΑΣΚΣ από δέκα λίθους από τραβερτίνη με τους κάτωθι όρους:

α) Τα μεγέθη των λίθων θα είναι της τάξεως των 15 εκατοστομέτρων και θα είναι τεμάχια που είχαν απομακρυνθεί κατά τη διάρκεια των ανασκαφών.

β) Η δειγματοληψία θα γίνει υπό την επίβλεψη συντηρητή της Ι' ΕΠΚΑ ύστερα από έγκαιρη συννενόηση.

γ) Τα αποτελέσματα της έρευνας θα κοινοποιηθούν στην ΙΓ' ΕΠΚΑ και στην Δ/νση Συντηρήσεως Αρχ/των για αρχειοθέτηση.

<u>Εσωτερική Διανομή</u>
Δ/νση Συντήρησης Αρχ/των

Ο Προϊστάμενος της Δ/νσης
κ.α.α.

Ακριβές Αντίγραφο
Ο Προϊστάμενος Γραμματείας

Permit from Greek Ministry of Culture, May 21, 1997

## GOVERNMENT OF GREECE
## MINISTRY OF CULTURE

DATE:            21 May 1997

TO:               The American School of Classical Studies

COPIED TO:     Ephoreia for Prehistoric and Classical Antiquities
                   The Delphi Museum

SUBJECT:       Sampling the limestone and porous rock deposits of the
                   archaeological site of Delphi

We have inspected the terms of K.N.5351/32 and UP.PO/ARCH/A3/2869/79/20-4-93. The ministerial decision is to provide permission for sampling and for transporting of the samples to America for analysis by Dr. John Hale, member of the American School of Classical Studies, of ten travertine rock samples taken from below the site, wherein:

a) The size of rock sample that will be taken is 15 centimeters (diameter), and samples will be taken from areas of the site that have already been excavated.

b) The rock samples will come under the supervision and maintenance of the Ephoreia for Prehistoric and Classical Antiquities in a timely manner.

c) The results of the investigation will be made known to the Ephoreia and to the Conservation Authority for their archives.

Signed and sealed,

Supervisor Secretariat

(Seal: Exact Copy)

Translation of Greek permit by Edward Tick, April 3, 2003

# New evidence for the geological origins of the ancient Delphic oracle (Greece)

**J.Z. de Boer**   Department of Earth and Environmental Sciences, Wesleyan University, Middletown, Connecticut 06459, USA
**J.R. Hale**   Department of Anthropology, University of Louisville, Louisville, Kentucky 40292, USA
**J. Chanton**   Department of Oceanography, Florida State University, Tallahassee, Florida 32306-4320, USA

## ABSTRACT

Ancient tradition linked the Delphic oracle in Greece to specific geological phenomena, including a fissure in the bedrock, intoxicating gaseous emissions, and a spring. Despite testimony by ancient authors, many modern scholars have dismissed these traditional accounts as mistaken or fraudulent. This paper presents the results of an interdisciplinary study that has succeeded in locating young faults at the oracle site and has also identified the prophetic vapor as an emission of light hydrocarbon gases generated in the underlying strata of bituminous limestone.

**Keywords:** Greece, normal faults, limestone, bitumen, travertine, ethylene.

## INTRODUCTION

The sanctuary of Delphi is located on the southern slope of Mount Parnassus overlooking the gorge of the Pleistos River (Figs. 1 and 2), where a settlement was established in the late Bronze Age, ca. 1600 B.C. (Müller, 1992). The famous oracle, originally sacred to the Earth goddess (Ge) and later to Apollo, made Delphi a major religious center. Greeks, Romans, and Anatolian rulers consulted the oracle for guidance concerning private affairs, colonizing ventures, wars, and changes of government. The oracle ceased to operate in A.D. 392, when it was closed by order of a Christian emperor.

Ancient authorities attributed the prophetic power of the Delphic oracle to three geological phenomena: a fissure in the bedrock, a gaseous vapor, and a spring. When French archaeologists excavated the temple a century ago, they encountered no obvious traces of a chasm or rising gases (Courby, 1927) and therefore rejected the tradition of intoxicating vapors as a myth. To explain how the myth might have originated, twentieth century investigators have pointed to prominent geological features in the vicinity of Delphi such as the Kastalia gorge to the east of the sanctuary (Oppé, 1904) or the Korykian Cave to the north (Fontenrose, 1978). Most recently, a fracture associated with a prominent east-west fault below the Phaedriades has been suggested as the origin of the ancient tradition of an oracular chasm (Piccardi, 2000).

Since 1995 we have been conducting a joint geological and archaeological project that incorporates field surveys with chemical analyses of spring water and mineral deposits. The new data collected in the course of this project indicate that both fracturing and gaseous hydrocarbon emissions occurred at the site of the Temple of Apollo. Both were directly connected to the operation of the Delphic oracle. These findings represent the first scientific evidence that establishes the accuracy of the ancient reports about a chasm and vapors at the oracle site.

## THE ORACLE

At Delphi the oracles were spoken by a local woman, called the Pythia, who acted as a medium for the god Apollo. Her state of trance could be induced only in the small enclosed chamber (*adyton*) below the floor level of the Temple of Apollo, adjoining a larger room for priests and consultants. Once inside the sunken chamber, the Pythia seated herself on a tripod and inhaled a gaseous emission rising from either a natural fissure or a spring (Holland, 1933).

During routine sessions, the Pythia appears to have been only mildly intoxicated (Fontenrose, 1978). She could recognize visitors in the outer room, understand their questions, and respond at times in poetic verse (Maurizio, 1995). According to the high priest Plutarch, however, if the Pythia was forced into the oracular chamber against her will, an abnormal and occasionally fatal state of delirium or frenzy might result.

Plutarch (1936) noted that in his day (end of the first century A.D.) the gaseous emission in the adyton was weak and unpredictable, but it had a sweet smell like perfume. Plutarch was aware that the vapors could reach the surface either as a free gas or in combination with spring water. In an attempt to account for a decline in the oracle's power over the 500 years before his own time, Plutarch theorized that the underlying rock might have run out of the vital essence that produced the gas. Alternatively, he suggested that the great earthquake of 373 B.C. (epicenter below the Gulf of Corinth, south of Delphi) had disrupted the flow of gas by closing the vents in the rock.

## TECTONIC SETTING

Delphi is located on the northern flank of the Corinth rift zone, which cuts an eastward-widening swath through central Greece (Fig. 1). Motion between the African, Anatolian, and Eurasian plates puts Greece in a tectonic vise. The opening of the Corinth rift zone is only

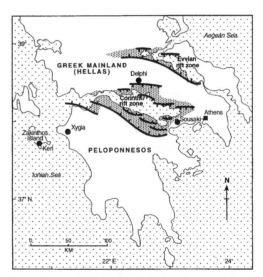

**Figure 1. Tectonic setting of Delphi and schematic representation of Corinth and Evvian rift zones. Keri, Xygia, and Sousaki represent sites of hydrocarbon gas emissions mentioned in text.**

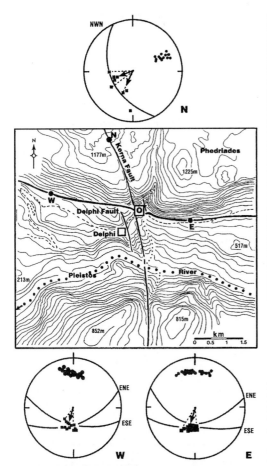

Figure 2. Topographic map of Delphi area and location of Kerna and Delphi faults. Stereographic projections (lower hemisphere) are of major fault surfaces (and slip directions) exposed north (*n* = 13), east (*n* = 22), and west (*n* = 18) of oracle site.

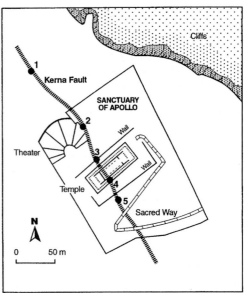

Figure 3. Trace of Kerna fault as outlined by line of five springs or related features in and directly outside sanctuary of Apollo. 1—Kerna Spring, 2—Theater Spring house, 3—travertine coating on Ischegaon retaining wall, 4—spring in temple foundation, near adyton, 5—former spring below sanctuary of Ge. Only Kerna Spring is active today.

one element in a general history of uplift and crustal stretching resulting from this confluence (Collier and Dart, 1991).

Each new episode of crustal extension widens the Corinthian Gulf as the peninsula of the Peloponnesos pulls farther south from the Greek mainland. In this highly fractured zone, characterized by normal faulting, long periods of gradually increasing strain are punctuated by major seismotectonic events (Ambraseys and Jackson, 1990; de Boer, 1992; Le Pichon et al., 1995; Armijo et al., 1996; Sorel, 2000). Energy released during such events causes earthquakes as well as frictional heating of the rock units along faults, both of which have proved crucial in the history of the Delphic oracle.

Records preserved in ancient sources show that the destruction of the Temple of Apollo in 373 B.C. may have been part of a more widespread tectonic event. In that year two small port towns, Helike and Boura, on the south shore of the Gulf of Corinth, disappeared below its waters during an earthquake (Mouyaris et al., 1992). Following this event, slip along an east-west–trending antithetic normal fault in the Delphic area destroyed the Temple of Apollo. Large surfaces of this fault are exposed both east and west of Delphi at short distances from the sanctuary (Fig. 2).

The southward-dipping Delphi fault extends below the ancient site of Delphi, but artificial terracing and erosional deposits obscure its trace in the vicinity of the temple. Detailed studies of this prominent tectonic feature have been published by Birot (1959), Pechoux (1992), and de Boer (1992).

During a field survey of the mountainside above Delphi, we discovered the exposed surface of a second major normal fault. This fault trends north-northwest and dips to the west (Fig. 2). The westward dip of the fault causes its surface trace to veer westward as it descends the cliffs to the sanctuary. Its southeastward continuation below the oracle site is shown by a more or less linear sequence of springs (Higgins and Higgins, 1996). We have named this structure the Kerna fault, after the Kerna Spring northwest of the temple (Figs. 2 and 3). The location of the fault as it passes under the Temple of Apollo is indicated by an ancient spring house built into the massive foundations, below and just to the south of the oracular chamber (Fig. 3).

The exact area of intersection of the Delphi and Kerna faults cannot be determined. Projections of fault trends, however, suggest that it is below the Temple of Apollo. The exceptionally dynamic geological situation at Delphi thus has its origins in the interaction between the Delphi and Kerna faults. Because cross-faulting happened to occur on a steep slope, the weakened mountainside underwent massive rock slides, forming the theater-like hollow backed by precipitous cliffs that

TABLE 1. HYDROCARBON GAS CONCENTRATIONS IN TRAVERTINE ROCK
FROM THREE AREAS IN THE SANCTUARY AT DELPHI

| Site | Methane | Ethane | Ethylene |
|------|---------|--------|----------|
| Kerna Spring | 0.13 | n.d. | n.d. |
| Ischegaon Wall | 1.16 | 0.04 | n.d. |
| Temple of Apollo | 2.82 | 0.03 | n.d. |

*Note:* Hydrocarbon concentrations are in ppm of volume of headspace; n.d.—not detected.

TABLE 2. HYDROCARBON GAS CONCENTRATIONS IN SPRING WATERS
FROM DELPHI AND THE ISLAND OF ZAKYNTHOS

| Location | Site | Methane | Ethane | Ethylene |
|----------|------|---------|--------|----------|
| Delphi | Kerna Spring | 15.3 | 0.2 | 0.3 |
| | Kastalia Spring | 4.9 | n.d. | n.d. |
| Zakynthos | Herodotus Spring | 109.4 | n.d. | 0.1 |
| | Xygia Spring | 169.8 | 1.6 | 0.4 |

*Note:* Concentrations are in nM/L of water; n.d.—not detected.

forms the spectacular setting for the oracle. When ancient authors mentioned a chasm in the inner sanctum of the temple, they were most likely describing a minor extensional fracture associated with the northwest-trending Kerna fault, at or near its intersection with the Delphi fault. A photograph taken during the French excavations shows subvertical extensional fractures in the bedrock below the temple foundation (Courby, 1927).

### EVIDENCE OF SPRINGS

Intersections of major fractures such as the Delphi and Kerna faults render the rock more permeable and provide pathways along which both groundwater and gases can rise. Springs were traditionally associated with the cult of Apollo at Delphi. Geological and archaeological evidence exists for as many as eight active springs in antiquity, although only the Kerna and Kastalia springs are still flowing today (de Boer and Hale, 2000). In the second century A.D., Pausanias (1935) observed a spring called Kassotis on the slope above the Temple of Apollo and stated that its waters plunged underground and emerged in the inner sanctum, where it made the women who drank from it prophetic. About A.D. 361, the Pythia claimed that she could give no more oracles because the temple had collapsed and the spring had fallen silent (Parke and Wormell, 1956).

Modern skepticism about gaseous emissions associated with springs at Delphi derives in part from an erroneous belief that only volcanic activity could produce vapors and gases (Amandry, 1950; Birot, 1959). In fact, the geological situation at Delphi is such that gases have been and continue to be produced in the bedrock underlying the oracle site. Although Paleocene flysch deposits (sandstone, shale, and conglomerate) occur at Delphi and in the Pleistos Valley, the local geology is dominated by a thick formation of Upper Cretaceous limestone. Some of its strata are bituminous, having a reported petrochemical content of as much as 20% (Aronis and Panayotides, 1960).

Seismotectonic activity increases the porosity and permeability of the rock units and can heat rock adjacent to faults to temperatures high enough to vaporize the lighter petrochemical constituents. Sulfur compounds, carbon oxides, and hydrocarbon gases can then be produced in the fault zones. Such emissions have been detected, for example, at Sousaki on the Isthmus of Corinth (east of Delphi) and at asphalt seeps on the island of Zakynthos (west of Delphi) (Dermitzakis and Alafousou, 1987). Higgins and Higgins (1996) suggested that the limestones at Delphi might have produced discharges of carbon dioxide, although this gas is not known to produce a euphoric trance.

### ANALYSES OF GASES

Conditions for the introduction of intoxicating vapors were thus present at Delphi, along with a geological setting that could concentrate them at the intersection of two major faults. To identify the specific gases involved, we took samples of water from the Kerna Spring on the slope above the Temple of Apollo, as well as samples of travertine deposited by ancient springs. The springs below the sanctuary originated in deep limestone deposits and emerged along the Kerna fault at warm temperatures (Fig. 3), depositing travertine as the water cooled. We hoped that the travertine preserved a record of incorporated gases.

We collected it from the surface of the ancient retaining wall (Ischegaon) directly north of the Temple, where a spring had deposited a stalactite mass of travertine on the masonry. Additional travertine samples were collected from fill inside the temple foundation, east of the inner sanctum, and from the Kerna Spring itself.

Using a headspace equilibrium technique (McAuliffe, 1971), we analyzed the water samples by gas chromatography. The travertine samples were placed in glass vials and stoppered; all air was removed by repeated vacuum evacuations and flushing with high-purity $N_2$ gas. The samples were then dissolved in $1M$ (molar) $H_3PO_4$ and the resultant headspace analyzed. Gases were identified by retention time and calibrated against known standards. The headspace to rock volume ratio in the vials was approximately 10 to 1. Detection limits were 0.01 ppm and 0.02 n$M$.

Hydrocarbon gases were detected within the travertine (Table 1). Methane ($CH_4$) in concentrations of 1–3 ppm was found in the headspace of vials containing dissolved samples collected from the Ischegaon wall and the temple, and ethane ($C_2H_6$) was present in all samples but those from the Kerna Spring. The vials were evacuated and several times filled with high-purity $N_2$; the headspace/rock ratio was 10. These results showed that at Delphi the spring waters brought light hydrocarbon gases to the surface in such volumes that some gas was trapped in the rapidly forming travertine.

The unusual nature of the spring water from the oracle site was emphasized by two comparisons. First, an analysis of water from the cold Kastalia Spring, located in a gorge ~100 m east of the temple, yielded only methane, at a level one-third that of the Kerna Spring waters. This finding suggests that the sites of hydrocarbon emissions at Delphi vary in output and are rather localized. Second, analysis of spring waters from the island of Zakynthos (Fig. 1) produced results that matched the types of gases from the Kerna Spring. The Keri site on Zakynthos is famous for its gas springs and tar pits (Dermitzakis and Alafousou, 1987). The latter deposits were used by ancient mariners to tar the hulls of their ships. The similarity of the composition of the spring waters sampled on Zakynthos and in Delphi confirms the link between hydrocarbon emissions and high concentrations of petrochemicals in the bituminous formations below these sites.

The modern Kerna Spring water yielded traces of ethylene ($CH_2$), another light hydrocarbon gas. Ethylene in particular was probably a significant component in the oracular sessions, because it has a sweet smell, which fits the description by Plutarch of vapors rising into the adyton.

First identified scientifically in 1865, ethylene was used as a surgical anesthetic in the early twentieth century. Under the influence of light doses, breathing is slow but regular, and the subject remains in full control of the body (Goodman and Gilman, 1996). Ethylene causes an excitation of the central nervous system.

The effect of low concentrations of ethylene is a sensation of floating or disembodied euphoria, with a reduced sense of inhibition. In some cases a more violent reaction may occur, including delirium and frantic thrashing of the limbs. Eventually, the anesthetic properties of ethylene can cause complete unconsciousness or even death. Plutarch (1936) described an event in which the Pythia died as a possible

**Figure 4. Sketch of Pythia in adyton, showing fractures from which gaseous vapors rose into narrow space during a mantic session.**

consequence of such an overdose. The effects of low-level intoxication wear off quickly once the subject has been removed from contact with the gas (Lockhardt and Carter, 1923).

## CONCLUSIONS

Contrary to current opinion among many archaeologists and some geologists, the ancient belief in an intoxicating gaseous emission at the site of the Delphic oracle was not a myth. An unusual but by no means unique combination of faults, bituminous limestone, and rising groundwater worked together at Delphi to bring volatile hydrocarbon gases to the surface. Ethane, methane, and ethylene all emerged, and modern research has shown that the effects of ethylene inhalation match the well-documented effects of the ancient prophetic vapors. The Temple of Apollo was constructed directly over an area of cross-faulting in order to enclose this unique occurrence and provide a setting for the oracle that was both dramatic and functional in terms of concentrating the gases in a constricted, poorly ventilated chamber above the source (Fig. 4). Our research has confirmed the validity of the ancient sources in virtually every detail, suggesting their testimony on geology is of more value than has recently been held to be the case.

## ACKNOWLEDGMENTS

Medical data concerning current research into the effects of ethylene inhalation were provided by Henry A. Spiller of the Poison Center at Kosair Children's Hospital, Louisville, Kentucky. We are grateful for the help and cooperation received from R. Kolonia, director of the Delphi Museum. We also thank the American School of Classical Studies, Athens, for assistance in securing permits for field work and sampling.

## REFERENCES CITED

Amandry, P., 1950, La mantique apollinienne à Delphes; essai sur le fonctionnement de l'oracle: Paris, E. de Boccard, 290 p.

Ambraseys, N., and Jackson, J., 1990, Seismicity and associated strain of central Greece between 1890 and 1988: Geophysical Journal International, v. 101, p. 663–708.

Armijo, R., Meyer, B., King, G.C.P., Rigo, A., and Papanastassiou, D., 1996, Quaternary evolution of the Corinth Rift and its implications for the late Cenozoic evolution of the Aegean: Geophysical Journal International, v. 126, p. 11–53.

Aronis, G., and Panayotides, G., 1990, Geological map of Greece: Delphi quadrangle: Athens Institute for Geology and Subsurface Research, scale 1:50 000.

Birot, P., 1959, Géomorphologie de la région de Delphes: Bulletin de Correspondance Hellenique, v. 83, p. 258–274.

Collier, R.E.L., and Dart, C.J., 1991, Neogene to Quaternary rifting, sedimentation and uplift in the Corinth Basin, Greece: Geological Society of London Journal, v. 148, p. 1049–1065.

Courby, F., 1927, Topographie et architecture: La terrasse du Temple: Fouilles de Delphes, v. 11, p. 65–66.

de Boer, J.Z., 1992, Dilational fractures in the Corinthian and Evvian rift zones of Greece: Their geometrical relation and tectonic significance in the deformational process of normal faulting: Annales Tectonicae, v. 6, p. 41–61.

de Boer, J.Z., and Hale, J.R., 2000, The geological origins of the oracle at Delphi, Greece, *in* McGuire, B., et al., eds., The archaeology of geological catastrophes: Geological Society [London] Special Publication, 171, p. 399–412.

Dermitzakis, M.D., and Alafousou, P., 1987, Geological framework and observed oil seeps of Zakynthos Island: Their possible influence on the pollution of the marine environment: Thalassographica, v. 10, p. 7–22.

Fontenrose, J.E., 1978, The Delphic oracle, its responses and operations: Berkeley, University of California Press, 476 p.

Goodman, L.S., and Gilman, A., 1996, The pharmacological basis of therapeutics: New York, McGraw-Hill, 1905 p.

Higgins, M.D., and Higgins, R., 1996, A geological companion to Greece and the Aegean: Ithaca, New York, Cornell University Press, 240 p.

Holland, L.B., 1933, The mantic mechanism at Delphi: American Journal of Archaeology, v. 37, p. 201–214.

Le Pichon, X., Chamot-Rooke, N., Lallemant, S., Noomen, R., and Veis, G., 1995, Geodetic determination of the kinematics of central Greece with respect to Europe: Implications for eastern Mediterranean tectonics: Journal of Geophysical Research, v. 100, p. 12 675–12 690.

Lockhardt, A., and Carter, J., 1923, Physiological effects of ethylene: American Medical Association Journal, v. 80, p. 765.

Maurizio, L., 1995, Anthropology and spirit possession: A reconsideration of the Pythia's role at Delphi: Journal of Hellenic Studies, v. 115, p. 69–86.

McAuliffe, C., 1971, Gas chromatographic determination of solutes by multiple phase equilibrium: Chemical Technology, v. 1, p. 46–51.

Mouyaris, N., Papastamatiou, D., and Vita-Finci, C., 1992, The Helice fault?: Terra Nova, v. 4, p. 124–129.

Müller, S., 1992, Delphes mycénienne: Un réexamen du site dans son contexte régional, *in* Bommelaer, J.-F., ed., Delphes: Centre de Recherche sur le Proche-Orient et le Grèce antique, Travaux, v. 12, p. 67–83.

Oppé, A.P., 1904, The chasm at Delphi: Journal of Hellenic Studies, v. 24, p. 214–240.

Parke, H.W., and Wormell, D.E.W., 1956, The Delphic Oracle, Volume II: The oracular responses: Oxford, UK, Blackwell, 436 p.

Pausanias, 1935, Description of Greece, Volume 4: Cambridge, Massachusetts, Harvard University Press, 605 p.

Péchoux, P.Y., 1992, Aux origines des paysages de Delphes, *in* Bommelaer, J.F., ed., Delphes: Strasbourg, France, University of Strasbourg II (Human Sciences), Publication 12, p. 1–38.

Piccardi, L., 2000, Active faulting at Delphi, Greece: Seismotectonic remarks and a hypothesis for the geologic environment of a myth: Geology, v. 28, p. 651–654.

Plutarch, 1936, The obsolescence of oracles: Moralia, Volume 5: London, W. Heinemann, 501 p.

Sorel, D., 2000, A Pleistocene and still-active detachment fault and the origin of the Corinth-Patras rift, Greece: Geology, v. 28, p. 83–86.

Manuscript received November 27, 2000
Revised manuscript received April 16, 2001
Manuscript accepted April 25, 2001

Printed in USA

# The Delphic Oracle: A Multidisciplinary Defense of the Gaseous Vent Theory

Henry A. Spiller,[1,*] John R. Hale,[2] and Jelle Z. De Boer[3]

[1]*Kentucky Regional Poison Center, Louisville, Kentucky*
[2]*Department of Anthropology, University of Louisville, Louisville, Kentucky*
[3]*Department of Earth and Environmental Sciences, Wesleyan University, Middletown, Connecticut*

## ABSTRACT

*Ancient historical references consistently describe an intoxicating gas, produced by a cavern in the ground, as the source of the power at the oracle of Delphi. These ancient writings are supported by a series of associated geological findings. Chemical analysis of the spring waters and travertine deposits at the site show these gases to be the light hydrocarbon gases methane, ethane, and ethylene. The effects of inhaling ethylene, a major anesthetic gas in the mid-20th century, are similar to those described in the ancient writings. We believe the probable cause of the trancelike state of the Priestess (the Pythia) at the oracle of Delphi during her mantic sessions was produced by inhaling ethylene gas or a mixture of ethylene and ethane from a naturally occurring vent of geological origin.*

*Key Words:   Delphi; Ethylene; Altered mental status; Oracle*

## INTRODUCTION

Oracles were used in the ancient world to gain insight to the future. Oracles were believed to have unique access to the gods of a particular religion and through this access were often able to see into the future. The most revered oracle in ancient Greece was located at the town of Delphi in the temple of Apollo, the god of prophecy. The prestige of this oracle made Delphi the most important, influential, and wealthy sacred place in the entire Greek world. For at least a thousand years, the pronouncements of the Delphic oracle offered divine guidance on issues ranging from the founding of colonies to declarations of war, as well as advice on personal issues. Rulers of Greece, Persia, and the Roman Empire made the arduous journey to this mountainous site. The ancient Greeks believed that the

*Corresponding author. Mr. Henry Spiller, Kentucky Regional Poison Center, P.O. Box 35070, Louisville, KY 40232-5070. Fax: 502/629-7277; E-mail: henry.spiller@nortonhealthcare.org

prophetic power of the Delphic oracle derived from the unique location of the temple at Delphi. According to classical authors such as Plutarch and Cicero, the priestess who spoke the prophecies (the Pythia) sat on a tripod that spanned a fissure or cleft in the rock deep within the temple of Apollo. A *pneuma* (breath, wind, vapor) rose from this chasm into the recessed inner sanctum, or "*adyton,*" where it intoxicated the Pythia and inspired her prophecies. During the last century, however, these ancient testimonies have been challenged and dismissed as unreliable, even as fraud. We use a combination of the ancient texts, geological evidence, and modern understanding of the properties of anesthetic gases to defend the argument that the prophesies of the Pythias in fact occurred after an intoxication from gases of geological origin.

## MODERN CONTROVERSIES

When French archaeologists began to dig at Delphi in the 1890s, they expected to find an elaborate marble temple, fine statuary, and an inner sanctum built on bedrock with a large cleft or fissure in the floor. To their disappointment, the excavations revealed only the foundations of the Temple of Apollo, along with parts of fallen columns. The center of the temple had no floor, but instead of revealing an expanse of fissured bedrock or the mouth of a cave it seemed to be built over a thick bed of natural clay.

A visiting English scholar was the first to express skepticism about the ancient traditions (1). Half a century later, an influential book was published by one of the leaders of the French team (2). Amandry maintained that there was no archaeological evidence in the temple itself to support the belief in a fissure or a gaseous emission (2). Moreover, he claimed that such an emission would be geologically impossible in the limestones of Mount Parnassus and stated that such vapors are only produced in volcanic areas.

The great authority of Amandry persuaded almost all historians, classicists, and archaeologists, except the Greeks themselves, that the ancient tradition recorded by Plutarch, Diodorus, and other writers was either a myth, a confusion, or a deliberate fraud. Most modern books on Delphi state categorically that there was not and could not ever have been an intoxicating gaseous emission inside the Temple of Apollo (3,4).

## HISTORICAL RECORD

The historical defense of the gaseous vent theory is based on evidence that includes ancient Greek and Latin texts as well as the archaeological remains of the temple and sanctuary. The literary texts include the testimony of ancient historians such as Pliny and Diodorus, philosophers such as Plato, poets such as Aeschylus and Cicero, geographers such as Strabo, the travel writer Pausanias, and even a priest of Apollo who served at Delphi—the famous essayist and biographer Plutarch. These writers consistently link the power of the oracle to natural features inside the temple, such as a fissure, a gaseous emission, and a spring. The geographer Strabo (64 B.C.–25 A.D.) wrote:

> They say that the seat of the oracle is a cave that is hollowed out deep down in the earth, with a rather narrow mouth, from which arises *pneuma* that inspires a divine frenzy; and that over the mouth is placed a high tripod, mounting which the Pythian priestess receives the *pneuma* and then utters oracles in both verse and prose.

The historian Diodorus of Sicily (first century B.C.) noted an additional element in his description of the Delphic oracle:

> They say that the water of the Cassotis spring plunges underground and emerges in the *adyton* of the temple, where it makes the women prophetic.

While the Temple of Apollo was run by men, the person who spoke the oracles was always a woman. She was given the title "Pythia." The Pythia served as a medium for Apollo, who was believed to take over her body and voice during her prophetic trances. The priest Plutarch described the supposed relationship between the god, the Pythia, and the natural forces by picturing the god Apollo as a musician, the Pythia as his musical instrument, and the *pneuma*, or vapor, as the plectrum or tool with which he drew sounds from the instrument. The Pythia was always a woman of the settlement of Delphi, but she could be old or young, rich or poor, well educated or illiterate. While she was serving as oracle, or mouthpiece of the god, the Pythia lived in the sanctuary, abstained from sexual intercourse, and fasted on or before the days scheduled for oracular sessions. During normal trances she heard the questions of visitors and gave coherent, if cryptic, replies in verse or ordinary speech. Occasionally she was seized with a violent delirium rather than a benign trance.

## THE SETTING OF THE ORACLE

The oracles at Delphi were delivered when the Pythia was placed on a tripod (somewhat like a modern barstool) in an enclosed chamber within the temple called the *adyton*, a Greek term meaning literally "do not enter." The Pythia alone remained in the *adyton*. Those consulting the oracle remained in a separate antechamber nearby. The one surviving depiction of the *adyton* seems to show that it had a low ceiling held up by a column or columns (Fig. 1). This part of the temple was recessed below the main floor level of the temple entrance. Visitors descended a long narrow ramp or staircase to reach the lower level of the waiting room and the Pythia's *adyton*.

Only nine times each year did the woman mount the tripod, enter the trance state, and speak for the god. These sessions were held on "Apollo's Day," the seventh day after each new moon in spring, summer, and fall. The oracle did not operate during the three months of winter. During days of oracle activity, the Pythia would initially be brought by priests of the temple from a secluded and protected residence and led through a series of purification and religious rituals in preparation for her performance. Eventually she was led down into the inner sanctum of the temple (the *adyton*).

### GEOLOGICAL EVIDENCE

The geological defense of the gaseous vent theory is supported by a series of associated facts: (1) the location of the temple directly above the intersection of two major faults, (2) the location of springs emerging from the ground inside and around the temple, (3) the limestone formations of the area containing petrochemicals, (4) the unique design of the temple, (5) the fact that the oracle did not function in the cold winter months, and (6) the documentation of the presence of hydrocarbon gases from the spring waters and travertine deposits.

The temple at Delphi is on a site intersected by major tectonic faults that are part of the Korinth rift zone, a region of crustal spreading (Fig. 2) (5). It sits on the mainland of Greece on the southern slopes of the Parnassos mountain range. Evidence for the recent geologic activity of one of these faults includes earthquakes and landslides. Evidence for intermittent seismic activity is important as it helps to explain: (1) the venting of the gases over an extended period of time, (2) the periodic changes in the intensity of the gaseous emissions, and (3) the eventual cause of their cessation. Neeft believes that an earthquake destroyed the main part of the Delphi settlement around 730 B.C. Delphi's highest period of prestige and wealth followed this event and lasted to the end of the fifth century B.C. Another earthquake severely damaged the temple in 375 B.C. and again in 23 A.D. Quakes in the area have been reported in 1580, 1769, and 1870. Early references to landslides may have been indicative of seismic activity in the region. Rockfalls thwarted attempts to ransack the temple complex by the Persians in 480 B.C., the Phocians in 354–352 B.C., and the Gauls in 279–278 B.C. Pechoux wrote:

> Time and again earthquakes had rumbled here, frightening away the plundering Persians and a

**Figure 1.** Only surviving depiction of the Pythia from the time when the oracle was active.

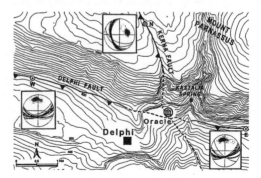

**Figure 2.** Intersection of the Kerna Fault and the Delphi Fault. The temple site is located directly above this intersection.

century later the plundering Phocians and a century later the plundering Gauls; it was the God protecting his shrine.

The oracle site is underlain by a limestone formation of late Cretaceous age, which contains layers rich in bitumen (oil). These limestone deposits formed some 100 million years ago in a shallow tropical sea. The tectonic collision between the Eurasian and African plates lifted these rocks above sea level to form the mountain range of the Parnassos. Friction along fault planes during slips heated and vaporized the lighter constituents in the bituminous layers forming hydrocarbon gases. The most likely path for the gases to follow upwards would be to rise through the fault lines dissolved in percolating ground water and to emerge eventually as springs. There were a number of such springs recorded at the Delphi site, including the Cassiotis that emerged in the *adyton*.

It is interesting to note that the mantic sessions of the Delphi oracle were never held during the winter months when the god Apollo was believed to have gone north to the land of the Hyperboreans. This suggests that the gas emissions at Delphi may have diminished during the colder periods when much of the water had accumulated on Mount Parnassos as snow and ice, and ground water temperatures were relatively low. As the ground water temperature rose in the spring, more of the gas it had incorporated was released.

Another key feature to support the argument for a geological vent is the unique design of the temple. Despite being one of the richest sites in ancient Greece, the temple had an earthen floor in its center with stone walls surrounding it. This unique construction, as opposed to the standard stone floor of the other major temples, suggests the design followed a physical need. The site certainly did not lack funds or engineers, as evidenced by the other major buildings at Delphi such as the gymnasium, theater, and Temple of Athena.

A final note of importance is the relationship of seismic activity and continued gas emissions. Because of changes in the solubility of calcium in enriched ground water the spaces in the fault zones would be slowly and inexorably filled with calcite. Such a process would inevitably clog or close the exit pathways for the trapped gases. To reopen such pathways brecciation is needed. Such a process commonly results from motion along a fault. Periodic seismic activity, as has been recorded in the area, is necessary to produce a ten-century-long venting of gas deposits. Additionally, seismic activity is also probably responsible for the final silencing of the gas

vents and of the oracle. Significant earthquakes shift the flow of ground water with its dissolved gases, frequently forcing it to emerge elsewhere along the fault.

A collapsed section of the ruined temple floor has been tentatively identified as the possible site of the *adyton* (26). It should be noted that three springs have been identified at the site whose location of emergence and flow all follow a pattern of northwest to southeast in a path that follows the geologic direction of the fault line and that crosses directly under the temple itself (Fig. 3).

Chemical analysis by gas chromatography was recently performed on water samples from the springs in and around the temple site and from travertine deposits in the *adyton* using a headspace equilibrium technique (5). The results of these samples have identified the trapped gases as primarily methane, ethane, and ethylene. Results showed the presence of methane and ethane in the travertine deposits with no ethylene detected. Evaluation of the spring water, however, showed a greater concentration of ethylene than ethane, with 0.3 and 0.2 nM/L, respectively. Ethylene is a significantly less stable molecule than ethane and methane, and may not have remained intact in the travertine deposits in the proportions that originally existed (11).

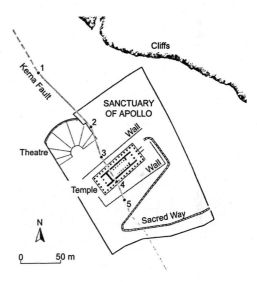

**Figure 3.** Location of fault line under temple and of the springs' emergence from ground.

*Table 1*

*Comparison of Historical Descriptions of Usual Response/Presentation of the Pythia with That of Mild Anesthesia via Inhalational Anesthetic Gases*

| Description of Pythia at Delphi from a "Normal" Mantic Session[a] | Description of Mild Anesthesia with Ethylene or Nitrous Oxide |
|---|---|
| Rapid onset of trance state | Full effects in 30 seconds to 2 minutes (Borne) |
| Calm response, willingly entered the *adyton*. Remained there for hours | Pleasant state of being, no sense of anxiety or asphyxiation. Happy to stay under influence of gas for long periods of time (Lockhardt) |
| Remain conscious | Remain conscious (Lockhardt) |
| Able to maintain seated position | Able to maintain seated position (James) |
| Can see others and hear questions | Responds to questions and write answers (James) |
| Tone and pattern of speech altered | Pattern of speech altered |
| Describe out of body experience—Feeling of being possessed by the god Apollo | Altered state—experienced religious revelations (James) |
| Free Association—images not obviously connected to questions | Free Association—random thought pattern not obviously connected to initial question (James) |
| Recovers rapidly | Complete recovery in 5–15 minutes from full operable anesthesia (Herb) |
| Amnesia of events while under influence | Amnesia of event while under influence (Lockhardt, James) |

[a] Based on evidence from Plutarch, Plato, Lucan, and other ancient authors, as well as depiction of Pythia on a vase from the fifth century.

The concentration of gases that were produced by the ground vent at the time the oracle was functioning is unknown. In all likelihood, the strength of the vent varied over the centuries due to geological conditions, and in fact Plutarch remarks on the waxing and waning of the oracle over time. The oracle was reported to have ceased to function sometime prior to the fourth century A.D. It is unclear that if 1800 years after the reported cessation of the vent, the gases present today reflect the same proportion as existed in ancient times.

## INTOXICATING PROPERTIES OF ETHYLENE

The third portion of the defense of the gaseous vent theory is that the volatile fumes produced an altered mental state that is similar to that produced by inhalation of anesthetic gases. The effects described in the Pythia are the same as the first stage of anesthesia, alternately referred to as the excitation or amnesia phase (Table 1).

Of the three proposed gases, all have the potential to produce an altered mental state, with ethylene > ethane > methane (6). In present day volatile inhalant abuse, hydrocarbon gases remain one of the primary sought-after substances for their intoxicating properties (7–10). In one report on volatile inhalant abuse, three of the top four volatile substances chosen for abuse were aliphatic hydrocarbons (7). Of the three gases available at

Delphi, ethylene is the most likely candidate to have produced the intoxicating vapor. Ethane, however, would be nearly equipotent and a mixture of the two would also produce significant altered mental status (6).

Ethylene is a simple aliphatic hydrocarbon gas ($C_2H_4$), with a sweet odor detectable at 700 ppm (11). It was one of the major inhalational anesthetic gases used in general anesthesia from the 1930s through the 1970s (12–16). Induction of full anesthesia with ethylene occurs rapidly (12,13,16–18). In less than 2 minutes after inhalation, levels of ethylene in the brain are capable of producing full anesthesia (16,17,19,21). Bourne found ethylene to be approximately 2.8 times as potent as nitrous oxide or ether (16). Some of the advantages of ethylene were its rapid onset and clearance, and the lack of respiratory and cardiovascular depressing effects. This is primarily due to low solubility and distribution outside of the vascular compartment. The major disadvantages were the rare fires or explosions in the operating room (22,23). It was replaced with safer, less explosive gases by the 1970s.

There are relatively few eyewitness descriptions of the Pythia in her intoxicated state. This is most likely because relatively few insiders knew the inner workings of the temple. The descriptions that do exist offer two distinct pictures of the Pythia in her intoxicated state. The first is the normal and "working state" of a calm relaxed woman able to respond to questions with visions that in many cases were random and not apparently associated to the subject at hand (Table 1). The second description is that of an apparently rare event of a delirious, ataxic, and combative woman, described as in a frenzy. Both descriptions are consistent with intoxication by ethylene and the early stages of anesthesia. This was a naturally occurring vent. Strict control of the flow of gas would not exist and therefore control of the depth of anesthesia would be crude, if at all. The likelihood of an adverse event is high with hundreds of subjects over the ten centuries using a poorly controlled source of gas. Lockhardt et al., in their description of the first human experiments with ethylene as an anesthetic gas, reported 10 of their 12 subjects had a very pleasant experience. However, two of the 12 had periods of excitement, confusion, and combative behavior.

An interesting modern illustration of the power of mild anesthesia to produce a religious visionary state comes from the philosopher William James who, during the late 19th and early 20th century, experienced religious mysticism and "extraordinary revelations" while experimenting with nitrous oxide. Descriptions of his "sessions" are similar to descriptions of the "normal" Pythia (24,25).

The most dramatic and detailed descriptions of the Pythia are during sessions gone awry. This situation, however, is not unlike the modern medical literature where case reports of difficult cases or adverse events get published in great detail, while the many "normal" cases go unrecorded and unpublished.

The most reliable documentation of the Pythia during her mantic sessions comes from the essayist Plutarch. His description of a mantic session gone awry correlates well with the effects that have been reported from the excitation phase of anesthesia: confusion, agitation, ataxia, and delirium. Plutarch described how a rich deputation from abroad had come to the temple for a consultation. The preparatory rituals indicated it was not the proper time but eventually the Pythia was forced to take her position on the tripod by the temple priests who were interested in satisfying the rich clients.

> She went down into the oracle unwillingly and half-heartedly; and with her first responses it was at once plain from the harshness of her voice that she was not responding properly; she was like a laboring ship and was filled with a mighty and baleful spirit. Finally she became hysterical and with a frightful shriek rushed toward the exit and threw herself down, with the result that not only the members of the deputation fled but also the oracle-interpreter Nicander and those holy men that were present.

Marcus Anneeaus Lucanus (Lucan) relates another story of a mantic session gone awry. It is the story of when the Roman Governor and General Appius Claudius consulted the oracle. Appius Claudius desperately needed a consultation concerning which side to join in the Roman civil war between Caesar and Pompey. He had traveled a great distance; however, it was not the seventh day and the oracle was not open for business. The Roman general forced the oracle to open for him and forced the Pythia down into the chamber against her will. Lucan reports:

> She initially dreaded the oracle recess of the inner shrine, she halted by the entrance and counter-feiting inspiration uttered feigned words from a bosom uninspired; with no inarticulate cry of indistinct utterances proved that her mind was not inspired with the divine frenzy. Her words that rushed not forth with the tremulous cry, her voice had not the power to fill the space of the vast cavern.

It is interesting to note that the ancient Greeks noted a change in her speech, both in tone and in pattern. The idea that lack of indistinct utterances proved her mind was not inspired, suggests recognition of a different clinical presentation during the time of "possession." Appius Claudius was not happy with the initial efforts of the Pythia and is reported to have shouted down to her:

> Profane wretch. I have come to inquire about the fate of this distracted world. Unless you stop speaking in your own voice and go down to the cave for true inspiration the gods whose oracles you are taking in vain will punish you—And so will I!

Lucan, then describes how, terrified by the Roman General, the Pythia entered the *adyton*.

> And her bosom for the first time drew in the divine power which the inspiration of the rock, still active after so many centuries, forced upon her. At last Apollo mastered the breast of the priestess. Frantic she careens about the cave, with her neck under possession, she whirls with tossing head through the void of the temple she scatters tripods that impede her random course.

What they appear to be describing are consistent with the excitation phase of general anesthesia. Lockhardt et al., when working with ethylene, describe a period of ataxia after ethylene intoxication and later describe two of their 12 subjects going through a period of excitement "so that restraint by holding down of the extremities was necessary" (12).

Another piece of supporting evidence is the report by Plutarch of a sweet odor like that of perfume that would drift to the outer sections of the temple from the *adyton*. Ethylene has a slight smell that is described as sweet with odor recognition at 700 ppm (11). That this odor was detectable in the outer sections of the temple, after diffusion over a large area, strongly suggests that greater concentrations existed in the enclosed *adyton* where the Pythia sat. The unique setup of the temple at Delphi, with a history of a recessed enclosed cell, would tend to concentrate the fumes around the Pythia allowing for a more significant exposure (Fig. 4). Also there is archeological evidence of efforts by the Greeks to concentrate the fumes by capping the vent and funneling it through a directed opening (26). It is suggested that the tripod of Pythia was then placed directly over this funneled gas jet.

Historically with ethylene, a mixture of 70–80% ethylene and 20–30% oxygen was used to produce full operable anesthesia. Mixtures of 20% ethylene and 80% oxygen were also used successfully (13). A concentration of less than 20% would be capable of producing an altered mental state while allowing her to remain conscious. This is something that could easily be produced using a directed vent in a small, enclosed chamber.

Another intriguing piece of evidence is offered by the sole representation of the Pythia existing from the period when the oracle functioned. It is interesting to note the unusual slumped posture of the Pythia, in a period when Greek human forms were proudly rendered with rigid erect posture (Fig. 1). The artists of the period did not attempt to portray her as raving and flailing, nor as erect and alert, but rather slumped over, as one would expect from a mildly anesthetized woman.

Another example from the ancient texts that suggest mild anesthesia is the report that the Pythia did not remember her utterances or other events after she had recovered. This is typical of the first stage of anesthesia, sometimes referred to as the stage of amnesia and analgesia (27). Luckhardt et al. also describe amnesia of the period under the influence of ethylene (12).

Several other interesting similarities exist but their value is unclear. The ritual to prepare the Pythia for her ordeal, for instance, involved fasting the day of the ordeal. Similar advice is given to patients undergoing anesthesia to fast the night before surgery. The well-known side effect of nausea and vomiting may have been learned by the ancient Greeks and incorporated into their ritual.

**Figure 4.** Sketch of the Pythia in the *adyton*, showing fractures from which gaseous emission rose into the small, enclosed structure. Based on archeological text and findings in the temple remains.

In conclusion we believe that: (1) the ancient texts all consistently refer to a gas or breath from a cavern as the source of the marvel at Delphi, (2) the geological studies show the existence of conditions for producing hydrocarbon gases in a manner consistent with the ancient texts, and (3) intoxication by one of the hydrocarbon gases, ethylene, produces effects consistent with the descriptions of behavior of the Pythia in the ancient texts. We believe the probable cause of the trancelike state used by the Pythia at the oracle of Delphi during her mantic sessions was produced under the influence of inhaling ethylene gas or a mixture of ethylene and ethane from a naturally occurring vent of geological origin.

## REFERENCES

1. Oppe, A.P. The Chasm at Delphi. J. Hellenic Stud. **1904**, *24*, 214–240.
2. Amandry, P. *La Mantique Apollinienne a Delphes*; Boccard: Paris, 1950; 215–230.
3. Fontenrose, J. *The Delphic Oracle*; University of California: Berkeley, 1978; 9–10.

4. Parke, H.W.; Wormell, D.E.W. *The Delphic Oracle, v. II: The Oracular Responses*; Blackwell: Oxford, 1956; 194–195.

5. De Boer, J.Z.; Hale, J.R.; Chanton, J. New Evidence for the Geological Origins of the Ancient Delphic Oracle (Greece). Geology **2001**, *29*, 707–710.

6. Dorsch, R.R.; De Rocco, A.G. A Generalized Hydrate Mechanism of Gaseous Anesthesia. Physiol. Chem. Phys. **1973**, *5*, 209–223.

7. Spiller, H.A.; Krenzelok, E.P. Epidemiology of Inhalant Abuse Reported to Two Regional Poison Centers. J. Toxicol. Clin. Toxicol. **1997**, *35*, 167–173.

8. Esmail, A.; Meyer, L.; Pottier, A.; Wright, S. Deaths from Volatile Substance Abuse in Those Under 18 Years: Results from a National Epidemiological Study. Arch. Dis. Child. **1993**, *69* (3), 356–360.

9. Parks, J.G.; Noguchi, T.T.; Klatt, E.C. The Epidemiology of Fatal Burn Injuries. J. Forensic Sci. **1989**, *34* (2), 399–406.

10. McGravey, E.L.; Clavet, G.J.; Mason, W.; Waite, D. Adolescent Inhalant Abuse: Environment of Use. Am. J. Drug Alcohol Abuse **1999**, *25*, 731–741.

11. Gibson, G.G.; Clarke, S.E.; Farrar, D.; Elcombe, C.R. Ethene (ethylene). In *Ethel Browning's Toxicity and Metabolism of Industrial Solvents*, 2nd Ed.; Snyder, R., Ed.; Elsevier: New York, 1987; 339–353.

12. Luckhardt, A.B.; Carter, J.B. Physiological Effects of Ethylene. J. Am. Med. Assoc. **1923**, *80*, 765–770.

13. Herb, I.C. Notes Taken from the Clinical Records. Anesth. Analg. **1923**, *Dec*, 230–232.

14. Herb, I.C. The Present Status of Ethylene. J. Am. Med. Assoc. **1933**, *101* (22), 1716–1720.

15. De Jong, R.H.; Eger, E.I. MAC Expanded: AD50 and AD95 Values of Common Inhalational Anesthetics in Man. Anesthesiology **1975**, *42*, 384–389.

16. Bourne, J.G. Uptake, Elimination and Potency of Inhalational Anaesthetics. Anesthesia **1964**, *Jan 19* (1), 12–32.

17. Cowles, A.L.; Borgstedt, H.H.; Gilles, A.J. Uptake and Distribution of Inhalation Anesthetic Agents in Clinical Practice. Anesth. Analg. **1968**, *47*, 404–414.

18. Cole, W.H.J. On Obtaining the Best Use from Methoxyflurane with Special Reference to Its Use with Ethylene. Med. J. Aust. **1968**, *1*, 7–9.

19. Cohen, E.N.; Chow, K.L.; Mathers, L. Autoradiographic Distribution of Volatile Anesthetics Within the Brain. Anesthesiology **1972**, *37*, 324–331.

20. Cowles, A.L.; Borgstedt, H.H.; Gillies, A.J. The Uptake and Distribution of Four Inhalation Anesthetics in Dogs. Anesthesiology **1972**, *36*, 558–570.

21. Salnitre, E.; Rackow, H.; Wolf, G.L.; Epstein, R.M. The Uptake of Ethylene in Man. Anesthesiology **1965**, *26*, 305–311.

22. Ngai, S.H. Explosive Agents—Are They Needed? Surg. Clin. N. Am. **1975**, *55* (4), 975–985.

23. MacDonald, A.G. A Short History of Fires and Explosions Caused by Anesthetic Agents. Br. J. Anesth. **1995**, *72* (6), 710–722.

24. James, W. Review of *The Anaesthetic Revelation of the Gist of Philosophy*. The Atlantic Monthly. http://www.theatlantic.com/issues/96may/nitrous/wmjgist.htm

25. Tymoczko, D. *The Nitrous Philosopher*. The Atlantic Monthly. http://www.theatlantic.com/issues/96may/nitrous/nitrous.htm

26. De Boer, J.Z.; Hale, J.R. The Geological Origins of the Oracle at Delphi, Greece. In *The Archeology of Geological Catastrophes*; McGuire, W.G., Griffiths, D.R., Hancock, P.L., Stewart, I.S., Eds.; Geological Society, London Special Publications: London, 2000; Vol. 171, 399–412.

27. Dripps, R.D.; Eckenhoff, J.E.; Vandam, L.D. *Introduction to Anesthesia*; WB Saunders: Philadelphia, 1977; 231–241.

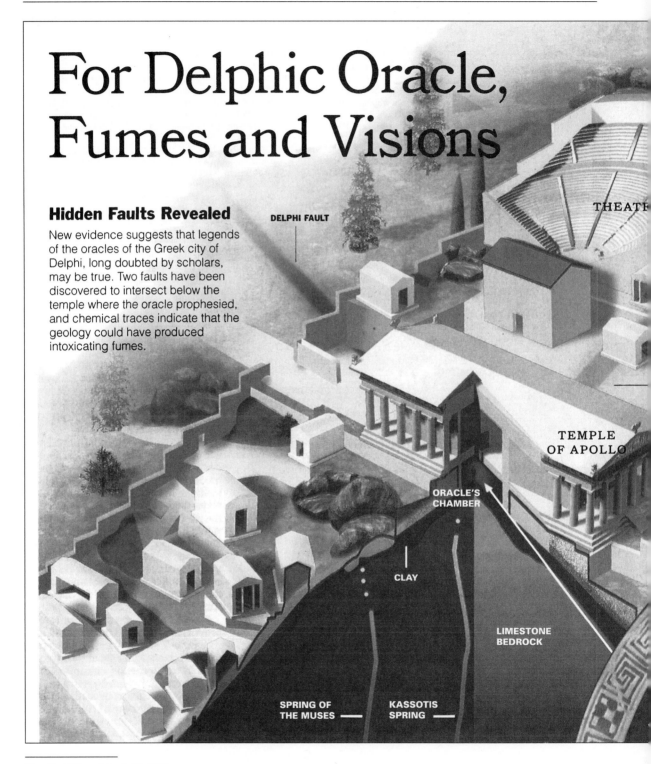

# For Delphic Oracle, Fumes and Visions

## Hidden Faults Revealed

New evidence suggests that legends of the oracles of the Greek city of Delphi, long doubted by scholars, may be true. Two faults have been discovered to intersect below the temple where the oracle prophesied, and chemical traces indicate that the geology could have produced intoxicating fumes.

DELPHI FAULT

THEATR

TEMPLE OF APOLLO

ORACLE'S CHAMBER

CLAY

LIMESTONE BEDROCK

SPRING OF THE MUSES

KASSOTIS SPRING

*New York Times*, March 19, 2002

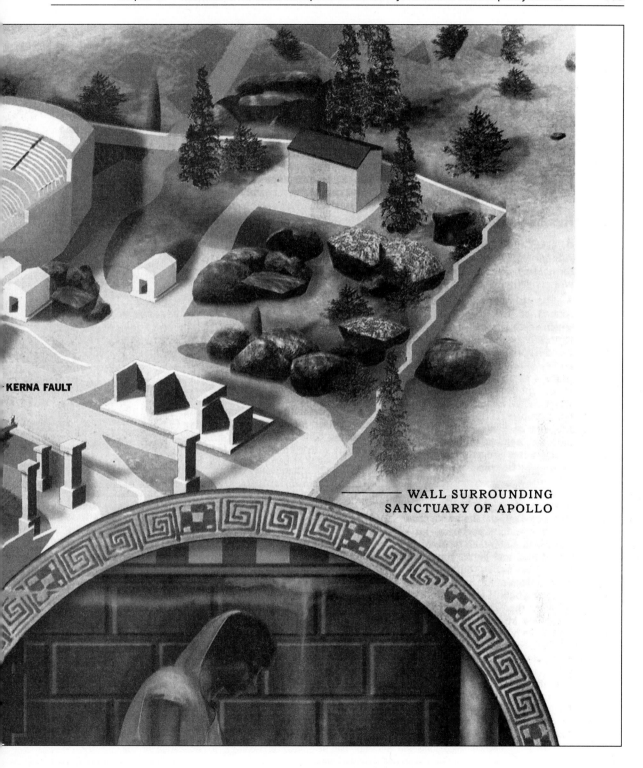

KERNA FAULT

WALL SURROUNDING
SANCTUARY OF APOLLO

**Inside The Chamber**

Not much is known about the women who served as oracles over the ages. In the illustration, one sits atop a tripod holding a bowl of water and a laurel leaf from Apollo's sacred tree. The most potent vapors from underground, scientists believe, were probably ethylene.

Illustrations by Frank Ippolito

### By WILLIAM J. BROAD

For at least 12 centuries, the oracle at Delphi spoke on behalf of the gods, advising rulers, citizens and philosophers on everything from their sex lives to affairs of state. The oracle was always a woman, her divine utterances made in response to a petitioner's request. In a trance, at times in a frenzy, she would answer questions, give orders and make prophecies.

Modern scholarship long ago dismissed as false the explanation that the ancient Greeks gave for the oracle's inspiration, vapors rising from the temple's floor. They found no underlying fissure or possible source of intoxicants. Experts concluded that the vapors were mythical, like much else about the site.

Now, however, a geologist, an archaeologist, a chemist and a toxicologist have teamed up to produce a wealth of evidence suggesting the ancients had it exactly right. The region's underlying rocks turn out to be composed of oily limestone fractured by two hidden faults that cross exactly under the ruined temple, creating a path by which petrochemical fumes could rise to the surface to help induce visions.

In particular, the team found that the oracle probably came under the influence of ethylene — a sweet-smelling gas once used as an anesthetic. In light doses, it produces feelings of aloof euphoria.

"What we set out to do was simple: to see if there was geological truth to the testimony of Plutarch and the others," said Dr. Jelle Zeilinga de Boer, a geologist at Wesleyan University, who began the Delphic investigations more than two decades ago.

As is often the case in science, the find was rooted in serendipity, hard work and productive dreaming. At one point, not unlike the oracle herself, the scientists were stimulated in their musings by a bottle of Dão, a Portuguese red wine.

The team's work was described last year in Geology, a publication of the

Geological Society of America, and at the annual meeting in January of the Archaeological Institute of America. It will also be reported in the April issue of Clinical Toxicology.

Over the years, scholarly doubt about the thesis has given way to wide acceptance and praise.

"I was very, very skeptical at first," said Dr. Andrew Szegedy-Maszak, a Wesleyan colleague and classicist, who specializes in Greek studies. "But they seem to have it nailed. I came to scoff but stayed to pray."

Near the Gulf of Corinth on the slopes of Mount Parnassus, the religious shrine was founded before 1200 B.C. and the temple eventually built there became the most sacred sanctuary for the ancient Greeks. They considered it the center of the world, marking the site with a large conical stone, the omphalos (meaning navel or center).

Originally a shrine to Gaea, the earth goddess, the temple at Delphi by the eighth century B.C. was dedicated to Apollo, the god of prophecy. His oracle spoke out, often deliriously, and exerted wide influence. One of her admired pronouncements named Socrates the wisest of men.

Before a prophetic session, the oracle would descend into a basement cell and breathe in the sacred fumes. Some scholars say her divine communications were then interpreted and written down by male priests, often in ambiguous verse. But others say the oracle communicated directly with petitioners.

With the rise of Christianity, the temple decayed. The Roman emperor Julian the Apostate tried to restore it in the fourth century A.D., but the oracle wailed that her powers had vanished.

French archaeologists began excavating the ruins in 1892, in time digging down to the temple's foundations. No cleft or large fissure was found. By 1904, a visiting English scholar, A. P. Oppé, declared that ancient beliefs in temple fumes were the result of myth, mistake or fraud.

The Oxford Classical Dictionary in 1948 voiced the prevailing view: "Excavation has rendered improbable the postclassical theory of a chasm with mephitic vapours."

Another round of myth-busting came in 1950 when Pierre Amandry, the French archaeologist who helped lead the temple excavations, declared in a book on Delphi that the region had no volcanism and that the ground was thus unable to produce intoxicating vapors.

Three decades later, in 1981, Dr. de Boer went to Delphi not to study old puzzles but to help the Greek government assess the region's suitability for building nuclear reactors. His main work was searching out hidden faults and judging the likelihood of tremors and earthquakes.

"A lucky thing happened," he recalled. Heavy tour traffic had prompted the government to carve in the hills east of Delphi a wide spot in the road where buses could turn around, exposing "a beautiful fault," he said. It looked young and active.

On foot, Dr. de Boer traced it for days, moving east to west over miles of mountainous terrain, around thorny bushes. The fault was plainly visible, rising as much as 30 feet. West of Delphi, he found that it linked up to a known fault. In the middle, however, it was hidden by rocky debris. Yet the fault appeared to run right under the temple.

"I had read Plutarch and the Greek stories," Dr. de Boer recalled. "And I started thinking, 'Hey, this could have been the fracture along which these fumes rose.' "

Dr. de Boer put the idea aside. Knowing little of the archaeological literature, he assumed that someone else must have made the same observation years earlier and come to the same conclusion.

In 1995, he discovered his mistake. While visiting a Roman ruin in Portugal, he met Dr. John R. Hale, an archaeologist from the University of Louisville, who was studying the Portuguese site. At sunset, the two men shared a bottle of wine, and the geologist began telling the archaeologist of the Delphi fault.

"I said, 'There is no such fault,' " Dr. Hale recalled. But Dr. de Boer convinced him otherwise. He cited both Plutarch, a Greek philosopher who served as a priest at Delphi, and Strabo, an ancient geographer. Each told of geologic fumes that inspired

Dr. John R. Hale

Dr. Jelle Zeilinga de Boer, above right, a geologist, and Dr. John R. Hale, below right, an archaeologist, helped solve the riddle of the intoxicating fumes of Delphi. They used rock and water samples, surveys of hidden faults and pharmacological analysis to reach their conclusion.

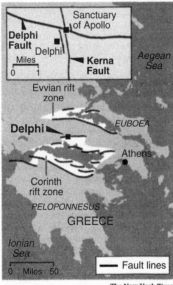

The New York Times

reports on the temple excavation and discovered to his surprise notations that the bedrock on which the temple was built was "fissured by the action of the waters."

The French archaeologists, expecting a yawning chasm, had apparently overlooked the importance of the small cracks.

"What I had been taught was wrong," Dr. Hale recalled. "The French had not ruled it out."

By 1996, the two men had traveled to Greece to resurvey the fault at Delphi and study the regional maps of Greek geologists. These revealed that underlying strata were bituminous limestone containing up to 20 percent blackish oils.

"I remember him throwing the map at me," Dr. Hale said of Dr. de Boer. " 'It's petrochemicals!' " No volcanism was needed, contrary to the previous speculation. Simple geologic action, Dr. de Boer insisted, could heat the bitumen, releasing chemicals into temple ground waters.

During a field trip in 1998, the vent notion grew more plausible still as the two men discovered a second fault, which they named Kerna after

divine frenzies, with Plutarch noting that the gases had a sweet smell. By the end of the evening, the geologist and archaeologist had decided to work together to find the truth.

Back in the United States, Dr. Hale tracked down the original French

George Ruhe for The New York Times

David R. Lutman for The New York Times

a well-known spring, going north-south under the temple. The intersecting faults now marked a provocative X.

As intriguing, the second fault appeared to be aligned with a series of ancient dry and modern wet springs, one directly beneath the temple.

The scientists found that the dry springs were coated with travertine, a rocky clue suggesting that the waters had come from deep below. When hot water seeps through limestone, it leaches out calcium carbonate that stays in solution until it rises to the surface and cools quickly. The calcium carbonate can then precipitate to form rocky layers of travertine.

Increasingly excited, the two men won permission from the Greek authorities to sample the travertine.

At this point, Dr. Jeffrey P. Chanton, a geochemist at Florida State University, joined the team. He analyzed the travertine samples gathered from dry springs near the temple and in its foundation, finding methane and ethane. Each can produce altered mental states. But a better candidate soon arose.

"A small light went off in my mind," Dr. de Boer recalled. Perhaps, he speculated, ethylene had been there as well.

Ethylene is significantly less stable than ethane and methane, so its absence in old rocks was understandable. Yet psychoactively, ethylene is quite potent, more so than ethane, methane or even nitrous oxide. From the 1930's to the 1970's, it was used for general anesthesia.

Dr. Chanton went to Greece, sampling an active spring near the temple.

The team waited. Days passed. Then his call came in. He had found ethylene, as well as methane and ethane. To all appearances, the ancient riddle had been solved.

In late 2000, Dr. Henry A. Spiller, the toxicologist who directs the Kentucky Regional Poison Center, joined the team to help with the pharmacological analysis.

"There's a fair amount of data on the effects of ethylene," Dr. Spiller said. "In the first stages, it produces disembodied euphoria, an altered mental status and a pleasant sensation. It's what street people would call getting high. The greater the dose, the deeper you go." Once a person stops breathing ethylene, he added, the effects wear off quickly.

Modern teenagers know of such intoxicants, including ones that in overdoses can kill. Experts say that youths who breathe fumes from gas, glue, paint thinner and other petrochemicals are toying with hydrocarbon gases.

Of late, Dr. Hale has been widening his focus, investigating other ancient Greek temples that he believes were built intentionally on geologically active sites.

And Dr. de Boer, now 67, is still concentrating on Delphi. On March 9, he and some students left for Greece to drill out rocky samples from the fault zones and illuminate them under a special light to try to establish dates of seismic activity.

Such geologic shocks, he said, may have influenced fume production over the ages, causing the intoxicating gases to wax and wane.

"You never know if it will work," he said of any research project shortly before the Delphi trip. "With the fumes, it did. With this, we don't know. But it's worth a try."

# Session 2E: Archaeology of Cult

NEW EVIDENCE FOR THE GEOLOGICAL ORIGIN OF THE DELPHIC ORACLE: ACTIVE FAULTS, GASEOUS EMISSIONS, AND ARCHITECTURAL ANOMALIES IN THE TEMPLE OF APOLLO: *John R. Hale*, University of Louisville, *Jelle de Boer*, Wesleyan University, *Jeffrey Chanton*, Florida State University, and *Henry Spiller, M.D.*, Kentucky Regional Poison Center

This report presents the results of a five-year interdisciplinary study of geological features and archaeological remains linked to the oracular cult at Delphi. The research team included a geologist, an archaeologist, a chemist, and a toxicologist. Beginning in 1996, the team identified and mapped geological faults at Delphi, conducted a field survey of ancient structures associated with those faults, and analyzed samples of spring water and travertine rock to detect both ancient and modern gaseous emissions. The study concluded with research into the physiological and psychological effects of the gases identified at Delphi.

Evidence derived from fieldwork and laboratory analysis indicates that the Apollo temple was built on bituminous limestone in a zone of cross-faulting, where the intersection of active faults provided pathways through which groundwater and gases rose to the surface. A previously undocumented NW–SE fault, designated the "Kerna" fault, runs through the sanctuary and under the temple. Its line coincides with an alcove on the south side of the cella and an elaborate conduit for spring water built into the foundation. Emissions of light hydrocarbon gases, including the sweet-smelling intoxicant ethylene, occurred along this fault. Ethylene can induce either a euphoric trance state or (more rarely) a violent delirium. The scientific evidence thus suggests that Plutarch, Pausanias, Diodorus, Strabo, and other ancient authors accurately recorded the unusual geological situation at Delphi, and understood its relation to the siting and design of the Apollo temple, the behavior of the Pythia, and other aspects of the oracular cult.

Conference Abstract, 103rd Annual Meeting of the Archeological Institute of America, 2002

# REFERENCES

[AAAS] American Association for the Advancement of Science [Internet]. Summer 2000. Professional ethics report XIII. [cited 2000 Dec 11]. Available from: http://www.aaas.org/spp/dspp/sfrl/per/per22 .htm.

[AAAS] American Association for the Advancement of Science [Internet]. 2002. Scientific association records programs: a beginner's guide. Association activities and the records they create. [cited 2002 June 4]. Available from: http://archives.aaas.org/guide/ guide4.php.

Abbott A. 1996. Scientists lose cold fusion libel case. Nature 380:369.

Adams JL. 1976. Conceptual blockbusting: a pleasurable guide to better problem solving. New York: W. W. Norton.

AgBioForum: A journal devoted to the economics and management of agrobiotechnology [Internet]. [cited 2003 Jan 3]. Available from: http://www.agbioforum .missouri.edu.

Agnew B. 1999. NIH eyes sweeping reform of peer review. Science 286:1074–6.

[AIP] American Institute of Physics. 1990. AIP style manual. 4th ed. New York: American Institute of Physics.

Alimenterics. Pylori-Chek Breath Test Kit. FDA 1999 Feb 4. Morris Plains (NJ): Alimenterics Inc.

Amato I. 1993. Pons and Fleischmann redux? Science 260:895.

Analytical Chemistry [Internet]. c2003. Author information. [cited 2003 Sep 22]. Available from: http:// pubs.acs.org/journals/ancham/.

Angell M, Kassirer JF. 1991. The Ingelfinger rule revisited. New England Journal of Medicine 325:1371–3.

Angell M, Kassirer JF. 1995. The Internet and the journal. New England Journal of Medicine 332:1709–10.

Angier N. 1992 May 10. Women still scarce on membership rolls of science academy. News and Observer (Raleigh, NC); Sect A:18.

Anholt RRH. 1994. Dazzle 'em with style: the art of oral scientific presentation. New York: Freeman and Company.

[Annals] Annals of Internal Medicine. [cited 2002 Oct 27]. Information for authors. Annals of Internal Medicine [Internet] 136(1). Available from: http://www.annals .org.

APA Style.org [Internet]. c2003. Washington: American Psychological Association; [cited 2003 Jan 29]. Available from: http://www.apastyle .org/aboutstyle .html.

[APS] American Physical Society [Internet]. 2002. Session W21-Cold Fusion (abstracts). [cited 2002 Dec 10]. Available from: http://www.aps.org/meet/MAR02/ baps/abs/S7810.html.

Aristotle. 1954. The rhetoric and the poetics of Aristotle. Rhys RW, Bywater I, translators. New York: Random House.

Arminen I. 2000. On the context sensitivity of institutional interaction. Discourse and Society 11:435–58.

Arroliga AC. 2002. Metaphorical medicine: using metaphors to enhance communication with patients who have pulmonary disease. Annals of Internal Medicine [Internet] 137(5) (Part 1). [cited 2003 Oct 5]. Available from: http://www.annals.org.

"arXiv.org" [Internet]. [cited 2003 Feb 3]. Available from: http://arXiv.org.

[ASCE] American Society of Civil Engineers [Internet]. c2000. Ethical standards for publications of ASCE journals. [cited 2003 Feb 22]. Available from: http://www.pubs.asce.org/authors/index .html#ethic.

[ASM] American Society for Microbiology. 1990. Call for abstracts. Anaheim (CA): American Society for Microbiology.

[ASM] American Society for Microbiology [Internet]. 2003. Call for abstracts. 103rd General Meeting; 2003 May 18–22; Washington, DC [cited 2003 Mar 15]. Available from: http://www.asmusa.org/mtgsrc/ gm2003CallforAbstracts.pdf.

Associated Press. 1992 Sep 5. Funding withheld for conference on genetics-crime link. News and Observer (Raleigh, NC); Sect A:7.

Associated Press. 1996 May 24. Quiz of adults discovers unscientific Americans. News and Observer (Raleigh, NC); Sect A:9.

Astrophysical Journal [Internet]. [cited 2003 Jan 3]. Author Information. Available from: http://www.journals .uchicago.edu/ApJ/.

Bacon F. 1605, 1859. The advancement of learning. The works of Francis Bacon. Volume 3. Spedding J, Ellis RL, Health DD, editors. New York: Longman.

Barker R. 1997. And the waters turned to blood. New York: Simon and Schuster.

Bazerman C. 1983. Scientific writing as a social act: a review of the literature of the sociology of science. In: Anderson P, Brochmann R, Miller C, editors. New essays in technical and scientific communication: research, theory, and practice. Farmingdale (NY): Baywood. p 156–84.

Bazerman C. 1985. Physicists reading physics: schema-laden purposes and purpose-laden schema. Written Communication 2(1):3–23.

Bazerman C. 1988. Shaping written knowledge: the genre and activity of the experimental article in science. Madison: University of Wisconsin Press.

Bazerman C. 1999. The languages of Edison's light. Cambridge (MA): MIT.

Bell GD. 1991. Anti-*Helicobacter pylori* therapy: clearance, elimination, or eradication? Lancet 337:310–11.

Bell HD, Walch KA, Katz SB. 2000. "Aristotle's pharmacy": the medical rhetoric of a clinical protocol in the drug development process. Technical Communication Quarterly 9(3):249–69.

Bellevue Literary Review [Internet]. [cited 2003 Feb 24]. Available from: http://www.blreview.org/index.htm.

Benditt J. 1995. Conduct in science. Science 268:1705.

Benson PJ. 1998. Changing moorings in scientific writing: suggestions to authors, allusions for teachers. In: Battalio JT, editor. Essays in the study of scientific discourse: methods, practice, and pedagogy. Stamford (CT): Ablex. p 209–25.

Berkenkotter C, Huckin TN. 1995. Genre knowledge in disciplinary communication: cognition/culture/ power. Hillsdale (NJ): Lawrence Erlbaum.

Biddle AW, Bean DJ. 1987. Writer's guide: life sciences. Lexington (MA): D. C. Heath and Company.

BioMed Central [Internet]. [cited 2003 Jan 18]. Available from: http://www.biomedcentral.com/start.asp.

BioMedNet. Elsevier Sciences [Internet]. [cited 2002 Dec 28]. Available from: http://www.bmn.com.

Bitzer L. 1968. The rhetorical situation. Philosophy and Rhetoric 1:1–14.

Blackman CF. 1994 Jan 27–29. Interaction of static and time-varying magnetic fields with biological systems. Ohlendorf Foundation Meeting, Margarita Island, Venezuela.

Blackman CF, Benane SG, House DH. 1985 June 16–20. A pattern of frequency dependent responses in brain tissue to ELF electromagnetic fields. The Annual Meeting of the Bio-electromagnetics Society, San Francisco, CA.

Blakeslee AM. 1994. The rhetorical construction of novelty: presenting claims in a letters forum. Science, Technology, and Human Values 19:88–100.

Blaser MJ. 1987. Gastric *Campylobacter*-like organisms, gastritis, and peptic ulcer disease. Gastroenterology 93:371–83.

Blaser MJ. 1996 Feb. The bacteria behind ulcers. Scientific American 104–7.

Bonk RJ. 1998. Medical writing in drug development: a practical guide for pharmaceutical research. New York: Pharmaceutical Products.

Boyd R. 1993. Metaphor and theory change: what is a metaphor a metaphor for? In: Ortony A, editor. Metaphor and Thought. London: Cambridge University Press. p 481–532.

Bradford A, Whitburn M. 1982. Analysis of the same subject in diverse periodicals: one method for teaching audience adaptation. Technical Writing Teacher 9:58–62.

British Medical Journal [Internet]. c2002. Advice to contributors. [cited 2002 Dec 31]. Available from: http://bmj.com/advice/sections.shtml.

Broad WJ. 2002 Mar 19. For Delphic oracle, fumes and visions. New York Times (national edition). Science Times D1, D4.

Brody AJ, Pelton MR. 1988. Seasonal changes in digestion in black bears. Canadian Journal of Zoology 66:1482–4.

Brody H. 1996 Sep. Wired science. Technology Review [Online]. Available from: http://web.mit.edu/afs/athena/org/t/techreview/www/tr.html.

Bronowski J. 1965. Science and human values. Rev. ed. New York: Harper and Row.

Brown FA. 1954 Apr. Biological clocks and the fiddler crab. Scientific American 190:34–7.

Brown FA. 1959. Living clocks. Science 130:1535–44.

Brown FA. 1960. Life's mysterious clocks. Saturday Evening Post 233:18–19, 43–44.

Brown FA. 1962. Responses of the planarian, Dugesia, and the protozoan, Paramecium, to very weak horizontal magnetic fields. Biological Bulletin 123:264–81.

Brummett B. 1976. Some implications of "process" or "intersubjectivity": postmodern rhetoric. Philosophy and Rhetoric 9:21–51.

Burke K. 1968. Counter-statement. Berkeley: University of California Press.

Burkholder JM, Glasgow HB. 1999. Science ethics and its role in early suppression of the *Pfiesteria* issue. Human Organization 58:443–55.

Burkholder JM, Glasgow HB, Springer J, Parrow MW. 2000. The life cycles of toxic *Pfiesteria* species and other estuarine dinoflagellates: toward verification of Pfiester's hypothesis. Poster. The CDC National Conference on Pfiesteria: From Biology to Public Health. Atlanta. GA, October 2000.

Burkholder JM, Lewitus AJ. 1994. Trophic interactions of ambush predator dinoflagellates in estuarine microbial food webs. Proposal submitted to the National Science Foundation. North Carolina State University, Department of Botany.

Burkholder JM, Noga EJ, Hobbs CH, Glasgow HB Jr, Smith SA. 1992. New "phantom" dinoflagellate is the causative agent of major estuarine fish kills. Nature 358:407–10; Nature 360:768.

Burkholder JM, Rublee PA. 1994. Improved detection of an ichthyotoxic dinoflagellate in estuaries and aquaculture facilities. Proposal submitted to the (U.S.) National Sea Grant College Program. North Carolina State University, Department of Botany.

Burrough-Boenisch J. 1999. International reading strategies for IMRD articles. Written Communication 16:296–316.

Butler D, Wadman M. 1999. Mixed response to NIH's web journal plan. Nature 399:8–9.

Byrne T, Hibbard J. 1987. Landward vergence in accretionary prisms: the role of the backstop and thermal history. Geology 15:1163–7.

[CAIRE] Committee on Assessing Integrity in Research Environments, Board of Health Sciences Policy, Division of Earth and Life Studies (The National Academies) [Internet]. 2002. Integrity in scientific research: creating an environment that promotes responsible conduct. Washington (DC): The National Academies Press. [cited 2003 Feb 27]. Available from: http://books.nap.edu/ books/0309084/ html/index.html.

Campbell JA. 1975. The polemical Mr. Darwin. Quarterly Journal of Speech 61:365–90.

Campbell KK, Jamieson KH. 1978. Form and genre in rhetorical criticism: an introduction. In: Campbell KK, Jamieson KH, editors. Form and genre: shaping rhetorical action. Falls Church (VA): Speech Communication Association. p 9–32.

Campbell KK. 1982. The rhetorical act. Belmont (CA): Wadsworth Publishing.

Carey J. 1992 Aug 10. What Barry Marshall knew in his gut. Business Week 68–9.

Carnap R. 1950. Logical foundations of probability. Chicago: University of Chicago Press.

[CBE] Council of Biology Editors. 1994. Scientific style and format: the CBE manual for authors, editors, and publishers. 6th ed. New York: Cambridge University Press.

[CBE] Council of Biology Editors. 1995. Report from the annual meeting—1994. In: [CBE] Views 18(1):14–16.

[CBE] Council of Biology Editors Task Force on Authorship. 2000. Who's the author? Problems with biomedical authorship, and some possible solutions. Report to the Council of Biology Editors (now Council of Science Editors) from the Task Force on Authorship February 2000. Science Editor 23(4):111–9.

[CDER and CBER] Center for Drug Evaluation and Research and Center for Biologics Evaluation and Research. 1995. *Guidance for industry: Content and format of Investigational New Drug Applications (INDs) for Phase 1 Studies of drugs, including well-characterized, therapeutic, biotechnology-derived products.*

[CFR] Code of Federal Regulations. 1999. Title 21. Sect 50, 56, 312, 314.

Chazin S. 1993 Oct. The doctor who wouldn't accept no. Reader's Digest 119–24.

Chiba N, Veldhuzen van Zanten SJO, Sinclair P, Fergusson RA, Escobedo S, Grace E. 2002. Treating *Helicobacter pylori* infection in primary care patients with uninvestigated dyspepsia: the Canadian adult dyspepsia empiric treatment—*Helicobacter pylori* positive (CADET-Hp) randomised controlled trial. British Medical Journal 324:1012–6. Retrieved July 21, 2000, from bmj.com.

Clay DE, Clapp CE, Linden DR, Molina JAE. 1989. Nitrogen-tillage-residue management: 3. observed and simulated interactions among soil depth, nitrogen mineralization, and corn yield. Soil Science 147:319–25.

Clayton CL, Mobley H. 1997. Helicobacter pylori protocols. Oxford; Cambridge (MA): Blackwell Healthcare Communications.

Close F. 1992. Too hot to handle: the story of the race for cold fusion. London: Penguin Books.

Cohen J. 1995. The culture of credit. Science 268:1706–11.

Cohen J. 1996 Jan 5. AIDS trials take on peer review. Science 271:20–1.

Cole KC. 1990 Jan 7. Science under scrutiny. New York Times; Sect EDUC:18.

Columbia Accident Investigation Board [Internet]. 2003 Aug. Report volume 1. 1st limited edition, Columbia Accident Investigation Board. Subsequent printing by NASA and the U.S. Government Printing Office, Washington DC. [cited 2003 Sep 23]. Available from: http://i.a.cnn.net/cnn/SPECIALS/2003/shuttle/CAIB.report.pdf.

Constantinides H. 2001. The duality of scientific ethos: deep and surface structures. Quarterly Journal of Speech 87:61–72.

Cornell News [Internet]. 2002 Sep 24. Cornell professor Paul Ginsparg, science communication rebel, named a MacArthur Foundation fellow. [cited 2003 Feb 3]. Available from: http://www.news.cornell.edu/chronicle/02/9.26.02/Ginsparg.html.

Couture B. 1993. Provocative architecture: a structural analysis of Gould and Lewontin's "The Spandrels of San Marco." In: Selzer J, editor. Understanding scientific prose. Madison: University of Wisconsin Press. p 276–309.

Cozman FG. 1999. Calculation of posterior bounds given convex sets of prior probability measures and likelihood functions. Journal of Computational and Graphical Statistics 8(4):824–38.

Cranor CF. 1993. Regulating toxic substances: a philosophy of science and law. New York: Oxford University Press.

Crease RP, Samios NP. 1989 Oct 1. The science of things that aren't so. News and Observer (Raleigh, NC); Sect D:1, 3.

Crismore A, VandeKopple WJ. 1988. Readers' learning from prose: the effects of hedges. Written Communication 5:184–202.

Dagani R. 1993 June 14. Latest cold fusion results fail to win over skeptics. Chemical & Engineering News 71:38–41.

Dagani R. 1995 May 15. New journal forges traditional peer review. Chemical & Engineering News 73:26–7.

Dautermann J. 1994. Negotiating meaning in a hospital discourse community. In: Spilka R, editor. Writing in the workplace: new research perspectives. Carbondale (IL): Southern Illinois University Press. p 98–110.

Davis RM. 1985. Publication in professional journals: a survey of editors. IEEE Transactions on Professional Communication 28(2):34–42.

Day M. 1999. The scholarly journal in transition and the PubMed Central. Ariadne [Internet] 21. [cited 2002 Dec 30]. Available from: http://www.ariadne.ac.uk/issue21/.

Day RA. 1998. How to write and publish a scientific paper. 5th ed. Phoenix (AZ): Oryx Press.

De Boer JZ, Hale JR. 1996. Proposal: a study of the evidence for geological faults and gaseous emissions at the sanctuary of Delphi, in order to test the validity of ancient testimonia concerning the origins of the oracle of Apollo.

De Boer JZ, Hale JR, Chanton J. 2001. New evidence for the geological origins of the ancient Delphic oracle (Greece). Geology 29:707–10.

Delmothe T, Smith R. 1999. Moving beyond journals: the future arrives with a crash. British Medical Journal 318:1637–9.

[DOC] U.S. Department of Commerce. 1994. Statistical abstract of the United States 1994. Washington: U.S. Government Printing Office.

[DOE] Department of Energy [Internet]. [updated 2002 May 31]. Grant application guide: instructions for application preparation. Washington: DOE Office of Science. [cited 2003 Mar 8]. Available from: http://www.science.doe.gov/ grants/App.html.

Dondis DA. 1973. A primer of visual literacy. Cambridge (MA): MIT Press.

Dong YR 1996. Learning how to use citations for knowledge transformation: non-native doctoral students' dissertation writing in science. Research in the Teaching of English 30:428–57.

Drazen JM, Curfman GD. 2002. On authors and contributors. New England Journal of Medicine [Internet] 347:55. [cited 2002 Nov 29]. Available from: http://content.nejm.org/content/vol347/issue1/index.shtml.

DuMez E. 2000. The role and activities of scientific societies in promoting research integrity. Professional ethics report XIII, American Association for the Advancement of Science [Internet]. [cited 2000 Dec 11]. Available from: http://www.aaas.org/spp/dspp/sfrl/per/per22.htm.

Ehninger D, Gronbeck BE, Monroe AH. 1988. Principles of speech communication. 10th brief ed. Glenview (IL): Scott, Foresman and Co.

[EPA] U.S. Environmental Protection Agency. 1998. What you should know about *Pfiesteria piscicida*. EPA Document #842-F-98-011. Washington: U.S. Government Printing Office. Also posted online as *Pfiesteria piscicida* Fact Sheet [updated 2002 Nov 27; cited 2003 Mar 7]. Available from: http://www.epa.gov/owow/estuaries/pfiesteria/fact.html.

[EPA] U.S. Environmental Protection Agency [Internet]. 1999 Dec 29. Consolidation of good laboratory practice standards. Sec. 806.81 Standard operating procedures. Federal Register 64(249). [cited 2003 Mar 14]. Available from: http://www.epa.gov/ EPA-TRI/1999/December/Day-29/tri33831.htm.

Fahnestock J. 1986. Accommodating science: the rhetorical life of scientific facts. Written Communication 3:275–96.

Fahnestock J. 1986. Accommodating science: the rhetorical life of scientific facts. Written Communication 15:330–50.

Fahnestock J. 1999. Rhetorical figures in science. New York: Oxford University Press.

Fallone CA, Veldhuyzen van Zanten SJO, Chiba N. 2000. The urea breath test for *Helicobacter pylori* infection: taking the wind out of the sails of endoscopy. Canadian Medical Association Journal 162:271–2.

Farrell TB, Goodnight GT. 1981. Accidental rhetoric: the root metaphors of Three Mile Island. Communication Monographs 48:271–300.

Feyerabend P. 1978. Against method. London: Verso.

Fisher WR. 1987. Human communication as narration: toward a philosophy of reason, value, and action. Columbia: University of South Carolina Press.

Fleischmann M, Pons S. 1989. Electrochemically induced nuclear fusion of deuterium. Journal of Electroanalytical Chemistry 261:301–8.

Flower L. 1993. Problem-solving strategies for writing. 4th ed. New York: Harcourt Brace Jovanovich.

Frayn M. 2000. Copenhagen. New York: Anchor Books.

Fry J. 2003. The disciplinary shaping of scholarly communication: patterns of computer-mediated communication across three diverse specialisms. Seminar, Center for Information Society Studies (CISS). February 12. North Carolina State University.

Fulbright MA, Reynolds SP. 1990. Bipolar supernova remnants and the obliquity dependence of shock acceleration. Astrophysical Journal 357:591–601.

Fuller S. 2000. Governing science before it governs us. Interdisciplinary Science Reviews 25:95–100.

Garfield E. 1995. Giving credit only where it is due: the problem of defining authorship. The Scientist 9:13.

Gastroenterology. [cited 2003 Feb 15]. Instructions to authors. Gastroenterology [Internet] 124(2). Available from: http://www2.gastrojournal.org/.

Gilbert GN, Mulkay M. 1984. Opening Pandora's box: a sociological analysis of scientists' discourse. Cambridge: Cambridge University Press.

Ginsparg P. 1994. First steps towards electronic research communication. Computers in Physics 8:390–6.

Glickman D. 1999. New crops, new century, new challenges: how will scientists, farmers, and consumers learn to love biotechnology and what happens if they don't? Remarks prepared for delivery before the National Press Club, Washington, DC, July 13, 1999. Release No. 0285.99.

Good GA. 2000. The assembly of geophysics: scientific disciplines in frameworks of consensus. Studies in History and Philosophy of Modern Physics 31B:259–92.

Goodbred SL Jr, Hine AC. 1995. Coastal storm deposition: salt-marsh response to a severe extratropical storm, March 1993, west-central Florida. Geology 23:679–82.

Gragson G, Selzer J. 1993. The reader in the text of "The Spandrels of San Marco." In: Selzer J., editor. Understanding scientific prose. Madison: University of Wisconsin Press. p 180–202.

Graham DY, Lew GM, Klein PD, Evans DG, Evans DJ Jr, Saeed ZA, Malaty HM. 1992. Effect of treatment of *Helicobacter pylori* infection on the long-term recurrence of gastric or duodenal ulcer. Annals of Internal Medicine 116:705–8.

Greek Ministry of Culture. 1997 May 21. Permit issued to John Hale for removal of porous stone from Delphi. Tick E, translator (2003).

Greenland PT. 1994. Essay review: the story of cold fusion. Contemporary Physics 35:209–11.

Griffith D. 1999. Exaggerating environmental health risk: the case of the toxic dinoflagellate *Pfiesteria*. Human Organization 58:119–27.

Gross AG. 1990. The form of the experimental paper: a realization of the myth of induction. The rhetoric of science. Cambridge (MA): Harvard University Press. p 85–96.

Gurak Laura J. 2000. Oral presentations for technical communication. Boston: Allyn and Bacon.

Hagmann M. 2000. Cancer researcher sacked for alleged fraud. Science 287:1901–2.

Hale JR, De Boer JZ, Chanton J, Spiller H. 2002. New evidence for the geological origin of the Delphic oracle: active faults, gaseous emissions, and architectural anomalies in the temple of Apollo. Conference abstract, 103rd Annual Meeting of the Archeological Institute of America. American Journal of Archeology 106:231-306.

Hall J. 1971. Decisions, decisions, decisions. Psychology Today 5(6):51–54, 86, 88.

Haller CR. 1998. Foucault's archeological method and the discourse of science: plotting enunciative fields. In: Battalio JT, editor. Essays in the study of scientific discourse: methods, practice, and pedagogy. Stamford (CT): Ablex. p 53–79.

Halloran SM. 1984. The birth of molecular biology: an essay in the rhetorical criticism of scientific discourse. Rhetoric Review 3:70–83.

Hamilton MJ, Kennedy ML. 1987. Genic variability in the racoon procyon lotor. The American Midland Naturalist 118:266–74.

Harmon JE. 1992. Current contents of theoretical scientific papers. Journal of Technical Writing and Communication 22:357–75.

Harnad S. 1990. Scholarly skywriting and the republication continuum of scientific inquiry. Psychological Science 1: 342–3.

Harnad S. 1997. The paper house of cards (and why it's taking so long to collapse). Ariadne [Internet] 8. [cited 2002 Dec 30]. Available from: http://www.ariadne.ac.uk/issue8/harnad/.

Harnad S, Hey J. 1995. Esoteric knowledge: the scholar and scholarly publishing on the net. In: Dempsey L, Law D, Mowat I, editors. Networking and the future of libraries. London: Library Association. p 110–6.

Harris AW, Misiewicz JJ. 1996. *Helicobacter pylori*. Oxford; Cambridge (MA): Blackwell Healthcare Communications.

Havelock EA. 1986. Orality, literacy, and Star Wars. Written Communication 3:411–20.

Hawes T, Thomas S. 1997. Tense choices in citations. Research in the Teaching of English 31:393-414.

Heatley RV. 1999. The *Helicobacter pylori* handbook. Oxford; Cambridge (MA): Blackwell Healthcare Communications.

Heisenberg W. 1971. Physics and beyond: encounters and conversations. New York: Harper and Row.

Hendrick SP, Reynolds SP. 2001. Maximum energies of shock-accelerated electrons in Large Magellanic Cloud supernova remnants. Astrophysical Journal 559:903–8.

Hendrick SP, Reynolds SP, Borkowski KJ. 2001. Maximum energies of shock-accelerated electrons in supernova remnants in the Large Magellanic Cloud. In: Holt SS, Hwang U, editors. Young supernova remnants: Eleventh Astrophysics Conference. Melville (NY): American Institute of Physics. AIP Conference Proceedings 565:387–90.

[Heredity]. Journal of Heredity. 2002. Information for contributors. Journal of Heredity 93(3): no page number [230+ p].

Herndl CG, Fennell BA, Miller CR. 1991. The accidents at Three Mile Island and the shuttle Challenger. In: Bazerman C, Paradis J, editors. Textual dynamics of the professions. Madison: University of Wisconsin Press. p 279–305.

Herrington AJ. 1985. Writing in academic settings: A study of the contexts for writing in two college chemical engineering courses. Research in the Teaching of English 19(4):331–61.

Heylin M. 2000 Oct 9. Journalism 101. Chemical Engineering News 46.

Hogan M. PubMed 1999. Central and beyond. HMSBeagle: The BioMedNet Magazine [Internet] 61:3. [cited 2002 Dec 28]. Available from: http://www.bmn.com.

Holton G. 1973. Thematic origins of scientific thought: Kepler to Einstein. Cambridge (MA): Harvard University Press.

Horn D. 1998. The shoulders of giants. Science 280:1354–5.

Hoshiko T. 1991 Oct 28. Facing ethical dilemmas: scientists must lead the charge. The Scientist 11, 13.

Houp KW, Pearsall TE, Tebeaux E. 1998. Reporting technical information. Boston: Allyn and Bacon.

Hsia HJ. 1977. Redundancy: is it the lost key to better communication? Audiovisual Communication Review 25:63–85.

Huizenga JR. 1992. Cold fusion: the scientific fiasco of the century. Rochester (NY): University of Rochester Press.

Huler S. 1990 Nov 12. How to get your research published: editors' thoughts. The Scientist 23.

Huyghe P. 1993 Apr. Killer algae. Discover 14(4):71–5.

Hyland K. 1996. Talking to the academy: forms of hedging in science research articles. Written Communication 13:251–81.

Hyland K. 1999. Disciplinary discourse: writer stance in research articles. In: Candlin CN, Hyland K, editors. Writing: texts, processes and practices. New York: Longman. p 99–121.

[ICH] International Conference on Harmonisation of Technical Requirements for Registration of Pharmaceuticals for Human Use [Internet]. 1996. Guidance for industry: structure and content of clinical study reports. [cited 2003 Oct 7]. Available from: http://www.fda.gov/cder/guidance/iche3.pdf.

[ICMJE] International Committee of Medical Journal Editors [Internet]. 2001. Uniform requirements for manuscripts submitted to biomedical journals. [cited 2003 Feb 22]. Available from: http://www.icmje.org.

Ingelfinger FJ. 1977. The general medical journal: for readers or repositories? New England Journal of Medicine 296:1258–64.

[ISI] Institute for Science Information[SM] [Internet]. 2003. Science Citation Index. [cited 2003 Mar 12]. Available from: http://www.isinet.com/isi/products/citation/sci/.

Issues in Science and Technology Online [Internet]. National Academy of Science, National Academy of Engineering, University of Texas at Dallas. [cited 2002 June 4]. Available from: http://www.nap.edu/issues/about.html.

Jacobs M, Storck W. 2000 May 8. Women in industry still hit glass ceiling. Chemical & Engineering News 78(9):36–7.

Jones SE, Palmer EP, Czirr JB, Decker DL, Jensen GL, Thorne JM, Taylor SF, Rafelski J. 1989. Observation of cold nuclear fusion in condensed matter. Nature 338:737.

Jorgensen-Earp CR, Jorgensen DD. 2002. "Miracle from mouldy cheese": Chronological versus thematic self-narratives in the discovery of penicillin. Quarterly Journal of Speech 88:69–90.

Kassirer JP, Angell M. 1991. On authorship and acknowledgements. New England Journal of Medicine [Internet] 325:1510–12. [cited 2003 Jan 4]. Available from: http://www.nejm.org/hfa/authorandacknowledge.asp.

Katz MJ. 1985. Elements of the scientific paper: a step-by-step guide for students and professionals. New Haven: Yale University Press.

Katz SB. 1992a. Narration, technical communication, and culture: *The soul of a new machine* as narrative romance. In: Secor M, Charney C, editors. Constructing rhetorical education. Carbondale (IL): Southern Illinois Press. p 382–402.

Katz SB. 1992b. The ethic of expediency: classical rhetoric, technology, and the Holocaust. College English 54:255–75.

Katz SB. 1993. Aristotle's rhetoric, Hitler's program, and the ideological problem of praxis, power, and professional discourse. Journal of Business and Technical Communication 7:37–62.

Katz SB. 2001. Language and persuasion in biotechnology communication with the public: How to not say what you're not going to not say and not say it. AgBioForum [Internet] 4:2. [cited 2003 Feb 3]. Available from: http://www.agbioforum.org.

Katz SB, Miller CR. 1996. The low-level radioactive waste siting controversy in North Carolina: toward a rhetorical model of risk communication. In: Herndl CG, Brown SC, editors. Green culture: environmental rhetoric in contemporary America. Madison: University of Wisconsin Press. p 111–40.

Kennicutt R. [cited 2003 Jan 3]. A brief guide to the *ApJ* editorial review process. Astrophysical Journal [Internet]. Available from: http://www.journals.uchicago.edu/ApJ/guide-ed_text.html.

Koshland D Jr. 1989 Aug 25. Minimizing fraud without stifling science. News and Observer (Raleigh, NC); Sect A:16.

Kuhn TS. 1970. The structure of scientific revolutions. 2nd ed. Chicago: University of Chicago Press.

Kuhn TS. 1979. Metaphor in science. In: Ortony A, editor. Metaphor and thought. London: Cambridge University Press. p 409–19.

Kuhn TS. 1996; org. 1970. The structure of scientific revolutions. 3rd ed. Chicago: University of Chicago Press.

Kynell-Hunt T, Savage GJ, editors. 2003. Power and legitimacy in technical communication. Vol. 1: The historical and contemporary struggle for professional status. Amityville (NY): Baywood Publishing Company.

Lakoff G, Johnson M. 1980. Metaphors we live by. Chicago: University of Chicago Press.

Lam SK. 1989. Duodenal ulcer relapse after eradication of *Campylobacter pylori*. Lancet 1:384.

Lancet. 1984. Spirals and ulcers. Lancet 1:1336.

Lancet. 1990. Information for authors. Lancet 336:1117.

Lane N. 2001. The new security environment. APS News Online [Internet]. January 2001. [cited 2002 June 18]. Available from: http://www.aps.org/apsnews/0101/010115.html.

Latour B. 1987. Science in action: how to follow scientists and engineers through society. Cambridge (MA): Harvard University Press.

Latour B, Woolgar S. 1979. Laboratory life: the social construction of scientific facts. Beverly Hills (CA): Sage.

Lawler A. 2000a. Silent no longer: model minority mobilizes. Science 290:1072–7.

Lawler A. 2000b. Relief, rebuke follow agreement on Lee. Science 289:1851, 1853.

Leary DE. 1990. Psyche's muse: the role of metaphor in the history of psychology. In: Leary DE, editor. Metaphors in the history of psychology. New York: Cambridge University Press. p 1–78.

Lewontin RC. 1993. Biology as ideology: the doctrine of DNA. New York: HarperPerrenial.

Liedes J. 1997. Copyright: evolution not revolution. Science 276: 223–5.

Lievrouw LA. 1990. Communication and the social representation of scientific knowledge. Critical Studies in Mass Communication 7:1–10.

Lim EM, Rauzier J, Timm J, Torrea G, Murray A, Gicquel B, Portnol D. 1995. Identification of *Mycobac-terium tuberculosis* DNA sequences encoding exported proteins by using *phoA* gene fusions. Journal of Bacteriology 177:59–65.

Locke S. 2001. Sociology and the public understanding of science: from rationalization to rhetoric. British Journal of Sociology 52:1–18.

Loffeld RJLF, Stobberingh E, Arends JW. 1989. Response to David Y Graham et al. Lancet 1:569–70.

Lorimer CG, Chapman JW, Lambert WD. 1994. Tall understorey vegetation as a factor in the poor development of oak seedlings beneath mature stands. Journal of Ecology 82:227–37.

Lyne J. 1998. Knowledge and performance in argument: disciplinarity and proto-theory. Argumentation and Advocacy 35:3–9.

MacArthur Fellows Programs [Internet]. 2002. MacArthur award announcement [cited 2003 Feb 3]. Available from: http://www.macfound.org/program/fel/2002fellows/ ginsparg_paul.htm.

Macrina FL. 2000. Scientific integrity. 2nd edition. Washington: ASM Press.

Maddox J. 1989. End of cold fusion in sight. Nature 340:15.

Madigan R, Johnson S, Linton P. 1995. The language of psychology: APA style as epistemology. American Psychologist 50:428–36.

Mallin MA, Burkholder JM, Larsen LM, Glasgow HB Jr. 1995. Response of two zooplankton grazers to an ichthyotoxic estuarine dinoflagellate. Journal of Plankton Research 17:351–63.

Marshall B. 2002. Helicobacter pioneers firsthand: accounts from the scientists who discovered Helicobacter 1892–1982. London: Blackwell.

Marshall B, Armstrong J, McGechie D, Glancy R. 1985. Attempt to fulfil Kochs postulates for pyloric campylobacter. Medical Journal of Australia 142:436–9.

Marshall BJ, Warren JR. 1984. Unidentified curved bacilli in the stomach of patients with gastritis and peptic ulceration. Lancet 1:1311–5.

Marshall BJ, Warren JR, Goodwin CS. 1989. Duodenal ulcer relapse after eradication of Campylobacter pylori. Lancet 1:836–37.

Marshall E. 1995. Dispute slows paper on "remarkable" vaccine. Science 268:1712–5.

Marshall E. 1996a. Trial set to focus on peer review. Science 273:1162–4.

Marshall E. 1996b. Peer review: battle ends in $21 million settlement. Science 274:911.

Marshall E. 2000a. Rival genome sequencers celebrate a milestone together. Science 288:2294–5.

Marshall E. 2000b. Storm erupts over terms for publishing Celera's sequence. Science 290:2042–3.

Maryland Department of Natural Resources [Internet]. 1999. Maryland guidelines for closing and reopening rivers potentially affected by Pfiesteria or Pfiesteria-like events. [cited 2003 Mar 29]. Available from: http://www.dnr.state.md.us/bay/cblife/algae/dino/pfiesteria/river_closure.html.

Matzat U. 1998. Informal academic communication and scientific usage of internet discussion of groups. IRISS '98 [Internet]: Conference Papers. [cited 2002 June 18]. Available from: http://www.sosig.ac.uk/iriss/papers/paper19.htm.

McDonald KA. 1995 April 28. Too many co-authors? The Chronicle of Higher Education 42:A35.

Medawar PB. 1964 Aug 1. Is the scientific paper fraudulent? Saturday Review 47:42–3.

Mehlenbacher B. 1994. The rhetorical nature of academic research funding. IEEE Transactions on Professional Communication 37:157–62.

Memory, JD. 2003 Feb 11. Science is working together. News and Observer (Raleigh, NC); Sect A:10.

[MEPS] Marine Ecology Progress Series. 2002 Mar 6. Editorial statement. Marine Ecology Progress Series 228: no page number.

Merton RK. 1973a. The normative structure of science. The sociology of science: theoretical and empirical investigations. Chicago: University of Chicago Press. p 267–8.

Merton RK. 1973b. Priorities in scientific discovery. The sociology of science: Theoretical and empirical investigations. Chicago: University of Chicago Press. p 286–324.

Miller CR. 1984. Genre as social action. Quarterly Journal of Speech 70:151–67.

Miller CR. 1992. Kairos in the rhetoric of science. In: Witte SP, Nakadate N, Cherry RD, editors. A rhetoric of doing. Carbondale: Southern Illinois University Press. p 310–27.

Miller CR, Halloran SM. 1993. Reading Darwin, reading nature; or, on the ethos of historical science. In: Selzer J, editor. Understanding scientific prose. Madison: University of Wisconsin Press. p 106–26.

Mobley HL, Hazel SL, Mendz G. 2001. Helicobacter pylori: physiology and genetics. Oxford; Cambridge (MA): Blackwell Healthcare Communications.

Monmaney T. 1993 Sept 20. Marshall's hunch. New Yorker 64–72.

Monroe J, Meredith C, Fisher K. 1977. The science of scientific writing. Dubuque (IA): Kendall/Hunt.

Montgomery SL. 1996. The Scientific Voice. New York: Guilford Press.

Moore B. 2000. Could cold fusion be for real? An interview with Dr. Eugene Mallove—Part 1. EV World [Internet]. [cited 2002 June 11]. Available from: http://www.evworld.com/archives/interviews2/mallove1.html.

Morris AJ, Nicholson GI, Perez-Perez GI, Blaser MJ. 1991. Long-term follow-up of voluntary ingestion of Helicobacter pylori. Annals of Internal Medicine 114:662–3.

Morton CC. 2000. Company, researchers battle over data access. Science 290:1063.

Myers G. 1985. The social construction of two biologists' proposals. Written Communication 2:219–45.

Myers G. 1991. Stories and styles in two molecular biology review articles. In: Bazerman C, Paradis J, editors. Textual dynamics of the professions. Madison: University of Wisconsin Press. p 45–75.

[NAS] National Academy of Sciences. 1989. On being a scientist. Washington: National Academy Press.

[NAS] National Academy of Sciences. 1995. On being a scientist: responsible conduct in research. 2nd ed. Washington: National Academy Press.

[NASA] National Aeronautics and Space Administration. 1993. NASA research announcement soliciting proposals for theory in space astrophysics. NRA 93-0SSA-06. Washington: National Aeronautics and Space Administration, Office of Space Science.

[NASA] National Aeronautics and Space Administration [Internet]. [updated 1999 June 23]. Instructions for responding to NASA research announcements for solicited basic research proposals (NRA 99-OSS-04; Appendix B). Washington: NASA Office of Space Science [cited 2003 Mar 9]. Available from: http://research.hq.nasa.gov/code_s/nra/current/NRA-99-OSS-04/apendixb.html.

[NASA] National Aeronautics and Space Administration [Internet]. 2003. Guidebook for proposers responding to a NASA Research Announcement (NRA). [cited 2003 Mar 11]. Available from: http://www.hq.nasa.gov/office/procurement/nraguidebook/proposer2003.pdf.

[NASA] National Aeronautics and Space Administration [Internet]. 2003. NASA accomplishments during 2002 and FY2004 budget request. [cited 2003 Oct 5]. Available from: http://www.nasa.gov/ pdf/ 2166main_04budget_detail_030227.pdf.

National Sea Grant. Sea Grant is competitive [Internet]. [cited 2003 Mar 12]. Available from: http://www.nsgo.seagrant.org/Competitive_FactSheet.html.

National Sea Grant College Program. 1994. Statement of opportunity for funding: marine biotechnology. Raleigh: North Carolina State University, UNC Sea Grant College Program.

Nature [Internet]. c2000. How to get published in Nature. [cited 2002 Nov 30]. Available from: http://www.nature.com/nature/submit/get_published/index.html.

Nature [Internet]. 2003a. *Nature* guide to authors. [cited 2003 Oct 4]. Available from: http://www.nature.com/nature/submit/gta/index.html.

Nature [Internet]. 2003b. *Nature's* other submitted material. [cited 2003 Oct 4]. Available from: http://www.nature.com/nature/submit/gta/Others/index.html.

Nature Biotechnology [Internet]. [cited 2003 Sep 22]. A guide to authors. Available from: http://www.nature.com/nbt/info/guide_authors/.

[NEJM] New England Journal of Medicine [Internet]. c2003. Help for authors. [cited 2003 Jan 4]. Available from: http://www.nejm.org/hfa/subinstr.asp.

Nelkin D. 1975. The political impact of technical expertise. Social Studies of Science 5:35–54.

Nelkin D. 1979. Controversy: politics of technical decisions. 2nd ed. Beverly Hills (CA): Sage Publications.

[Neuroscience] Society for Neuroscience. 1994. Preliminary program for 24th annual meeting, Nov 13–14; Miami Beach, FL.

[NHEERL] National Health and Environmental Effects Research Laboratory. 1997. NHEERL/97/01 Guidelines for operating procedures. Research Triangle Park, NC: Office of Research and Development, U.S. Environmental Protection Agency.

[NHEERL] National Health and Environmental Effects Research Laboratory. 2000 June 21. Standard operating procedure: staining with acridine orange and micronucleus analysis. SOP No. NHEERL-H/ECD/ GCTB/ADK/00/22/000.

[NIAID] National Institute of Allergy and Infectious Diseases, National Institutes of Health [Internet]. 2003a [updated 2003 Jan 14]. NIAID funding: how to write a grant: write to your audience. Bethesda (MD): NIAID Funding and Council News Center. [cited 2003 Mar 9]. Available from: http://www.niaid.nih.gov/ ncn/grants/write/write_d1.htm.

[NIAID] National Institute of Allergy and Infectious Diseases, National Institutes of Health [Internet]. 2003b [updated 2003 Mar 5]. NIAID funding: grant application basics: other factors play a role in review. Bethesda (MD): NIAID Funding and Council News Center. [cited 2003 Mar 9]. Available from: http:// www.niaid.nih.gov/ ncn/grants/basics/basics_b4 .htm.

[NIAID] National Institute of Allergy and Infectious Diseases, National Institutes of Health [Internet]. 2003c [updated 2003 Mar 5]. NIAID funding: grant application basics: most reviewers scan each application. Bethesda (MD): NIAID Funding and Council News Center. [cited 2003 Mar 9]. Available from: http:// www.niaid.nih.gov/ncn/grants/basics/basics_c2.htm.

[NIAID] National Institute of Allergy and Infectious Diseases, National Institutes of Health [Internet]. 2003d [updated 2003 Mar 5]. NIAID funding: how to write a grant: research plan section g. Literature cited. Bethesda (MD): NIAID Funding and Council News Center. [cited 2003 Mar 10]. Available from: http://www.niaid.nih .gov/ncn/grants/write/write_p1.htm.

[NIAID] National Institute of Allergy and Infectious Diseases, National Institutes of Health [Internet]. 2003e [updated 2003 Mar 5]. NIAID funding: how to write a grant: more common problems cited by peer reviewers. Bethesda (MD): NIAID Funding and Council News Center. [cited 2003 Mar 9]. Available from: http:// www.niaid.nih.gov/ncn/grants/write/write_d6.htm.

Nightingale V. 1986. Community as audience—audience as community. Australian Journal of Communication 9 & 10:31–41.

[NIH] National Institutes of Health. 1993. Helpful hints: preparing an NIH research grant application. Bethesda (MD): National Institutes of Health, Division of Research Grants.

[NIH] National Institutes of Health [Internet]. 2000. An introduction to extramural NIH. [cited 2002 Nov 27]. Available from: http://grants.nih.gov/grants/ intro2oer.htm.

[NIH] National Institutes of Health [Internet]. [updated 2001 Aug 24]. A straightforward description of what happens to your research project grant application (R01/R21) after it is received for peer review. Bethesda (MD): NIH Center for Scientific Review. [cited 2003 Mar 9]. Available from: http://www.csr.nih.gov/review/ peerrev.htm.

[NIH] National Institutes of Health [Internet]. 2003a [updated 2003 Feb 3]. OER: peer review policy and issues. Bethesda (MD): NIH Office of Extramural Research. [cited 2003 Mar 9]. Available from: http://grants1.nih.gov/grants/peer/peer.htm.

[NIH] National Institutes of Health [Internet]. 2003b [updated 2003 Feb 3]. Summary of the FY 2004 President's Budget. [cited 2003 Mar 9]. Available from: http://www.the-aps.org/pa/action/news/ fy04presbud.pdf.

[NIH] National Institutes of Health, Consensus Development Panel on *Helicobacter pylori* in Peptic Ulcer Disease. 1994. *Helicobacter pylori* in peptic ulcer disease. Journal of the American Medical Association 272:65–9.

[NLM] National Library of Medicine [Internet]. 2001. National Library of Medicine recommended formats for bibliographic citation. Supplement: Internet formats. (K. Patrias, ed.) Bethesda (MD): National Library of Medicine, US Dept of Health and Human Services. [cited 2003 Jan 20]. Available from: http://www.nlm .nih.gov/pubs/formats/internet.pdf.